AF527867

Advances in Ethnobotany

Advances in Ethnobotany

Santosh Kumar Jha
Astt. Professor
P.G. Department of Plant Science
J.N.R.M. Port Blair
Andaman and Nicobar Islands (India)

SATISH SERIAL PUBLISHING HOUSE
403, Express Tower, Commercial Complex
Azadpur, Delhi-110033 (India)
Phone : 011-27672852, Fax : 91-11-27672046
e-mail : info@satishserial.com, hkjain1975@yahoo.com
Website : www.satishserial.com

Published by :

SATISH SERIAL PUBLISHING HOUSE

403, Express Tower, Commercial Complex, Azadpur, Delhi-110033 (INDIA)
Phone : 011-27672852 Fax : 91-11-27672046
E-mail : info@satishserial.com, hkjain1975@yahoo.com

ISBN 978-93-86200-27-3

Composed, Designed & Printed in India

Dedicated to

My parent

(Sri Sushil Jha and Smt. Chanchala Devi)

whose upbringing makes me able to do this and

My mentors

Late Dr. Bijendra Pathak and

Prof. (Dr.) Usha Singh

(my research supervisor) who shaped and

oriented me to do this noble work

AMITY INSTITUTE FOR HERBAL AND BIOTECH PRODUCTS DEVELOPMENT
An Institution of Ritnand Balved Education Foundation – Thiruvananthapuram

Prof. (Dr.) P. PUSHPANGADAN,
M.Phil., Ph.D., FAS, FBRS, FES, FIAT, FNRS, FNSE, FNESA, FNAASc, FNASc, FRSC (UK)
(Padma Shri Awardee, UN Equator Initiative Laureate and Borlaug Awardee)
(Former Director, NBRI & CIMAP, Lucknow & JNTBGR & RGCB, Trivandrum)
Hon. Director General & Senior Vice President, RBEF

Foreword

The book "Advances in Ethnobotany" is an attempt to integrate herbal research edited by Dr. Santosh Kumar Jha. Plants in India have been used for medicinal purposes long before recorded history. India is - a - mega-diversity centre harboring a multitude of medicinal plant species each presumably studded with as yet unknown genetic and chemical variations of economic importance. The book provides comprehensive, extensive and realistic information on a large number of medicinal plants grown, found and used by tribes and various ethnic groups in different parts of India. This book is divided into three sections. In first section ethnobotanical works of Veligonda, Shivalik hills, and some other research works of North eastern states are described. A chapter is dedicated on research methodology in ethnobotany. Another chapter describes ethnobotany of orchidaceae. The second section deals with Pharmacological activities of *Cyperus rotundas,* bioactive metabolites from plant associated endophytic fungi, berries: source of natural antioxidant, phytochemistry of *Murraya koenigii* and several other herbal drugs are explained in very lucid manner. The third section focuses on utilization and conservation of plants in which medicinal and economic importance *of Rhododendron* in Sikkim Himalaya, an efficient micropropagation protocol for *Melissa officinalis,* utilization of plant biodiversity for treatment and cure of diseases and many other relevant topics arc discussed in length.

The World Health Organization estimated that 80% of people worldwide rely on herbal medicines for some part of their primary health care. Also, modern pharmacopoeia still contains at least 25% drugs derived from plants while many others are synthetic analogues built on prototype compounds isolated from plants. India and China are the largest users of herbal medicines. But there is a great concern about the safety and efficacy of herbal use. The quality assurance of these products is very important.

Fuelling renewed public interest in these alternative systems of medicine are reasons such as increased side effects, lack of curative treatment for several chronic diseases, high cost of new drugs, microbial resistance and emerging diseases.

This book having 25 chapters, which will be equally helpful for students of pharmacognosy, pharmacology as well as environmentalist and conservationist. The editor Dr. Santosh Kumar Jha has attempted to organise the book in such a way that it will be useful to everyone interested in the field of medicinal botany, pharmacy, ethno-medicine, ethnobotany, herbal therapy, naturopathy and green health therapy.

I congratulate Dr. Santosh Kumar Jha for his effort in bringing out this volume which I hope will be welcomed by all those who are interested in the study of ethnobotany of India.

P. Pushpangadan

Preface

Ethnobotany, in totality, is virtually a new field of research and if this field is investigated thoroughly and systematically, it will yield results of great value to the ethnologist, archeologist , anthropologist, botanist and ultimately to pharmacologists and phytochemists. In the recent years, there has been tremendous growth of interest in plant based drugs, pharmaceuticals, perfumery- products, cosmetics and aroma compounds used in food flavours, fragrances and natural colours in the world. Hence, ethnobotanical research provides new vistas for applications, improvement and popularization of traditional medicines.

The field of medicine has long been divided between the so called 'rationalist' and 'vitalistic' approaches. The same dichotomy exists today among herbal practitioners. But as herbal medicine moves increasingly into mainstream acceptance, it is being placed more and more under the rationalist microscope. Today ethnobotany is in the midst of Renaissance. This revival reflects increasing concern about the disappearance of rain forests and the tribal culture inhabiting them. India is well known for her traditional knowledge and wisdom. Traditional medicine is a synthesis of therapeutic experience of generations of practicing physicians of an indigenous system of medicine.

The present publication "Advances in Ethnobotany" provides comprehensive and realistic information of an extensive and large number of medicinal plants grown, found and used by tribes and various ethnic groups in different parts of India including Andaman and Nicobar Island. The book is divided into three sections. In first section ethnobotanical works of Veligonda, Shivalik hills, and some other research works of north eastern states are described. A chapter is dedicated on research methodology involves in ethnobotanical researches. The second section deals with pharmacological activities of *Cyperus rotundus*, bioactive metabolites from plant associated endophytic fungi, berries: source of natural antioxidant, phytochemistry of *Murraya koenigii* and several other works on herbal drugs are explained in a very lucid manner. The third section focuses on utilization and conservation of plants in which medicinal and economic importance of *Rhododendron* in Sikkim Himalayas, an efficient micropropagation protocol for *Melissa officinalis*, utilization of plant biodiversity for treatment and cure of diseases and many other relevant topics are discussed at length. This book will be very much helpful for post graduate students, researchers, scientists and doctors working in the field of Indian system of medicine and academicians working in the field of medical botany, Herbal technology and ethnobotany as well as other branches of life sciences.

The editor has attempted to organise the book in such a way that it will be useful to everyone interested in the field of medicinal botany, pharmacy, ethno-medicine, ethnobotany, herbal therapy, naturopathy and green health therapy. This book will be equally helpful to students of pharmacognosy, pharmacology as well as to the environmentalists and conservationists. The editor has endeavor here is to make people aware of potential of medicinal plants from all angles so that these life saving plants can be protected from total destruction.

I express my sincere thanks to all the contributors for providing me with excellent papers. I had a lot of discussion with them during editorial process. I have tried my best to edit those papers keeping their original version and ideas intact. However the onus of technical contents and views of any chapter rests with their respective author(s).

Sincere thanks are also due to my wife Smt. Kalpana Jha who had been a source of inspiration to pursue the work in all the circumstances. My sons Sankalp Jha and Sarthak Jha deserves special appreciation for cooperating me in various ways. Actually the book utilized their time which I could have spent with them.

I am also thankful to the publishers. The publishers 'Satish Serial Publishing House Delhi' deserves special appreciation for their painstaking and enthusiastic approach to complete this strenuous task well in time. While every effort has been made to avoid mistakes there could be still some errors. I do hope the readers will bear with me for any shortcomings in this account.

Editor

About the Editor

Dr.Santosh Kumar Jha is a Practicing ethnobotanist. He has been working on Ethnobtany of Bihar and Andaman and Nicobar Islands. He has six years of experience in teaching Ethnobotany, Medical botany, Cytology, Genetics and Ecology at UG and PG level. He has Post-graduated from Lalit Narayan Mithila University, Darbhanga and secured his Ph.D. degree from Nalanda Open University, Patna. He has supervised more than a dozen master dissertations on Ethnobotany. Presently, he has been working as an Assistant Professor in Jawaharlal Nehru Government College, Port Blair. He has ten research papers to his credit in various National and International journals. He has immensely contributed to the development of course material for Post Graduate students in Ethnobotany and Cytogenetics of Nalanda Open University. He is a life member of Indian Science Congress Association, Kolkata, Medicinal and Aromatic Plant Association of India and Andaman Science Congress Association, Port Blair. He is also a member of editorial board of three reputed research journals.

Contents

SECTION-III : UTILIZATION AND CONSERVATION

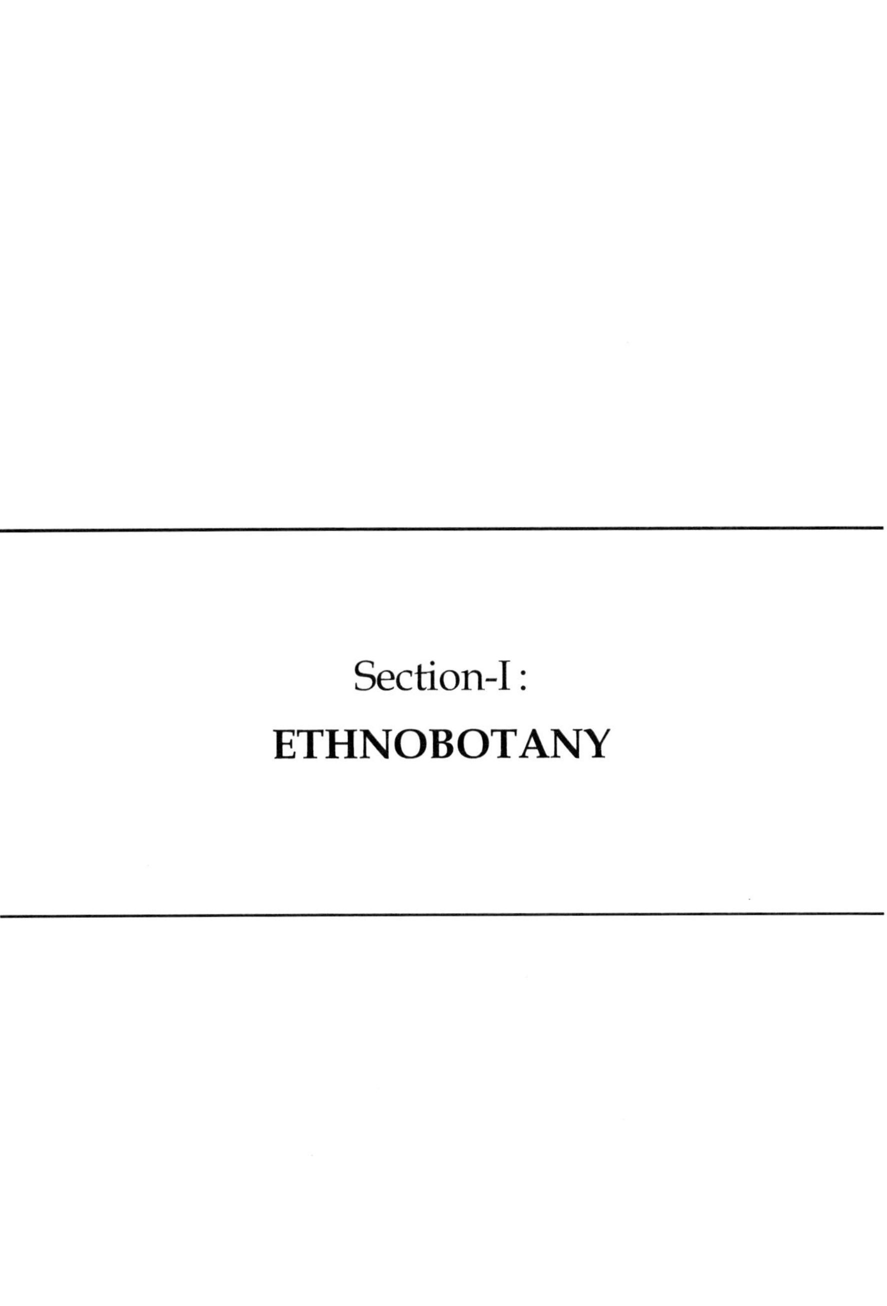

Section-I :

ETHNOBOTANY

Chapter 1

Ethnobotanical Plants of Veligonda Hills, Southern Eastern Ghats, Andhra Pradesh, India

Basha S.K.M., Sivakumar Reddy P.

NBKR Medicinal Plant Research Institute;Vidya Nagar; SPSR Nellore District. A. P., Research and Development Centre; Bharathiyar University; Coimbatore; Tamil Nadu

ABSTRACT

This paper deals with ethnobotany of veligonda hills of Eastern Ghats. The forests of Eastern Ghats in Andhra Pradesh are inhabited with 33 tribal groups. They contain valuable information regarding therapeutic properties of their surrounding crude drugs for different human and veterinary ailments which was recorded and critically analyzed with the help of literature as well as field observations. Based on these observations 226 plants species have been listed alongwith their common name and family.

Keywords: Ethnoarchaeology; ethnobryology; ethnoecology; eastern ghats; venkatagiri range.

Introduction

Plants have been used both in the prevention and cure of various diseases of humans and their pets. With the advent of human civilization, many systems of therapy have been developed primarily based on plants. Ayurveda, Homeopathy, Sidda, Unani, etc. are our traditional systems of medicines. The plant-based traditional medical systems continue to provide the *primary health care* to more than three-quarters of the world's populace. The World Health Organization has estimated that over 80% of the global population relies chiefly on traditional medicine (Akerele 1992).

Indigenous herbal treatment is a part of the culture and dominant mode of therapy in most of the developing countries. These traditional phytoremedies, with a considerable extent of effectiveness, are socially accepted, economically viable and mostly are the only available means.

Still, one-third of the modern pharmaceutical preparations have botanical origin. International trade on medicinal plants is, therefore, increasing rapidly mainly as result of intensified adoption of crude extracts for self-medication by the general public in the developed countries. In India, the use of plants for medicinal treatment dates back to 5000 years. It was officially recognized that 2500 plant species have medicinal value while over 6000 plants are estimated to be explored in traditional, folk and herbal medicine (Huxley 1984).

Ethnobotany

Ethno botany deals with the relationship between human societies and plants. It has been recognized as a multidisciplinary science comprising of many interesting and useful aspects of plant science, history, anthropology, culture and literature. Its importance stems from the varied economic uses of plants among the primitive human societies which are equally beneficial to modern man. It has also brought to light numerous little known or unknown uses of plants (Jain 1981).

Hershberger (1896) of Pennsylvania University was to first coin the term ethnobotany to denote the study of plants used by primitive and aboriginal people. Ethno botanical research can provide a wealth of information regarding both past and present relationships between plants and the traditional societies. Investigations into traditional use and management of local flora have demonstrated the existence of extensive local knowledge of not only about the physical and chemical properties of many plant species, but also the phenological and ecological features in the case of domesticated species. In addition to its traditional roles in economic botany and exploration of human cognition, ethno botanical research has been applied to current areas of study such as biodiversity prospecting and vegetation management. It is hoped that, in the future, ethno botany may play an increasingly important role in sustainable development and biodiversity conservation (Rajasekaran & Warren 1994). In an interaction with the traditional areas of science, ethnobotany gives out several interrelated and interdisciplinary subjects link *ethnomedicine, ethnoarchaeology, ethnobryology, ethnoecology, ethnoagriculture, ethnonarcotics, ethnopharmacology,* etc.

Ethnobotanical investigation has led to the documentation of a large number of wild plants used by tribal for meeting their multifarious requirements (Anonymous 1990). Studies on ethnobotany were initiated by Janaki-Ammal as an official programme in the Economic Botany Section of Botanical Survey of India (Howrah) in 1954. From 1960, Jain started intensive field studies among tribal areas of central India (Jain 1963 a-e; 1964 a-c; 1965 a-b). These publications in early sixties triggered ethno botanical activities among botanists, anthropologists and ayurvedic medical practitioners. An AICRP on Ethnobiology came into operation from 1982 at NBRI, Lucknow, and four other centres (Shillong, Howrah, Coimbatore and Port Blair) of Botanical Survey of India (Jain & Mitra 1997). Mudgal (1987) provided a synoptic account of ethnobotanical works in India. Binu *et al.* (1992) compiled the ethno botanical work carried out in India. Later, Lalramnghinglova & Jha (1999) reviewed work on ethnobotany of the World with special reference to India. An important prerequisite for proper utilization of raw materials of the country is the survey of its natural resources and the preparation of an inventory. It is necessary that we should have full knowledge regarding the occurrence, frequency, distribution and phenology of various plants for their proper utilization. The forests of Andhra Pradesh have great potentiality both from the economic and botanical points of view. The State has one of the timber and non-timber rich forests in India.

Study Area

Eastern Ghats are one of the nine major floristic zones of India possessing rich and diversified plant wealth due to undulated topography and availability of rich humus content. The forests of Eastern Ghats in Andhra Pradesh are inhabited with 33 tribal groups (Tribal welfare department, Government of Andhra Pradesh-2011). They have valuable information regarding therapeutic properties of their surrounding crude drugs for different human and veterinary ailments which was recorded and critically analyzed with the help of literature as well as field observations. Based on these observations some potential drug yielding plants, which have limited distribution, were selected for scientific evaluation. Most of the enumerated taxa were reported as endemic and endangered (Nair & Sastry, 1998) as they have been over exploited for different purposes.

Eastern Ghats - An overview

The Eastern Ghats cover on area about 75, 000 sq. km. traversing the coromandel between 11^0 30′ - 22^0 N latitudes and 76^0 50′ - 86^0 30′

E longitudes. Its northern boundary is marked by river Mahanadi basin while the southern boundary is Cauvery and west lives tips bastar, Telangana, Karnataka plateaus and Tamilnadu uplands. They pass mainly through three states viz. Orissa, Andhra Pradesh and Tamilnadu.

In Andhra Pradesh the Eastern Ghats pass through Srikakulam, East Godavari, West Godavari, Khammam, Krishna, Guntur, Mahaboobnagar, Prakasam, Kurnool, Kadapa, Nellore and Chittoor districts. The Eastern Ghats do not form continuous range like Western Ghats but assemblage of discontinuous ranges of hills with plateaus, escarpments, butters, tors, narrow basins and gorges with elevation ranging from few meters to more than 1600m. Mahanadi, Godavari, Pennar and Cauvery are main rivers which rise in Western Ghats have cut extensively through the Eastern Ghats to escape in to the Bay of Bengal, hence, they do not form a continuous range. Based on the climates, topographic, geographical features the Eastern Ghats of Andhra Pradesh can be divided in to the following regions.

1. Northern Eastern Ghats: The stretching extreme north of the state *i.e.* Simhachalam and Rampa hills.
2. Southern Eastern Ghats: These Ghats stretching between the South of the river, the cannery through Papi hills, Kondapalli range, Nallamalais, Yerramalais, Palakonda, Veligonda range, Horseley hills, Seshachalam hills, Nagari hills, etc.

The altitudes in the Eastern Ghats of Andhra Pradesh range from 300 - 1500m above MSL. The altitudes more than 1000m above MSL in central parts of the north Eastern Ghats and 300 – 600m, and above in Southern Eastern Ghats, while in Nallamalais the highest peak rises between 600-800m above MSL.

Veligonda Hills

Veligonda and adjoining hill ranges spread along about 170 Km North to South in Kadapa and Nellore Districts and stretching a little further into Prakasam District. Geographically these hill ranges lie in between 79^0 E to 79^0 30′ E and 13^0 45′ N to 150 15′ N. The altitude in general ranges up to 1000 m. The forests are in general dry deciduous type. Veligonda and adjoining hill ranges comprising of Palakonda, Seshachalam, Lankamala and terminal part of Nallamalais from mid region of Southern Eastern Ghats.

The Veligonda range which separates the Nellore district from Kadapa and Kurnool is the back bone of the Eastern Ghats, starting from Nagari promontory in Chittor district. It runs in a northerly direction along the western borders of the Nellore district, rising to the elevation of 3,626 feet at Penchalakona in Rapur Thaluk. Veligonda hill ranges have high altitudinal and deep valleys. Among the Velugondas range of hills the Durgam in Venkatagiri range and Penchalakona are the most prominent and 914 meters above mean sea level.

Climate

Mean Annual Rain Fall

The rain fall ranges from700-1000mm through south west and NorthEast monsoons.

Mean Annual Temperature.

Maximum temperature is 36-47°C.

Soils

Red sandy soil, Mixed red and black soil.

Vegetation types in Eastern Ghats in Andhra Pradesh

The forest area of the State extends to about 63,814 sq km which constitutes 23.2 % of the total land area (Andhra Pradesh State of forest report- 2013). The vegetation in Eastern Ghats is determined by climate, edaphic factors and biotic factors along with altitude. The wide variations in climate and topography of the Eastern Ghats have resulted in various types of forest growth. The vegetation in Eastern Ghats are classified based on the concept of Champion and Seth (1968). The following types of vegetation are found in Eastern Ghats of Andhra Pradesh.

I. Tropical moist deciduous vegetation

These are typical deciduous forest with high annual rainfall of over 1000mm and mixed with evergreen species only along the patches of selected habitats. This type of forests found in Eastern Ghats of Andhra Pradesh like, Ananthagiri, Maredumilli, and East Godavari District. It has the following prominent sub types *viz.*, a) Forests dominated with Sal (*Shorea robusta*), b) Those completely devoid of Sal or non Sal forests, c) Riparian forests found along with river banks, streams and in the low hilly areas.

The vegetation is characterized by dense foliage at top canopy with abundant large climbers and epiphytes with scattered bamboo growth. The deciduous period is very less i.e., March-April. These can be sub divided into following categories.

a) Northern Tropical moist deciduous Sal forests

This type of forest sub type found at Donubai area, Srikakulam, Vijayanagaram and Seshachalam hills of Chittoor and Kadapa Districts. The most dominant tree species are *Shorea robusta* along with other codominate species like *Xylia xylocarpa, Haldinia cordifolia, Anogeissus latifolia, Terminalia alata, Lannea coromandelica, Madhuca longifolia, Albizzia procera, Syzygium cumini, Pterocarpus marsupium* mixed with species of middle canopy like *Cleistanthus collinus, Dillinia pentagyna*. The lower canopy with shrubby species *Alstonia venenata, Cipadessa baccifera, Woodfordia fruiticosa, Helictres isora,* etc. and ground is covered with some herbs, like *Desmodium pulchellum, Curcuma neilghierensis, Globba merantina, Tephrosia tinctorea,* etc. along with grass species like *Arundinella setacea, Apluda muitica,* etc. A few evergreen species like, *Syzygium cumini, Memecelon umbellatum, Diospyros malabarica, D. melanoxylon,* etc.

b) Southern Tropical moist deciduous (non-Sal) forests

These are found in the districts of Vizayanagaram (Punyagiri area), Visakhapatnam (Gudem), East Godavari (Rampa), West Godavari (Polavaram), Kurnool (Nallamalais), which contain the dominant species like *Anogeissus latifolia, Dalbergia latifolia, Mangifera indica, Pterocarpus marsupium, Terminalia alata, Sterculea urens* and *Xylia xylocarpa,* represents top canopy mixed with middle canopy species like *Bridelia retusa, Careya arborea, Grewia tilifolia, Glochidion zeylanicum, Holarrhena pubescens, Litsea glutinosa, Mallotus philippensis, Polyalthea cerasoides* along with bamboo breaks commonly of *Dendrocalamus strictus* and occassionally of *Bambusa arundinacia* and teak also found in some plantations. The middle and lower canopy with the species of *Cissus vitigenea, Gardenia gummifera, Helectres isora, Ixora arborea, Nyctanthus arbor-tristis, Woodfordia fruiticosa,* etc. The low shrubby layer mixed with tall grasses such as *Apluda mutica, Themeda triandra, Chlorophytum tuberosum, Pimpinella tirupatiensis,* etc. along with lianas

strictus and occassionally of *Bambusa arundinacia* and teak also found in some plantations. The middle and lower canopy with the species of *Cissus vitigenea, Gardenia gummifera, Helectres isora, Ixora arborea, Nyctanthus arbor-tristis, Woodfordia fruiticosa,* etc. The low shrubby layer mixed with tall grasses such as *Apluda mutica, Themeda triandra, Chlorophytum tuberosum, Pimpinella tirupatiensis,* etc. along with lianas of *Bauhinia vahlii, Entada pursaetha, Toddalia asiatica,* etc with ground species.

c) Southern tropical moist deciduous riparian forest

These are common along with river banks (Godavari), with semi evergreen species like *Barringtonia acuminata, Ficus racemosa, Homonia riparia, Terminalia arjuna, Mimosa pudica, Syzygium cumini,* etc. The dominant grass in this vegetation type is *Saccharum spontaneum.*

II. Tropical dry deciduous forests

This type represents typical deciduous forest growing in larger areas along with the northern, middle and southern Eastern Ghats. The upper canopies in these forests are uneven with mixture of species mostly typical deciduous trees which become leafless during dry seasons. Shrubs and grasses grow as undergrowth in a limited density in frequent forest fires. Bamboo and woody climbers being exposed, moist areas along low stream banks are the suitable habitats for epiphytes. These forests are classified into following sub types:

a) Teak - bearing dry deciduous forests

These are distributed mostly in Eastern Ghats of Visakhapatnam, East Godavari, West Godavari, Khammam districts and Rayalaseema region. The teak is associated with *Anogeissus latifolia, Boswellia serrata, Cassia pinnata, Chloroxylon swietenia, Garuga pinnata, Pterocarpus marsupium, Terminalia alata, T .chebula, T. bellirica,* etc., mixed with shrub species like *Canthium dicoccum, Chomelia asiatica, Erythroxylon monogynum, Holarrhena pubescens, Helecteris isora,* etc., covered by climbers like *Bauhinia vahli, Cissus pallida, Mucuna pruriens, Ventilago maderaspatana.* The ground layer is covered gregariously with bamboo bushes, *Dendrocalamus strictus* and other tall grasses like *Curcuma pseudomontana* and *Desmodium cimicina.*

b) Non – Teak dry deciduous forests

These forests are found in the districts of Rayalaseema and Nellore, interestingly some parts of these forests are dominated by endemic species like *Pterocarpus santalinus, Shorea tumbaggaia* on the hill tops of Seshachalam (Tirumala) hills and *Syzygium alternifolium* is also sub dominant species in these areas. The other endemics like *Boswellia ovalifoliolata, Cycas beddomei, Pimpinella tirupetiensis, Rhynchosia beddomei* and *Actinodaphne madraspatana* are not uncommon in above area. This can be termed as gaps in Seshachalam hills growing along with some common elements like *Terminalia pallida* (endemic tree) *T. alata, Bridelia retusa, Pinus roxburghii* (exotic) *Acacia auriculiformis* (introduced), mixed with dry deciduous elements like *Erythroxylum monogynum, Ziziphus maurintiana,* etc. The climbers are *Celastrus paniculata, Cayratia padata, Marsdenia tenacissima.* The common grasses found in these forests are *Cymbopogon coloratus* and *Heteropogon contortus.* On the hill tops gregarious patches of *Phoenix loureirii* are also found.

III. Mixed dry deciduous forests

These forests are found in drier localities in Rayalaseema region (Anantapuram, Chittoor, Kadapas and Kurnool) of Eastern Ghats. In these forests a mixed type of vegetation is seen. These forests are classified into following sub types:

a) Southern - mixed dry deciduous forests

These are more common in drier localities and subjected to extreme biotic interference like grazing, fires and collection centers of NWFP, found mostly in all districts of Rayalaseema region of Eastern Ghats. The floristic components comprises *Gardenia gummifera,* different species of *Terminalia, Albizia* and Acacias, *Pterocapus marsupium, Hardwickia binata, Randia uliginosa, Balanites eagiptica,* etc. along with gregarious growth of *Phoenix loureirii* and *Chloroxylon swietenia.* The hill slopes are found with *Boswellia serrata, Commiphora caudata, Terminalia arjuna* (trenches of hill slopes). The climbers like *Pterospermum hexapetalum, Decalepis hamiltonii, Gymnema sylvestre,* etc., found in open areas of Nallamalais, Seshachalam and Yerramalais of the Eastern Ghats.

b) Northern mixed dry deciduous forests

These forests are not frequent but present in northern and southern corners like in Orissa and Tamilnadu states respectively.

IV. Dry evergreen forests

This type of forests occur in coastal plains like Vijayanagaram (Poolbagh), Visakhapatnam (Madugula), Srikakulam (Pathapatnam), Nellore (Sriharikota) and Chittoor (Mamandur) district. Sriharikota Island is located in Nellore district of southern part adjoining the Pulicat Lake. This island and its surroundings support dry evergreen vegetation. The common species found in these forests are *Albizia amara, Manilkara hexandra, Sapindus emarginatus* and *Strychnos nux-vomica*. The climbers like *Strychnos minor, Pyrenacanthus volubilis,* and *Derris scandens.*

V. Thorny – Scrub forests

These are degraded deciduous forests due to biotic interference, over exploitation, and frequent fires and are widely distributed in arid and semiarid parts of Eastern Ghats are especially in forest peripheries. Hence the climax was changed to thorny scrub forests (secondary in origin). Due to frequent forest fires some fire resistant spiny species like *Lantana camara, Zizyphus oenoplea* are established as invaders. In some places *Hyptis suaveolens* and *Cassia alata* are also found as invaders. These forests supports the growth of grass species for short period of rainy season and vegetation termed as dry Savannah forests (*Infact Savannah are native to African countries).

Typical species are *Chloroxylon swietenea, Terminalia alata, Atalantia monophylla, Capparis zeylanica, Cadaba fruiticosa, Zyziphus mauritania, Z. xylopyrus, Lantana camara, Euphorbia antiquorum, E. tirucalli, Flacourtia indica, Dodonaea viscosa, Cassia auriculata, Dichrostachys cineria,* etc. with stunted growth bearing elements of *Terminalia alata, Anogeissus latifolia, Pterocarpus marsupium,* etc. The grass species like *Apluda mutica, Themeda triandra, Cymbopogon* sp. etc., are also common.

Ethnology in Eastern Ghats of Andhra Pradesh

The tribal people of Eastern Ghats of Andhra Pradesh originated long back of prehistoric era. The stone-age culture of these tribals is evidenced by micro and mega-lithic sites at upper Godavari (Sanapati and Sahu, 1966). As Andhra Pradesh is drained by major rivers mainly in Eastern Ghats, the tribals also settled along the river streams which pass through the Ghats. These tribes have their own styles of culture, language, heritage, customs, religious practices, food habits etc., and mainly live deep in forests, unable to contact with civilized people. They mainly depend on minor forest produce, hunting and rarely on agriculture for their livelyhood. Out of 33 types of tribe of Andhra Pradesh, 27 communities are confined to these isolated hills and adjacent plains.

Yanadis

The Yanadis are more primitive aboriginal and concentrated mainly in Chittoor, Kadapa, Nellore, Ongole and Guntur districts of Andhra Pradesh. The tribe is said to be the direct descendents of Paleolithic people. Chenchus and Yanadis both are from one parental stock and are believed to be originated in Nallamalai hill tracts (Raghavaiah, 1962). Yanadis speak only Telugu language with a characteristic dialect and accent. They don't have any special functions, ceremonies or celebrations particular to them. They are integrated with Hindu social system and practices.

The Yanadi lead a carefree, life with contentment and unbridled merriment. Their diet chiefly consists of vegetable food and animals, wild fowls and other birds of food value. They even dig rat holes and use them in menu. However, the best satiating food for them is fish. Honey gathering in forests and plains is also a common practice among them.

Two sub-tribes are recognized within Yanadis, based on their occupation are Manchi yanadi, the superior type and the Challa yanadis, the inferior type and carry different names including Garapa Yanadis, Chettu yanadis, Kappa yanadis, based on their habitation and the food taken. The Yanadis are short statured with dark skin colour, platyrrhine nose, long head, prominent chin, thick lips and scanty hair both on head and body. They reside in huts usually constructed adjacent to a water source. Yanadis living in and around forests keep themselves busy in collecting and selling minor products.

Table 1. Ethnobotanical Plants of Veligonda Hills

S.No	Botanical Name	Vernacular Name	Family	Habit
1	*Abrus precatorius* L.	Gurivinda	Fabaceae	Climber
2	*Abuliton indicum* (L.) *Sweet.Hort.Brit.*	Duvvenakaya /Tutturubenda	Malvaceae	Shrub
3	*Acacia leucophloea (Roxb.)* Willd.	Tella tumma	Mimosaceae	Tree
4	*Acacia tora(Roxb.) Craib.*	Korinteega	Mimosaceae	Climber
5	*Achyranthes aspera* L.	Uttareni	Amaran-thaceae	Herb
6	*Actinopteris radiata (Koenig ex Sw.)*	Mayuri shika	Actinopteri-daceae	Herb
7	*Adiantum caudatum* L.	Raja hamsa	Adiantaceae	Herb
8	*Aegle marmelos* (L.)	Maredu/Bilva	Rutaceae	Shrub
9	*Aerva lanana* (L.)	Pindikura	Amaran-thaceae	Herb
10	*Ageratum conyzoides* L.	Goat weed	Asteraceae	Herb
11	*Alangium salvifollium* (L.f.)	Udaga / Ankolamu	Alangiaceae	Tree
12	*Albizia amara (Roxb.)*	Cheekireni	Mimosaceae	Tree
13	*Albizzia odoratissima* (L.f.) *Benth*	Chinduga	Mimosaceae	Tree
14	*Alstonia scholaris* L.	Edakulapala	Apocynaceae	Climber
15	*Andrographis paniculata (Burm.f.)* Wall.	Nelavemu	Acanthaceae	Herb
16	*Anisomelea malabarica* (L.)	Moga-Bira	Lamiaceae	Shrub
17	*Annona squamosa* L.	Sitapalem	Annonaceae	Tree
18	*Annona reticulate* L.	Ramapalam	Annonaceae	Tree
19	*Anogeissus latifolia* (Roxb.ex Dc.)	Chirimanu / Elama	Combre-taceae	Tree
20	*Argemeone mexicana* L.	Kusuma / Brahmadandi	Pepepave-raceae	Herb
21	*Aristolochia braceteolata* Lam.	Gadidagadapa	Aristolo-chiaceae	Herb
22	*Aristolochia indica* L.	Easwari	Aristolo-chiaceae	Herb

23	*Asparagus racemosus* Willd.	Sathavari	Liliaceae	Herb
24	*Atalantia monophylla* (L.)	Munukudu	Rutaceae	Shrub
25	*Atylosia scarabaeoides* (L.) *Benth.*	Adavikandi	Fabaceae	Climber
26	*Azadirachta indica* A.Juss. In Mem. Mus. Natl.	Vepa	Meliaceae	Tree
27	*Azima tetracantha* Lam.	Tella uppili	Salvado-raceae	Shrub
28	*Bacopa monnieri* (L.) *Pennel*	Brahmi	Scrophul-riaceae	Herb
29	*Basella alba* L.	Bachali	Basellaceae	Climber
30	*Bauhinia racemosa* Lam.	Are fibres	Caesalpin-aceae	Tree
31	*Blumea mollis(D.Don)* Merr.	Kukkapogaku	Asteraceae	Erect aromat-ic herb
32	*Boerhavia diffusa* L.	Attamamidi	Nyctagin-aceae	Herb
33	*Bombax ceiba* L.	Adavi Buruga	Malvaceae	Tree
34	*Borassus flabellifer* L.	Tati	Araceae	Tree
35	*Boswelia ovalifoliata* Bal. & Henry	Sambrrani	Burseraceae	Tree
36	*Boswelia serrata* Roxb.	Sambrani	Burseraceae	Tree
37	*Buchnania axilaris* (Desr.)	Sara	Anacardi-aceae	Tree
38	*Butea monosperma* (Lam)	Moduga	Fabaceae	Tree
39	*Caesalpinia bonduc* (L.) *Roxb.*	Gacha	Fabaceae	Shrub
40	*Calophyllum inophyllum* L.	Ponna	Calophy-llaceae	Tree
41	*Canavalia ensiformis* (L.) *DC*	Adavi thamba	Fabaceae	Climber
42	*Capparis sepiaria* L.	Nalla uppili	Capparaceae	Shrub
43	*Capparis zeylanica* L.	Adonda	Capparaceae	Shrub
44	*Carissa carundus* L.	Vaka	Apocynaceae	Small Tree
45	*Cardiospermum halicacabum* L. Sp.	Buddakakara	Sapindaceae	Climber

46	*Careya arborea* Roxb.	Budda darimi	Barringtoni-aceae	Tree
47	*Carallumma adsendens* Haw.	Kundelu Kommulu	Asclepiad-aceae	Herb
48	*Carmona retusa* (Vahl)	Nomuchettu / Barranki	Boraginaceae	Shrub
49	*Cassia absus* L.Sp.	Chanupala vittulu	Caesalpin-aceae	Herb
50	*Cassia fistula* L.Sp.	Rela	Caesalpin-aceae	Tree
51	*Cassia italica* (Mill.) Spreng.	Nelatangedu	Caesalpin-aceae	Herb
52	*Cassia Montana Meyne ex.* Roth.	Pyditangedu	Caesalpin-aceae	shrub
53	*Cassia occidentalis* L.Sp.	Kasintha	Caesalpin-aceae	shrub
54	*Cassine glauca (Rottb.)* Kuntz.	Nerdhi	Celastraceae	
55	*Cassytha filiformis* L.	Sitamma savaralu	Lauraceae	Climber
56	*Catunaregam spinosa* (Thung.)	Manga	Rubiaceae	Shrub
57	*Cayratia pedata* (Lam.)	Adavi gummaditeega	Vitaceae	Climber
58	*Centella asiatica* (L.)	Saraswathi	Apiaceae	Herb
59	*Chionanthus zeylanica* L.Sp.	Punagani	Oleaceae	Tree
60	*Chloroxylon swietenia* DC. Prodr.	Billudu	Meliaceae	Tree
61	*Christella dentata* (Forssk.)	Downy wood fern	Thelipteri-daceae	Herb
62	*Cipadessa baccefera* (Roth) Miq. In Ann. Mus	Ranaberi	Meliaceae	Shrub
63	*Cissampelos pareira* L.Var.hirsuta	Visha boddi	Menisper-maceae	Shrub
64	*Cissus quadrangularis* L.	Nalleru	Vitaceae	Herb
65	*Cissus vetigenia* L.Sp.Pl.	Adavi gummidi	Vitaceae	Climber
66	*Citrullus colocynthus* (L.)	Papara	Cucurbit-aceae	Climber

67	*Cocculus hirsutus* (L.) Diels in Engl.	Dusari Teega	Menisperm-aceae	Climber
68	*Coccinia grandis* J.Voigt.	Kakidonda	Cucurbit-aceae	Climber
69	*Cochlospermum religiosum* (L.) Alston	Konda gogu	Cochlosper-maceae	Tree
70	*Coldenia procumbens* L.	Hamsapadu	Boraginaceae	Herb
71	*Commifera caudate* (White & Arn.) Engl.	Kondamamidi	Burseraceae	Small Tree
72	*Corallocarpus epigaeus* (Rott.)	Mukkudonda	Cucurbit-aceae	Climber
73	*Cordia dichotoma Forst.f*	Bankamanu / Nakkera	Boraginaceae	Tree
74	*Costus speciosus* (Koen.)	Adavi allam/ Chengalva cost	Costaceae	Herb
75	*Crateva religiosa G.* Forst.	Varuna	Capparaceae	Tree
76	*Crotalaria retusa* L.	Sanapusphi	Fabaceae	Herb
77	*Curculigo orchioides Gaertn.,Fruct.*	Nelathati	Hypoxid aceae	Herb
78	*Cycas beddomi Dyer.*	Peritha	Cycadaceae	Tree
79	*Cymbopogon colorattus* (Hook.f.Stapf.	Boda Gaddi		
80	*Cymopogon flexuosus* (L.) Rendle	Nimma gaddi	Poaceae	Herb
81	*Dalbergia latifolia* Roxb.	Jittagi / Iridi	Fabaceae	Tree
82	*Dalbergia paniculata* Roxb.Pl.Cor.t.	Pacchari	Fabaceae	Tree
83	*Datura metal* L.	Nalla ummetta	Solanaceae	Herb
84	*Datura stromonium* L.	Ummetta	Solanaceae	Shrub
85	*Decalepis hamiltonii* Wight & Arn	Maredu kommulu	Asclepedi-aceae	Shrub
86	*Decaschistia crotonifolia* Wight & Arn	Adavigogu	Malvaceae	Shrub
87	*Deccannia pubscens* (Roth)	Konda manga	Rubiaceae	Tree
88	*Derris scandens* (Roxb.)	Nalla teega	Fabaceae	Climber
89	*Desmdium triflorum* (L.) Dc.	Munta mandu	Fabaceae	Herb

90	*Dillenia pentagyna roxb.*	Chinna kalinga	Dilleniaceae	Tree
91	*Dioscoria pentaphylla* *L. Crinum viviparum* (Lam.)R.Ansari &V.J.Nair	Injedigadda	Dioscore aceae	Climber
92	*Diospyros ebenum J.*Koenig.	Nalla uti	Ebenaceae	Tree
93	*Diospyros melanoxylon* Roxb.	Tumki	Ebenaceae	Tree
94	*Dodonea viscosa* (L.) Jacq.Enum.	Bandaru	Sapindaceae	Shrub
95	*Eclipta prostrata* (L.)	Gunta galijeru	Asteraceae	Herb
96	*Echinops echinatus*	MullaBanthi	Asteraceae	Herb
97	*Ehretia pubescens* Benth.in Royle.	Pakki	Boraginaceae	Tree
98	*Enicostema axillare* (Lam.)	Gulividi	Gentianaceae	Herb
99	*Entada pursaetha DC.*	Gila teega / Konda chinta	Mimosaceae	Climber
100	*Erythroxylum monogynum* (Roxb.)	Devadari	Erythroxy-laceae	Shrub
101	*Euphorbia hirta* L.	Nanabala	Euphorbi-aceae	Herb
102	*Ficus benghalensis* L.Sp.	Marri	Moraceae	Tree
103	*Ficus microcarpa* L.f.	Kondajuvvi	Moraceae	Tree
104	*Gardenia gummifera*.L.f.	Bikki	Rubiaceae	Tree
105	*Gardenia resinifera* Roth.	Erribikki	Rubiaceae	Tree
106	*Givotia moluccana* (L.)	Tella poliki	Euphorbia-ceae	Tree
107	*Gmelia asiatica* L.	Adavi Gummadi	Verbenaceae	Shrub
108	*Gloriosa superba* L.Sp.Pi.	Nabhi / Nagetigadda	Liliaceae	Climber
109	*Glycosmis pentaphylla (Retz) DC.*	Gonji	Rutaceae	Shrub
110	*Grewia tiliifolia* Vahl.	Adavicham-anthi	Tiliaceae	Shrub
111	*Guazuma tomentosa* Kunth. *(Guazuma ulmifolia* Lam.*)*	Rudhracksha	Tiliaceae	Tree
112	*Guidonia esculenta* (Roxb.) Baill.	Chilukadudhi	Flacourti-aceae	Shrub

113	*Gymnema sylvestre* (Retz)	Podapatri	Asclepedi-aceae	Shrub
114	*Gyrocarpus asiaticus* Willd.	Taniki / Nalla poliki	Hernandi-aceae	Tree
115	*Habenaria apetala*		Orchidaceae	Herb
116	*Hardwickia binata* Roxb.	Api	Fabaceae	Tree
117	*Haldinia cordifolia* (Roxb)	Rudra ganapa	Rubiaceae	Tree
118	*Hedtitus peberula* (G.Don) Arn.	Chiruveru	Rubiaceae	Herb
119	*Hedyotis corymbosa* (L.)	Vermela-vemu	Rubiaceae	Herb
120	*Hedyotis herbacea* L.	Chiriveru	Rubiaceae	Herb
121	*Heliotropium indicum* L.	Nagadanthi	Boraginaceae	Herb
122	*Helicters isora* L.	Gooba thada	Sterculiaceae	Shrub
123	*Hemidesmus indicus* (L.) *var.inducus*	Sugandhapala	Periplocaceae	Herb
124	*Hemionitis arifolia (Burm.f.)* Moore	Rama bhanam	Hemionitid-aceae	Herb
125	*Hugonia mystax* L.	Kakibeera	Linaceae	Shrub
126	*Hibiscus platanifolius* (Willd.)	Kondagogu	Malvaceae	Tree
127	*Hiptage benghalensis* (L.) Kurz	Madhavi tega	Malphigi-aceae	Climber
128	*Holarrhena antidysenterica* (Roxb. ex Fleming)	Kola musthi / pala/kodisapala	Apocynaceae	Tree
129	*Holostemma ada-kodein Schultes*	Tella jilledu / Peyyi baddu	Asclepedi-aceae	Climber
130	*Hybanthus enneaspermus* (L.) Muell. Arg. Fragm.	Ratna purusha	Violaceae	Herb
131	*Ichnocarpus frutescens* (L.) R.Br.	Palateega	Apocynaceae	Climber
132	*Indigofera asphelthoides* Vahl.ex.	Sivavemu	Fabaceae	Herb
133	*Impatines leschenaulti* (DC.)Wall.ex.Wight&Arn		Balsamin-aceae	Herb
134	*Ixora pavetta Andr.* Bot.Repos.t.	Korivi/Papidi	Rubiaceae	Tree
135	*Jasminum auriculatum* Vahl.	Adavimalli	Oleaceae	Climber

136	*Justicia adhatoda* L.	Addasaram	Acanthaceae	Shrub
137	*Lawsonia inermes* L.	Gorintaku	Lythraceae	Small Tree
138	*Lannea coromandelica* (Houtt.) Merr.	Gumphena	Anacardi-aceae	
139	*Lantana camara* L.	Phallikampa	Verbanaceae	shrub
140	*Leonotis nepetiifolia* (L) R.Br.Prodr	Ranabheri	Lamiaceae	Herb
141	*Lepisanthes tetraphylla* (Wall.) Radf.	Sali kunkudu	Sapindaceae	Tree
142	*Leptadenia reticulate* (Retz.)	Mukkupala-teega	Asclepiad-aceae	Climber
143	*Limnophila indica* (L.)	Sambrani	Scrophulri-aceae	Herb
144	*Limonia acidissima* Groff.	Velaga	Rutaceae	Tree
145	*Listea glutinosa* (Lour.) C.B.Rob.	Pulusumamidi	Lauraceae	Tree
146	*Lygodium flexuosum* (Linn.)	Mekasannu	Schizaeaceae	Climber
147	*Madhuca longifolia* (Koen.) Macbr.	Ippa	Sapotaceae	Tree
148	*Mallotus philippensis* (Lam.) Mull.Arg.	Sinduri	Euphorbi-aceae	Tree
149	*Manikara hexandra* (Roxb.)	Pala	Sapotaceae	Tree
150	*Mimusops elengi* L.	Pogada	Sapotaceae	Tree
151	*Mimosa pudica* (L.)	Aathipathi	Sapotaceae	Herb
152	*Momordica charantia* L.Var.muricata Willd	Buddakakara	Cucurbit-aceae	Climber
153	*Moringa concanensis*	Adavi munaga	Moringaceae	Tree
154	*Murraya paniculata* (L.) *Jack.*	Naramusti	Rutaceae	Tree
155	*Naraveliazeylanica*(L.)DC	Korivi kattaku	Ranunkul-aceae	Climber
156	*Neptunia oleraceae* Lour.	Neruthaluvapu	Fabaceae	Shrub
157	*Ochna obtusata* DC.	Errijambi	Ochnaceae	shrub
158	*Orabanche cernua* Loefl.	Pogakumalle	Orabanch-aceae	
159	*Olax scandens* Roxb.	Mekabanda	Olacaceae	Climber

160	*Opilia amentacea* Roxb.	Nallameka-banda	Opilaceae	Climber
161	*Pavonia xylanica* (L.)Cav.	Adavi puttudu/ Chiru benda	Malvaceae	Herb
162	*Pentatropus capensis* (L.f.)	Yedupullateega	Asclepiad-aceae	Climber
163	*Pergularia daemia* (Forssk)	Dushtapaku	Asclepiad-aceae	Climber
164	*Phoenix sylvestris* (L.) Roxb.	Eetha	Arecaceae	Tree
165	*Phyllanthus amarus* Schum & Thonn	Nelausiri	Euphorbi-aceae	Herb
166	*Phyllanthus embllica* L.	Nelli/Usiri	Euphorbi-aceae	Tree
167	*Physalis minima* L.	Budama	Solanaceae	Herb
168	*Pimpinella tirupathensis* L.	Adavi kottimeera	Apiaceae	Herb
169	*Piper Sylvester*	Toka mereyalu	Piperaceae	Climber
170	*Pithacalobium dulce* (Roxb.) Benth.	Simachintha	Fabaceae	Tree
171	*Plumbago zylanica*	Tella chitramulam	Plumbagin-aceae	Herb
172	*Pongamia pinnata* (L.) *Pierre*	Kanuga	Fabaceae	Tree
173	*Polyalthiya cerasoides* (Roxb.)	Dudduga	—	Tree
174	*Pouzolzia zeylanica* (L.) Benn.	Uchchagadda	Urticaceae	Herb
175	*Premna tomentosa* Willd	Narava/Namari	Verbanaceae	Tree
176	*Pterocarpus marsupium* Roxb.	yegisa	Fabaceae	Tree
177	*Pterocarpus santalilnus* L.F.	Rakta chandanam	Fabaceae	Tree
178	*Pterospermum xulocarpum* (Gaertn.)	Tada	Sterculiaceae	Tree
179	*Pueraria tuburosea Roxb.ex* Willd.	Chenchu gadda /Bhoochakra	Fabaceae	Climber
180	*Rivea gtoicraterufirnus* (Desr.)	Boddi teega	Convolvul aceae	Shrub

181	*Rhynchosia minima* (L.) Dc.	Adavichikkudu	Fabaceae	Tree
182	*Salvadora persica* L.	Nalla uppili/ Varagogu	Salvador-aceae	Tree
183	*Santalum album* L.	Chandanam , Srigandham	Santalaceae	Tree
184	*Schefflera stellata* (Gaertn.) Harms	Reval, Ededdula	Araliaceae	Shrub
185	*Scilla hyacinthina* (Roth)	Nakkeragadda	Liliaceae	Herb
186	*Shorea roxburghii* G.Don Gen.Syst	Jalari	Dipteeroc-arpaceae	Tree
187	*Shorea tumbeuggaia* Roxb.	Tamba / Guggilam	Dipteroca-rpaceae	Tree
188	*Solanum Melanogena* L.*varinsanum* L.	Chiruvanga	Solanaceae	Shrub
189	*Solanum surrattense* Burm. F.	Errivanga	Solanaceae	Tree
190	*Solanum trilobatum* L.	Mulla mushti	Solanaceae	Climber
191	*Soymida febrifuga* (Roxb)	Somi	Meliaceae	Tree
192	*Spondias pinnata* (L.f.)Kurz	Adavimamidi	Anacardiaceae	
193	*Sphaeranthus indicus* L.	Bodasaram	Asteraceae	Herb
194	*Sterculia urens* Roxb.	Thapasi	Sterculiaceae	Tree
195	*Strychnos calubrina* L.	Nagamusti	Loganiaceae	Climber
196	*Strychnos potatorum* L.f.	Musthi	Loganiaceae	Tree
197	*Strychnos potatorum* L.F.Suppl.	Chilla	Loganiaceae	Tree
198	*Suregada angustifolia* (Baill.ex Muell.Arg)	Sapranchi	Euphorbi-aceae	Shrub
199	*Syzygium cumini* (L.)	Neredu	Myrtaceae	Tree
200	*Syzygium alternifolium* (Wight) Walp.	Mogi	Myrtaceae	Tree
201	*Tamarindus indica* L.	Chinta	Fabaceae	Tree
202	*Tarenna asiatica* L.	Kommi	Rubiaceae	Shrub
203	*Terminalia arjuna* (DC.) Wight & Arn)	Arjuna / Tella maddi	Combret-aceae	Tree
204	*Terminalia bellirica* (Gaertn.)	Thandra / tani	Combret-aceae	Tree

205	*Terminalia pallida* Brandis	Tella karaka	Combret-aceae	Tree
206	*Terminallia chebula* Retz.	Karaka	Combret-aceae	Tree
207	*Thespisia populnia* (L.) Correa	Gangaravi	Malvaceae	Tree
208	*Tinospora cordifolia* (Willd.) Hook.f. &Thoms	Tippa teega	Menisper-maceae	Climber
209	*Tribulus terrestris* L.	Palleru	Zygophyll-aceae	Herb
210	*Trichosanthes cucumeria* L.	AdaviPotla	Cucurbit-aceae	Climber
211	*Trichosanthes tricuspidata* Lour.	Papara	Cucurbit-aceae	Climber
212	*Tridax procumbens* L.	Gaddi chamanthi	Asteraceae	Herb
213	*Triumfetta rhomboidea* Jacq.		Tiliaceae	Shrub
214	*Tylophora indica* (Burm.f.)	Kakkupala	Asclepedi-aceae	climber
215	*Vanda spathulata* L.	Nusti bhadhanika	Orchidaceae	Herb
216	*Vanda roxburghii* Nicolson in Salda	Veduru bhadhanika	Orchidaceae	Herb
217	*Ventilago denticulata* Willd.	Surati / Surudu	Rhamnaceae	Climber
218	*Vernonia anthelmintica* (L.)	Adavi jeelakarra	Asteraceae	Herb
219	*Vettiveria zizanioides* (L.)	Vattiveru	Poaceae	Herb
220	*Viscum articulatum* Burm.f.	Badanika	Viscaceae	Shrub
221	*Vitex altissima* L.f.syppl.	Nemaliadugu	Verbanaceae	Tree
222	*Walsura trifolia* (A.Juss)	Valudu	Meliaceae	Tree
223	*Wattakaka volubilis* (L.f.)	Kallisi	Asclepedi-aceae	Climber
224	*Wrightia tinctoria* (Roxb.) R.Br.	Reppala	Apocynaceae	Tree
225	*Ziziphus mauritiana* Lam. Encycl.	Regu	Rhamnaceae	Tree
226	*Ziziphus xylopyrus* (Retz.)	Gotti	Rhamnaceae	Tree

Table 2. List of Threatened, Endangered and Vulnerable medicinal plants of Veligonda Hills.

Threatened Species	Endangered Species	Vulnerable Species
Decalpis hamilitonii	*Cristella dentate*	*Alstonia scholaris*
Cocholospermum	*Lygodium flexuosum*	*Centella asiatica*
Pavonia somifera	*Costus speciosus*	*Coralloacarpus epigaeus*
Tylophora indica	*Vanda spathulata*	*Schefflera stellata*
Plumbago indica	*Gloriosa superba*	*Careya arborea*
Purenia tuberosa	*Hemidesmus indicus*	*Cucumeria aliangium*
Strychnos potatorum	*Hybanths enneaspermus*	*Entada pursaetha*
	Hemionitis arifollila	*Dalbergia latifolia*
	Glycosmis pentaphylla	*Dellenia pentagyna roxb*
	Moringa concanensis	*Cardiospermum helicabum*
	Strychnos nux-vomica	*Habenaria apetala*
	Vernonia anthelmintica	*Maeruva oblongifolia*
	Piper sylevestre	*Holostemma ada-kodein*
	Vanda roxburghii	*Madhuca langifollia*
		Pterocarpus marsupirm rox
		Soymida febrifuga

References

Bartle L. 1997. Why do we need new crops as alternative sources for industrial and medicinal material. In *"Domestication, Production and Utilization of New Crops"*, edited by Smart J & Haq N (Colorilene Printers, Bangladesh).

Bhandary M.J. and Chandrasekhar K.R. 2003. Sacred groves of Dakshina Kannada and Udupi districts of Karnataka. *Curr. Sci.* 85 1655-1656.

Gamble J.S. 1915-36. *Flora of Presidency of Madras.* Adlard and Son Lrs Lonswn 1 55.

Gamble J.S. and CEC Fischer, 1967. *Flora of the Presidency of Madaras.* 1-3 (Calcutta).

Hooker J.D. 1897. *The Flora of British India,* 1-7 (Reeve, L.Co.Ltd. Ashford Kant, London).

Jain S.K. and RR Rao, 1977. *Field and Herbarium Methods.* (Today and Tomorrow Publishers, New Delhi).

Lakshminarayana K. and Venkaiah M. 1998. Biodiversity in the sacred groves of the North Costal Districts of Andhra Pradesh. *National symposium on conservation of Eastern Ghats,* 52-58.

Prakasa Rao, J. and Tarakeswara Naidu, 2010. Diversity of Vascular Plants in the sacred gove of Uppa from Eastern Ghats, Visakhapatnam District in Andhra Pradesh. *Journal of Indian Botany Society*(3) 406-411.

Ranga Rao T.R. 1901. The Yanadis of Nellore district. *Bull. Madras Govt. Museum* 4(2) 87-113.

Ravi Prasad Rao B. 1998. Sacred Groves in Southern Eastern Ghats, India. Paper presented in *International Conference on Conservation of Tropical Species. Communities and Ecosystems.*

Savithramma N., Basha SKM, Rao K.N. 2002. Ethnobotanical wealth of Sriharikota A.P. in *Cultural Ecology of Indian Tribes* Eds.,(Raj publication, New Delhi) 200-235.

Surya Narayana B. and Sreenivasa Rao A. 2002. *Flora of Nellore District,* Andhra Pradesh, (Publication Maharashtra).

□□□

Chapter 2

Appraisal of Ethno Medicinal Plants Used to Treat Human and Livestock Ailments in Shiwalik Hills of Northwest Himalaya, India

Neetu Sharma*

Extension Specialist, KVK, Kangra-176001, CSKHPKV, Palampur, H.P., India, Email: neepradu@gmail.com

ABSTRACT

Different communities throughout the world have specialized and profound knowledge on the use of medicinal plants for various diseases. However, the detailed information on the respective use may extinct in near future as this knowledge is passed only orally among generations in most of the communities. Therefore the present study was conducted to document important ethno medicines used for the treatment of various ailments and diseases by traditional practitioners/local people belonging to Kangra and Hamirpur districts of Himachal Pradesh. Validated questionnaires were applied to elders/informants, traditional healers, local practitioners, medicine men and women from ten villages each from the selected districts. The study demonstrated the use of 39 medicinal plant species for curing various ailments. The most frequently utilized plant part was the leaf followed by root. Basic processes/preparations used by healers for drug preparations included decoction, infusion, extract, powder, juice, paste and poultice. The ever-increasing demand for these species has increased overexploitation from wild habitats. Thus it is recommended to develop farming techniques of commercially viable species and disseminate knowledge among stakeholders.

Keywords: Ethno medicine, Medicinal plants, Traditional Knowledge.

Introduction

Plants are one of the most important sources of medicines. The application of plants as medicines dates back to prehistoric period. In India the references to the curative properties of some herbs in the Rig-veda seems to be the earliest records of use of plants in medicines

(Mohd. et al., 2012). Ethnobotanical studies are often significant in revealing locally important plant species especially for the discovery of crude drugs. Right from its beginning, the documentation of traditional knowledge, especially on the medicinal uses of plants, has provided many important drugs of modern day (Flaster, 1996). The World Health Organization (WHO) has a keen interest in documenting the use of medicinal plants by native people from different parts of the world (Mahwasane et al., 2013). About 60 percent of the world's population and 80 percent of developing countries' populations rely on traditional medicine, mostly plant drugs for their primary healthcare needs (Gaur and Tiwari, 1987). The reliance of this large population could be attributed to cultural acceptability, efficacy against certain type of diseases, relatively good accessibility to the plants and economic affordability of the herbal material as compared to modern medicine and the extensive local knowledge and expertise among the local communities (Omwenga *et al.*, 2012).

The Indian Himalayan Region (IHR) is a mega hot spot of biological diversity (Myers 2000). The flora includes about 8,000 species of angiosperm (40% endemic), 44 species of gymnosperm (16% endemic),600 species of pteridophyte (25% endemic),1737 species of bryophyte (33% endemic), 1,159 species of lichen (11% endemic) and 6,900 species of fungi (27% endemic) (Singh and Hajra, 1996; Samant et al., 1998).These include some 1748 species of medicinal plant with various traditional and modern therapeutic uses (Samant et al. 1998), 675 species of wild edible plants (Samant and Dhar 1997), 118 species of medicinal plants yielding essential oils, 279 species of fodder, 155 sacred plants (Samant and Pant 2003) and 121 rare-endangered plants (Nayar and Sastry 1987, 1988, 1990). Being a hilly state, Himachal Pradesh has rich plant diversity due to varying degree of agro climatic zonation from subtropical to extreme cold. The state is a bucket of large variety of medicinal herbs. There are about 3500 known plant species recorded in the state, about 500 are reported on the medicinal value (Chauhan 2003). This plant diversity is used by the local people for various purposes like, traditional healthcare, cultural, and religious acts. Unfortunately this indigenous knowledge has been passed from one generation to another orally for centuries without the aid of writing as traditional healers do not keep written records for the fear of misuse. As a result, much of the valuable knowledge of the practitioners used to be burnt in the funeral pyre with the death of such persons. Thus there is a great need to document this indigenous knowledge.

Documentation can also play a key role towards conservatory aspects of such species of medicinal plants that could have been proven scientifically to be effective in the management of given diseases (Omwenga *et al.*, 2015).

Methodology

The State of Himachal Pradesh (30°22´40"-33°12´40"N to 75 ° 45´55"- 79 ° 04´20" E) includes parts of the Trans and Northwest Himalaya and covers 55,673 km^2: 9% of the Indian Himalayan Region. Topographically, the state can be divided into three zones -The Shiwaliks or outer Himalayas, The inner Himalayas or mid-mountains and Alpine zone or The greater Himalayas. Among the three zones of the state, the present study was carried out in Kangra and Hamirpur districts of the Shiwaliks or Outer Himalayas. The Kangra district is situated in the Western Himalayas between 31°2 to 32°5 N and 75° to 77°45 E. The district Hamirpur of Himachal Pradesh is situated in between 76°18'-76°44' East longitudes and 31°-32°30' North latitudes. Hamirpur district has wide phyto- diversity of broad-leaved and deciduous plants and has been regarded as a veritable emporium of plant- genetic –resources of medicinal plants. Data collection was initiated with a prior reconnaissance survey and interactions with the village headman and the people in groups, so as to build confidence with them and to get acquainted with area. During the surveys, participatory interview tools including group discussions, informal meetings, questionnaire surveys, and field observations were used for data collection. The information was collected by asking questions in native local language in order to minimize biased information. For accurate and best possible information, knowledgeable elders/ informants, traditional healers, local practitioners, *vaids* were contacted, including medicine men and women from ten villages each from the selected districts. They were asked to give their knowledge about the plants they use against a disease, plant parts harvested, method of preparing the medicine along with form and the mode of administration. Field visits were undertaken along with the respondents in order to identify plant species of medicinal importance.

Results

The present study revealed the use of 39 medicinal plant species. These plant species belonging to 31 families are being used extensively for curing number of ailments like fever, headache, weakness, cough, cold, whooping cough, eye infection, tooth infection, diarrhoea, constipation, acidity, boils and wounds, diabetes, sore throat, blisters in mouth, eye burning, jaundice, stomach pain, rheumatism, sprain, arthritis, bleeding from nose etc. Out of the total 39 plant species, 11plants are also used in ethno veterinary practices. They have been used against various diseases/ailments viz. foot and mouth, mastitis, joint pains, arthritis, broken horns, fractured bones, diphtheria and sprain.

Concerning the use of the different plant parts the healers commonly harvest the leaves (67 preparations), the roots (21 preparations), the fruit (13 preparations) and the bark (12 preparations). Sharma (2015), in her survey on ethno medicinal plants used for traditional healthcare systems of Shiwalik Hills, Himachal Pradesh, also observed that the most frequently used herbal material against different ailments was the leaf .

Different parts of the plant such as bark, leaf, seed, root, etc. have been used in different ways for preparing drugs for different health conditions. The common preparations suggested for internal application mentioned in the chapter are: **decoction** (*simmering of the thicker and less permeable plant parts such as the roots, bark, fruit and seed, for easy extraction of their medicinal constituents*), **infusion** (*a simple way of extracting active medicinal constituent of plants through the medium of hot water. The volatile components of the aerial parts like roots leaves, bark, flowers, fruits, etc. are extracted*), **extract** (*pounded plant parts are steeped in hot or cold water, and finally are sieved out through a clean cloth*), **powder** (*dry plant parts are pounded into powder*) and **juice** (*a fresh product obtained from crushed plant parts*). Externally applied preparations are **paste** (*a soft, wet mixture obtained by grinding plant parts so as to be used as a thin layer application*) and **poultice** (*a soft, moist mass made of plant parts , that is spread on a cloth and then applied to/placed (when it is hot) on* the *skin*).

In majority of the respiratory, digestive and urinary disorders/ ailments, a decoction of plants was prescribed for diseases like cough, fever, constipation, diarrhoea and dysentery while gargling with decoction was recommended for sore throat. Massage of seed oil or application of rhizome paste/leaf juice /decoction/ poultice on joints was recommended for ailments like arthritis and joint pains. Further, it was revealed that pulpy portion of some of the plant leaves or leaf paste was used as a bandage for boils, mumps, wounds and cuts etc. whereas other skin diseases were cured by external application of the paste of leaf, latex or sticky sap. It was further observed that the medicine men believe to use a combination of plants for best results to promote the activity of curing the diseases. The herbal drugs prepared with the traditional methods through slow grinding and mixing processes conserves all the natural substances within it in the 'naturally balanced form' without losing any essential component and maintains the activity and purity of the drug.

Discussion

This chapter provides comprehensive information on the uses of medicinal plants in traditional health care system of Himachal Pradesh. During the survey, it was found that besides traditional herbal healers, most of the elderly person, both men and women in the villages had sound knowledge and deep understanding about medicinal use of some plants. This indicates that the traditional societies, though convinced of the efficacy of local medicines are using the medicinal plants to a good extent but still face a major constraint to its wider use. The constraint is the effort involved in preparation of the medicines and availability of ingredients, since large number of medicinal plants has become quite rare and not easily available. According to the All India Trade Survey of prioritized medicinal plants, demand for some high-value medicinal plants has increased 50%, whereas availability has declined by 26% (Anonymous, 2001). The ever-increasing demand for these species in the pharmaceutical industries has increased habitat degradation and levels of overexploitation from wild habitats. If the overexploitation of entire medicinal plants continues, many species may decrease in, and ultimately disappear from their natural habitats. Thus to ensure the long-term conservation of medicinal plants, agro-techniques of the commercially viable medicinal plants need to be developed and disseminated among the stakeholders. Further, capacity-building of farmers, exposure visits and awareness camps on various aspects of medicinal plants, to enhance their skills in cultivation, conservation and marketing should be organised.

Table 1. Medicinal plants used to treat various diseases/ ailments.

Scientific name **Family *Local name	Disease/ ailment	Part Used	Use
Abrus precatorius L. Leguminosae Ratti	Mouth ulcer	Leaf	Green leaf juice is applied on the ulcers.
	Cough	Leaf	Decoction of young leaves is given orally.
	Stomachache	Seed	Seeds (about 100g) along with ghee or butter are prescribed.
	Induce abortion		Seed Poultice of seeds is suggested.
Acacia catechu Willd. Mimosaceae Khair	Dysentery	Bark	Decoction of 15gms of bark in 1 litre water is prepared and served at regular intervals.
	Rheumatism	Root	Paste of fresh root is applied on the joints once a day for a week.
	Cold and Cough		Bark The decoction of bark mixed with milk is taken.
	Skin diseases	Bark	Bark decoction is served for cutaneous diseases, also used in washing the wounds.
	Leprosy	Whole	Plant Decoction of root, leaf, flower, bark and fruits is given orally as well as for external dressing.
	Urinary problem		Stem One cup decoction prepared by boiling stem in water is given four times a day.
	Foot and Mouth Disease (Cattle)	Bark	Bark boiled with water is fed to cattle.
Acacia nilotica (L.) Willd. ex Del Leguminosae Kikar	Conjunctivitis	Leaf	Paste prepared by grinding the leaves of the plant is applied over the eyes.

	Sore throat	Gum	Use of gum is suggested.
	Eczema	Gum	Gum of the plant is mixed with the pulp of *Mangifera indica*. The mixture so obtained can be applied on the skin.
	Healthy teeth	Twigs	Twigs are used for scouring teeth.
	Prevention of uterus prolapse	Bark	Sittings are done in water boiled with bark (200gm each) of *Acacia nilotica* and *Ficus religiosa*.
Achyranthes aspera L. Amaranthaceae Puthkanda	Dry Cough	Seed	Powder obtained from roasted seeds is mixed with honey and served with luke warm water before going to bed.
	Jaundice	Root	Decoction of root is prescribed.
	Joint pains	Root	Root paste is applied on joints to relieve pain.
	Stomach troubles	Root	Decoction of roots is advised.
	Asthma	Whole Plant	The whole plant is crushed and taken with honey twice a day for a month.
	Blisters in mouth	Leaf	Leaf juice gives relief.
	Bone fracture (Cattle)	Root	A piece of fresh root is ground and the paste is applied to cure bone fracture.
Aconitum heterophyllum Wall. ex Royle Ranunculaceae Patish	Fever	Tuber	Powdered mixture of 1g tuber and flowers along with a glass of water or milk acts as a tonic especially in combating debility after fevers.
	Cough	Root	The root powder mixed with honey is prescribed.
	Cold	Rhiz-ome	Fresh rhizome is inhaled.
	Tonsillitis	Leaf & Seed	External application of concoction of crushed leaves and seeds on the throat is suggested.
	Headache	Root	Paste of ground tuber is applied on forehead.
	Stomach ache	Roots	Oral administration of powder of dried roots

Acorus calamus Wall. Bachh Araceae Barein	Weakness	Rhiz-ome	An application prepared by mixing ground rhizome of *Acorus calamus, Zingiber officinale* with ghee and powdered sugar is served to the patient early in the morning.
	Chest congestion	Root	Roots are macerated in the form of paste and applied to the chest of the patient.
	Cough & fever	Rhiz-ome	Infusion of the rhizome is given (especially to children).
	Wound healing	Leaf	Leaf paste is applied to wound.
	Arthritis	Rhiz-ome	Rhizome paste is applied on painful joints.
	Indigestion	Root	Black peppers (100), little amount of *Zingiber officinale* and the root of *Acorus calamus* together are ground and tablets prepared by this mixture are given.
	Toothache	Root	Chewing the roots alleviates toothache.
	Abdominal pain and diarrhoea	Whole plant	The plant juice is administered orally to treat abdominal pain and diarrhoea.
	Joint pains (cattle)	Rhiz-ome	Rhizomes mixed with mustard oil are used.
Adhatoda vasica Nees Acanthaceae Basunti	Cough	Leaf	A decoction of the leaves is served.
	Wound healing	Leaf	A poultice of the leaves is applied to wounds.
	Rheumatic pains	Leaf	Poultice of leaves is also helpful in relieving rheumatic symptoms when applied to joints.
	Cough (cattle)	Leaf	Decoction of leaves is prescribed.
Aegle marmelos (L.) Correa Rutaceae Bael	Diarrhoea	Fruit	50 gm mixture prepared from equal quantity of pulp and powdered green mango stone is consumed.

	Stomach disorders	Fruit	Squash prepared by mixing the fruit pulp with sugar and water is very useful.
	Constipation	Leaf	Juice of leaves when taken with black pepper helps to relieve constipation.
	Intestinal worms	Leaf	Leaf extract is consumed twice a day to get rid of intestinal worms.
Agave Americana L. Agavaceae Ramban	Toothache	Root & Leaf	A poultice made from the root and the leaves is often used to treat toothache.
	Digestive disorders	Leaf & Flo-wer Stalk	The base leaves and flower stalks are cooked as vegetable and is believed to cure digestive disorders.
	Bone fracture (cattle)	Leaf	Leaf fibres used to tie the fractured bone.
Ajuga bracteosa Wall. ex Benth. Lamiaceae Neelkanthi	Fever	Leaf	Leaf extraction is used to cure fever.
	Joint pains	Leaf	External application with leaf paste is beneficial.
	Burns and Boils	Leaf	Leaf powder is prescribed for burns and boils.
Allium sativum L. Liliaceae Lahsoon	Earache	Cloves	2-4 garlic cloves are heated with a teaspoon of mustard oil for few minutes and then this oil is applied around or in the ear.
	Skin diseases	Cloves	Juice is used as a rubifacient in skin diseases.
	Wounds	Cloves	Juice diluted with water is used for washing wounds.
Aloe barbadensis Miller Liliaceae Kavarein	Skin diseases	Leaf	The sticky sap from the succulent leaves is applied to skin itches, cuts and burns.
	Boils and wounds	Leaf	Cut leaf into two parts and sprinkle turmeric powder on it. Place it on the boil and bandage it.

	Arthritis	Leaf	Two teaspoon juice of the leaf , mixed with 6-8 drops of lemon juice and ½ teaspoon of cumin powder is applied on joints.
	Constipation	Leaf	Half teaspoon of paste prepared from ground leaves (6-8), salt (a pinch) and turmeric (a pinch) is taken.
	Eczema	Leaf	Juice extracted from leaves is applied on affected part as an ointment till problem persists.
	Fractured bones	Pulp	Poultice of pulp is applied on fractured bones
Amaranthus spinosus L. Amaranthaceae Chulai	Pimples	Leaf	Leaf juice with a pinch of turmeric powder is applied on the face twice daily.
	Gonorrhoea	Whole plant	Plant decoction is prescribed.
	Gastric ulcer	Root	10g of fresh root boiled in water and one cup of the decoction is taken daily.
Annona squamosa L. Annonaceae Sitaphal	Burns	Fruit & Leaf	The fruit and leaf juice heated over boiling water mixed with honey is rubbed over burnt area.
	Cold	Leaf	Leaf decoction is served.
	Skin diseases	Latex	The latex is applied on the skin to treat skin diseases like eczema and inflammation.
Artemisia sieversiana Willd. Compositae Charmara	Induce abortion	Leaf	Decoction of leaves is given to the pregnant ladies as an abortifacient.
	Wound	Leaf	Paste prepared from the leaves is applied topically on wounds to cure pain and swelling.
Asparagus racemosus Willd. Asparagaceae Satavari	Constipation	Rhiz-ome	Rhizome of wild asparagus in boiled and this liquid is served with little sugar.

	Weakness	Root	Dried root (15g) is ground to a fine powder and is taken (5g powder) with a glass of warm milk.
	Increase lactation	Root	Root powder of 10gm *Asparagus racemosus* L. along with water is administered orally with a glass of milk twice a day for 7 days after delivery.
Bauhinia variegata L. Leguminosae Karalen	Sore throat	Bark	Crush two inches dried bark and prepare its decoction in ½ glass of water. Gargle with the decoction.
	Skin problems	Bark	Bark paste is used as a liniment.
Benincasa hispida (Thunb.) Cogn. Cucurbitaceae Petha	Jaundice	Fruit	About 50 ml of fruit juice is given once a day for three days consecutively.
	Eye infection	Shoots	New vegetative apical shoots are crushed and the sap is applied directly on the eyes.
	Healthy teeth	Stem	Peeled stem is considered good for scouring teeth.
Bombax ceiba L. Bombacaceae Simbal	Constipation	Bark	Bark powder mixed with coriander powder, jaggery and water is served.
	Urinary disorder	Bark	Decoction (about 10 ml) prepared from bark of *Bombax ceiba* and flowers of *Butea monosperma* mixed in the ratio of 2:1 is served.
	Mastitis	Bark	Ground mixture of bark of *Bombax ceiba* and seeds of *Glycine max* is formed into a paste adding water and is fed to the cattle.
Butea monosperma (Lamk.) Taub Leguminosae Palash	Eye infection	Root	Extract the juice from 2-3 roots and strain it. Put 1-2 drops in eyes.

	Scabies	Seed	Seeds (20g) are ground with lemon juice to make a fine paste and the paste so obtained is applied regularly till the problem persists.
	Whooping cough	Seed	Oral consumption of seed ash along with honey is suggested.
	Ulcers and septic sore throat	Gum	Fresh gum from bark is applied to ulcers and septic sore throat.
	Cough & cold	Leaf & Flo-wer	Extraction of leaves and flowers is used.
Calotropis gigantea (L.) Ait. f. Asclepiadaceae Aak	Cough	Root/ Flo-wer	Place 100g of root/flowers on fire. After drying, grind into fine powder. Take one teaspoon of powder in the morning and evening with warm water.
	Indigestion	Root	One cup decoction prepared from one handful root powder in one litre water is given twice a day.
	Arthritis	Leaf	Mix juice of 5-6 leaves in equal quantity of *Sesamum* oil. Apply on affected part.
	Arthritis (cattle)	Leaf	Leaves of *Calotropis gigantea* and bulb of *Allium sativum* fried with mustard oil is rubbed on infected part.
Cannabis sativa L. Cannabinaceae Bhaang	Arthiritis	Seed	The affected part is massaged with the lukewarm seed oil. The treatment is undertaken at bedtime once a day.
	Wounds	Leaf	Leaf paste is used for dressing wounds and sores.
	Dandruff	Leaf	Leaf juice is applied on the scalp.
	Wasp bite	Leaf	Leaf paste is used.
Cassia angustifolia Vahl. Caesalpiniaceae Sanai	Cold	Leaf	Add 5g fresh leaves to 15 ml of water in a very small earthen container. Heat it on slow fire till water evaporates and then inhale.

	Constipation	Leaf	Boil 10-12 leaves in 40ml. of water till 10ml. solution is left. Strain and drink the solution till recovery.
Cassia fistula L. Caesalpiniaceae Gurlakdi	Constipation during pregnancy	Pod (Pulp)	Boiled and strained mixture of 1 teaspoonful of ground *Trachyspermum ammi,* 2 cm piece of *Cassia fistula* and 1 teaspoonful of ground *Foeniculum vulgare* in a glass of water is given 2-3 times a day.
	Ringworm	Tuber	Paste of the scrubbed tuber is applied on the infected part of the body for few days.
	Bleeding from nose	Pulp	The pulp of the plant and *gulkand* (rose petals preserve) is mixed (20gms each) together and boiled in 25 ml of water until it dries up. Served to control bleeding from nose.
	Muscular pains	Leaf	Ash of burnt leaves is mixed with mustard oil. The prepared paste is applied on the aching part.
	Stomach-ache	Fruit	Oral consumption of fruit decoction along with jaggery is helpful.
	Diarrhoea	Leaf	Soak dried leaves in water overnight. Boil it next morning. Mix 1 teaspoon of sugar and strain the liquid. Drink 1 teaspoon.
	Intestinal parasite	Leaf	About 10-15 ml leaf juice is suggested to be consumed adding a pinch of rock salt for three times a day after meals.
Curcuma longa L. Zingiberaceae Haldi	Wounds	Rhiz-ome	Paste of fresh rhizome mixed with warm water is given to heal up internal wounds.
	Mastitis	Rhiz-ome	Rhizome of *Curcuma longa* is mixed with oil of *Brassica campestris* and rubbed to cure mastitis.

	Broken horns	Rhizome	External application of rhizome paste is prescribed.
	Yoke galls (cattle)	Rhizome	Rhizome of *Curcuma longa is dried* over flame followed by grinding and mixing with oil of *Brassica compestris*. The paste so formed is daubed over the infected part.
Cuscuta reflexa Roxb. Convolvulaceae Akashbel	Abortion	Stem	One teaspoonful paste of stem with little lime is given once in the morning for 4-5 days to induce abortion at the early stages of pregnancy.
	Joint pains	Stem	Stem decoction is used.
	Diarrhoea	Twigs	Oral administration of juice from twigs (200ml), twice daily till cured.
Datura stramonium L. Solanaceae Datura	Arthritis	Leaf	Take 8-10 leaves and put in any hot oil till they turn black. Strain oil and use for massaging.
	Whooping cough	Leaf, Stem	Powder of dried leaves (4-5) and stems of *Datura stramonium* are smoked till disease is cured.
	Pain killer	Seed	Seed roasted in *Brassica compestris* oil is used for massage.
	Skin disease	Leaf, Seed	Leaf and seed paste is applied to cure the skin disease.
Ficus benghalensis L. Moraceae Bar, Bargad	Wounds	Leaf	Paste of the leaves is applied externally to abscesses and wounds to promote suppuration.
	Pimples	Leaf	Paste of young aerial roots is applied on pimples.
	Diarrhoea, Dysentery	Bark	Decoction of bark is beneficial.
	Healthy teeth and gums	Twigs	Young twigs when used as tooth brush, strengthen gum and teeth.
	Joint and muscular pains	Latex	External application of latex relieves such pains.
	Cracked or inflamed soles	Latex	Milky latex is applied externally.

Ficus religiosa L. Moraceae Pipal	Heart palpitation and cardiac weakness	Leaf	Leaves are soaked in water overnight and distilled the next morning. Then stored. About 15 mg of this solution can be administered thrice daily.
	Arresting excessive bleeding	Leaf	50 ml of juice or 1 tablespoon of dry leaf powder is served with water.
	Asthma	Fruit	The dried, powdered fruit if taken with water for a fortnight gives relief.
	Wound	Bark	Paste of bark is used for wound healing.
	Mumps and boils	Leaf	Mildly heated leaf smeared with ghee or butter is bandaged over the affected part.
	Diphtheria (animal)	Leaf	Infusion prepared by boiling leaves of *Ficus religiosa* and rhizome of *Zingiber officinale* with water is fed to the ailing cattle.
Leucas aspera (Willd.) Link Lamiaceae Gumabuti	Eye infection	Flow-er	A drop of juice extracted from 5-6 flowers is put in the eyes till the problem persists.
	Jaundice	Leaf	Grind to extract juice of 5-6 leaves, add 4-5 drops of limejuice, a pinch of calcium carbonate (lime) and ¼ th teaspoon of turmeric powder. Mix with ½ glass of milk or 1teaspoon of honey. Take once daily.
	Tonsillitis	Leaf	Grind and extract juice from 5-6 leaves of the plant. Put two drops in each ear.
	Scabies	Leaf	Extraction of leaves is applied on skin.
Murraya koenigii Spreng. Rutaceae Gandala/ Gandhela	Stomach pain	Leaf	Prepare a paste by grinding 15-20 *Murraya koenigii* leaves and mix it with 1 cup of buttermilk. Drink it in the morning before consuming anything.

	Cuts and wounds	Leaf	Leaf paste is applied as bandage.
	Healthy teeth and gums	Stem	Stem used for scouring teeth and for healthy gums.
Phyllanthus emblica L. Euphorbiaceae Amla	Indigestion and Diarrhoea	Shoot	Tender shoots taken with butter milk cure these problems.
	Burning sensation on skin	Fruit	Paste of fruits is used to get relief from the burning sensation.
	Gonorrhoea	Fruit	Fruit powder (2-5 grams) mixed with water is served for drinking.
	Mouth ulcers	Fruit	Gargles with fruit juice diluted with water are prescribed.
Pinus roxburghi Sarg. Pinaceae Cheer	Sprain in animals	Wood	Boiled the leaves of *Cannabis sativa* with ash of *Pinus roxburghi* and black salt and used externally.
	Broken horns	Resin	Resin is used to join the broken parts.
Spilanthes acmella L. Compositae Akarkara	Throat infection	Flow-er	Gargles with boiled solution of *Spilanthes acmella* flower, *Syzygium cumini* leaves and *Mangifera indica* leaves in water are prescribed.
	Toothache	Flow-er	Chew the flower to get relief in tooth ache.
	Head ache	Whole plant	Paste of whole plant is applied on forehead.
Taraxacum officinale F.H. Wigg Asteraceae Dudali	Incontinent bladder and kidney disorders	Whole plant	Infusion of whole plant is taken regularly as tea.
	Jaundice	Root	5 g of root powder given with milk twice a day for one week.
	Constipation during pregnancy	Leaf	Vegetable prepared from *Taraxacum officinale* and *Spinacia oleracea* leaves is given.

	Joint pains	Root	External application of root paste is used in the treatment of joint pains.
	Cuts and Injuries	Leaf	Fresh leaves are used as bandages on cuts and injuries.
Terminalia chebula Retz. Combretaceae Harad	Constipation during pregnancy	Fruit	Decoction of one fruit each of *Terminalia chebula* and *Terminalia bellirica* and two fruits of *Emblica officinalis* is administered once a day till problem persists.
	Diarrhoea	Fruit	Ground dry fruits (2) mixed in water (200ml) are given twice daily for two days to cure diarrhoea.
Tinospora cordifolia (Willd.) Miers ex Hook.f. & Thoms. Menis-permaceae Giloen	Constipation	Leaf	2-3 leaves dipped in 2 teaspoons of ghee are consumed.
	Diabetes	Stalk	One cup juice extracted from crushed stalk (leaves and stem) is served daily.
	Diarrhoea	Leaf	Ground leaves (10-15) are added to a glass of water and is consumed once after having meals at night.
	Jaundice	Leaf	Juice extracted by grinding 5-6 leaves is consumed adding 2 teaspoons of honey. It is continued till recovery.
	Weakness	Stem	Half cup decoction prepared from 10g stem of *Tinospora cordifolia* adding 300 ml of water is consumed.
	Dysentery with abdom-inal pain	stem	Half cup of the decoction prepared from stem is taken internally thrice a day.
	Ease in Delivery	Stem	Stem juice is given at the time of delivery.
	Debility (animals)	Root	Ground root is served with water to curing debility

Viola odorata L. Violaceae Banafshah	Fever	Flow- er	Decoction of flowers is prepared adding water and is served to the patient.
	Swollen tonsils and sore throat	Flow- er	Gargles with the infusion of a spoonful of dried flowers per cup of water.
	Cold & Cough during pregnancy	stem	Two tea spoons decoction of *Viola serpens*, ground *Elletaria cardamomum* and *Glycyrrhiza glabra* is given to pregnant woman.
Vitex negundo L. Verbenaceae Banna	Ear ache	Leaf	Leaves are boiled in mustard oil, filtered and are used as ear drop.
	Sprain	Leaf	About 10gm *Vitex negundo* leaves with equal amounts of leaves of *Prunus cerasoides, Murraya koenigii* leaves and henna leaves are crushed and the prepared application is tied as poultice on the affected area every night till recovery.
	Boils and wounds	Leaf	Leaves are wrapped in a cloth and hot fomentation with the wrapped leaves is applied on the affected area once a day for 3-4 days.
	Catarrhal and Headache	Leaf	Dried leaves are smoked.
	Acute rheumatism	Leaf	Leaf decoction is employed in medicinal baths.
	Common cold (running nose)	Leaf	Pounded leaves are boiled in water. After cooling, the strained mixture (quarter to half of the mixture) is taken three times a day.

References

Anonymous. 2001. *Appendix I, II and III to Convention on International Trade in Endangered Species of Wild Fauna and Flora*. Arlington VA: US Fish & Wildlife Services, pp 32.

Chauhan, N.S. 2003. Important medicinal and aromatic plants of Himachal Pradesh. *Indian Forester,* 129: 979-998.

Flaster T.1996. Ethnobotanical approaches to the discovery of bioactive compounds. Progress in new crops. In *Proceedings of the Third National Symposium* ASHS Press, Alexandria, pp 561-565.

Gaur, R.D. & Tiwari, J.K.1987. Some little known medicinal plants of Garhwal Himalaya. An ethnomedicinal study .In: *Medicinal and poisonous plants of the Tropics* (Leeuwenberg, A.G.M. eds.). Pudoc. Wageningen, Netherlands, pp 139-142.

Mahwasane, S.T., Middleton, L., & Boaduo, N. 2013. An ethnobotanical survey of indigenous knowledge on medicinal plants used by the traditional healers of the Lwamondo area, Limpopo province. *South Afr. J. Bot.,* 88: 69–75.

Mohd M., Taqi A. K.,& Firoz M.2012. Medicinal Plants of Rural India: A Review of Use by Indian Folks. *Indo Global Journal of Pharmaceutical Sciences,* 2(3): 286-304.

Myers, N., Muttermeier, R.A., Muttermeier C.A., da Fonseca A.B.G. ,& Kent J. 2000.Biodiversity hotspots for conservation priorities, 403:853–8

Nayar, M.P., & Sastry, A.R.K. 1987, 1988, 1990.*Red Data Book of Indian Plants,* Vol. I–III. Calcutta: Botanical Survey of India.

Omwenga, E.O., Okemo, P.O., & Mbugua, P.K. 2012. In vitro antimicrobial and preliminary photochemical screening of some Samburu anti-diarrhoeal medicinal plants-Kenya. *South Asian J. Exp. Biol.,* 2: 217–226.

Omwenga, E.O., Hensel, A., Shitandi, A., Goycoolea, F.M. 2015. Ethnobotanical survey of traditionally used medicinal plants for infections of skin, gastrointestinal tract, urinary tract and the oral cavity in Borabu sub-county, Nyamira county, Kenya. *Journal of Ethnopharmacology,* 176:508–514.

Samant, S.S. ,& Dhar, U.1997. Diversity, endemism and economic potential of wild edible plants of Indian Himalaya. *International Journal of Sustainable Development & World Ecology,* 4:179–191.

Samant,S.S., Dhar, U.,& Palni, L.M.S. 1998.*Medicinal Plants of Indian Himalaya: Diversity Distribution Potential Values*. Nainital: Gyanodaya Prakashan.

Samant, S.S., & Pant, S. 2003. Diversity,distribution pattern and traditional knowledge of Sacred Plants in Indian Himalayan Region. *Indian Journal of Forestry*, 26(3):201–13.

Sharma, N. 2015 Application of Ethno Medicinal Plants in Traditional Healthcare Systems Prevalent in Shiwalik Hills, Himachal Pradesh, India. In: *Recent Advances in Ethnobotany* (Kumar, S. ed.), Deep Publications, New Delhi, pp.171-179.

Singh, D.K., & Hajra, P.K.1996. Floristic diversity. In: *Biodiversity Status in the Himalaya* (Gujral, G.S.,; Sharma, V. eds.), New Delhi: British Council, pp.23–38.

❑❑❑

Chapter 3

Ethnobotanical Use of Plants in Banjara Community of Marathwada

Krishna S. Rathod[1], Nikhil B. Rathod[2], Rafuddin Naser*[3]

[1,2] Research Scholar Pacific University Udaipur. [3] Department of Botany, Maulana Azad College, Dr. Rafiq Zakaria Campus, Aurangabad (M.S.) 431001, krishnasrathod55@gmail.com, rnilu75@gmail.com, mrnaser2000@yahoo.com

ABSTRACT

The paper deals with ethnobotanical uses of plants by Banjara community. In this paper more than 30 important medicinal plants along with their botanical and local name and their ethnomedicinal uses have been described.

Keywords: Banjara, Gor Vansha, Choko, Osmanabad

Introduction

The word *"Banjara"* must have evolved from Prakrit, Hindi and Rajasthani words "Bana/Ban or Vana/Van" meaning Forest or Moorlands and "Chara" meaning 'Movers'. The Banjara are (together with the Domba) are sometimes called as the "Gypsies of India". The word Banjara is a deprecated, colloquial form of a word of Sanskrit origin. The Sanskrit compound-word *vana chara* means "forest wanderers" was given to them presumably because of their primitive role in the Indian society as forest wood plant collectors and distributors. The traditional knowledge available with the ethnic people plays an important role in quick and proper identification of natural resources. The traditional knowledge systems of the folk, oral tradition, and also published and unpublished literature are the important sources of locating potential bio resources. Unfortunately, due to lack of written documents, most of the traditional knowledge about medicinal plants and their uses survived only by words of mouth from generation to generation and are being gradually lost. Banjara is also known as 'GOR' or Gorvamshi Gor which is one of the ancient vanshas in the world. Its history dates back to 5-6 thousand years. The references about Gor Vansha are found in Greek civilization up to Harappa and Mohenjodaro. Even proof has been found regarding the presence of a civilization here 4500 years prior to Harappa civilization. This civilization is the Gor civilization (Motiraj Rathod, 1970). The Gorvanshi are witness to the hunting stage of early humans dating back to thousands of years. Later in the state of domestication of animals Banjaras was an important community in rearing the cows and protecting them. They took their ethnic name from this occupation. 'Go' means cow and 'ra' means the protector, hence 'Gor.' References regarding people engaged in trading by loading their goods on oxen since 4500 years ago are found. As they were engaged in trade they came to be known as Banjara. (Devilal Paliwal).

With traditional Banjara women in the region

Gor Banjaras were worshippers of nature since ancient times. The sun, wind, fire, water and the earth were their gods. By applying a thin layer of cow dung mixed in water a small place on the right side of the house is cleaned. A design of the earth and its four parts showing four continents are drawn with the help of jowar flour in the cleaned place". It is called 'Choks', A bronze pot filled with water is kept in the center of the Choko and a small stick of neem leaves is kept inside it (Ramsingji Bhanavat). Gor vanshiyas have an independent dialect. It is known as Gor Boli or Gormati Boli. This dialect is still spoken since pre-Indus period in Gor provinces of

Afghanistan, Baluchistan, Sindh, Punjab, Rajasthan, Gujarat, Kathewad, Harappa and Mohenjodaro. This belt is regarded as the original region of Indus culture. If we make a comparative study of the Gor dialect spoken in India and the above-mentioned tribal belt then it can be said that the Gor dialect might have been the original dialect. Further, various languages might have originated from the Gor dialect.

Study Area

Marathwada region represents large population of Banjara people. They live in forest or hilly areas they are so much familiar with nature.

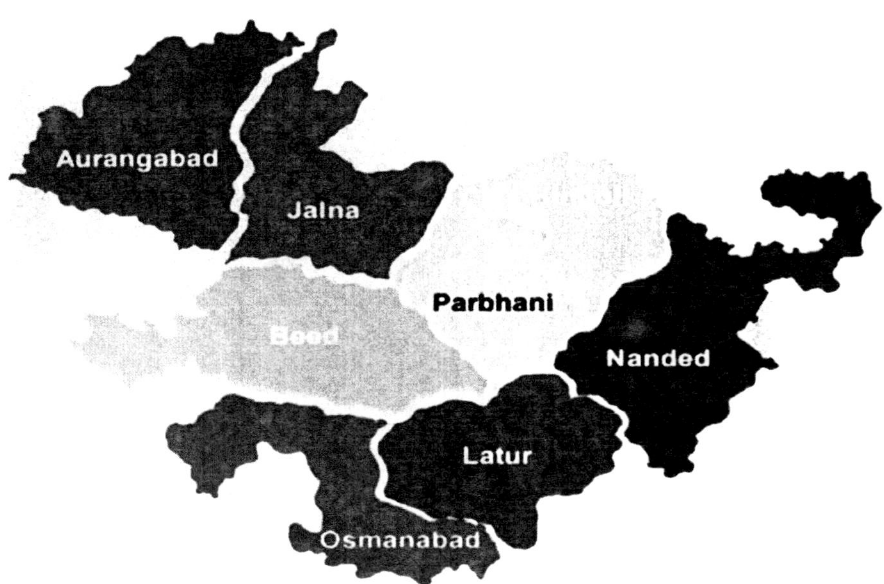

Map of Marathwada

They use traditional herbals formulation to treat common ailments despite the availability of the modern pharmaceutical drugs in the nearby towns and cities. Plant or plant parts used by Banjara community in India or Marathwada for the treatment of human diseases and other purpose, have not been studied by any botanist although Naik (1998), Vaikos & Naser (2002), Pokale 1978, & other studied general and ethenobotinacal use of plants from the area. Banjara's utilize plants for food ceremonies, festivals and medicines.

Material and Methods

Ethnobotanical survey was conducted in 8 districts of Marathwada, i.e. Aurangabad, Beed, Jalna, Nanded, Parbhani, Latur, Hingoli and Osmanabad. The investigations were carried out using a questionnaire (Jain 1987, Miguel 1998). The people of all ages, as well as the senior persons were consulted. Around 55 plants were claimed for various purposes like agriculture implements, fiber yielding, sacred colour yielding construction, timber, gum & fodder, spices, vegetable and food mat & plant making fuel etc. The local names parts used and mode of administration of dose was noted during the interaction. These plants are listed in Table 1, in alphabetical order. Herbarium specimens were authenticated through the Botanical Survey of India. The voucher specimens are deposited in the herbarium of Botany Research laboratory, Maulana Azad College, Rauza Baugh, Aurangabad, Maharashtra Suitable voucher specimen number is also given and mention here.

Author discussing with a Banjara Bhagat about the traditional uses of plants in Marathwada Region

Botanical Name	Local Name	Family & voucher specimen No.	Use
Acacia Leucophloea (Roxb) Willd.	Hiwar	Mimosaceae MACH 00048	Fresh plant twigs are tied at the marriage place of Banjara Community as an indication that the boy or girls is getting married for the first time. The wood is used in making agricultural implements, like the plough. Preferred for firewood.
Adhatoda zeylanica Medic.	Adhulsa	Acanthaceae MACH 0125	The dried leaf is smoked as a cigarette. The leaves of this plant are also utilized as green manure and for yielding a yellow dye
Aegle marcelos (L) *Correa*	Bel	Rutaceae MACH 00226	This is generally considered as sacred tree by the Banjara Community, as its leaves are offered to Lord Shiva during worship. According to Hindu mythology, the tree is another form of Lord Kailashnath.
Agave americana Linn.	Ghaypat	Liliaceae MACH 0315	In Banjara community the leaves are useful for yielding fibers, which are suitable for making rope, matting, coarse cloth and are used for embroidery of leather. These fibres are important to the economy of banjara community of Marathwada. The plants are grown around the field for fencing as well as for preventing soil erosion.
Ailanthus excels Roxb	Maha-rukh	Bersuraceae MACH 00425	Wood/timber is used by community for constriction of house
Alangium Salvifolium (L.F.)	Ankul	Alangiaceae MACH 00511	In banjara community of Marathwada the wood is used for musical instruments and furniture. The twigs are used for brushing the teeth

Albizia lebbeck (L) Benth.	Sipo, Shirish	Mimosaceae MACH 00589	In Marathwada region especially in Banjara Community, the tree is used to produce timber.
Allium Cepa L	Piaj/ kanda	Alliaceae MACH 620	Steam and leaves are eaten as a vegetable all over the Marathwada region.
Azadirachta indica A. Juss.	Kadul-imb, Neem	Meliaceae MACH670	During most occasions of celebrations and weddings the banjaras adorn their surroundings with the neem leaves and flowers as a form of decoration and also toward off evil spirits and infections. Neem seeds are grinded and converted into powder. They are soaked overnight in water the sprayed on the corp. Neem cake is often used as a fertilizer by banjara peoples.
Bahunia Purpuria. L	Lal Kachna/ khariwal / Rakta Kanchan	Caesal-piniaceae MACH 00980	The bark of Bauhinia is used to make rope and stems of smaller lianescent species are used for binding. Some Bauhinia species are being aused for bindingand also used for yielding edible gum. Wood is used for agricultural implements.
Balanites aegyptiaca (L) Del	Hingan bet	Balanitaceae MACH 00990	B. *Aegyptiaca* has fine grained dense and heavy heartwood; it is easily worked and takes a good polish. The wood is durable and resistant to insects, In Banjara community this plants used for making tool handles and domestic items such as spoons.
Bambusa arundifolia Ret.	Bamboo	Poaceae MACH 01015	Bamboo is used in manufacturing a number of products, like building materials, carpentry, farming, hunting and finish apparatus, fuel and lighting, household, domestic and personal items by banjara community. The culms of bamboo are largely used in constructional work, tool handles, weapons, furniture,

			musical instruments, handicrafts, stakes, pots, etc.
Biscofia javanica Blume.	Boke/ Bishop Wood	Euphorbiaceae MACH 01089	The tree is a potential source of long fibres for pulp and paper production. Timber : Bishop wood is medium weight and moderately hard. The heartwood is purplish - brown to reddish brown and is sharply differentiated from the narrow, pale brown to pale reddish brown sapwood. It is used for general construction (beams, posts), bridges, decking, sleepers, mining props, flooring, interior finish, veneer, plywood, implements, carving etc. Tannia or dyestiff : A red dye obtained from the bark is used to stain rattan baskets.
Bombax ceiba L.	Simel/ silk cotton tree	Bombacae MACH 01125	Banjara Marathwada region use of its wood for preparation of musical instruments such as Dholak and Tambura and also to make spoon for kitchen. These are full of cotton like fibrous stuff. It is for the fiber that villagers gather the essential fruit and extract the cotto substance. This substance is used for filling economically priced pillows, quilts, sofas etc. The fruit is cooked and eaten and also pickled. Holi, the festival of color is also the blooming time of this tree and its flower is used to prepare eco-friendly color. Besides this, a pole of debarked of this trees stem or branch is used to as main pillar to be burnt in *Holi dahan* by Banjara community.
Biscofia javanica L.	Hadsan	Acanthaceae MACH 01130	The young soft leaves are cooked and eaten as a vegetable. The dark red, dense wood is used as a building material. The bark is used as a source of red dye. The wood is susceptible

			to Lyctus and dry wood, termite attack, longhorn and ambrosia beetles as well as wood rotting fungi. Fresh wood smells of vinegar. It is used for general construction (beams, posts), bridges, decking, sleepers, mining props, flooring, interior finish, veneer, plywood, implements, carving etc. A red dye obtained from the bark is used to stain baskets. The bark is also employed in the toughening of nets and ropes.
Blumea lacera (Burm f)	Bhamu-rda Burandu, Kakr-onda.	Asteraceae MECH 01140	The plant is used to drive away fleas and other insects
Boerhaavia diffusa L.	Khapar-khuti, Punar-nava, Tambadi vasu.	Nyctagin-aceae MECH 01160	*Boerhavia diffusa* can be used as a fodder for livestock.
Brassica campestris L.Var.	Sorso	Brassicaceae MECH01192	Edible oil is obtained from the seed, it is best when cold pressed
Brassica nigra (L) Koch.	Kalasorso	Brassicaceae MECH01214	Semi-drying oil is obtained form the seed, as well as being edible it is used as a lubricant. The plant is often grown as a green manure, it is very fast producing bulk, suitable for digging into the soil in about 8 weeks. It used to harbor the pests and diseases of the cabbage family so is probably best avoided where these plants are grown in a short rotation and especially if club root is a problem. Mustard oil (allyl isothiocyanate) is used in commercial cat and dog repellent mixtures.

Butea Monosperma (Lam.) Taub.	Palas	Fabaceae MECH 01318	The flower are used to prepare a traditional Holi colour . It is also used a s dyeing color for fabric. Fresh twigs a are tied on horns of bullocks on occasion of Pola festival
Cajans cajan (L) Hutch.	Tur	Fabaceae MECH 01416	Community, split pigeon peas called toor dal in Marathi are one of the most popular pulsed, being an important source of protein in a mostly vegetarian diet. In regions where it grows, fresh young pods are eaten as a vegetable in dishes such as sambar.
Chloroxylon Swietenia D.C.	Behru Halda	Meliaceae MECH 01518	The wood produced by the tree is often a golden colour with a reflective sheen. It is used for small luxury items and as a veneer in wooden furniture. It is one of the best known satinwoods.
Cochlosp-ermum religiosum (L.) Alst. Handb.	Ganeri	Cochlospe-rmeaceae MECH 01590	Gum is used for book binding, calico printing, cosmetic industry cigar paste, seed oil is used in, soap making, seed cake is used, as manure and cattle feed, floss covering seed, is used for stuffing in pillows, mattresses, cushions life belts, wood is used as, fuel wood
Commiphora wightii Arnott.	Guggul	Burseraceae MECH 01600	Dried resin is burnt as dhop (inscence sticks) during religious ceremonies for worshipping of gods and goddesses.
Corchorus olitorius Linn.	Banpat	Tiliaceae MECH 01615	Banjara use the fibers form Corchorus (known as jute) for production of rope and are the most widely cultivated vegetable fiber after cotton.
Cordia dichotoma Forst. F.	Bhokar	Boragin-aceae MECH1628	In banjara community, the fruits are used as a vegetable, raw, cooked, or pickled. The immature fruits are pickled and are also used as a vegetable fodder. The leaves also yield good fodder.

Cordia gharaf (Forsk.) Ehrenb. & Asch.	Gondhan	Boragin-aceae MECH 01648	Wood used in making agricultural implements and furniture.
Coriandrum Sativum Linn.a	Dhaniya, Kotham-bir, Kothmir	Apiaceae MECH 01689	Banjara peoples use fresh leaves and the dried seeds for cooking. The dry fruits are known as coriander or coriandi seeds Corinder seed is a spice in garam masala and Banjarian curries.
Crateva adansonii DC	Waiwa-rna	Capparid-aceae MECH 01700	It is as animal fodder; and also as ornamental purpose. People use leaves as food, sauces, condiments, spices, flavourings.
Curcuma longa L.	Zingiber-aceae	*Haldi* MECH 01819	In curry powder, Curcua is added to the mix of spices which usually consist of : coriander, cumin seeds, fenugreek seeds, chilli, mustard seeds, black peppercorns and salt. Turmeric can be added to hot water and then strained (with touch of ginger and lemon juice); it can also be added to milk and simmered.
Cymbopogon flexuosus Wats	Tikhadi /Gauti Chai	Poaceae MECH1880	The oil is used in cosmetic industries and medicines. It is also used as household disinfectant.
Diospyros melanoxylon Roxb.	Tendu-ptta, Temb-urni, Tendu	Ebenaceae MECH1898	Tendu patta (Diospyros melanoxylon) leaves are used for wrapping the tobacco and making "beedis" in banjara community
Embilica Officinalis Gaertn.	Aavala	Euphorbi-aceae MEC01914	In banjara community, the fruit is pickled with salt, oil, and spices. Aamla is eaten raw or cooked into various dishes. In Marathwada region of banjara, tender arieties are used to prepare dal (a lentil preparation), and amle ka murabba, a sweet dish indigenous to the Banjara community of Marathwada

			(Wherein the berries are soaked in sugar syrup for a long time till they are imparted sweet flavor); it is traditionally consumed after meals.
Erythrina suberosa L.	Pangara	Papilion-aceae MECH01990	In Hinduism, tree is Indra's garden in heaven (Swarga) is held to be *E.stricta*. The same is found is Tibetan Buddhism, where the man do ra ba growing in Sukhavati is identified as an Indian Coral tree. The concept of the Five Trees of Paradise is also found in Christian Gnosticism.
Ficus beng-alensis L.	Vad	Moraceae MECH 02000	In Hinduism, the leaf of the banyan tree is said to be the resting place for the god Krishna.
Ficus religiosa L.	Pipel	Moraceae MECH02080	Peepal tree has the great importance in Banjara community of India especially among the Buddist who regard peepal tree as the personification of Buddha. Lord Buddha attained enlightenment mediating under the Peepal tree. It is regarded as the sacred tree and the person used its leaves for the religious purposes.
Guzuma ulmifolia Lamk.	Rudraksh	Sterculi-aceae MECH 02098	The wood of the Guazuma ulmifolia is utilized for posts, interior carpentry, light construction, boxes, crates, shoe horns, tool handles and charcoal. The wood is found to be very unproblematic to work with. The sapwood has a brown color (light) and the heartwood is pink to brown
Jatropha Curcas L.	Parsi errand Magali errand	Euphorbi-aceae MECH02114	As animal fodder. The cake can also be used as feed in digesters and gasifiers to produce biogas.
Lagerstr-oemia Indica L.	Saoni	Lythraceae MECH02160	The timber of some species has been used to manufacture bridges, furniture and railway sleepers.

Lantana Camera L.	Ghaneri	Lamiaceae MECH02169	Banjara community uses this plant for household furniture, such as tables and chairs (made from the stalks), or the small branches are bundled together to make brooms. It has become popular in gardens for its hardy nature. it is not affected by pests or disease, has low water requirements, and is tolerant of extreme heat.
Lavandula bipinnata O Ktze.	Deepmal, Asmani gulgota	Lamiaceae MECH02178	Herbs, lavender has few insect pests: the green capsid bug or cuckoo spit, can be an aesthetic problem, however this may be unsightly but are harmless.
Limonia acidissim Linn.	Kawat, Kauth	Rutaceae MECH02198	The rind of the fruit is so thick and hard that it can be carved and used as a utensil such as a bowl or ashtray. The bark also produces an edible gum. The tree has hard wood which can be used for woodworking.
Mahauca longifolia (J.) Macb.	Mahawa /ahawda	Sapotaceae MECH02231	Banjara use heartwood which is reddish brown, strong, hard and durable; it is used in construction, naves and felloes of cartwheels, door and window frames. The sweet, fleshy corolls are a rich source of sugars, vitamins, calcium and essential oil and eaten raw or cooked, used in manufacture of country liquor and vinegar. The flower are used to produce an alcoholic drink in banjara community. Several parts of the tree, including the bark, are used for their medicinal properties. It is considered holy by many tribal communities because of its usefulness
Mangi fora indica	Anacard-iaceae	*Amba* MECH02390	Food : Mango should always be washed to remove any spa residue, before handling, some seeding mangos are so fibrous that they

			cannot be sliced. Instead, they are massaged, the stem-end is cut off, and the juice squeezed form the fruit into the mouth. Thin slices, seasoned with turmeric, are dried, and sometimes powdered, and used to impart an acid flavor to chutneys, vegetables and soup, green or ripe mangoes may be used to make relish.
Melia azardira-chta L.	Bakan-neem	Meliaceae MECH02415	As pesticide, neem oil plays an important role in the biological control of insects. While the Indian have been using the neem for centuries, the west has come to recognize the potential of neem oil in the control of insects in the last decade only.
Oryza sativa L. (Dhan)	Rice, Tandul, Bhat, Chawal	Poaceae MECH02448	Rice is most often consumed a whole grains, boiled or steamed in water, fice flour lacks gulten and so is usually consumed as noodles, as the absence of gluten results in poor quality bread.
Oxalis corniculata L.	Ambushi Omlika	Oxalidaceae MECH02459	The leaves of wood sorrel are quite edible, with a tangy taste of lemons. A drink can be made by infusing the leaves in hot water for about 10 minutes, sweetening and then chilling. The entire plant is rich in vitamin C. Any wood sorrel is safe in low dosages, but if eaten in large quantities over a length of time can inhibit calcium absorption by the body.
Phyllanthus acidus (.)	Skeel, Rai Awla	Euphorbi-aceae MECH02490	Varous parts of the plant are used for food. In India and Indonesia, the cooked leaves are eaten. While the fruit is eaten fresh, and is sometimes used as flavoring for other dishes in Marathwada, it is generally regarded as too hard to be eaten by

			itself in its natural form and is processed further. It is candied in sugar or pickled in salt, used in chutney, relish or preserves. The tree is used as fuel wood. The wood is fairly hard, strong, tough and durable if seasoned. The seasoned wood is also used for making utensils and other small objects.s
Pongamia pinnata	Karanj	Asteraceae MECH02509	Karanja is commonly used as a fuel. Its wood is susceptible to insect attack, so wood is not considered as quality timber. But it may be used in agricultural implements, tools and combs.
Portulaca oleracea Linn.	Ghol	Portulace-aceae MECH 02515	The leaves and stems are also edible. They can be pounded into a mush and eaten raw, cooked as a vegetable or added to salads.
Prosopia cineraria L.	Shami, Sondad	Mimosaceae MECH2540	The wood of *P. cineraria* is a good fuel source, and provides excellent charcoal. The leaves and pods are consumed by livestock and are a beneficial forage.
Punica granatum	Dalimb	Myrtaceae MECH02590	The major use of wild pomegranate is for the making of anardana, which is used in making chutney and as a souring agent in various preparations. The wood of the tree is very hard and durable. It is generally used in making agricultural implements.
Santalum album L.	Chandan	Santalaceae MECH02640	It has use as wood and oil in religious practices. It also feature as a construction material in temples and elsewhere.
Schleichera oleosa (Lour) Oken.	Kesumb	Sapindaceae MECH02670	The timber is suitable where hardness, bending strength and toughness are required. The wood is commonly used for oil and sugar mills, cart wheels, and agricultural implements.

Some Photographs of Banjara Communities

2/27/2012

References

Jain S.K. 1991. *Dictionary of Indian folk medicine and Ethnobotany.* Deep Publications, New Delhi.

Naik V.N. 1998. *Marathwadyatil Samanya Vanaushadhi* (Marathi) Amrut Prakashan, Aurangabad.

Naser Rafiuddin and Vaikos N.P., Plants used in ethnoveterinary practices in Aurangabad district Maharashtra I, In: Proceedings of Plant Resource Development, by AM Mungikar and As Bhutkar (Eds), Dr. B.A.M.University, Aurangabad 2002, pp-202-210.

Pardeshi V.N. 1983. *Flora of Godavari Yalley an Sourthern Plains of Aurangabad* Ph. D. Thesis, Dr. Babasaheb Ambedkar Marathwada University, Aurangabad.

Pokale D.S. 1982. *Flora of Northern Hills of Aurangabad district;* Ph.D. thesis. Dr. Babasaheb Ambedkar Marathwada University Aurangabad.

Pokle D.S.&V.N. Pardeshi, 1978. *Some Noteworthy Plants from Aurangabad district.* Nat Sci J. Dr. Dr. Babasaheb Ambedkar Marathwada University Aurangabad. 17-71-75.

Chapter 4

Plants of Ethnomedicinal Importance From Sikkim Himalayan Region, India

Jitendra Kumar Shukla*[1], C.S. Purohit[2], Sandhya Thapa[1] and Preeti Dhakal[1]

[1]Regional Centre of Institute of Bioresources and Sustainable Development, 5th Mile, Near Metro Point, Tadong, Gangtok-737102, Sikkim (*E-mail: shuklajk2000@yahoo.co.in), [2]Botanical Survey of India, Arid Zone Regional Centre, P.O. Pal, AIIMS road, Subhash Nagar, Jodhpur-342014, Rajasthan (E-mail: chandansinghpurohit@yahoo.com)

ABSTRACT

The present communication deals with the collection and compilation of the information, obtained by several forays undertaken by different botanists, gathered information & discussion with local people and review of the relevant literature. In total, 15 medicinal plants with their local name(s), ethno-botanical importance, uses and their distribution in Sikkim is described.

Keywords: Ethno-botany, medicinal plants, distribution, Sikkim Himalaya.

Introduction

Ethno botany as an academic discipline initially named 'Aboriginal Botany' (Powers, 1874), received its widely accepted name from Harshberger in 1895 (Harshberger, 1896; Cotton, 1996; Hidayati *et al.,* 2015). Ethno botanical research deals with the relationship between ethnic communities and plants. Even before the emergence of ethno botany and ethno biology as disciplines, various societies and individuals explored the relationship between humans and plants and animals (Ford, 2011). Oldest system of medicine has evolved as a constant effort of mankind toward, and as a remedy to the suffering since the evolution of humanity. Use of biological resources the for health care has developed based on available resources, experiences and knowledge of people in different cultures. Old systems of medicine are believed to be the written compilation of contemporary folk knowledge, and thus folk knowledge can be considered as the precursor of all traditional medicinal systems. For instance, traditional Indian medicinal systems such as the Ayurveda, Siddha and Yoga, acknowledge folk medicinal knowledge as the root source of information (Hidayati *et al.*, 2015; Ravishankar and Shukla, 2007). Traditional medicine is practiced non-formally by traditional healers as local healing traditions, and formally through well known medical systems. Despite the diverse traditional systems, they share a common aim of health, healing and factors of well-being. Health is not only related to physical healing, but also to lifestyle practices, mind-body-nature balance, nutrition and livelihoods. In the present scenario there is renewed interest in the potential of traditional medicine and health care systems. With the rapid increase of communicable diseases (such as HIV, malaria, tuberculosis, pneumonia, diarrheal diseases), chronic diseases (such as diabetes, ischemic heart diseases) and new & dreaded diseases, the importance of ethno medicines is increasing throughout the world. One-third of the global population mainly in economically developed and transition economies do not have access to essential modern health care (Bodeker *et al.*, 2005). Recurring ailments such as infectious, chronic, and lifestyle-related disease are drawing interest of people towards the use of ethno medicine. More over 70-80% of the world population use herbal medicines , mainly in the developing countries, for primary health care because of better cultural adaptability, better compatibility with human body and lesser side effects (Shukla *et al.*, 2001). The need to provide primary health care against the increasing and upcoming new ailments requires an intensification of research on novel medicines at one end with search for new medicinal crops which can provide optimum cure for novel/ existing diseases. There is an urgent need to create greater awareness amongst the population as a whole particularly amongst the farmers about the medicinal and economic values of medicinal plants (Mohammed *et al.*, 2004). As this system primarily consists of biological resources, it is also important to ensure their sustainability. Ensuring sustained availability of biological resources simultaneously with the access to quality health care for all individuals of society is a necessity. Sustainability and approaches to health care are closely related, especially in rural communities of developing countries, as health care in these communities is primarily delivered by native healers or community health workers or householders using biological resources. Because of low reach of modern doctors and health care facilities, and the knowledge & experience possessed by knowledgeable local healers, it becomes important to involve the latter more actively in health care delivery systems. Further, protecting such knowledge from erosion and misappropriation also need to be addressed. There is a fast increasing consumption of plant resources by the natural product industry in India. Studies suggest that 960 medicinal plant species are the source of 1,289 botanical raw drugs that are traded in Indian markets and are used by around 9,000 herbal industries in the country. Of this, 81 percent (780) of species in active trade are entirely or largely sourced from the wild (Ved and Goraya, 2008). From the wild collection more than 70 per cent is collected through destructive practices as stem, wood, bark, roots or as whole plant (FRLHT, 1999; 2009; Barata et al., 2016). In addition to this, the demand for plant resources for modern industries including pharmaceuticals, botanicals and cosmetics is also on the rise (Laird and Wynberg, 2008). This rise in demand, along with various other factors such as loss of forests, encroachment and conversion, destructive practices such as grazing and overharvesting have led to the destruction of genetic diversity and habitats of several valuable natural resources. It is estimated that in India, around 300 plants are in various threat categories. Cultivation of such resources is not yet a viable economic option due to preference for wild sourcing given lower costs, in addition to a general lack of information on agro-techniques (Hamilton, 2004; UNU-IAS Policy Report, 2012).

Materials and Methods

The methods employed in this study were designed with the purpose of providing baseline information on the medicinal and domestic uses of plant species in local systems and among ethnic communities of Sikkim, through a survey, field visits, and personal observations about the uses of these plants. Extensive formal and informal interviews were conducted with local users and herbal medicinal practitioners. Some local markets were also surveyed during 2012 to 2015 regarding the uses and availability of medicinal plants. Samples of most of the species were brought to the laboratory and identified on the basis of available scientific studies (Hooker, 1872-1897; Cowan and Cowan, 1929; Biswas, 1956; Nadkarni, 1954; Polunin and Stainton, 1984, Grierson and Long, 1983-1999; Bhakuni, 1990; Noltie, 1994-2000; Sinha and Sinha, 2001).

Ethno-botanical plants of Sikkim- Hooker (1872-1897) made the first floristic study of Sikkim, which was followed by a comprehensive work on medicinal plants of Darjeeling and Sikkim by Biswas (1956). Over 400 plants possessing therapeutic properties have been recorded from the region (Srivastava and Kapaki 1990); many of which are used in the traditional medicine of ethnic groups in this region. The Indian Himalaya cover approximately 4,19,873 km^2 area (Rodgers & Panwar, 1988) extending form Jammu & Kashmir in the North-West to the Arunachal Pradesh in the East. It has an altitude ranging from 300-8000m above mean sea level (amsl) with vegetation comprises of tropical, sub-tropical, temperate, sub-alpine and alpine types. The Himalayan range is home to the planet's highest peaks, including the highest, Mount Everest. The Himalayas include over a hundred mountains exceeding 7,200 m (23,600 ft) in elevation. Sikkim state lies between 27°46′-28°7′48″ N latitude and 88°05″-88°55′25″ E longitude in the lap of Eastern Himalayas below Khanchendzonga with an area of 7299 sq km. The state contains over 4000 species of flowering plants, including 600 species of orchids and more than 100 species of medicinal plants (Singh and Chauhan, 1998). The elevation in Sikkim Himalayan region ranges from 300m in the south to more than 6000m in the north with perpetual snowy mountain ranges. Depending on the elevation, the climate of the state is divided into tropical in lower hills (300-900m), sub tropical in the mid hills (900-1800m), temperate (1800-2700m), sub alpine (2700-4000m) in hills and alpine (4000->5000m) in very high hills. The average rainfall of the state varies from 1200mm (at 300m elevation) to 4500mm (at 2000m elevation) and over 80% of it rains during Monsoon season (June to September). Depending on the season and elevation, the average maximum temperature varies from 13°C to 35°C while the minimum temperature ranges from 0°C to 23°C (Sundriyal et al., 2004). The unique geographical position and wide range of topography, high fertile soil, sufficient rainfall and presence of large number of perennial stream makes the state of Sikkim one of the treasure houses of bio-diversity in the country. Great topographical variation and wide arrays of climatic zones favours the rich growth of diversified flora. This region is also the habitat of many endemic floras which have become rare, threatened or endangered. Sikkim has vast reserve of medicinal plants and rich culture of folk medicine. The rich flora of Sikkim has a number of raw drugs described in Ayurvedic texts. There are about 420 plants which are used by the tribal people for various diseases in Sikkim Himalayan region, out of which few are utilized on commercial basis. Brief particulars of the important medicinal plants commercially utilized in Sikkim are given below.

Aconitum ferox, Aconitum heterophyllum, Acorus calamus, Aloe barbadensis, Angelica archangelica, Artemisia vulgaris, Asparagus racemosus, Bergenia ciliata, Cinnamomum tamala, Costus speciosus, Dendrobium nobile, Hedychium spicatum, Mallotus philippensis, Nardostachys jatamansi, Oroxylum indicum, Picrorhiza kurrooa, Podophyllum hexandrum, Rubia sikkimensis, Swertia chirayita, Saussurea lappa, Smilax lanceifolia, Taxus baccata, Valeriana hardwickii, Zanthoxylum armatum, etc.

A large number of Medicinal and Aromatic Plants (MAPs) have been distributed within different forests along an altitudinal gradient. Most of the MAPs are used in the Indian Systems of Medicines, and pharmaceutical. These wild plants are sometimes an important local medicine on one hand and supplement/substitute food beside routinely available resources for native communities. These medicinal plants are important source of perpetual health and of income to local residents. The present work draws an attention to the some of the major medicinal plants utilized by local healers in Sikkim Himalayan region. The present compilation aims to sum up the information on the traditional knowledge of plants, which are used for different diseases and ailments, and which play a vital role in health care among the ethnic group of Sikkim. Alphabetical list is given below including family, vernacular name, ethno-botanical uses and distribution in Sikkim.

(i) *Aconitum heterophyllum* Wall. ex Royle (Ranunculaceae) Bikh

Taxonomic characters: It is a tall herb with tuberous roots. Based on morphology and anatomy, several forms of *A. heterophyllum* are recognized viz. white, yellow, black and red, amongst which the white variety is considered best in Ayurveda.

Ethno-botanical Uses: *A. heterophyllum* is used both externally and internally for various ailments. The plant is used as an anti-pyretic, analgesic, anti-periodic, aphrodisiac, astringent, anti-venom, anti-inflammatory, anti-rheumatic, vermifuge, etc. It is used in the treatment of stomach disorders (diarrhoea, dyspepsia, dysentery, anorexia, piles, etc.); to reduce obesity; to treat cough and headaches caused by eating excessive amounts of oily foods; thirst associated with fever; yellowish sclera; nausea; vomiting; and throat pain (Butola and Badola, 2008).The underground stem and root are used in traditional system of medicine. The white tuberous roots are plumpy with a pale yellow colour, acrid, bitter, themogenic, expectorant, alexeteric, stomachic, digestive, antiperiodic and tonic. The root powder with honey is prescribed for cough irritations, bronchitis and as an anthelmintic (Ukani *et al.*, 1996). Roots are anti-fertility agent, tonic, stomachic, anti-periodic, used in hysteria, piles and throat diseases, dysentery, diarrhoea, stomach disorders, malarial fever, vomiting, helminthiasis, haemorrhoids, haemorrhages, internal inflammatory conditions, to relieve body-ache, fever, cold, cough, nose discharge, general debility and recommended for diseases in children etc. (Prajapati and Kumar, 2003; Pradhan and Badola, 2008; Das *et al.*, 2012). Roots contain alkaloids namely heteratisine, heterophyllisine, heterophylline, heterophyllidine, atisine, atidine, F-dihydroatisine (Pelletier *et al.*, 1968), hetisine, hetidine, hetisinone, isoatisine (Ukani *et al.*, 1996) and benzoylheteratisine (Hussain and Hore, 2007).

Distribution: Sikkim- Kupup, Thangu

Specimen examined: Sikkim, North district, Thangu, 13950ft, SK Jana 18399 (BSHC-30772).

(ii) *Artemisia vulgaris* L. (Asteraceae) Tite Patti

Taxonomic characters: *Artemisia vulgaris* is an aromatic perennial shrub, sparsely pubescent; middle stem leaves are elliptic or ovate-elliptic, 1 or 2 pinnatipartite or sect. inflorescence a dense, narrow panicle.

Ethno-botanical Uses: The plant mainly contain coumarin, glucosides, sesquiterpene lactones, polyacetylenes, essential oils, diterpenes and triterpenes, flavonoids aglycones and falvonoids di-O-glycosides (Nikolova *et al.*, 2004; Lee *et al.*, 1998), a germacrene derivativery (Sosa *et al.*, 2001; Pires *et al.*, 2009). Many diseases have traditionally been treated with this plant which includes inflammatory diseases, gastric ulcers, cancer and infections by bacteria, fungi & viruses (Carvalho, 2011). Many authors have reported the antibacterial (Mitra, 2014a, Chen, 1989), antihypertensive (Tigno, 2000), hepatoprotective (Gilani, 2005), antioxidant (El-Tantawy, 2015; Temraz and El-Tantawy, 2008), hypolipidemic, antiinflammatory (El-Tantawy, 2015) and antispasmodic (Khan and Gilani, 2009) properties of *A. vulgaris*. The essential oil of this plant has been widely studied (Bagci *et al.*, 2010; Blagojevic, 2006; Bora and Sharma, 2011; Govindaraj *et al.*, 2008). Phenolic compounds present in the aerial parts of plant are important as antioxidant, and for their health benefits

(Costa, 2013; Fernandez-Panchon *et al.*, 2008). The above ground part of plant is used as antispasmodic, haemostatic in nervous affection, menstrual disorders, leucorrhoea, haemorrhagic dysentery, vomiting, colic, rheumatism and in asthma (fig. 1a). The roots are used as tonic and antispasmodic. Local healers of Sikkim use *Artemisia* to treat nose bleeding, asthma, diseases of brain, sprain, high blood pressure, skin diseases, mouth ulcer, allergies, gout and rheumatism to ward off mosquito, insect & pests and in religious ceremony (Bharati and Sharma, 2012; Hussain and Hore, 2007; Pal and Palit, 2011; Pradhan and Badola, 2008). Leaf extract is used on cuts and bruises and considered to possess detergent effect & used as cleansing agent (Das *et al.*, 2012). Anti gastric ulcer activity was observed in the leaves of *A. vulgaris* (Mitra, 2014b). Singh *et al.* (2012) observed antimicrobial activity in essential oil of *A. vulgaris*. Saju *et al.* (2012) observed inhibition of mycelia growth of *Pestalotiopsis* sp. by using combined aqueous extracts of *A. vulgaris* and *Schima wallichi*.

Distribution: *A. vulgaris* is an aromatic perennial shrub of temperate regions of Europe, Asia, North Africa, and North America. In India it is found in W. Himalayas, Sikkim, Khasi hills, Darjeeling, Manipur and W. Ghats from Konkan southwards. In Sikkim it is found in Kyongnosla, Namchi, Rumtek, Toong, Geyzing, Pelling, Jorethang and Gursey.

(iii) *Astilbe rivularis* Buch.-Ham. ex D. Don (Sexifragaceae) Buriokhati

Taxonomic characters: It is a perennial, erect, hairy shrub with large pinnate leaves. Leaflets are ovate long pointed toothed with bristle on lower surface of midrib (fig. 1b). Flowers are greenish yellow about 2mm in diameter born on pyramid shaped terminal cluster.

Ethno-botanical Uses: Major traditional uses of its dried rhizomes are mainly for the treatment of headache and infertility by the tribal community of the hills. For several decades, numerous medicinal preparations of rhizomes of *A. rivularis* have been used as potential phyto-therapeutics among different ethnic races (Chettri *et al.*, 2005). Local healers use its root for body ache, back pain, bone fracture, joint pain, peptic ulcer and to control dysentery & diarrhoea (Shrestha, 1988; Sundriyal *et al.*, 1998; Shrestha and Dhillion, 2003; Manandhar, 1991, 1992). Its rhizome is a sold in market, which is chewed as areca nut and used to relieve pain. The juice of the plant is applied to sprains and swelling. Rajbhandari *et al.* (2009) mentioned the extract of A.

rivularis with some other herbs showed potent anti-herpes and anti-influenza viral activity. Mandal et al. (2009) observed free radical scavenging effect and anti-lipid peroxidation in purified solvent fractions of *A. rivularis*. The physiological mechanism of anti-inflammatory effect of plant extract is not well understood. High amount of polyphenols, flavonoids, coumarin berginin, triterpenoids, acetyl-â peltoboykniolic acid, peltoboykniolic acid and astilbic acid are present in its rhizome.

Distribution: Sikkim, North district: Lachung, Zema, Logbridge, Rabong, Lachen, Tholung, Kishong; East district: Lungthung, Burtuk, Penangla, Zuluk, Nathula, Pangthang; West district: Vareng, Ribdi, Okhrey, Sombaria; South district: Damthang, Ravangla, Rabong, Kewzing;

Specimen examined: North Sikkim: Lachung, 07.09.2002, S Rathore 24318, Acc. No. (BSHC)- 34921; Zema to Logbridge, 02.10.1997, P Singh & SS Dash 19922, Acc. No. (BSHC)- 23327.

(iv) *Bergenia ciliata* (Haw.) Sternb.f. (Sexifragaceae) Pakhanbed

Taxonomic characters: *Berginea ciliata* is an important medicinal herb of perennial nature with creeping root stock. *B. ciliata* is the source of Pashanbheda (name in Hindi language) after *B. ligulata.* Its leaves are basal, sub orbicular to round, with bristly margins, short stalked and turn bright red in autumn (fig. 1c). Flowering ranges from February to April. Flowers are white or pink-purple in dense or in spreading clusters.

Ethno-botanical Uses: Local healers of Sikkim use it in bone fracture, body ache, back pain, sprain and in joint & chest pain. In other usages it has been reported to be useful in dissolving kidney and bladder stones, useful in fever, diarrhoea, abnormal leucorrhoea, piles, pulmonary affections, cough & cold, asthma, boils, ophthalmia and has many medicinal properties such as antibacterial, anti-inflammatory, anticancer, antidiabetic (Chauhan *et al.,* 2012, Chopra *et al.,* 1956; Manandhar, 2002).

Distribution: It is found in Afghanistan, South Tibet and Bhutan. In India it is found in Meghalaya, West Bengal, Arunachal Pradesh and Sikkim. In Sikkim Himalayas it is found in Kyongnosla, Changu, Karponanag, Lachen to Thongu, Nathang, Prekchu-Tsokha, Pangolakha-Subaney Dara and Gangtok under both natural and in domesticated conditions (Hafidh *et al.*, 2009; Grierson and Long, 1987; Chauhan *et al.*, 2012).

Specimen examined: West Sikkim, Bakhim, 2800 m, 04. 3. 1996, *S. K. Jana* 18084 (BSHC); North Sikkim, Changu - Lachung road, 12. 4. 1986, *DCS Raju & S. Singh* 5380 (BSHC).

(v) *Costus speciosus* (Koenig ex Retz.) J.E. Smith (Zingiberaceae) Betlauri

Taxonomic characters: It is a fleshy herb with long leafy spirally twisted stem 1-2 m in height and horizontal rhizomes, leaves are simple and spirally arranged, oblanceolate to oblong acute, glabrous, sheathing stem. Flowers are white with orange centre, having calyx with oval lobes, corolla with unequal lobes, bracts are bright red (Fig. 1d).

Ethno-botanical Uses: *C. speciosus* is an important medicinal and ornamental plant used to cure different diseases. The plant possess many pharmacological activities such as antibacterial, antifungal, antioxidant, antihyperglycemic, antiinflammatory, analgesic, antipyretic, antidiuretic, larvicidal, antistress and estrogenic (Srivastava *et al.*, 2013; Pawar and Pawar, 2014). The rhizomes are bitter, astringent, acrid, cooling, aphrodisiac, purgative, anthelmintic, depurative, febrifuge, expectorant, tonic, improve digestion, and is a stimulant that clears toxins (Nahak and Sahu, 2011; Pawar and Pawar, 2014). It also has anti-fertility and anabolic properties. Rhizomes are also given in diseases as pneumonia, rheumatism, dropsy, urinary diseases, jaundice, and leaves are given in mental disorders. Due to its excess use and poor conventional propagation method, it is becoming one of the rare and threatened plants. Recently it is used in drug industry as a natural source of diosgenin which is a steroidal sapogenin used for synthesis of sex hormones, cortisone and oral contraceptives (Sarin *et al.*, 1974; Kumar *et al.*, 1984; Sharma and Chattopadhyay, 1983). Diosgenin content up to 3.37% has been reported in rhizome of *C. speciosus* (Singh *et al.*, 2013). Ethno medicinally root powder is used as 1 teaspoon twice a day for 3 days to treat rheumatism, asthma and sexual dysfunction. Decoction of root powder is given to children to treat diarrhea, dysentery, stomach pain and root extract is used to get relief from ear ache. Rhizome is used to treat dropsy and oedema (Ramawat, 2008). In Sikkim, people of Lepcha tribe use rhizome mixed with sugar to treat veneral diseases;

being pungent, it is used as a substitute to zinger. Its juice if taken before breakfast cures urinary tract infection (Pradhan and Badola, 2008). As an ethno-veterinary medicine the rhizomes are used for healing of wound, to wash injuries and to cure fever (Bharati and Sharma, 2009).

Distribution: It is widely distributed in Assam, Khasi and Jaintia Hills, sub-Himalayan tracts of U.P. & H.P. and W. Ghats.

(vi) *Curcuma caesia* Roxb. (Zingiberaceae) Kalo hardi

Taxonomic characters: *Curcuma caesia* is commonly known as Kali-haldi or Kalo-haldi is an ethno medicinally important plant of South East Asia. The plant is erect ranging up to 1.0 m in height, differentiated into underground tuberous rhizome and an erect aerial shoot with leaves and flowers. Rhizome is tuberous with camphoraceous odour, and covered with adventitious roots, root scars, and warts; moreover, it shows longitudinal circular wrinkles on the surface giving the look of nodal and internodal zones. Leaves are broad oblong lanceolate and glabrous with deep purple colour in the middle of lamina. Petiole is ivory colour, ensheathing one another forming a pseudo stem (fig. 1e). Inflorescence is 15–20 cm long spike. Flowers are smaller than bracts, pale yellow with reddish border (Palewal *et al.,* 2011).

Ethno-botanical Uses: In the traditional system of medicine, fresh and dried rhizomes of *C. caesia* are used in treating leucoderma, asthma, tumours, piles, bronchitis, bruises etc. *C. caecia* possess therapeutical activities like antioxidant, antibacterial, antipyretic, larvicidal, insecticidal, antimicrobial, wound healing and antihyperglycemic (Gantait *et al.,* 2011; Satyavam and Warjeet, 2012; Dewangan *et al.,* 2014). In Sikkim *C. caecia* is traditionally used in the treatment of loss of appetite, food poisoning and jaundice. The rhizomes are used by the tribal community of Madhya Pradesh state for the treatment of pneumonia, cough & cold in children, fever and asthma in adults. The powder of herb is used by tribal women as a face-pack (Paliwal *et al.,* 2011).

Distribution: This herb is available throughout north-east, central India, hills of East Godavari, West Godavari, and Andhra Pradesh. In Sikkim it is found in East Sikkim.

(vii) *Hedychium spicatum* Smith (Zingiberaceae) Sathi, Saro

Taxonomic characters: It is a perennial, erect herb with leafy stem of about 150 cm height. Leaves are alternate, broadly ovate-lanceolate, glabrous above and sparsely pubescent beneath with clasping stalks. The flowers are fragrant growing along the stem in a dense spike (fig. 1f). Calyx is membranous, three-lobed, ovate, obtuse and smaller than bracts. Corolla tube is longer than calyx. Petals are white with an orange-yellow or red base; filaments of stamen are red. Capsules are globular, three-valved with an orange-red lining; seeds are black with red aril (Gaur, 1999).

Ethno-botanical Uses: Its rhizome is used in Ayurvedic and Unani medicine. A perfumed coloured powder (called "Abir", used during festivals in India) is prepared from the rootstock. Dried and powdered roots are used for perfuming tobacco. The roots are also used medicinally as carminative and stimulant. The leaves are woven into mats (Handa, 2000). Aqil and Ahmad (2003) reported antimicrobial activity in the fruit extract of *H. spicatum*. Seeds are believed to cause abortion. Dried rhizome of the plant contains essential oil, starch, resins, organic acids, glycoside, albumem and saccharine. The rhizome yields essential oil which has p-methoxycinnamate as major chemical constituent (Sarin, 2008). The essential oil also contain ethyl ester of p-methyl cinnamic acid, d-sabirene cineole, sesquiterpenes and pentadecane mentyl paracumarine (Board NIIR, 2005), cineole, γ-terpinene, limonene, β-phellanderene and β-terpineol (Hussain and Hore, 2007); α-pinene, β-pinene, limonene, 1:8 cineole, linalool, camphor, linalyl, acetate, terpineol, bornneol, carophyllene, r-cadinene, humulene, terpineolene, p-cymene in *H. spicatum* var. *acuminatum* (Garg et al., 1977). The essential oil can be used in perfumes, soaps, hair oils and in cosmetics. The essential oil from rhizomes are used in the treatment of respiratory disorders, fevers, tranquilizer (Dixit and Varma,1979), hypotensive, antispasmodic, CNS depressant, analgesic (Rastogi and Mehrotra, 1980), antimicrobial (Bisht et al., 2006), antioxidant (Joshi *et al.,* 2008), antifungal (Aqil and Ahmad,2003), pediculicidal (Jadhav *et al.,* 2007), anthelmintic activity (Sravani and Paarakh, 2011), antimalarial (Misra, 1991), hypocholesterolemic (Malini and Vanithakumari, 1990), analgesic, antihistaminic, anti-inflammatory (Tandan *et al.,* 1997), ulcer protective, useful in respiratory disorders (Ghildiyal *et al.,* 2012) and cytotoxic (Suresh *et al.,* 2013) activities. The rhizomes are useful in the treatment of asthma and bronchitis (Singh, 1983). Rhizome powder is sprinkled as an antiseptic agent and also used in various aches and pains (Thakur *et al.*, 1989). The rhizomes have insect repelling properties and used for the preservation of cloth in some part of Uttarakhand. Decoction of rhizome with deodar saw dust is taken for tuberculosis (Gaur, 1999). Rhizomes are used in vomiting, stomach and liver disorder, local inflammations, nausea, asthma, bronchitis, hiccups, local pain and to counteract the bad mouth taste & smell (Board NIIR, 2005). In Sikkim Himalayan region the root of *H. spicatum* are used in the preparation of tooth powder due to pungent and bitter taste. In Mizoram the rhizome is used as carminative, tonic, stimulant, expectorant, in liver problem, vomiting, stomachache, inflammatory & pains and in snakebite (Rai and Lalramnghinglova, 2010). Reddy *et al.* (2009a) isolated labdane diterpenes as intestinal α-glucosidase inhibitor from antihyperglycemic extract of *H. spicatum*. Reddy et al, (2009b) isolated two diterpenes from rhizome extract showing cytotoxic activity against colon, skin, lung and breast cancer. Rawat *et al.* (2011) observed potential antioxidant activity in the rhizomes of H. spicatum. Tandan *et al.* (1997) reported significant analgesic and anti-inflammatory activity in rhizome of *H. spicatum*.

Distribution: *H. spicatum* commonly known as Kapur Kachari, found in temperate and sub-temperate zones in Himalayas between 1500 to 2700 m (Naithani, 1985).

(viii) *Heracleum nepalense* D. Don (Umbelliferae) Chimphing

Taxonomic characters: *Heracleum nepalense* is a small shrub up to 2m high; stem solitary, pubescent; Inflorescence umbel, 8-30 flowered; Petals white, occasionally pinkish, outer flowers in umbel conspicuously radiant (Fig. 1g).

Ethno-botanical Uses: *H. nepalense* is known for its anti-diarrhoeal, anti-septic, anti-influenzal and anti-spasmodic activities (Dhar *et al.*, 1968). Various phytochemicals have been isolated from the root of *H. nepalense* such as furocoumarins (bergapten, byakangelicin, allo-imperatorin etc), comrarins, steroids namely α- sitosterol and alkaloids (Anonymous, 1959; Gupta *et al.*, 1975; Sun *et al.*, 1980; Asolkar *et al.*, 1992; Bose *et al.*, 2007). The plant is used in veterinary medicine. The roots of the plant are used as digestive, carminative and anti-diarrhoeal; fruits are pickled, used in cure of typhoid, nausea, thyroid trouble and during vomiting; seeds are used in fever and influenza (Dash *et al.*, 2006; Chettri *et al.*, 2005; Yonzone *et al.*, 2012a; Bose *et al.*, 2007; Saha *et al.*, 2011). *H. nepalense* exhibits a dose-dependent immune-stimulant effect (Dash *et al.*, 2006).

Distribution: *H. nepalense* is generally found beside stream banks in Nepal and Sikkim (Gurung, 1999).

(ix) *Litsea cubeba* (Lour.) Pers. (Lauraceae) Siltimbur

Taxonomic characters: *Litsea cubeba* is a fast growing, multipurpose tree of about 8 to 10m height.

Ethno-botanical Uses: As a multipurpose tree it is used for feeding muga silk worms, production of fast growing timber (Chaudhury, 1981), in aromatherapy and treatment of many diseases (Bhuinya *et al.*, 2010). It's a bark, leaf, root and fruits are used in traditional medicines (Mao *et al.*, 2000) (Fig. 1h). In traditional Chinese medicine it is used for the treatment of rheumatic diseases, common cold and stomach ache (Feng *et al.*, 2009). Many pharmacological activities of *L. cubeba* has been reported such as anticancer (Ho *et al.*, 2010), antibacterial (Wang and Liu, 2010), antioxidant (Hwang *et al.*, 2005), anti-inflammatory, immunological activities (Choi and Hwang, 2004) and as a therapeutic agent in arthritis (Lin *et al.*, 2013). Fresh fruits of *L. cubeba* yield an aromatic essential oil which is used as a flavor enhancer in foods, cosmetics, cigarettes; as a raw material in the manufacture of citral, vitamins A, E, and K, ionone, methyl ionone, and perfumes; and as an antimicrobial and insecticide (Luo *et al.*, 2005). In Sikkim Himalayan region the raw fruit of *L. cubeba* are chewed twice to thrice a day to control diabetes (Chhetri *et al.*, 2005; Gurung, 2002) and also used as carminative (Chanda *et al.*, 2007). Fresh leaves and mature fruits are used in the treatment of cholera, hysteria, vermifuge, gastritis, indigestion and stomach colic (Yonzone *et al.*, 2012a). The tree bark is used for treating foot and mouth disease of cattle in *Khasi* and *Jantia* tribes of Meghalaya (Mao *et al.*, 2009). Sharma *et al.* (2015) found essential oil of *L. cubeba* as an effective antifungal agent for controlling the growth of phyto-pathogenic fungi.

Distribution: It is native of Southeast Asian countries including India, China, Bhutan, Nepal, Myanmar, Vietnam, Korea, Taiwan and Indonesia. In India, it is found mainly in the Himalayas and reported from Arunachal Pradesh, Assam, Manipur, Meghalaya, Mizoram, Nagaland, Sikkim, Uttaranchal, Uttar Pradesh and West Bengal.

(x) *Oroxylum indicum* Vent. (Bignoniaceae) Totala

Taxonomic characters: Oroxylum indicum, also known as Sonapatha or Shyonaka is commonly used herbal medicine in Ayurvedic system. It is a small tree of about 8-15 m height with light brown bark having corky lenticels. Leaves are 3-7 cm long, 2-3 pinnate having opposite, pinnae, stout cylindrical rachis with 2-4 pairs of leaflets. Flowers are numerous, in large erect racemes. Capsules are 30-60 cm long and 5-10 cm broad, straight, tapering to both ends, like sword case and semi-woody (Fig. 1i). Seeds are numerous and winged (Deka *et al.,* 2013; Kirtikar & Basu, 2001; Ayurvedic Pharmacopoeia of India).

Ethno-botanical Uses: Roots, leaves and stems are used either as a single drug or as component of drug preparations for treating various ailments and as a tonic and Rasayana drug. It possesses anticancer, antioxidant, hepatoprotective, immunomodulaory properties, contains flavonoids like chrysine, baicalein and Oroxylin-A. Various other properties like gastro-protective, antibacterial, antioxidant and analgesic of *O. indicum* have also been reported (Dev *et al.,* 2010; Ali *et al.,* 1998; Yan *et al.,* 2011). *O. indicum* is vulnerable taxa for Sikkim (IUCN criteria). Its flowers are used as vegetable. Bark and seeds are used for the treatment of fever, pneumonia, throat complications (Idrisi *et al.,* 2010) and bark decoction or juice in diabetes (Chhetri *et al.,* 2005). Unbroken pods are used in local ceremony (Pradhan and Badola, 2008). Flower ash is applied to treat burns, bark ash and pods are used to heal cut. Leaves are used as fodder. Flower and root bark improves appetite, used in vomiting, asthma, bronchitis, etc. (Chettri *et al.,* 2005). Under ethano-veterinary uses the seeds and root bark of *O. indicum* are used on wounds. Seeds pounded with mustard oil are applied on cracked nipple of milking cattle (Bharati and Sharma, 2009).

Distribution: O. indicum is found throughout southeast and south Asian countries. It is found in Fujian, Guangdong, Guangxi, Guizhou, Sichuan, Taiwan, Yunan, Myanmar, Vietnam, Laos, Thailand, Cambodia, Malaysia, Philippines, Indonesia (Java, Sumatra), Nepal, Bhutan, Bangladesh, Pakistan and Sri Lanka (Theobald, 1981; Zhang and Santisuk, 1998; Kochummaen, 1978). It is native to the Indian sub-continent, in the Himalayan foot hills (Theobald, 1981). In India, it is found in the Himalayan foothills and Eastern and Western Ghats, and it is cultivated throughout India in forest areas (Sasidharan, 2004; Dinda *et al.,* 2015). In Sikkim it usually grows in open forests and slopes at 300–1200 m altitude and is also cultivated.

(xi) *Panax sokpayensis* Sharma & M.K. Pandit (Araliaceae) Ajambari

Taxonomic characters: *Panax sokpayensis* is a perennial herb of 20 to 150 cm height with horizontal rhizome varying in length from 2 cm to up to 40 cm depending upon age of plant. Rhizome has persistent coin shaped scars (stem scars) at nodes separated by short and thick internodes. New stem emerge out each year which vary in height depending upon age of plant. Stem culminate in whorl of 4-5 palmate compound petiolate leaves (Fig. 1j). Inflorescence is umbel and originates from the centre of leaf whorl. Fruits are globose to sub-globose containing one seed or bilobed containing two seed or trilobed containing three seeds. Fruit on maturation turn red and black form lower and upper portion, respectively.

Ethno-botanical Uses: *Panax* spp. are among the most searched after medicinal specie with importance in pharmacological, food and nutraceutical field. Commonly known a ginseng, these have anti-stress, anti-aging and aphrodisiac properties (Cheng *et al.*, 2007; Lu *et al.*, 2009; Kim *et al.*, 2013; Biswas *et al.*, 2015). About 150 naturally obtained ginsenosides have been isolated from roots, leaves, stems, fruits and flower heads of ginseng (Christensen, 2009). Many biological activities, such as anti-inflammatory, anti-tumor, and anti-dementia effects (Yang *et al.*, 2014; Shi *et al.*, 2013; Sun *et al.*, 2008; Mochizuki *et al.*, 1995), have been shown by ginsenosides. *P. sokpayensis* is found growing in the submontane region of West Sikkim at an altitude ranging between 1,700 m and 2,300 m (Sharma and Pandit 2009). However Sharma and Pandit (2011) delineated the difference between the *P. sikkimensis*, *P. bipinnatifidus* & *P. sokpayensis* and considered that *P. sikkimensis* is a synonym of *P. bipinnatifidus*. *Panax* as such are used in the traditional medicine in the Sikkim Himalayan region. Rhizomes of these species are harvested from the wild and are exploited for commercial and domestic use. Locally the rhizomes are used as stimulant, in general debility, weakness, headache, vomiting, blood pressure, as carminative, tonic, expectorant, gastritis, menstrual disorder, asthma, liver complain and for longevity & vitality (Hussain and Hore, 2007; Yonzone *et al.*, 2012a).

Distribution: Sikkim, West Sikkim at an altitude of 2000 to 2500m above mean sea level.

(xii) *Picrorhiza kurroa* Royle ex. Benth. (Scrophulariaceae) Kutki

Taxonomic characters: *Picrorhiza kurroa* has a long, creeping rootstock of bitter taste growing in moist rocky crevices and sandy soils. The leaves are oval with serrate margins. Flowers appear from June to August and are white or pale purple, borne on a tall spike.

Ethno-botanical Uses: *P. kurroa* is a well-known herb of Ayurvedic system of medicine. Traditionally it has been used to treat liver disorders and ailments of upper respiratory tract, to reduce fevers, to treat chronic diarrhoea, dyspepsia and scorpion sting. In Sikkim Himalayan region the dried roots are used in malarial fever, cold, cough, as cathartic, purgative and in dyspepsia (Pradhan and Badola, 2008; Chettri *et al.*, 2005). *P. kurroa* is source of glycosides, picroside I and II, and kutkoside which are extracted from dried rhizomes and roots of 3 to 4 year old plants (Rastogi *et al.*, 1949; Kitagawa *et al.*, 1969; Weinges *et al.*, 1972; Jia *et al.*, 1999). *P. kurroa* is used as hepato-protective, diuretic, finds use as antiperiodic, stomachic, cathartic and cholagogues (Hussain, 1984; Kirtikar and Basu 1984). Although *P. kurroa* propagates mainly through roots or suckers in nature, indiscriminate collection of its underground parts has led to considerable depletion of natural populations. Its propagation through seeds is also poor (Chandra *et al.*, 2006). Moreover, there is an acute lack of attention towards its cultivation practices, resulting in the enlistment of plant as endangered plant. Therefore, it is utmost important to explore the different methods of propagation and conservation of *P. kurroa.*

Distribution: It is an endemic, herbaceous plant of alpine Himalayas from west to east, ranging from 3000 to 4500 m above mean sea level (Thakur *et al.*, 1989). It is endemic to alpine Himalayas and grows in the inner ranges from Kashmir to Sikkim (Chandra *et al.*, 2006).

Specimen examined: Sikkim, West district, 10.06.2004, SS Rathore 28943 (BSHC-38033).

(xiii) *Podophyllum hexandrum* (Podophyllaceae) Bankakari

Taxonomic characters: *Podophyllum hexandrum* is perennial herb, growing in the lower alpine regions of Sikkim Himalaya. *P. hexandrum* grows up to 1.5 feet in high with deeply lobed leaves, fleshy erect stems, rising from the soil. The plant forms a pair of crow-foot (palmate) leaves with cuneate base and serrate margins. Flowers are solitary and erect, during fruiting season the plants can be easily identified due to the reddish fruit colour.

Ethno-botanical Uses: It has been used traditionally as well as in modern times as a purgative and emetic, for cure of necrotic wounds and inhibitor of tumour growth. Traditionally *P. hexandrum* had been used for the treatment of septic wound, cold, burning sensation, constipation, erysipelas, plague, mental disorders, allergic and inflammatory conditions of the skin, cancer of brain, bladder and lung, Hodkins disease and non- Hodkins lymphoma (Singh and Shah, 1994; Beutner and Vonkrogh, 1990; Wani *et al.*, 2012). The rhizome contains a resin, known as Indian Podophyllum Resin, which contains podophyllol, podophyllin and podophyllotoxin as well as a sticky resin-quercetin. Podphyllotoxin is the most active cytotoxic natural product. It is used as starting compound for the synthesis of anticancer drug etoposide (Vepeside), an FDA approved anticancer drug and teniposide. Podophyllotoxin inhibits microtubule assembly. These drugs are used for lung cancer, testicular cancer, neuroblastoma, hepatoma and other tumours. It also shows antiviral activities by interfering with some critical viral processes (Giri and Narasu, 2000). Podophyllotoxin also finds use in the treatment of leukaemia, dermatological disorders like warts, rheumatoid arthritis and psoriasis. The roots are used as emetic, cholagogue, blood purifier, purgative and alterative. It is considered as a cardiac tonic in small doses. It also finds use as a stimulant in peristalsis, allergy and skin inflammations. It also has numerous applications in modern medicine by virtue of its free radical scavenging capacity. An extract of *P. hexandrum* has been shown to provide approximately 80% whole-body radioprotection in mice. Total synthesis of podophyllotoxin is an expensive process and availability of the compound from natural resources is an important issue for pharmaceutical companies that manufacture these drugs. The Indian *P. hexandrum* is superior to its American counterpart, *P. peltatum* in terms of its higher podophyllotoxin content (>5%) in dried roots (Purohit *et al.*, 1999; Giri and Narasu, 2000; Fay and Ziegler, 1985; Drew *et al.*, 1987; Thakur, 1993). Presently the *P. hexandrum* population is scarce due to excess collection from nature. It need attention for conservation and in depth study for propagation, cultivation techniques, podophyllotoxin production, a detailed phytochemical diversity study, particularly of marker podophyllotoxin and its glycosides (Qazi *et al.*, 2011).

Distribution: Sikkim, North district- Thila, Thangu and Yongdi.

Specimen examined: Sikkim, North district- Shingba Rhododendron Sanctuary, N 27^{0}45'27.7; E 88^{0}43'27.2, 13.09.2014, CS Purohit 38279.

(xiv) *Swertia chirayita* (Roxb. ex Fleming) Karsten (Gentianaceae) Chirowto

Taxonomic characters: Plants of *Swertia chirayita* are described as annual (Kirtikar and Basu, 1975; Anonymous, 1976; Bentley and Trimen, 1983; Chauhan, 1999) or biennial (Garg, 1987; Grierson and Long, 1999; Shreshtha and Joshi, 1992; Nautiyal and Nautiyal, 2004; Joshi and Dhawan, 2005; Rijal, 2009) or triennial (Pradhan and Badola, 2012) or pluri-annual i.e flowering once in the third year (Edwards, 1993; Shah, 2008; Raina *et al.,* 2013). In the first and second year of growth the plant of *S. chirayita* remain in rosette form and in later part of second year to third year, plant rises up (erect stage) to bear flower and fruits. Erect stage of plant remains for 6-7 months from June to December (fig. 1k).

Ethno-botanical Uses: Commercially the plants of *S. chirayita* are harvested during erect stage. The plants of *S. chirayita* are very specific to its altitudinal and cool environment requirements. *S. chirayita* is one of the nationally and internationally renowned medicinal plants of high demand. *S. chirayita* has drawn attention since ancient times due to its medicinal importance because of the presence of bitter glycosides, xanthones, triterpenoids, iridoid and flavonoides (Negi *et al.,* 2011; Kumar and Staden, 2016). Significant biological activities of *S.chirayita* includes hepatoprotective, antihepatotoxic, antimicrobial, anti-inflammatory, anticarcinogenic, antileprosy, hypoglycemic, antimalarial, antioxidant and CNS depressant (Negi *et al.,* 2011). In Sikkim Himalayan region the plant of *S. chirayita* is used in traditional medicine and as home remedy for ailments such as malarial fever, indigestion, intestinal worms in children, liver diseases, cough, cold, asthma, headache, boils and scabies (Joshi and Joshi, 2008; Pradhan and Badola, 2008; Pradhan and Badola, 2012). In Sikkim Himalayan region the cultivation of *S.chirayita* in and around its niche environment has been taken up by farmers in a very sporadic and scattered manner. Farmers hesitate to take up its cultivation due to its slow growing nature (Basnet, 2001), long gestation/cultivation period of three years (Abrol *et al.,* 2012: Badola and Pal, 2002), poor germination of seed and difficult to handle seedlings because of very small size, etc. (Joshi and Dhawan, 2005; Raina, *et al.*, 1994; Chaudhuri *et al.*, 2007).

Distribution: *S. chirayita* is a plant of temperate Himalayas growing at an altitude of 1200-3000 m above mean sea level (msl) from Kashmir state of India to Bhutan and also in the Khasi hills of Meghalaya state at 1200-1500 m above msl (Kirtikar and Basu, 1975; Clarke, 1885; Blatter, 1984; Tandon *et al.*, 2010). In Sikkim it is found in Kyongnosla, Namchi, Rumtek, Tong, Geyzing, Pelling, Jorethang, Gursey and sporadically cultivated in Uttrey and Yuksam.

Specimen examined: Sikkim, West district, Hilley-Varsey route, 22.09.2002, SK Rai & S Pradhan 25411 (BSHC-28796).

(xv) *Tupistra nutans* Wall. ex Lind. (Liliaceae) Nakima

Taxonomic characters: *Tupistra nutans* is a glabrous herb with radical, oblanceolate leaves, flower are dull brown, born on 10-15 cm spike. It is naturally found as well as cultivated in subtropical and temperate zone of Sikkim Himalayan region.

Ethno-botanical Uses: Locally the powdered roots, and flower decoction are taken to control diabetes (Hussain and Hore, 2007). Inflorescence is powdered and mixed with water and taken to relieve body pain (Pradhan and Badola, 2008; Idrisi *et al.*, 2010) (Fig. 11). Saha *et al.*, 2011 reported use of inflorescence in food poisoning which are boiled, cooked and consumed for at least 7 days. Root decoction is helpful in curing urinary troubles, insomnia, constipation and rheumatic pain. Use of spike as vegetable is helpful in gastrointestinal problem, constipation and is advised as tonic. Flowers are used for consumption as curry (Chettri *et al.*, 2005). Dry powder of inflorescence is eaten to treat diabetes, and as a tonic to relieve body pains (Purohit *et al.*, 2014).

Distribution: In Sikkim it is found distributed in Tinding, Yoksum, Phodong, Kabi, Rangit, Tumlong, Luybik, Rungit, Bakhim, Rungbee, Rayang valley, Dickling.

Conclusion

Ethno medicinal use of plant is an effort of mankind over-million of years of search for eternal health, longevity and remedy to relieve discomfort & pain; which compelled early man to explore their surroundings and try & develop these natural resources as therapeutic agents. The systematic record of use of ethno medicinal plants, their incorporation into regular system of medicine, refinement and further development became part of the Materia Medica of many cultures all over the world. The exploration of vast literature, the phytogeographic location and ethnic knowledge of people of Sikkim Himalayan region indicates that the area is not only rich in the medicinal plant diversity, but also in the traditions of folk-medicine. Despite gradual socio-cultural transformation, local communities possess substantial knowledge of plants and their uses. In Sikkim Himalayan region the reliance of people on folk medicines for health care is more likely to be associated due to the lack of modern medicines, area remoteness and the belief of people on effectiveness of traditional folk medicine. More exploration needs are felt in the remote villages of Sikkim Himalayan region. In order to safeguard and further perpetuate the ethno medicinal plant associated knowledge, it should be well documented, preserved and patented. Cures for many of the dreaded diseases such as cancer, AIDS, etc. may lie hidden in the treasure of these Himalayan folk medicines. A concerted effort by scientists, farmers, financers and political leaders is called for, to benefit from the knowledge and resources nurtured on the lap of this part of Himalayas, over thousands of years. This, however furthermore requires detail quantitative assessment of resource availability, productivity potential, sustainable harvesting and domestication methodology, assessment of market value of potentially promising species and most importantly the equitable benefit sharing regimes. Farmers should be involved in the cultivation of medicinal plants at least in their barren and fallow land. The methods for cultivation, collection, processing and preparation of crude drugs have to be established on a scientific basis. Appropriate research should be carried out by the apex institutions in the hills to develop agro-techniques for the cultivation of medicinal plants on priority basis. Intensive training on cultivation, conservation, and processing techniques for ethno medicinal plant species, needs to be carried out. There is also a lack of phyto-therapeutic evidence for most of the species. It is advisable that phytochemical and pharmacological studies be carried out in order to confirm the validity of properties attributed to these species; this is particularly important for species with market potential beyond the local level. This will pave the path for the scientific establishment of these medicinal plants. With setting up management plans for the extraction and collection of ethno medicinal species, these can also provide income to local people.

Validation of therapeutic value of ethno medicinal plants by apex institutions so that they can be used in development of novel drug or as a remedy to ailment is highly recommended. Renewed interest of developing and developed countries in the natural resources has opened new horizons for the exploration of natural sources with the perspectives of safety and efficacy. The development of these traditional systems of medicines with the perspectives of safety, efficacy and quality will help not only to preserve this traditional heritage but also rationalize the use of natural products in the health care. The ethno medicinal plant species of Sikkim Himalayan region may be explored with the modern scientific approaches for better leads in the health care.

Fig. 1. Ethanomedicinally used plants of Sikkim Himalayan region: leaf bearing branch of *Artemisia vulgaris* (a), *Astilbe rivularis* (b), *Bergenia ciliate* (c), flower bearing branch of *Costus speciosus* (d), *Curcuma caesia* (e), flowering in *Hedychium spicatum* (f), flowering and fruiting in *Heracleum nepalense* (g), fruit bearing branch of *Litsea cubeba* (h), mature pods of *Oroxylum indicum* (i), fruit bearing *Panax sokpayensis* (j), flowering branch of *Swertia chirayita* (k), inflorescence of *Tupistra nutans* (l).

Acknowledgements

The authors are thankful to Department of Biotechnology, Government of India, New Delhi for financial assistance in the form project: Establishment of Regional Centre of Institute of Bioresources and Sustainable Development vide sanction order No. BT/01/08/ IBSD/Sikkim dated 23/03/2009. Authors are also grateful to the Director, Institute of Bioresources and Sustainable Development, Imphal for providing facilities.

References

Abrol, E., Vyas, D. & Koul, S. 2012. Metabolic shift from secondary metabolite production to induction of anti-oxidative enzymes during NaCl stress in *Swertia chirata* Buch.-Ham. *Acta Physiologiae Plant*, 34: 541-546.

Ali, R.M., Houghton, P.J., Raman, A. & Hoult, R.S. 1998. Antimicrobial and antiinflammatory activities of extracts and constituents of *Oroxylum indicum* (L.) Vent. *Phytomedicine*, 5(5): 375-381.

Anonymous. 1959. *The Wealth of India*. Vol. V: H-K, CSIR, New Delhi. pp. 350-351.

Anonymous. 1976. *The Wealth of India* – a dictionary of Indian raw materials and industrial products. Raw materials. Vol. X, Sp–W. Publications and Information Directorate, Council of Scientific and Industrial Research, New Delhi. Pages 78–81.

Aqil, F. & Ahmad, I. 2003. Broad spectrum antibacterial and antifungal properties of certain traditionally used Indian medicinal plants. *World Journal of Microbiology and Biotechnology*, 19(6): 653-657.

Asolkar, L.V., Kakkar, K.K., & Chakre, O.J. 1992. Second supplement to glossary of Indian. *Medicinal Plants with Active Principles* Part-I (A-K). CSIR (PID), New Delhi. pp. 350-351.

Ayurvedic Pharmacopoeia of India, Part 1st, Government of India, Ministry of Health and Family Welfare, Vol.3: 183-184.

Badola, H.K. & Pal, M. 2002. Endangered medicinal plant species in Himachal Pradesh. *Current Science*, 83: 797–798.

Bagci, E., Kursat, M. & Civelek, S. 2010. Essential oil composition of the aerial parts of two *Artemisia* species (*A. vulgaris* and *A. absinthium*) from East Anatolian region. *Journal of Essential Oil Bearing Plants*, 13(1): 66–72.

Barata, A.M., Rocha, F., Lopes, V. & Carvalho, A.M. 2016. Conservation and sustainable uses of medicinal and aromatic plants genetic resources on the worldwide for human welfare. *Industrial Crops and Products*. (In press) doi:10.1016/j.indcrop.2016.02.035

Basnet, D.B. 2001. Evolving nursery practices and methods of cultivation of high value medicinal plant *Swertia chirayita* Ham. *Environment and Ecology*, 19:935-938.

Bentley, R. & Trimen, H. 1983. Medicinal plants (Vol. 3). Dehradun, India: International Book Distributors.

Beutner, K.R. & Vonkrogh, G. 1990. Current Status of Podophyllotoxin for the treatment of Genital warts. In *Seminars in Dermatology*, 9(2): 148-151.

Bhakuni, D.S. 1990. Drugs from Plants. *Science Reporter*, 27(8): 12.

Bharati, K.A. & Sharma, B.L. 2009. Studies on ethno-veterinary uses of plant resources of Sikkim. *Indian Forester*, 135(5): 691-696.

Bharati, K.A. & Sharma, B.L. 2012. Plants used as ethnoveterinary medicines in Sikkim Himalayas. *Ethnobotany Research and Applications*, 10: 339-356.

Bhuinya, T., Singh, P. & Mukherjee, S.K. 2010. *Litsea cubeba* - Medicinal Values - Brief Summary. *J. Trop. Med. Plants*, 11(2): 179-183.

Bisht, G.S., Awasthi, A.K. & Dhole, T.N. 2006. Antimicrobial activity of *Hedychium spicatum*. *Fitoterapia*, 77(3): 240-242.

Biswas, K. 1956. *Common Mdcicinal plants of Darjeeling and Sikkim Himalaya.* West Bengal: Bengal Government Press.

Biswas, T., Singh, M., Mathur, A.K. & Mathur, A. 2015. A dual purpose cell line of an Indian congener of ginseng—*Panax sikkimensis* with distinct ginsenoside and anthocyanin production profiles. *Protoplasma, 252(2):* 697-703.

Blagojevic, P., Radulovic, N., Palic, R. & Stojanovic, G. 2006. Chemical composition of the essential oils of Serbian wildgrowing *Artemisia absinthium* and *Artemisia vulgaris*. *Journal of Agricultural and Food Chemistry*, 54(13): 4780–4789.

Blatter, E. 1984. Beautiful flowers of Kashmir (Vol. II). Dehradun, India: International Book Distributors.

Board, N. I. I. R. 2005. *Compendium of medicinal plants*. National Institute of Industrial Re - Materia medica, vegetable. Pp. 444. ISBN: 8186623809, 9788186623800

Bodeker, G., Neumann, C., Lall, P. & Oo, Z.M., 2005. Traditional medicine use and health worker training in a refugee setting at the Thai-Burma border. *Journal of Refugee Studies*, 18(1), pp.76–99.

Bora, K.S. & Sharma, A. 2011. The genus *Artemisia*: A comprehensive review. *Pharmaceutical Biology*, 49(1): 101–109.

Bose, S. K., Dewanjee, S. & Mandal, S.C. 2007. Antibacterial activity of methanol extract of roots of *Heracleum nepalense* D Don. on bacteria causing diarrhoea. *Oriental Pharmacy and Experimental Medicine*, 7(3): 286-289.

Butola, J.S. & Badola, H.K. 2008. Propagation conditions for mass multiplication of three threatened Himalayan high value medicinal herbs. *Plant Genetic Resource Newsletter*, 153: 43-47.

Carvalho, I.S., Cavaco, T. & Brodelius, M. 2011. Phenolic composition and antioxidant capacity of six *Artemisia* species. *Industrial Crops and Products*, 33(2): 382–388.

Chanda R., Mohanty, J.P., Bhuyan, N.R., Kare, P.K. & Nath, L.K. 2007. Medicianal plants used against gastrointestinal tract disorders by the traditional healers of Sikkim Himalayas. *Indian Journal of Traditional Knowledge*, 6(4): 606-610.

Chandra B., Palni L.M.S. & Nandi S.K. 2006. Propagation and conservation of *Picrorhiza kurrooa* Royle ex Benth.: an endangered Himalayan medicinal herb of high commercial value. *Biodiversity and Conservation*, 15: 2325–2338. DOI 10.1007/s10531-005-0770-z

Chaudhuri, R.K., Pal, A. & Jha, T.B. 2007. Production of genetically uniform plants from nodal explants of *Swertia chirata* Buch.-Ham. ex Wall.- and endangered medicinal herb. *In Vitro Cellular & Developmental Biology-Plant*, 43: 467-472. DOI: 10.1007/s11627-007-9095-9

Chaudhury, S.N. 1981. *Muga Silk Industry*, Directorate of Sericulture and Weaving. Government of Assam, Guwahati, Assam, India, 1-31.

Chauhan, N.S. 1999. *Medicinal and aromatic plants of Himachal Pradesh*. Indus Publishing Company, New Delhi, p.632.

Chauhan, R., Ruby, K. & Dqivedi, J. 2012. *Bergenia ciliate* mine of medicinal properties: a review. *Int. J. Pharm. Sci. Rev. Res.*, 15(2): 20-23.

Chen, C., Lin, C. & Namba, T. 1989. Screening of Taiwanese crude drugs for antibacterial activity against *Streptococcus* mutans. *Journal of Ethnopharmacology*, 27(3): 285–295.

Cheng, A., Lou, Y., Mao, Y., Lu., S., Wang, L. & Chen, X. 2007. Plant terpenoids: biosynthesis and ecological functions. *Journal of Integrative Plant Biology*, 49(2): 179–186.

Chettri, N., Sharma, E. & Lama, SD. 2005. Non-timber forest produces utilization, distribution and status in a trekking corridor of Sikkim, India. *Lyonia*, 8(1): 89-101.

Chhetri, D.R., Parajuli, P. & Subba, G.C. 2005. Antidiabetic plants used by Sikkim and Darjeeling Himalayan tribes, India. *Journal of Ethnopharmacology*, 2: 199-202.

Choi, E.M., & Hwang, J.K. 2004. Effects of methanolic extract and fractions from *Litsea cubeba* bark on the production of inflammatory mediators in RAW264.7 cells. *Fitoterapia* 75: 141–148.

Chopra R.N., Nayar S.L. & Chopra I.C. 1956. *Glossary of Indian Medicinal Plants*, Council of Scientific and Industrial Research, New Delhi.

Christensen, L.P. 2009. Ginsenosides chemistry, biosynthesis, analysis, and potential health effects. *Adv Food Nutr Res.*, 55: 1–99.

Clarke, C.B. 1885. Gentianaceae. In: Hooker, J.D. (ed.), *The Flora of British India*. Vol. IV. L. Reeve and Co., Ltd., London. Pp. 121-130.

Cowan, A.M. & Cowan, J.M. 1929. *The trees of north Bengal. Bengal Secretariat Book* Dept. Calcutta: Bengal Government Press.

Costa, A.G.V., Garcia-Diaz, D.F., Jimenez, P. & Silva, P.I. 2013. Bioactive compounds and health benefits of exotic tropical red-black berries. *Journal of Functional Foods*, 5(2): 539–549.

Cotton, CM. 1996. Ethnobotany: principles and applications. John Wiley & Sons, Chichester, UK. Pp. 424.

Das, T., Mishra, S.B., Saha, D., & Agarwal, S. 2012. Ethnobotanical survey of medicinal plants used by ethnic and rural people in eastern Sikkim Himalayan region. *Afr. J. Basic Ap. Sci.*, 4: 16-20.

Dash, S., Nath, L.K., Bhise, S., Kar, P. & Bhattacharya, S. 2006. Stimulation of immune function activity by the alcoholic root extract of *Heracleum nepalense* D. Don. *Indian journal of Pharmacology*, 38(5): 336.

Deka, D.C., Kumar, V., Prasad, C., Kumar, K., Gogoi, B.J., Singh, L., & Srivastava, R.B. 2013. *Oroxylum indicum* - a medicinal plant of North East India: an overview of its nutritional, remedial, and prophylactic properties. *Journal of Applied Pharmaceutical Science*, 3(4): S104-S112.

Dev, L R., Anurag, M. & Rajiv, G. 2010. *Oroxylum indicum*: A review. *Pharmacognosy Journal*, 2(9): 304-310.

Dewangan, M.K., Dwivedi, C., Sivna, S., Rao, P, Yadav, R., Chandrakar, K., Shashank, P.L., Singh, S. & Sinha, D. 2014. Medicinal value of

Curcuma cassia roxb: an overview. *Critical Review in Pharmaceutical Sciences,* 3(4): 1-9.

Dhar, M.L., Dhar, M.M., Dhawan, B.N., Mehrotra, B.N. & Ray, C. 1968. Screening of Indian Plants for biological activity: I. *Indian J. Exp. Biol.,* 6: 232-247.

Dinda, B., Silsarma, L., Dinda, M. & Rudrapaul, P. 2015. *Oroxylum indicum* (L.) Kurz, an important Asian traditional medicine: From traditional uses to scientific data for its commercial exploitation. *Journal of Ethnopharmacology*. 161: 255-278.

Dixit, V.K. & Varma, K.C. 1979. Effect of essential oils of rhizomes of *Hedychium spicatum* on central nervous system. *Indian J. Pharmacol.,* 1(2): 147-149.

Drew, S.E., Conway, S.J., Jenning, P. & Helliwell, K. 1987. Determination of the aryltetralin lignan content of podophyllum resins and roots/ rhizomes. *J Pharm Pharmacol.,* 39: 738–739.

Edwards, D.M. 1993. The marketing of non-timber forest product from the Himalaya: The trade between East Nepal and India. *Rural Development Forestry Network Paper 15b,* Overseas Development Institute, London. Pp. 24.

El-Tantawy, W.H. 2015. Biochemical effects, hypolipidemic and anti-inflammatory activities of *Artemisia vulgaris* extract in hypercholesterolemic rats. *Journal of Clinical Biochemistry and Nutrition,* 57(1): 33.

Fay, D.A. & Ziegler, H.W. 1985. Botanical sources differentiation of *Podophyllum* resin by high performance liquid chromatography. *Journal of Liquid Chromatography,* 8(8): 1501–1506.

Feng, T., Xu, Y., Cai, X.H., Du, Z.Z. & Luo, X.D., 2009. Antimicrobially active isoquinoline alkaloids from *Litsea cubeba. Planta Medica,* 75, 76–79.

Fernandez-Panchon, M.S., Villano, D., Troncoso, A.M. & GarciaParrilla, M.C. 2008. Antioxidant activity of phenolic compounds: From in vitro results to *in-vivo* evidence. *Critical Reviews in Food Science and Nutrition,* 48(7), 649–671.

Ford, R.I. 2011. History of ethnobiology. In: Ethnobiology. Anderson, EN, Pearsall, D.M., Hunn, E.S., Turner, N.J. & Hoboken, N.J. (eds.), Wiley-Blackwell. Pp.15–26.

Foundation for Revitalisation of Local Health Traditions (FRLHT), 1999. The Key Role of Forestry Sector in Conserving India's Medicinal Plants: Conceptual and Operational Features. Bangalore: FRLHT.

Foundation for Revitalisation of Local Health Traditions (FRLHT), 2009. Overview. *Sci. Med.*, 57(4): 745–756.

Gantait, A., Barman, T. & Mukharjee, P. 2011 Validated method for determination of curcumin in turmeric powder. *Indian Journal of Traditional Knowledge*, 10(2): 247-250.

Garg, S. 1987. Gentianaceae of the north west Himalaya (a revision). *In:* Bioscience International, (ed.), Monograph 17. New Delhi: Today and Tomorrow's Publication Co., p. 183-184.

Garg, S.N., Shawl, A.S. & Gulati, B.C. 1977. Essential oil of *Hedychium spicatum* var. *acuminatum*. *Indian Perfumer*, 21(2): 79-82.

Gaur, R.D. 1999. Flora of District Garhwal North West Himalaya. Transmedia Srinagar, (Garhwal). P. 715.

Ghildiyal, S., Gautam, M.K., Joshi, V.K. & Goel, R.K. 2012. Pharmacological evaluation of extracts of *Hedychium spicatum* (Ham-ex-Smith) rhizome. *Ancient science of life*, 31(3): 117.

Gilani, A., Yaeesh, S., Jamal, Q. & Ghayur, M. 2005. Hepatoprotective activity of aqueous-methanol extract of *Artemisia vulgaris*. *Phytotherapy Research*, 19(2): 170–172.

Giri, A. & Narasu, M.L. 2000. Production of podophyllotoxin from *Podophyllum hexandrum*: a potential natural product for clinically useful anticancer drugs. *Cytotechnology*, 34: 17–26.

Govindaraj, S., Kumari, B.D.R., Cioni, P.L. & Flamini, G. 2008. Mass propagation and essential oil analysis of *Artemisia vulgaris*. *Journal of Bioscience and Bioengineering*, 105(3): 176–183.

Grierson, A.J.C. & Long, D.G. 1983-1999. *Flora of Bhutan* 1(1-3) & 2(1-2). Royal Botanical Garden, Edinburgh.

Grierson A.J.C. & Long D.G. 1987. *Flora of Bhutan* (Vol. 1, Part 3). Royal Botanic Garden, Edinburgh. Pp. 492.

Grierson, A.J.C. and Long, D.G. 1999. Flora of Bhutan, In: Long, D.G. (ed.), Published by Royal Botanical Garden Edinburgh, UK and the Royal Government of Bhutan, 2: P. 626.

Gupta B.D., Banarjee S.K. & Handa K.L. 1975. Coumarins from *H. nepalense* and *H. wallaichi*. *Phytochemistry*, 14: 2533-2538.

Gurung, B. 2002. The medicinal plants of Sikkim Himalaya Maples, Chakung, West Sikkim (India).

Gurung, G. 1999. The medicinal plants of Sikkim Himalaya, Subash Publication, Sikkim, Pp 156-158.

Hafidh, R.R., Abdulamir, A.S., Jahanshiri, F., Abas, F., AbuBakar, F. & Sekawi, Z. 2009. Asia is the mine of natural antiviral products for public health. *The Open Complementary Medicine Journal*, 1: 58-68.

Hamilton, A.C. 2004. Medicinal plants, conservation and livelihoods. *Biodiversity and Conservation*, 13, pp.1477-1517.

Handa, K.L. 1952. A modified method for the assay of *Swertia chirata*. *Indian Journal of Pharmaceutical Sciences*, 14: 87-88.

Handa, N. 2000. *Wild Flowers of India*. Books Today. Pp.66.

Harshberger, JW. 1896. The purpose of ethnobotany. *Bot Gaz.*, 21: 146-54.

Hidayati, S., Franco, F.M. & Bussmann, W. 2015. Ready for phase 5 - current status of ethnobiology in Southeast Asia. *Journal of Ethnobiology and Ethnomedicine*, 11: 17. DOI 10.1186/s13002-015-0005-7

Ho, C.L., Jie-Pinge, O., Liu, Y.C., Hung, C.P., Tsai, M.C., Liao, P.C., Wang, E.I., Chen, Y.L. & Su, Y.C. 2010. Compositions and in vitro anticancer activities of the leaf and fruit oils of *Litsea cubeba* from Taiwan. *Natural Product Communications*, 5: 617-620.

Hooker, J.D. 1872—97. *Flora of British India*. London. (Reprinted in 1982 by Bishen Singh Mahendra Pal Singh, Dehra Dun, Vols. I—VII.)

Hussain, A. 1984. Conservation of genetic resources of medicinal plants in India. In: Jain S.K. (eds), *Conservation of Tropical Plant Resources*. Botanical Survey of India, Howrah, Pp. 110-117.

Hussain, S. & Hore, D.K. 2007. Collection and conservation of major medicinal plants of Darjeeling and Sikkim Himalayas. *Indian Journal of Traditional Knowledge*, 6(2): 352-357.

Hwang, J.K., Choi, E.M. & Lee, J.H. 2005. Antioxidant activity of *Litsea cubeba*. *Fitoterapia* 76: 684-686.

Idrisi, M.S., Badoal, H.K. & Singh, R. 2010. Indigenous knowledge and medicinal use of plants by local communities in Rangit Valley, South Sikkim, India. NeBIO, NECEER. Imphal, 1(2): 34-45.

Jadhav, V., Kore, A. & Kadam, V.J. 2007. *In-vitro* pediculicidal activity of *Hedychium spicatum* essential oil. *Fitoterapia*, 78(7): 470-473.

Jia, Q., Hong, M.F. & Minter, D. 1999. Pikuroside: a novel iridoid from *Picrorhiza kurroa*. *J. Nat. Prod.*, 62: 901-903.

Joshi, K. & Joshi, A. 2008. *Swertia* L. (Gentianaceeae) in Nepal Himalaya: checklist, phytogeography, ethnobotany and conservation status. *Ethnobotanical Leaflets*, 12: 361-372.

Joshi, P. & Dhawan, V. 2005. *Swertia chirayita* – an overview. *Current Science,* 89: 635–640.

Joshi, S., Chanotiva, C.S., Agarwal, G. & Prakash, O. 2008. Terpenoid compositions and antioxidant and antimicrobial properties of the rhizome essential oils of *Hedychium spicatum. Chem. Biodivers.,* 5(2): 299-309.

Kala, C.P. 2005. Ethnomedicinal botany of the Apatani in the Eastern Himalayan region of India. *Journal of Ethnobiology and Ethnomedicine,* 1(1): 1-8.

Khan, A. & Gilani, A.H. 2009. Antispasmodic and bronchodilator activities of *Artemisia vulgaris* are mediated through dual blockade of muscarinic receptors and calcium influx. *Journal of Ethnopharmacology,* 126(3): 480–486.

Kim H.J., Kim P. & Shin CY. 2013 A comprehensive review of the therapeutic and pharmacological effects of ginseng and ginsenosides in central nervous system. *J Ginseng Res.,* 37: 8–29.

Kirtikar, K.R. & Basu, B.D. 1975. *Indian medicinal plants.* Vol. III; 2nd ed. Dehradun, India: Bishen Singh Mahender Pal Singh. p. 1663–1668.

Kirtikar, K.R. & Basu, B.D. 1984. *Indian Medicinal Plants.* Vol. III. Bishen Singh Mahendra Pal Singh, Dehradun, pp. 1824–1826.

Kirtikar, K.R. & Basu, B.D. 2001. *Indian Medicinal Plants,* Oriental Enterprises, Dehradun, 4: 1105–1107.

Kitagawa, I., Hino, K., Nishimura, T., Mukai, E., Yosioak, I., Inouye, H. & Yoshida, I. 1969. Picroside I: a bitter principle of *Picrorhiza kurrooa. Tetrahedron Lett.,* 43: 3837–3840.

Kochummen, K.M. 1978. Bignoniaceae, Tree Flora of Malay. Part-3, Longmann Malaysia Sdn. Bhd.. Wisma Damansara, Kuala Lumpur, pp. 36–44.

Kumar, S. Shukla, R.S., Singh, K.P., Paxton, J.D. & Husain, A. 1984. Glyceollin : A Phytoalexin in Life Blight of *Costus speciosus. Phytopathology,* 74: 1349-1352.

Kumar, V. & Staden, J.V. 2016. A review of *Swertia chirayita* (Gentianaceae) as a traditional medicinal plant. *Frontiers in Pharmacology,* 6: 1-14. DOI: 10.3389/fphar.2015.00308

Laird, S. & Wynberg, R. 2008. Access and Benefit-Sharing in Practice: Trends in Partnerships Across Sectors. Technical Series no. 38, Montreal: Secretariat of the Convention on Biological Diversity.

Lee, S.J., Chung, H.Y., Maier, C.G.A., Wood, A.R., Dixon, R.A. & Mabry, T.J. 1998. Estrogenic flavonoids from *Artemisia vulgaris* L. *J. Agric Food Chem.*, 46: 3325-3329.

Lin, B., Zhang, H., Zhao, X.X., Rahman, K., Wang, Y., Ma, X.Q., Zheng, C.J., Zhang, Q.Y., Han, T. & Qin, L.P. 2013. Inhibitory effects of the root extract of *Litsea cubeba* (lour.) pers. on adjuvant arthritis in rats. *Journal of Ethnopharmacology*, 147: 327-334.

Lu, J.M., Yao, Q. & Chen, C. 2009 Ginseng compounds: an update on their molecular mechanisms and medical applications. *Curr. Vasc. Pharmacol.*, 7:293–302.

Luo, M., Jiang, L.K., & Zou, G.L. 2005. Acute and genetic toxicity of essential oil extracted from (Lour.) Pers. *Journal of Food Protection*, 68(3): 581-588.

Malini, T. & Vanithakumari, G. 1990. Rat toxicity studies with â-sitosterol. *J Ethnopharmacol.* 28(2): 221-34.

Manandhar, N.P. 2002. *Plants and people of Nepal*. Timber Press, Portland, Oregon. pp. 599. ISBN: 0-88192-527-6.

Manandhar, N.P., 1991. Medicinal plant-lore of Tamang tribe of Kabhrepalanchok District, Nepal. *Economic Botany*, 45: 58–71.

Manandhar, N.P., 1992. Folklore medicine of Dhading District, Nepal. *Fitoterapia*, 63: 163–177.

Mandal, P., Misra, T.K. & Basu, P.K. 2009. *In vitro* antioxidant potential *of Astilbe rivularis* rhizome. *Canadian Journal of Pure and Applied Sciences*, 3(1): 649-654. ISSN: 1715-9997.

Mao, A.A, Wetten A., Far, M.F. & Caligari, P.D.S. 2000. In vitro propagation of *Litsea cubeba* (Lours.) Pers., a multipurpose tree. *Plant Cell Report*, 19: 263-267.

Mao, A.A., Hynniewta, T.M., & Sanjappa, M. 2009. Plant wealth of Northeast India with reference to ethnobotany. *Indian Journal of Traditional Knowledge*, 8(1): 96-103.

Misra, S.B. 1991. Antimalarial activity of traditional plants against erythrocytic stages of *Plasmodium berghei*. *Int. J. Pharmacog.* 29:19-23.

Mitra, P.K. 2014a Antibacterial activity of an isolated compound (av-2) from the leaves of titeypati (*Artemisia vulgaris* Linn.). *Mintage Journal of Pharmaceutical & Medical Science*, 3(3): 12-14.

Mitra, P.K. 2014b. Screening the efficacy of forty medicinal plants north east Himalayas of Indian subcontinent in indomethacin induced gastric ulcer in albino rats. *Mintage Journal of Pharmaceutical & Medical Sciences*. 3(2): 14-17.

Mochizuki, M., Yoo, Y.C., Matsuzawa, K., Sato, K., Saiki, I., Tono-Oka, S., et al. 1995. Inhibitory effect of tumor metastasis in mice by saponins, ginsenoside Rb2, 20(R)- and 20(S)-ginsenoside-Rg3, of red ginseng. *Biol Pharm Bull.*, 18(9):197–202.

Mohammed, S., Kasera, P.K. & Shukla, J.K. 2004. Unexploited plants of potential medicinal value from the Indian Thar desert. *Natural Product Radiance*. 3(2): 69-74.

Nadkarni, A.K. 1954. *Indian Materia Medica*. Popular Book Depot, Bombay, India.

Nahak, G. & Sahu, R.K. 2011. Free radical scavenging activity of rhizome of *Costus speciosus* (koen) j.e. sm. *International Journal of Institutional Pharmacy and Life Sciences*, 1(3): 61-69.

Naithani, B.D. 1985. *Flora of Chamoli*. Vol. II. Botanical Survey of India, Howarth.

Nautiyal, M.C. & Nautiyal, B.P. 2004. *Agro-techniques for high altitude medicinal and aromatic plants*. Bishen Singh Mahendra Pal Singh, Dehradun, India.

Negi, J.S., Singh, P. & Rawat, B. 2011. Chemical constituents and biological importance of *Swertia*: A review. *Current Research in Chemistry*, 3: 1-15.

Nikolova, I., Gevrenova, R., Ivancheva, S. & Gevrenova, S. 2004. High performance liquid chromatographic separation of surface flavonoid aglycones in *Artemisia annua* L. and *Artemisia vulgaris* L. *J. Serbian Chem Soc.*, 69: 571-574.

Noltie, H.J. 1994-2000. *Flora of Bhutan* 3(1-2). Royal Botanical Garden, Edinburgh.

Pal, S. & Palit, D. 2011. Traditional knowledge and bioresource utilization among Lepcha in North Sikkim. *NeBIO*. 2(1): 13 -17

Paliwal, P., Pancholi, S.S. & Patel, R.K. 2011. Pharmacognostic parameters for evaluation of the hizomes of *Curcuma caesia*. *J Adv Pharm Tech Res.*, 2: 56-61.

Pawar, V.A. & Pawar, P.R. 2014. *Costus speciosus*: An Important Medicinal Plant. *International Journal of Science and Research*, 3(7): 28-33.

Pelletier, S. W., Aneja, R., & Gopinath, K. W. 1968. The alkaloids of *Aconitum heterophyllum* Wall.: isolation and characterization. *Phytochemistry*, 7(4): 625-635.

Pires, J. M., Mendes, F. R., Negri, G., Duarte Almeida, J. M., & Carlini, E. A. 2009. Antinociceptive peripheral effect of *Achillea millefolium* L. and *Artemisia vulgaris* L.: both plants known popularly by brand names of analgesic drugs. *Phytotherapy Research,* 23(2): 212-219.

Powers S. 1874. *Aboriginal botany.* Proc Calif Acad Sci. 5: 373–9.

Pradhan B.K. & Badola H.K. 2008. Ethnomedicinal plant use by Lepcha tribe of Dzongu valley, bordering Khangchendzonga Biosphere Reserve, in North Sikkim, *India. Journal of Ethnobiology and Ethnomedicine,* 4(22). doi:10.1186/1746-4269-4-22.

Pradhan, B.K. & Badola, B.K. 2012. *Swertia chirayita,* a high value critically endangered medicinal herb of Himalaya- an overview on its potential, conservation status and policies. In: Panda S. & Ghosh, C. (eds.), *Diversity and Conservation of Plants and Traditional Knowledge – A Felicitation Volume of Prof. Abhaya Prasad Das.* Bishen Singh Mahendra Pal Singh, Dehradun, India. Pages 333-355.

Prajapati, N.D. & Kumar, U. 2003. *Agro's Dictionary of Medicinal Plants.* Agrobios (India), Jodhpur .Pp. 7.

Purohit, C.S., Agrawala, D.K. & Das, K. 2014. Ethno-botanical plants of Sikkim and their status- A review. Sanjeev Kumar, (Ed.), Deep Publication. 256-277.

Polunin, O. & Stainton, A. 1984. *Flowers of the Himalaya.* New Delhi: Oxford University Press.

Purohit, M.C., Bahuguna, R., Maithani, U.C., Purohit, A.N. & Rawat, M.S.M. 1999. Variation in podophylloresin and podophyllotoxin contents in different populations of *Podophyllum hexandrum. Current Science,* 77(8): 1078-1080.

Qazi, P.H., Raashid, A. & Shawl, S.A. 2011. *"Podophyllum hexandrum"* – a versatile medicinal plant. *International Journal of Pharmacy and Pharmaceutical Sciences,* 3(5): 261-268.

Rai, P.K., & Lalramnghinglova, H. 2010. Ethnomedicinal plant resources of Mizoram, India: Implication of traditional knowledge in health care system. *Ethnobotanical Leaflets,* 14(3): 274-305.

Raina, R., Johri, A.K. & Srivastava, I.J. 1994. Seed germination studies in *Swertia chirayita* L. *Seed Research,* 22: 62-63.

Raina, R., Patil, P., Sharma, Y.P. & Rana, R.C. 2013. Reproductive biology of *Swertia chirayita* - a temperate critically endangered medicinal plant. Cryologia: *International Journal of Cytology, Cytosystematics and Cytogenetics,* 66: 12-22.

Rajbhandari, M., Mentel, R., Jha, P.K., Chaudhary, R.P., Bhattarai, S., Geali, M.B., Karmacharya, N., Hipper, M. & Lindequist, U. 2009. Antiviral Activity of Some Plants Used in Nepalese Traditional Medicine. *Evidence-Based Complementary and Alternative Medicine*, 6(4): 517–522.

Ramawat, K.G. 2008. *Herbal Drugs: Ethnomedicine to Modern Medicine.* Springer Science & Business Media, Pp. 402.

Rastogi R.P., Sharma V.N. & Siddiqui S. 1949. Chemical examination of *Picrorhiza kurrooa* Benth. *Indian J. Sci. Ind. Res.*, 8B: 172.

Rastogi, R.P. & Mehrotra, B.N. 1980. *Compendium of Indian Medicinal Plants,* New Delhi: Central Drug Research Institute.

Ravishankar, B. & Shukla, V.J. 2007. Indian systems of medicine: a brief profile. *Afr J Tradit Complement Altern Med.*, 4(3): 319–37.

Rawat, S., Bhatt, I.D. & Rawal, R.S. 2011. Total phenolic compounds and antioxidant potential of *Hedychium spicatum* Buch. Ham. ex D. Don in west Himalaya, *India. Journal of Food Composition and Analysis*, 24(4): 574-579.

Reddy, P.P., Rao, R.R., Shashidhar, J. & Sastry, B.S. 2009b. Phytochemical investigation of labdane diterpenes from the rhizomes of *Hedychium spicatum* and their cytotoxic activity. *Bio. Org. Med. Chem. Let.*, 19(21): 6078-81.

Reddy, P.P., Tiwari, A.K., Rao, R.R., Madhusudhana, K., Rao, V.R.S., Ali, A.Z., Banu, K.S. & Rao, J.M. 2009a. New Labdane diterpenes as intestinal á-glucosidase inhibitor from antihyperglycemic extract of *Hedychium spicatum* (Ham. Ex Smith) rhizomes. *Bioorganic & medicinal chemistry letters.* 19(9): 2562-2565.

Rijal, D.P. 2009. Taxonomic study of some medicinally important species of *Swertia* L. (Gentianaceae) in Nepal. *Botanica Orientalis – Journal of Plant Science,* 6: 18-24.

Rodgers, W.A. & Panwar, H.S. 1988. *Planning a wildlife protected area network in India. Vol.1.* Report. Wildlife Institute of India, Dehradun.

Saha, J., Sarkar, P.K., & Chattopadhyay, S. 2011. A survey of ethnomedicinal plants of Darjeeling hills for their antimicrobial and antioxidant activities. *Indian Journal of Natural Products and Resources,* 2(4): 479-492.

Saju, K.A., Mech, S., Deka, T.N. & Biswas, A.K. 2012. In vitro evaluation of biocontrol agents, botanicals and fungicides against *Pestalotiopsis* sp. infecting large cardamom (*Amomum subulatum* Roxb.). *Journal of Spices and Aromatic Crops,* 20(2): 89-92.

Sarin, Y.K. 2008. *Principal crud herbal drugs of India- An illustrated guide to important, largely used and traded medicinal raw materials of plant origin.* Bishen Singh Mahendra Pal Singh, Dehradun. pp. 176-177.

Sarin, Y.K., Bedi, K.L. & Atal, C.K. 1974. *Costus speciosus* rhizome as a source of Diosgenin. *Current Science,* 43(18): 569-570.

Sasidharan, N. 2004. Biodiversity Documentation for Kerala—Flowering Plants. vol. 6. Forest Research Institute, Peechi, Kerala, India, pp. 321–324.

Satyavam, D. A. & Warjeet S L. 2012. Investigation of the structure - non linearity Relationship of zederone from rhizome of *Curcuma caecia* Romb. *Indian Journal of Chemistry,* 51(B): 1738-1742.

Shah, I.A. 2008. *Studies on seed germination and seed set in Swertia chirayita (Roxb. ex. Flem.) Karst.* MSc (forestry) thesis, Dr.Y.S. Parmar University of Horticulture & Forestry, Nauni, Solan (HP), India.

Sharma, A.K. & Chattopadhyay, S. 1983. Relative Amounts of Nuclear DNA in populations of *Costus speciosus* (Koen.) Sm. *Current Science,* 52(14): 653-658.

Sharma, D., Yami, H., Sharma, D. & Shukla, A.K. 2015. Antifungal activities of essential oils from four commonly used ethno-medicinal plants. *Asian Journal of Ethnopharmacology and Medicinal Foods,* 25.

Sharma, S.K. & Pandit, M.K. 2009. A New Species of *Panax* L.(Araliaceae) from Sikkim Himalaya, India. *Systematic botany,* 34(2): 434-438.

Sharma, S.K. & Pandit, M.K. 2011. A morphometric analysis and taxonomic study of *Panax bipinnatifidus* Seem.(Araliaceae) species complex from Sikkim Himalaya, India. *Plant Systematics and Evolution,* 297(1-2): 87-98.

Sharma, S.K., Bisht, M.S. & Pandit, M.K. 2010. Synaptic mutation-driven male sterility in *Panax sikkimensis* Ban.(Araliaceae) from Eastern Himalaya, India. *Plant Systematics and Evolution,* 287(1-2), 29-36.

Shi, J, Xue, W., Zhao, W.J. & Li, K.X. 2013. Pharmacokinetics and dopamine/acetylcholine releasing effects of ginsenoside Re in hippocampus and mPFC of freely moving rats. *Acta Pharmacologica Sinica,* 34(2): 214–20.

Shreshtha, J.N. & Joshi, S.D. 1992. Tissue culture techniques for medicinally important herbs- *Orchis incarnate* and *Swertia chirayita.* Banko Janakari *(a Journal of Forestry Information for Nepal),* 3: 24-26.

Shrestha, P.M. & Dhillion, S.S. 2003. Medicinal plant diversity and use in the highlands of Dolakha district, Nepal. *Journal of Ethnopharmacology,* 86: 81–96.

Shrestha, P.M. 1988. Contribution to the ethonobotany of the Tamangs of Kathmandu Valley. *Contribution to Nepalese Studies*, 15: 247–266.

Shukla, J.K., Kasera, P.K. and Chawan, D.D. 2001. Conservation and multiplication of endangered plants: 1. *Leptadaenia reticulata* (Retz.) Wight. & Arn. *Ancient Science Life*, 20: 65-69.

Singh, B.R., Singh, V., Singh, R.K., Toppo, S., Haque, N. & Ebibeni, N. 2012. Comparative evaluation of antimicrobial effect of *Artemisia vulgaris* essential oils extracted from fresh and dried herb. Medicinal Plants-International *Journal of Phytomedicines and Related Industries*, 4(2): 76-82.

Singh, I., Gautam, Y. & Vimala, Y. 2013. Detection and isolation of diosgenin from *Costus speciosus* callus raised from non-germinal seeds. *Int. J. Chem. and Life Sciences*, 2(10): 1240-1242.

Singh, J. & Shah, N.C. 1994. Podophyllum: A Review. *Curr. Res. Med. Arom. Plants*, 16, 53-83.

Singh, P. & Chauhan, A.S. 1998. An overview of the plant diversity in Sikkim state. In: Rai, S.C., Sundriyal, R.C. & Sharma, E. (eds.), *Sikkim: Perspectives for Planning and Development*, Sikkim Science Society, Bishen Singh, Mahendra Pal Singh Pvt. Ltd., Dehradun. Pp.219-232.

Singh, R.S. 1983. *Vanausadhi Nidarshika (Ayurvedic Pharmacopeia)*. Uttar Pradesh Hindi Sansthan, Lucknow.

Sinha, R.K. & Sinha, S. 2001. *Ethnobiology*. Surabhi Publicaions, Jaipur.

Sosa, S., Tubaro, A., Kastner, U., Glasl, S., Jurenitsch, J. & Della Loggia, R. 2001. Topoical anti-inflammatory activity of a new germacrene derivative form *Achillea pannonica*. *Planta Med.*, 67: 654-658.

Sravani, T. & Paarakh, P.M. 2011. Evaluation of anthelmintic activity of rhizomes of *Hedychium spicatum* Buch. Ham. *Int. J. Res. Pharm. Sci.*, 2: 66-68.

Srivastava, S., Singh, P., Jha, K.K., Mishra, G., Srivastava, S. & Khosla, R.L. 2013. Antiinflammatory, analgesic and antipyretic activities of aerial parts of *Costus speciosus* Koen. *Indian J Pharm Sci.*, 75(1): 83–88. doi: 10.4103/0250-474X.113532

Sun, H.D., Lin, Z.Q. & Nij, F.D. 1980. Study of the Chinese drugs of umbelliferae II. Chemical constituents of the roots of *H. nepalense* D. Don. *Yunnan Zhiwu Yanjiu*, 2: 224-227.

Sun, J., Song, X. & Hu, S. 2008. Ginsenoside Rg1 and aluminum hydroxide synergistically promote immune responses to ovalbumin in BALB/c mice. *Clin Vaccine Immunol.*, 15(2):303–7.

Sundriyal, M., Sundriyal, R.C. & Sharms, E. 2004. Dietary use of wild plant resources in the Sikkim Himalaya, India. *Econ Bot.*, 58: 626-638.

Sundriyal, M., Sundriyal, R.C., Sharma, E. & Purohit, A.N. 1998. Wild edibles and other useful plants from the Sikkim Himalaya, India. *Oecologia Montana*, 7: 43-54.

Suresh, G., Poornima, B., Babu, K.S., Yadav, P.A., Rao, M.S.A., Siva, B., Prasad K.R., Nayak, V.L. & Ramakrishna, S. 2013. Cytotoxic sesquiterpenes from *Hedychium spicatum*: Isolation, structure elucidation and structure–activity relationship studies. *Fitoterapia*, 86: 100-107.

Tandan, S.K., Chandra, S., Gupta, S. & Lal, J. 1997. Analgesic and anti-inflammatory effects of *Hedychium spicatum*. *Indian Journal of Pharmaceutical Sciences*, 59(3): 148-150.

Tandon, P., Kumaria, S. & Kayang, H. 2010. Conservation of medicinal and aromatic plants of northeast India. In: Ahmad, A., Siddiqi, T.O. and Iqbal, M. (eds.), Medicinal Plants in Changing Environment, Capital Publishing Company, New Delhi, India. Pp. 203-212

Temraz, A. & El-Tantawy, W.H. 2008. Characterization of antioxidant activity of extract from *Artemisia vulgaris*. *Pakistan Journal of Pharmaceutical Sciences*, 21(4): 321–326.

Thakur, R.S. 1993. Plant drugs: Emerging areas of modern drug research. In: Govil, J.N., Singh, V.K. and Hashmi, S. (eds.), *Glimpses in Plant Research Vol. XI*, Today and Tomorrow's Publishers, New Delhi. pp. 371–375.

Thakur, R.S., Puri, H.S. & Hussain, A. 1989. *Major Medicinal Plants of India*. CIMAP, Lucknow, pp. 404–407.

Theobald, W.L. 1981. Bignoniaceae. In: Dassanayake, M.D. & Fosberg, F.R. (eds.), A Revised Handbook to the Flora of Ceylon. Amerind Publishing Co., Pvt., Ltd., New Delhi, India.

Tigno, X., de Guzman, F. & Flora, A. 2000. Phytochemical analysis and hemodynamic actions of *Artemisia vulgaris* L. *Clinical Hemorheology and Microcirculation*, 23(2–4): 167–175.

Ukani, M.D., Mehta, N.K., & Nanavati, D.D. 1996. *Aconitum heterophyllum* (Ativisha) in ayurveda. *Ancient science of life*, 16(2): 166.

United Nations University-Institute of Advanced Studies Policy Report, 2012. *Biodiversity, Traditional Knowledge and Community Health:*

Strengthening Linkages. In: Unnikrishnan, P.M. and Suneetha, M.S. (eds.), Policy Report .Pp. 82.

Ved, D.K. & Goraya, G.S. 2008. *Demand and Supply of Medicinal Plants in India*. Dehra Dun: Bishen Singh, Mahendra Pal Singh.

Wang, H. & Liu, Y. 2010. Chemical composition and antibacterial activity of essential oils from different parts of *Litsea cubeba*. *Chemistry and Biodiversity*, 7: 229–235.

Wani, S.H., Ashfaq, M., Shsh, K.W. & Singh, D. 2012. Phytochemical screening of methanolic extracts of *Podophyllum hexandrum* royle and *Rheum emodi* wall. *Journal of Current Chimical & Pharmaceutical Sciences*, 2(2): 125-128. ISSN 2277-2871.

Weinges, K., Kloss, P. & Henkels, W.D. 1972. Natural products from medicinal plants. XVII. Picroside-II, a new 6-vanilloyl-catapol from *Picrorhiza kurooa* Royle and Benth. *Justus Liebigs Annalen der Chemie*, 759: 173.

Yan, R., Cao, Y., Chen, C., Dai,, H., Yu, S., Wei, J., Li, H. & Yang, B. 2011. Antioxidant flavonoids from the seed of *Oroxylum indicum*. *Fitoterapia*, 82: 841–848.

Yang, W., Hu, Y., Wu, W., Ye, M. & Guo, D. 2014. Saponins in the genus *Panax* L. (Araliaceae): A systematic review of their chemical diversity. *Phytochemistry*, 106: 7-24.

Yonzone, R., Rai, S. & Bhujel, R.B. 2012a. Ethnomedicinal and aromatic plant diversity and resources of Darjeeling district of Eastern Himalaya in India. *International Journal of Advances in Pharmaceutical Research*, 3(4): 859-871.

Yonzone, R., Rai, S., & Bhujel, R.B. 2012b. Genetic diversity of ethnobotanical and medicinal plants resources of Darjeeling district, West Bengal, India. *International Journal of Advances in Pharmaceutical Research*, 3: 713-729.

Zhang, Z.Y. & Santisuk, T. 1998. Bignoniaceae. In: Wu, Z.Y. & Raven, P.H. (Eds.), Flora of China, Vol 18. Science Press, Beijing and Missouri Botanical Garden Press, St. Louis, USA, pp. 213–215.

□□□

Chapter 5

Traditional Medicines and Current Practices: A Review of Medicinal Plants Used to Treat Cancer

Santosh Kumar Jha*

*P.G. Department of Plant Science, J.N.R.M. Port Blair, India, E-mail: yesbotany@rediffmail.com, Mobile: 9531855587

ABSTRACT

The article provides a general outline on descriptions of various types of cancers and their management from medicinal and ethnobotanical perspective underlying its scientific principles involved in treating these conditions with the use of natural products. The article reviews the available literature regarding researches on anti-cancerous traditional and ayurvedic herbs and also includes how these plants are used to control and cure various types of cancers.

Keywords: Inflammation, Chemotherapy, Leukaemia, Carcinogens.

Introduction

Cancer is a dreadful disease caused by abnormal and uncontrolled cell division. Cancer, after cardiovascular disease, is the second leading cause of death (Kutluk and Kars, 1998; Turkistan, 2005). Out of about 10 million people diagnosed of cancer every year about 6 million die of the disease (Pinar, 1998). Deaths from cancer worldwide are projected to continue rising with an estimated 12 million deaths by 2030(Wang *et al.*, 2007). In 2012, there were an estimated 14.1 million cases around the world, of these 7.4 million cases were in men and 6.4 million women. This number is expected to increase to 24 million by 2035 (Ferlay, 2013).Cancer is considered to be more of a developed world issue, in fact rate of all cancers (excluding non-melanoma skin cancer) was 1.8 times higher in more developed compared to less developed countries (Ferlay, 2013).

The major causes of cancer are smoking, dietary imbalances, hormones and chronic infections leading to chronic inflammation (Ames and Gold, 1995). The most frequent types of cancer worldwide in order of the number of global deaths are: among men- lung, stomach, liver, colorectal, oesophagus and prostate; and among women- breast, lung, stomach colorectal and cervical (Abiodun *et al.*, 2010). The most prevalent cancer among females is breast cancer, accounting for about 23% of total cancer cases; in males, the most prevalent is lung cancer, which accounts for 17% of total cancer cases (Jemal et al., 2011).Treatment of cancer usually involves a combination of surgery, radiation therapy, and chemotherapy but despite these therapeutic options, cancer remains associated with high mortality. Various cancer and cancer-related conditions have been treated by local herbalists for ages (Sofowora, 1984) and many plants have been reported as useful in the management of such conditions. As chemotherapy destroys the normal cells along with cancer cells, biological active components from plants are significant and important source of new drugs that are likely to lead to new drugs and better treatments for cancer. Recently the scientific world has experienced an upsurge of interest in the therapeutic potential of medicinal plants as a source of promising anticancer agents. However, the application of plant-based compounds for the treatment of cancer can be traced back to 1950s. Plants have produced many anticancer drugs such as taxanes and vincristine and still serve as a veritable source of new products through the use of standard bioassay methods (Noble *et al.*, 1959; Wani *et al.*, 1971). An estimate suggests that 50% of breast cancer and 37% of prostate cancer patients use herbal products (Richardson, 2001). Some of the very first anticancer agents derived from plants are *Vinca* alkaloids, vinblastine, vincristine, and cytotoxic podophyllotoxins. Statistical data suggest that 16 plant-derived anticancer drugs have been subjected to clinical trials so far (Belayachi *et al.*, 2013). Landmarks of these clinical trials are flavopiridol, isolated from the Indian tree *Dysoxylum binectariferum*, and meisoindigo, isolated from the Chinese plant *Indigofera tinctoria*, which have been documented to have less toxicity than conventional chemotherapeutic anticancer drugs (Saklani and Kutty, 2008).

The broad aim of this article is to provide a general outline on descriptions of cancers and their management from medicinal and ethnobotanical perspective underlying its scientific principles involved in treating these conditions with the use of natural products. This article reviews the available literature regarding researches on anti-cancerous traditional and ayurvedic herbs and also includes how these plants are used to control and cure various types of cancers. The review is written with an intention to raise awareness and encourage uses of phytochemicals obtained from medicinal plants for combating cancer and suggesting an integrated approach in tumour management and treatment.

Causes of Cancer

Major causes of cancer are -Viruses such as *Epstein-Barr-Virus* (EBV), *Hepatitis-B-Virus* (HBV), *Human Papilloma Virus* (HPV). Environmental and occupational exposure such as ionizing, radiation UV radiation, and exposure to chemicals including vinyl chloride, benzene and asbestos. Life style factors such as high-fat, low fiber diets, tobacco, ethanol etc., medication and such as alkylating agents and immuno-suppressants, Genetic factors such as inherited mutations, cancer causing genes, defective tumor suppressor genes are major causes of cancer.

Table 1. Selected Known Causes of Cancer and the Associated Cancer Types

Category	Suspected carcinogens	Associated cancer
A. Radiation	Ionizing Radiation (Therapeutic or Diagnostic) Non-ionizing Radiation (Ultraviolet light)	Leukaemia, thyroid, breast, lung, liver, bone cancer, Skin cancer
B. Occupational Carcinogens	Aromatic amines (Dyes) Asbestos, Benzene Chromium, Nickel dust Vinyl chloride (PVC)	Bladder Cancer, Lung, pleural, peritoneal cancer, Acute myeloidleu-kaemia myeloid leu Lung cancer, Nasal sinuses, Lung cancer, Liver cancer
C. Therapeutic Agents 1. Chemotherapy	Cyclophosphamide Chlorambucil & Melphalan	Bladder cancer, leukaemia Acute nonlymphocytic leukaemia
2. Hormones	Androgens Diethylstil-bestrol Oestrogens	Prostate cancer Vaginal, endometrial, breast cancer, Endometrial, breast, liver cancer
3. Other drugs Non-	Phenacetin Azathioprine & Cyclosporine	Kidney, bladder cancer Hodgkin's lymphoma

D. Infectious Agents		
1. Bacteria	*Helicobacter pylori*	Gastric cancer
2. Viruses	Human Epstein-Barr virus (EBV) papilloma virus (HPV) Hepatitis B or C virusHuman immunodeficiency virus (HIV) Human T-lymphotropic virus type 1 (HTLV-1)	Cervical cancerBurkitt's lymphoma, Hodgkin's lymphoma Liver cancer Kaposi's sarcoma, non-Hodgkin's lymphoma Adult T-cell leukaemia/ lymphoma
3. Parasites: Flukes	*Schistosomahaematobium* *Clonorchissinensis* *Opisthorchisviverrini*	Bladder cancer Cholangiocarcinoma
E. Miscellaneous		
1. Tobacco	Cigarette smoking	Lung, bladder cancer
2. Alcohol	Tobacco chewing	Oropharyngeal cancer
3. Dietary factors	Excessive alcohol consumption Low-fibre/ High fat dietInadequate folic acid intake, excess red or processed meat consumption High Nitrosamine intake Aflatoxins (food contaminant)	Oropharyngeal, laryngeal, oesophageal, liver cancer Colorectal cancer Colon cancer Gastric cancer Liver cancer
4. Lifestyle risk factors	Obesity (BMI >30 kg/m2) Physical inactivity	Males: esophagus, colon cancer Females: gallbladder, endometrial, breast (postmenopausal) cancer Colon, breast cancer

Adapted from Colledge *et al.*, 2010, Munjal *et al.*, 2012, Longo *et al.*, 2012, Goldman *et al.*, 2012.

Cancer-causing genes

There are four main types of gene involved in cell division. Most tumours have faulty copies of more than one of the genes.

1. **Oncogenes** - Oncogenes are genes which under normal circumstances play a major role i.e. to convey the cells for dividing.

When oncogenes are activated, they speed up growth rate of cells.

2. ***Tumour suppressor genes-*** These genes were discovered in 1979 by Professor Sir David Lane. These genes make proteins which normal function is just opposite to that of oncogenesis. One of the most important tumour suppressor genes is called p53.

3. ***Suicide genes*** - Apoptosis or cell suicide is a highly complex and an important process. Cells usually have the ability to commit suicide whenever something goes wrong, to prevent damage to their neighbours. There are several genes involved in this process. If any suicide gene is damaged then a faulty cell can keep dividing and becomes cancerous.

4. ***DNA-repair genes*** - The DNA in every cell in the body is under constant assault from a variety of directions. But cells contain many different proteins whose job is to repair the damaged DNA. Most of DNA damage is repaired immediately, with no ill effects. But if the DNA damage occurs in a gene that makes a DNA repair protein, ability of the cell to repair itself will be reduced, and that can allow errors to accumulate in other genes over time.

Cancer therapy — a practical dilemma

To find out any practical solution to combat cancer is of paramount importance. An alternative solution to allopathic system of medicine embodied with severe side effects is the use of medicinal plant preparations to arrest the insidious nature of the disease. Many herbs and plant products have been evaluated in clinical studies and are currently being investigated phytochemically to understand their tumouricidal actions against various cancers. So, cancer patients who already got crippled with this disease, further burdened by drug-induced toxic side effects have now turned to seek help from the complementary and alternative system of medicine hoping for a better cure. Medicinal plants based therapy has been found to be able to cure these chronic diseases better, which were previously not amenable to treat by western medical practices. This traditional Indian system of medicine with its evolution through centuries has always fascinated practitioners and researchers for its applications in cancer treatment on a scientifically proven research background.

Drugs used to treat Cancer

Drugs used to treat most cancers are those that can block cell signalling, including growth factor signalling (e.g., epidermal growth factor); prostaglandin production (e.g., COX-2 i.e.cyclooxygenase-2); inflammation (e.g., inflammatory cytokines: NF-kB i.e.Nuclear factor-kappa B, TNF, IL-1, IL-6, chemokines); drug resistance gene products (e.g., multi-drug resistance); cell cycle proteins (e.g., cyclin D1 and cyclin E); angiogenesis (e.g., vascular endothelial growth factor); invasion (e.g., matrix metalloproteinases); antiapoptosis (e.g., bcl-2, bcl-XL, XIAP i.e.Activated protein-1, survivin, FLIP); and cellular proliferation (e.g., c-myc, AP-1, growth factors).

Plant based Drugs

Around 25 centuries ago, Hippocrates proclaimed, 'Let food be thy medicine and medicine be thy food.' According to a recent report by Newman *et al.*, as many as 65% of formally synthetic hypertension drugs are plant based .Out of 121 prescription drugs which are in use today for cancer treatment, 90 are derived from plants. Almost 74% of these, including taxol, were discovered by investigating a folklore claim. Between 1981 and 2002, 48 out of 65 drugs approved for cancer treatment were natural products, based on natural products, or mimicked natural products in one form or another.

Recently, a greater emphasis has been given towards the researches on complementary and alternative medicine that deals with cancer management. Several studies have been conducted on herbs under a multitude of ethno botanical grounds. For example, Hartwell has collected data on about 3000 plants, those of which possess anticancer properties and subsequently been used as potent anticancer drugs. Ayurveda, a traditional Indian medicine of plant drugs has been successful from very early times in using these natural drugs and preventing or suppressing various tumours using various lines of treatment.

Table 2. List of herbs commonly used as anticancer agents in Ayurvedic system

S.No.	Name of the herb	Family	Use
1.	*Amorphopallus campanulatus* B.	Araceae	The mature tuber is first burnt and then mixed with butter and jaggery and applied for tumour destruction.
2.	*Baliospermum montanum*	Euphorbiaceae	The paste comprising of *Baliospermum montanum, Plumbago zeylanica, Euphorbia neriifolia, Calotropis procera,* jaggery, *Semecarpus anacardium* applied over the tumours.
3.	*Barleria prionitis* L.	Acanthaceae	The *Barleria prionitis* oil prepared with whole plant is indicated for external application during acute stages of cyst in blood vessels.
4.	*Basella rubra* L.	Basellaceae	The plant and leaves are ground with sour buttermilk with salt for preparing a poultice and indicated for *arbuda.*
5.	*Curcuma domestica* Valeton	Zingiberaceae	The *Curcuma domestica* powder in combination with *Symplocos racemosa, Soymida febrifuga,* is mixed with honey and this is used as an external remedy.
6.	*Ficus bengalensis* L.	Moraceae	Application of mixture of *Ficus bengalensis* and *Saussurea lappa* pacify tumour growth on bone.
7.	*Flacourtia romantchi* L. Herit	Flacourtiaceae	The paste of *Flacourtia romantchi, Cassia fistula, Capparis sepiaria,* is recommended for *kaphaja* tumours.
8.	*Madhuca indica* Gmel.	Sapotaceae	This paste is prepared from the barks of *Madhuca indica, Syzygium cumini, arjuna Terminalia arjuna* and *Salix caprea* and prescribed for local application
9.	*Moringa oleifera* Lam.	Moringaceae	The paste of *Moringa oleifera* seeds, *Solanum xanthocarpum, Sinapis dichotoma, Holarrhena antidysenterica*

			and *Nerium odorum* roots prepared with buttermilk is used for *arbuda* tumours.
10.	*Oroxylum indicum* (L.) *Benth*	Bignoniaceae	The drug *Oroxylum indicum* prescribed in treatment of *granthi* .
11.	*Pandanus odoratissimum* L.f.	Pandanaceae	A paste of *Pandanus odoratissimum* with sugar was applied externally.
12.	*Prosopis cineraria*	Fabaceae	This paste made up of *Prosopis cineraria* seeds, *Raphanus sativa, Moringa oleifera,* barley and mustard with sour buttermilk was applied locally for disintegrating cysts.
13.	*Pterospermum acerifolium* Willd.	Sterculiaceae	The flowers of *Pterospermum acerifolium* mixed with sugar to be applied locally
14.	*Vitis vinifera* Dunal.	Vitaceae	The mixture of *Terminalia chebula,* grape juice and *sugar cane juice* has been used. Resveratrol, a natural product derivative from grape juice has been proved to possess cancer chemopreventive activity.

Table 3. Scientific evidences on herbs used in Ayurveda proven to have anticancer property

S.No.	*Name of the herb*	*Family*	*Indications*	*References*
1.	*Abrus precatorius* L.	Fabaceae	Yoshida sarcoma (rats) Fibrosar-coma (mice)Ascites tumour cells	Subba Reddy and Sirsi
2.	*Albizzia lebbeck* Benth.	*Mimosaceae*	Sarcoma 180 (mice)	Dhar *et al.*
3.	*Allium sativum* L.	*Liliaceae*	Sarcoma (rat)	Hu *et al.*
4.	*Aloe vera* (L.) Burm.f.	*Liliaceae*	Yoshida AH-130 ascite hepatoma (pleural tumour) human neuroecto-dermal tumours	Corsi *et al.* Pecere *et al.*

5.	*Alstonia scholaris* R.Br.	*Apocyanaceae*	HSI human sarcoma benzo(a)pyrene induced for stomach carcinoma	Dhar *et al.*, Jagetia *et al.*
6.	*Amura rohitaka*		Leukaemia	Prasad and Deshpande, Rabi and Gupta
7.	*Anacardium occidentale* L.	*Anacardiaceae*	Hepatoma 129	Dhar *et al.*
8.	*Asparagus racemosus* Willd.	*Liliaceae*	Human epidermoid carcinoma	Dhar *et al.*
9.	*Bacopa monniera* (L.) Penn.	*Scrophulariaceae*	Walker carcin osarcoma 256	Bhakuni *et al.*
10.	*Berberis cristata* L.	*Acanthaceae*	Human epidermal carcinoma of the nasopharynx *N*-nitrosodiethylamine induced carcinogenesis	Bhakuni *et al.*, Anis *et al.*
11.	*Boswellia serrata* Roxb.ex Coelbr.	*Burseraceae*	Human epidermal carcinoma of the nasopharynx Leukaemia and brain tumours	Dhar *et al.* Hostanska *et al.*
12.	*Calotropis gigantea* (L.) R.Br.	*Asclepidaceae*	Human epidermal carcinoma of the nasopharynx	Bhakuni *et al.*, Dhar *et al.*
13.	*Curcuma longa* L.	*Zingiberaceae*	Fibrosarcoma, Preclinical and clinical trials review	Sriganth and Premalatha Aggarwal *et al.*
14.	*Datura metel* L.	*Solanaceae*	Human epidermal carcinoma of the nasopharynx	Dhar *et al.*
15.	*Erythrina suberosa* Roxb.	*Fabaceae*	Sarcoma 180	Dhar *et al.*
16.	*Euphorbia hirta* L.	*Euphorbiaceae*	Freund virus leukaemia	Dhar *et al.*
17.	*Gynandropis pentaphylla* DC.	*Capparridaceae*	Hepatoma 129	Dhar *et al.*

18.	*Heliotropium indicum* L.	*Boraginaceae*	P-388 lymphocytic leukaemia	Pal *et al.*
19.	*Hygrophila spinosa* T. *Anders*	*Acanthaceae*	Dalton's lymphoma, Ehrlich ascites carcinoma and Sarcoma-180	Maiti Mazumdar *et al.*
20.	*Ixora undulate*	*Rubiaceae*	P-388 lymphocytic leukaemia	Dhawan *et al.*
21.	*Juniperus indica*	Cupressaceae	Human epidermoid carcinoma of the nasopharynx	Dhawan *et al.*
22.	*Luffa cylindrical M.V.Roem.*	*Cucurbitaceae*	Schwartz leukaemia	Bhakuni *et al.*
23.	*Melia azedarach* L.	*Meliaceae*	Walker carcin osarcoma 256	Bhakuni *et al.*
24.	*Moringa oleifera* Lam.	*Moringaceae*	Human epidermoid lymphocytic leukaemia, Skin papillomagenesis	Dhawan *et al.* Bharali *et al.*
25.	*Nerium indicum* Mill.	*Apocyanaceae*	Erlish ascites carcinoma	Pal *et al.*
26.	*Nigella sativa* L.	*Ranunculaceae*	Lewis lung carcinoma, Colon cancer	Dhar *et al.* Salim and Fukushima
27.	*Ocimum sanctum* L.	*Lamiaceae*	Skin and liver tumours	Dubey *et al.*
28.	*Paederia foetida* L.	*Rubiaceae*	Human epidermoid carcinoma of the nasopharynx	Dhar *et al.*
29.	*Picrorrhiza kurroa* - Royle Ex Benth	*Plantaginaceae*	Hepatic cancers	Dhar *et al.*
30.	*Plumbago zeylanica* L.	*Plumbaginaceae*	Hepatoma	Parimala and Sachdanandam
31.	*Rubia cordifolia* L.	*Rubiaceae*	P-388, L-1210, B-16 melanoma, colon 388, Lewis lung carcinoma, mammary carcinoma	Itokawa *et al.*

32.	*Taxus buccata*	Taxaceae	Cytotoxic against various tumours	Melado *et al.*
33.	*Vinca rosea* L.	*Apocynaceae*	P-1534, carcinoma of the breast, cervix, kidney, lung and ovary	Rastogi and Mehrotra
34.	*Withania somnifera* L. *Dunal*	*Solanaceae*	Various tumours	Dhar *et al.*

Table 4. Therapeutic enhancement potential of ayurvedic herbs on cancer chemotherapy/radiation

S. No.	Name of the herb	Family	Chemotherapy/ayurvedic herb intervention studies
1.	*Allium sativum* L.	*Liliaceae*	Water-soluble derivative of garlic, S-allylmercaptocysteine (SAMC), inhibited proliferation and cell cycle progression in two human colon cancer cell lines, SW-480 and HT-29, similar to the effects of sulindac sulfide (SS), a well-known colon cancer chemopreventive agent. Co-adminis-tration of SS with SAMC enhanced the growth inhibitory and apoptotic effects of SS, suggesting the usefulness of SAMC alone or in combination with SS or other chemopreventive agents.
2.	*Aloe vera* (L.) Burm.f.	*Liliaceae*	In a randomised double-blinded clinical trial, comparing mild soap and aloe vera gel against incidence of radiation therapy induced skin reactions; the median time of five weeks was taken to show any skin changes in the aloe/soap treatment versus three weeks in the soap only treatment. The protective effect of adding aloe to the soap regimen increases during long time radiation exposure.In another clinical trial involving patients with advanced solid tumours, for whom no other standard effective therapy was available, combination of pineal indole melatonin (MLT) plus Aloe vera extracts produced some therapeutic benefits,at least in terms of stabilization of disease and survival when compared to MLT alone treatment.

3.	*Alstonia scholaris* R.Br.	*Apocyanaceae*	The *Alstonia scholaris* extract pre-treatment increased the effect of radiation as by enhancement of cell killing in HeLa and KB cells, followed by HL60, MCF7, and HePG2 cells. In in vivo studies, with Ehrlich ascites carcinoma bearing mice the pre-treatment of extract caused increased life span of animals when compared with untreated irradiated group. The combination treatment of *Alstonia scholaris* extract with cyclophosphamide was also found to be most effective against Ehrlich ascites carcinoma as it caused the highest tumour regression and enhanced the mean and average survival time when compared with cyclophosphamide alone treated group.
4.	*Curcuma longa* L.	*Zingiberaceae*	When radiation and curcuma were applied together as synergical therapy, curcuma showed a radiation sensitising effect in HeLa, K-562 and IM-9 cell lines. Curcumin, the active constituent from *Curcuma longa* also enhances the anticancer potential of Cisplatin and reduces its nephrotoxicity in fibrosarcoma bearing rats.
5.	*Heliotropium indicum* L.	*Boraginaceae*	In a Phase I study consisting of Solid tumour patients who have undergone prior chemotherapy/ radiation therapy, Indicine N-oxide, an alkaloid from *Heliotropium indicum* have shown some improvement against skin melanoma and ovarian carcinoma.
6.	*Moringa oleifera* Lam.	*Moringaceae*	Pre-treatment with the leaf extract of *M. oleifera* exhibits significant radiation protection to the bone marrow chromosomes in mice and this could be useful to overcome side effects of radiation therapy.
7.	*Nigella sativa* L.	*Ranunculaceae*	In mice bearing Ehrlich ascites carcinoma, thymoquinone (TQ), the main constituent o f the *Nigella sativa* oil, significantly enhanced the therapeutic efficacy of ifosfamide by improving its antitumour effect and reducing its nephrotoxicity. Furthermore, mice treated with ifosfamide in combination with TQ showed less body weight loss and mortality rate compared to IFO single therapy.

8.	*Ocimum sanctum* L.	*Lamiaceae*	Orientin and Vicenin, two water-soluble flavonoids isolated from the leaves of *Ocimum sanctum* have shown significant protection to the human lymphocytes against the clastogenic effect of radiation, radiation lethality and chromosomal aberrations in vivo. This radioprotection associated with their antioxidant activity may have clinical potential in cancer therapy.
9.	*Taxus buccata*	*Taxaceae*	In a Phase II study, the triplet regimen based on taxol (active constituent of *Taxus buccata*), ifosfamide, and carboplatin has proved active, safe, and easy to deliver on an outpatient basis for patients with advanced stage IIIB-IV non-small-cell lung cancer. Another combination of Herceptin withs Taxol significantly improves the overall response rate, increases the time to progression and the overall survival in breast cancer patients. These effects are more pronounced in patients characterized with HER/ 2 +++ over expression.Taxol also exerts a weak radiosensitising effect on breast and cervical carcinoma cells on the basis of an optimal Taxol/ radiation scheduling.
10.	*Withania somnifera* L. Dunal	*Solanaceae*	*W. somnifera* when administered for 4 days before paclitaxel treatment and continued for 12 days caused significant reversal of neutropenia of paclitaxel in mice. It can be used as an adjuvant during cancer chemotherapy for the prevention of bone marrow depression associated with anticancer drugs. The active component, withaferin A isolated from the extract showed significant antitumour and radiosensitising effects in experimental tumours in vivo, without any noticeable systemic toxicity. In Ehrlich ascites carcinoma mice, the extract showed dose dependent inhibition on tumour growth and increased the survival rate. Combination of radiation therapy with extract increased tumour cure and tumour-free survival.

Table 5. Anti cancerous activity of medicinal plants described in Siddha system.

S.No.	Botanical name/ Siddha view	Family	Tamil name	Phytochemicals	Action against specific cancer	(Mudhaliar, 2006) *Taste* (S/T/P)*	Pacifies (V,P,K)**
1.	*Allium sativum*	Liliaceae	*Poondu*	Allin, allicin, allicnase,s-allylc cysteine,diallyl disulphide, methylallyl trisulphide(Sabnis,2001).	Carcinoma of mammary gland, Hepato carcinoma (Sabnis,2001).	S- *Karpu* T- *Veppamn* P- *Karpu*	*Kapham*
2	*Aloe vera*	Liliaceae	*Kumari*	Emodin, aloien, glycoproteins, polysaccharides(Sabnis,2001).	Anti angiogenic activity (Sabnis,2001).	S- *Kaippu* T-*Seetham* P- *Ennipu*	*Kapham, Pitham*
3.	*Andrographis paniculata*	Acanthaceae	*Nilavembu*	Andrographiolide, Andrographiside (Trivedi,2001).	Squamous cell carcinoma of nasopharynx, Lymphatic leukemia(Trivedi,2001).	S- *Kaippu* T- *Veppam* P- *Karppu*	*Pitham, Kapham*
4.	*Bacopa monnieri*	Scropularaciae	*Brami*	Saponins-mannitol, Hersaponin, Monerin (Premalatha,2005)	Carcinosarcoma (Premalatha,2005)	S- *Thuvarpu, kaippu* T- *Veppam* P- *Karppu*	*Pitham, Kapham*
5.	*Boerrhavia diffusa*	*Boraginaceae*	*Mukirratai*	Punarnavine, Boeravionones G& H (Merina et al,2012).	Malignant melanoma, breast cancer (Merina et al,2012).	S- *Kaippu* T- *Veppam* P- *Karppu*	*Kapham, Pitham*
6.	*Corcus sativus*	Iridaceae	*Kunguma poo*	Crocetin(Sabnis,2001).	Skin tumours (Sabnis,2001).	S- *Kaippu* T- *Veppam* P- *Karrpu*	*Kapham, Pitham*
7.	*Curcuma longa*	Zinziberaceae	*Manjal*	Tumerone,Curcumine (Baatout et al,2004).	Oral cancer and Stomach cancer (Agarwal et al,2003)	S- *Karpu, kaippu* T- *Veppamn* P- *Karpu*	*Kapham*

8.	*Emblica officinalis*	*Euphorbiaceae*	*Nelli*	Polyphenols, flavones, tannins (Merina *et al.*, 2012).	Lymphoma, Anti metastatic activity, melanoma (Merina *et al.*, 2012).	S- *Pullipu, thuvarpu, ennipu* T- *Seetham* P- *Ennipu*	*Vatham, Pitham, Kapham*
9.	*Glycyrrhiza galbra*	Leguminaceae	*Athimathuram*	Licochadcone & flavanoides (Fuy *et al.*, 2004)	Prostate cancer (Fuy *et al.*, 2004)	S- *Ennipu* T- *Seetham* P- *Ennipu*	*Pitham*
10.	*Linum usitatissimum*	Linaceae	*Allisi vithai*	Cynogenetic glycosides, Lignans (Sakarkar, 2011)	Breast cancer (Sakarkar, 2011)	S- *Ennipu, thuvarpu* T- *Seetham* P- *Ennipu*	*Pitham*
11.	*Momordica charantia*	*Cucurbitaceae*	*Paagal*	Charantin, sitosterol, Ascorbigin (Sabnis, 2001).	Colon cancer, Leukemia (Sabnis, 2001).	S- *Kaippu* T- *Veppam* P- *Karppu*	*Kapham, Pitham*
12.	*Ocimum sanctum*	*Laminaceae*	*Thulasi*	Orientin & Vicenin (Premalatha, 2005)	Radioprotective antioxidant (Premalatha, 2005)	S- *Karppu* T- *Veppam* P- *Karppu*	*Kapham*
13.	*Picrorrhizia kurroa*	Scrophulariaceae	*Kadugurhogni*	Picrosides. (Sakarkar, 2011)	Liver cancer (Sakarkar, 2011)	S- *Kaippu, karppu* T- *Veppam* P- *Karrpu*	*Kapham*
14.	*Plumbago indica*	Plumbginaceae	*Kodiveli*	Plumbagin, Apigenin, Glycosides (Parimala, 1993).	Prostate cancer (Hafeez *et al.*, 2012)	S- *Karpu* T- *Veppam* P- *Karpu*	*Kapham*
15.	*Rhus succedanea*	Anacardiaceae	*Karkadagasingi*	Hirokiflavone, Robustaflavone (Ruchen *et al.*, 2009)	Brain, breast, colon, lung, liver cervical and prostate cancer Shirin *et al.*, 2001)	S- *Thuvarpu* T- *Veppam* P- *Karppu*	*Kapham, Pitham*
16.	*Rubia cordifolia*	Rubiaceae	*Manjitti*	Rubiadin (Karnick, 1996).	Melanoma, Sarcoma, Lung carcinoma, Lymphatic leukemia. (Sabnis, 2001).	S- *Karrpu, kaippu* T- *Veppam* P- *Karrpu*	*Kapham*

17.	*Semicarpus anacardium*	*Anacardiaceae*	*Seerankottai*	Semicarpin, anacardin (Premalatha,1999)	Leukemia, Melanoma, Glioma, Hepato carcinoma (Premalatha, 1999)	S - *Kaippu* T - *Veppam* P - *Karppu*	*Pitham, Kapham*
18.	*Taxus buccata*	Taxaceae	*Thalisabathari*	Taxanes,Taxol, Cepholomannine (Baatout *et al.*, 2004).	Leukemia, Breast cancer. Sarcoma, Cancer of larynx, ovary and colon (Fuy *et al.*, 2004)	S - *Karppu* T - *Veppam* P - *Karppu*	*Kapham*
19.	*Tinospora cardifolia*	*Menispermiaceæ*	*Seenthil*	Tinosporin (Matthew, 1999)	Breast cancer, leukemia and cervical cancer (Premalatha, 2005).	S - *Kaippu* T - *Veppam* P - *Karppu*	*Kapham, Pitham*
20.	*Vitex negundo*	*Verbanaceae*	*Nochi*	Chrysoplenetin (Awale *et al.*, 2011)	Human pancreatic cancer. Effective against myelo suppression and anaemia during chemotherapy (Merina *et al.*, 2012).	S - *Karppu, Kaippu* T - *Veppam* P - *Karppu*	*Kapham*
21.	*Vitis vinefera*	Vitaceae	*Thiratchai*	Resveratrol (Jang *et al.*, 1997).	Breast cancer, Prostate cancer, Non–hodgkinslymphoma (Sakarkar, 2011).	S - *Thuvarpu* T - *Seetham* P - *Karppu*	*Pittham, Kapham*
22.	*Withania somnifera*	Solanaceae	*Ammukura*	Withanolides, Withaniferin (Devi, 1996)	Antitumor and Radio sensitizing effect (Devi, 1996)	T - *Veppam* P - *Karppu*	*Kapham, Pitham*
23.	*Zingiber officinale*	Zingiberaceae	*Ennchi*	Gingerenone A, gingeols Zingerone, oleoresin (Sabnis, 2001).	Leukemia, Skin cancer (Sabnis, 2001).	S - *Karrpu* T - *Veppam* P - *Karrpu*	*Kapham*

* S – *Suvai*,(Taste) T – *Thanmai*,(Character) P – *Pirivu* (Division) ; ** V – *Vatham*, P – *Pitham*, K – *Kapham*; *Enippu* (Sweet), *Pulippu* (Sour), *Kaippu* (Bitter), *Karppu* (Pungent), *Thuvarppu* (Astrigent), *Veppam* (Hot), *Seetham* (Cold)

Table 6. Anti cancerous activity of medicinal plants described in Unani system.

S.No.	Botanical Name	Family	Unani name	Mode
1.	*Artemisia absinthium* L.	Asteraceae	Afsantīn	Shafi G. *et al.* (2012)
2.	*Cuscuta reflexa* Linn.	Convolvulaceae	Aftīmūn	Suresh V. *et al.* (2011)
3.	*Phyllanthus emblica* L.	Euphorbiaceae	Āmla	Sultana S. *et al.* (2008)
4.	*Withania somnifera* (L.) Dunal	Solanaceae	Asgand	Nema R. *et al* . (2013)
5.	*Glycyrrhiza glabra* Linn.	Fabaceae	Asl-us-Sūs	Rahman S. *et al* . (2007)
6.	*Semecarpus anacardium* Linn.	Anacardiaceae	Balādur	Sugapriya D. *et al.* (2008)
7.	*Terminalia bellirica* (Gaertn.) Roxb.	Combretaceae	Balela	Dinesh M.G. *et al.* (2014)
8.	*Achyranthes aspera* Linn.	Amaranthaceae	Chirchita	Subbarayan P.R. *et al.* (2010)
9.	*Cedrus deodara* (Roxb.) G.Don	Pinaceae	Deodār	Gaidhani S.N. *et al.* (2013)
10.	*Tinospora cordifolia* (Thunb.) Miers	Menispermaceae	Gilo	Verma R. *et al.* (2011)
11.	*Nymphaea alba* Linn.	Nymphaeaceae	Gul-i-Nīlofer	Khan N. *et al.* (2005)
12.	*Terminalia chebula* Retz.	Combretaceae	Halela	Ahuja R.*et al.* (2013)

13.	*Solanum nigrum* Linn.	Solanaceae	'Inab-us-Sā'lab	Jeong J.B. *et. al.* (2007)
14.	*Jatropha curcas* Linn.	Euphorbiaceae	Jangli Arand	Oskoueian E. *et.al.* (2011)
15.	*Bauhinia variegata* (L.) Benth.	Fabaceae	Kachnār	Balakrishnan S.K. *et.al.* (2014)
16.	*Picrorhiza kurroa*	Plantaginaceae	Kutki	Kumar M.H. *et.al.* (2014)
17.	*Catharanthus roseus* (L.) G. Don	Apocyanaceae	Sadābahār	Ruskin S.R. *et.al.* (2014)
18.	*Aloe vera* (L.) Burm.f.	Liliaceae	Sibr Zard (Elva)	Naveena. *et.al.* (2011)
19.	*Acorus calamus* Linn.	Acoraceae	Waj Turki	Sreejaya S.B. *et.al.* (2013)
20.	*Berberis aristata* Linn.	Berberidaceae	Zarishk	Pai K.S. *et.al.* (2012)

Table 7. Some other Medicinal plants claimed to have anti-cancer properties.

Botanical name	**Family**	**Part used**	**Parts used and their main active components**
Acorus calamus	Araceae	Roots and rhizome	β-asarone (1, 2, 4-trimethoxy-5-prop-1-enyl-benzene) , calamusin
Acronychia baueri	Rutaceae	Bark	Triterpene lupeol and the alkaloids melicopine, acronycine, and normelicopidine
Agave americana	Agavaceae	Leaf	Steroidal saponin, alkaloid, coumarin, isoflavonoid, hecogenin
Agropyron repens	Poaceae	Rhizomes	Rhizome contains essential oil, polysaccharide and mucilage
Agrimonia pilosa	Rosaceae	Herb	Agrimonolide, flavonoid, triterpene, tannin and coumarin
Ailanthus altissima	Simaroubaceae	Bark	Triterpene, tannin, saponin and quercetin-3-glucoside
Akebia quinata	Lardizabalaceae	Fruit	Flavonoid and saponin
Allium cepa	Liliaceae	Bulb	Quercetin, allicin
Allium sativum	Liliaceae	Bulb	allicin, methyl allyl trisulfide, and diallyl trisulfide S-allylcysteine and S-allylmercapto-L-cysteine

Aloe barbadensis	Liliaceae	Leaf juice	saponins, salicylic acids
Alpinia galanga	Zingiberaceae	Rhizomes	Kaempferide and flavone
Aphanamixis polystachya	Meliaceae	Bark	Sterol, saponins
Aristolochia contorta	Aristolochiaceae	Root and fruit	Lysicamine and oxaaporphine
Asparagus racemosus	Liliaceae	Root	Shatavarin I-IV
Aster tataricus	Asteraceae	Whole plant	Triterpene, monoterpene and epifriedelanol
Azadirachta indica	Meliaceae	Bark	nimbin, nimbinene, azadirachtin, azadirachtol, azadirachnol, desacetynimbinene, nimbandiol, nimbolide, quercetin, beta-sitosterol, n-hexacosanol, nimbiol and nimocin.
Bacopa monnieri	Scropulariaceae	Whole plant	Brahmine ,herpestine, saponins
Bauhinia variegata	Caesalpinaceae	Root	flavonolglycoside, triterpene saponin, phenanthraquinone, flavonoids.
Butea monosperma	Fabaceae	Bark	Cajanin and isoformononetin;isobutrin
Cajanus cajan	Fabaceae	Leaves	cajanuslactone

Chelidonium jajus var. *asiaticum*	Papaveraceae	Herb	Alkaloids (sanguinarine, chelerythrine, berberine)
Chimaphila umbellate	Ericaceae	Whole plant	Ericolin, arbutin, urson and tannin
Coix lachryma jobi	Poaceae	Seed	Trans-ferulyl stigmasterol
Dryopteris crassirhizoma	Polypodiaceae	Rhizomes	Filicinic and filicic acids, aspidinol and aspidin
Echinops setifer	Asteraceae	Whole plant	Echinopsine
Erythronium americanum	Liliaceae	Whole plant	Alpha-methylenebutyrolactone
Euonymus alatus	Celastraceae	Whole plant	Triterpene, euolatin, steroid and sesquiterpene alkaloid
Eupatorium cannabinum	Asteraceae	Whole plant	Sesquiterpene, lactone, pyrrolizidine alkaloidand flavonoid
Fragaria vesca	Rosaceae	Leaf and fruit	Flavonoid, tannin, borneol and ellagic acid Asia, Europe
Fritillaria thunbergii	Liliaceae	Whole plant	Alkaloid and peimine
Galium aparine	Rubiaceae	Cleaver	Iridoid, polyphenolic acid, tannin, anthraquinoneand flavonoid

Hydrastis canadensis	Ranunculaceae	Whole plant	Isoquinoline alkaloids (hydrastine, berberine, berberastine, candaline), resin and lactone
Junchus effuses	Juncaceae	Whole plant	tridecanone, effusol, juncanol, phenylpropanoid and a-tocopherol
Lantana camara	Verbenaceae	Whole plant	Alkaloids (camerine, isocamerine, micranine,lantanine, lantadene)
Larrea tridentate	Zygophyllaceae	Whole plant	Resin
Lonicera japonica	Caprifoliaceae	Whole plant	Tannins, saponins and carotenoids
Olea europrae	Oleaceae	Leaf and oil	Oleic acid and polyphenol
Panax quinquefolium	Araliaceae	Roots	Ginsenoside, sesquiterpene, limonene vitamins (B1, B2, B12)
Phaleria macrocarpa		Fruits	Gallic acid
Polygonatum multiflorum	Liliaceae	Whole plant	Saponin, flavonoid and vitamin A
Potentilla chinensis	Rolsaaceae	Whole plant	Gallic acid and tannin
Pygeum africanum	Boraginaceae	Bark	Phytosterol, triterpene and tannin

Pyrus malus	Rosaceae	Bark and fruit	Quercetin, catechin, flavonoid, coumaric and gallic acids, and procyanidin
Rhus chinensis	Anacardiaceae	Leaf	Tannin, apigenin and glycoside; seed contains bruceosides (A, B),
Rubus idaeus	Rosaceae	Leaf	Flavonoid and tannin; fruit contains vitamins (A, B, C) and ellagic acid
Scilla natalensis	Hyacinthaceae	Bulb	Bulb
Scrophularia nodosa	Scrophulariaceae	Aerial parts	Iridoid, flavonoid and phenolic acid
Smilax chinensis	Liliaceae	Rhizomes	Tannin, saponins and flavonoid
Tabebuia spp.	Bignoniaceae	Bark	Quinine, bioflavonoid and co-enzyme Q
Thuja occidentalis	Cupressaceae	Whole plant	Flavonoid, tannin, volatile oil and mucilage
Thymus vulgaris	Lamiaceae	Whole plant	Volatile oil, flavonoid and tannin
Trifolium pratense	Fabaceae	Flower	Glucosides (trifolin, trifolitin, trifolianol), flavonoid
Vitex rotundifolia	Verbenaceae	Whole plant	Camphene, pinene and diterpene

Phytochemicals and Cancer – An overview

Plant materials have been used for the treatment of malignant diseases for centuries. Recent phytochemical examination of plants which have a suitable history of use in folklore for the treatment of cancer had induced often resulted in the isolation of principles with antitumour activity. An intensive survey of plants, micro organism and marine animals for antitumour activity began in the later 1950s mainly because the United States National Cancer Institute (NCI) initiated and funded a major screening programme.

Phytochemicals obtained from Soybean such as genistein (4',5,7-tribydroxy isoflavone) inhibits the growth of transplantable human prostate carcinoma. Epidemiological studies have consistently shown that regular consumption of fruits and vegetables is strongly associated with reduced risk of developing chronic diseases such as cancer as the phytochemical extracts from them exhibit strong antioxidant activity.

In the screening of Yemeni plants used in folk medicine for the anticancer potential, the methanolic extracts of *Dendrosicyos socotrana, Withania aduensis, Withania riebeckii, Dracena cinnabari* and *Buxus hildebrandlii* exhibited the highest toxicity on all tumor cell lines. The four varieties of muscadine grape extract had the ability to inhibit the activitiy of matrix metalloproteinases implying that those could be good inhibitors of carcinogenesis. The limonoids isolated from the methanol extract of *Khaya senegalensis* proved to have good anticancer activity. The leaf extract of *Ashwagandha* selectively killed tumor cells and thus it was a natural source for safe anticancer medicine. The fruit of deer berry (*Vaccinium stamineum*) exhibited the anticancer capability of human lung and leukemia cancer cells. Polyphenolic extracts from *Vaccinium macrocarpon* inhibited the growth and proliferation of breast, colon, prostate, lung, and other tumors as do flavonols, proanthocyanidin, oligomers, and triterpenoids isolated from the fruits of the same.

Morinda citrifolia showed cancer preventive affect in both clinical practice and laboratory animal models. An alcoholic extract of *Biorhythms sensitivum* for antitumor activity could inhibit the solid tumor development in mice induced with Dalton's lymphoma ascites (DLA) cells and increased the life span of mice bearing Ehrlich ascites carcinoma (EAC) tumors. Edible fruits and berries served the source for novel anticancer agents, given that extracts of those foods have demonstrated cytotoxic activity against tumor cell lines. Nimbolide, a triterpenoid extract from the flowers of the neem tree was found to have antiproliferative activity against some cancer cell lines. *Semecarpus anacardium* Linn. (nut milk) extract exerts its anticancer effect through quenching - reactive oxygen species. The cytotoxic activities of two medicinal herbs *Linum persicum* and *Euphorbia cheradania* that are native to Iran showed cytotoxic activity on tumor cell lines. The Pomegranate extracts inhibits the growth of breast cancer cells. Brassinosteroids, steroid plant hormones are promising leads for potential anticancer drugs. The *Careya arborea* bark significantly reduced the solid tumor volume induced by DLA cells. The methanol extract of *Bauhinia racemosa* stem bark exhibited antitumor effect in EAC bearing mice. The antitumor activity of the ethanol extract of *Indigofera aspalathoides* was established.

The extract of some medicinal herbs such as *Anemarrbena asphodeloides* (Root), *Artemisia argyi* (leaf), *Commiphora Myrrh* (Resin), *Duchesnea indica* (Aerial Plants), *Gleditsia sinessis* (Fruit), *Ligustrum lucidum* (fruit), *Rheumpalmatum* (Root and Rhizome), *Rubia cordifolia* (Root), Salvia Chinesis (Aerial parts), *Scutellaria barbata* (Aerial Parts), *Uncaria rhychopylla* (Stem), *Vaccaria segetails* (seed) showed anticancer effects invitro and those effects were markedly greater on cancer cells compared with normal cells.

Phytoconstituents extracted from a large number of plants belonging to the genus Hypericum are known to possess potent anticancer nature. Cytotoxic activity of *Sarris cernuss* extract on human colon and breast carcinoma cultures was proved. The natural antioxidant gallic acid (GA) isolated from the fruits of an Indonesian medicinal Plant, *Phaleria Macrocarpa* was proved to be a potent anticancer compound. The rhizome *Zingiber officinalis,* one of the most widely used species of the ginger family is a common condiment for various foods and beverages. The pungent vallinoids i.e., 6-gingerol and 6-paradol, shogaolsand zingerone attributed to the anticancer properties of ginger.

The antineoplastic activity of methanolic extracts of five medicinal plants that are native to Iran including *Galium mite, Ferula Angulata*, *Stachys obtuscrena, Grsium bracteosum,* and *Echinophora Cinerea* was investigated and proved to have anti tumor activity. *Panax ginseng* and its extracts have long been used for medical purposes and there is increasing interest in developing ginseng products as cancer preventive agents. Purified bioactive compounds derived from medicinal mushrooms were potentially important for new source of anticancer agents.

The Saponins from the plant *Clematis manshrica* has obvious antitumor effects against various transplanted tumor on mice. The Embelin derivatives such as 1,4 - benzoquinone derivative 5-0 ethyl embelin(1) and 5-0 methyl embelin are promising antimitotic and anti cancer molecules. Sesquiterpenes the class of naturally occuring molecules that are 15-carbon isoprenoid compounds typically found on plants and marine life. They have therapeutic potential in decreasing the progression of cancer.

The anticancer activity from *Platycodon grandiflorum* has been proved and established. The methanol extract of stem bark of *Dillenia pentagons* appears to be more active against Dalton's lymphoma. *Limonium Vulgare,* Artemisia *Maritima* and *Salicornia europaea* showed antineoplastic activities. The extracts of *Ononis spinosa, Trifolium fragiferum* and *Trifolium repen* showed tumor growth inhibiting activities. Methanol extract *Ledum groelandicum* Retzius (Labrador tea) leaf twig extract showed anticancer activity. The anti-neoplastic activity of guduchi (*Tinospora cordifolia*) on Ehrlich ascities carcinoma has also been proved.

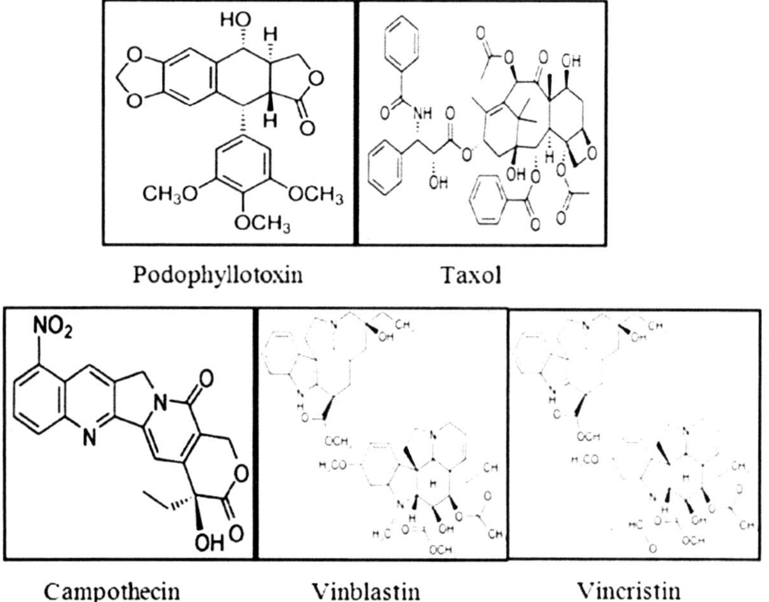

Fig. 1. Structures of Some Plant-Derived Anti-Cancer Agents in Clinical Use.

References

Aggarwal B.B. Nuclear factor-kappa B, the enemy within.*Cancer Cell* (2004) 6: 203-208.

Aggarwal B.B., Kumar A., Bharti A.C. 2003. Anticancer potential of curcumin: preclinical and clinical studies. *Anticancer Res* 23: 363-398.

Ahuja R., Agrawal N., Mukerjee A. 2013. Evaluation of anticancer potential of *Terminalia chebula* fruits against Ehrlich Ascites Carcinoma induced cancer in mice. *J Scient Innovat Res.* 2(3): 549-554.

Anis K.V., Rajeshkumar N.V., Kuttan R. 2001. Inhibition of chemical carcinogenesis by berberine in rats and mice. *J Pharm Pharmacol.* 53: 763–768.

Anonymous, Yugi vaithya chinthamani, 1998. Chennai, India; Directorate of Indian Medicine and Homeopathy Publications.

Awale S. *et al.*, 2011. Identification of Chrysoplenetin from Vitex negundo as a potential cytotoxic agent PANC-1 human cancer cells. *Phytotherapy Research,* 25(12): 1770–1775.

Baatout S., Derradji H., Jacquet P., Ooms D., Michaux A., Mergeay M. 2004. Effect of curcuma on radiation-induced apoptosis in human cancer cells. *Int J Oncol.* 24: 321–329.

Baatout S., Derradji H., Jacquet P. *et al.* 2004. Effect of curcuma on radiation-induced apoptosis in human cancer cells. *Int J Oncol,* 24: 321–329.

Badary O.A. 1999. Thymoquinone attenuates ifosfamide-induced Fanconi syndrome in rats and enhances its antitumour activity in mice. *J Ethnopharmacol,* 67: 135–142.

Balakrishnan S.K., Bhat K.I. 2014. Apoptosis and flowcytometric studies of *Bauhinia variegata* bark extract. *Asi J Pharmaceut Cl Res.* 7(1).

Bhakuni D.S., Dhar M.L., Dhar M.M., Dhawan B.N., Gupta B., Srimal R.C. 1971. Screening of Indian plants of biological activity, Part-III. *Indian J Exp Biol.* 9: 91–102.

Bharali R., Tabassum J., Azad M.R. 2003. Chemomodulatory effect of *Moringa oleifera* Lam., on hepatic carcinogen metabolising enzymes, antioxidant parameters and skin papillomagenesis in mice. *Asian Pac J Cancer Prev.* 4: 131–139.

Bhakuni D.S., Dhar M.L., Dhar M.M., Dhawan B.N., Mehrotra B.N. 1969. Screening of Indian plants of biological activity, Part-II. Indian *J Exp Biol.* 7: 250–262.

Bhishagratha K.L. 1991. Sushruta samhita. Varanasi: Choukhamba Orientalia.

Bilal Bin Hafeez, Weixiong zhong *et al.*, 2012. Plumbagin inhibits prostate cancer development in TRAMP mice via targeting PKCå, Stat3 and neuroendocrine markers. *Oxford journals, Carcinogenesis,* 33(3): 644-651.

Bray F., Ren J.S., Masuyer E., Ferlay J. 2013. Global estimates of cancer prevalence for 27 sites in the adult population in 2008. *Int J Cancer,* 132(5): 1133–1145.

Corsi M.M., Bertelli A.A., Gaja G. The therapeutic potential of *Aloe vera* in tumour-bearing rats. 20: 115–118.

Coussens L.M., WERB Z. 2002. Inflammation and cancer. *Nature.* 420: 860-867.

Davis L., Kuttan G. 1998. Suppressive effect of cyclophosphamide-induced toxicity by *Withania somnifera* extract in mice. *J Ethnopharmacol.* 62: 209–214.

Devi P.U. 1996. *Withania somnifera* Dunal (Ashwagandha): potential plant source of a promising drug for cancer chemotherapy and radiosensitisation. *Indian J. Exp Biol.* 34: 927–932.

Dhar M.L., Dhar M.M., Dhawan B.N., Mehrotra B.N., Ray C. 1968. Screening of Indian plants for biological activity, Part-I. *Indian J Exp Biol.* 6: 232–247.

Dhawan B.N., Dubey M.P., Mehrotra B.N., Rastogi R.P., Tandon J.S. 1980. Screening of Indian plants for biological activity, Part-IX. *Indian J Exp Biol.* 18: 594–606.

Devi P.U. 1996. *Withania somnifera* Dunal (Ashwagandha) a potential plant source of a promising drug for cancer chemotherapy and radio sensitisation. *Indian J Exp Biol,* 34: 927–932.

Dinesh M.G., Rajasekaran S., Suneel R., Chandrasekaram K., Kalaivani R. 2014. *Terminalia bellerica* leaf extracts induce apoptosis in HepG2 cells and regulates cell cycle progression by inducing G2/M cell cycle arrest. *Indian J Res Pharmacy Biot.* 2(1): 1044-1057.

Dubey N.K. 1997. Cytotoxicity of the essential oil of *Cymbogon citrus* and *Ocimum gratissimum. Indian J Pharm Sci.* 59: 263-269.

Eferl R., Wagner E.F. 2003. AP-1: a double-edged sword in tumorigenesis. *Nat. Rev. Cancer.* 3: 859-868.

Estrov Z., Shishodia S., Faderl S. *et al.*, 2003. Resveratrol blocks interleukin-1betainduced activation of the nuclear transcription factor NF-kappaB, inhibits proliferation, causes S-phase arrest, and induces apoptosis of acute myeloid leukemia cells. *Blood.* 102: 987-995.

Fu Y., Hsieh T.C. *et al.*, 2004. Licochalcone -A novel flavanoid isolated from licorice root (Glycyrrhiza glabra) arrests prostate cancer cells. Biochemical and Biophysical Research Communications, 10:332(1): 322(1): 263-270.

Gaidhani S.N., Singh A., Kumari S., Lavekar G.S., Juvekar A.S., *et al.* 2013. Evaluation of some plant extracts for standardization and anticancer activity. *Indian J Traditional Knowledge.* 12(4): 682-687.

Garg A., Aggarwal B.B. 2002. Nuclear transcription factor-kappa B as a target for cancer drug development. *Leukemia.* 16: 1053-1068.

Gupta Y.K., Sharma S.S., Rai K., Katiyar C.K. 2001. Reversal of paclitaxel induced neutropenia by *Withania somnifera* in mice. *Indian J Physiol Pharmacol.* 45: 253–257.

Hong R.L. Spohn W.H., Hung M.C. 1999. Curcumin inhibits tyrosine kinase activity of p185neu and also depletes p185neu. *Clin. Cancer Res.* 5: 1884-1891.

Hostanska K., Daum G., Saller R. 2002. Cytostatic and apoptosis-inducing activity of boswellic acids toward malignant cell lines *in vitro. Anticancer Res.* 22: 2853–2862.

Hu X., Cao B.N., Hu G., He J., Yang D.Q., Wan Y.S. 2002. Attenuation of cell migration and induction of cell death by aged garlic extract in rat sarcoma cells. *Int J Mol Med.* 9: 641–643.

Itokawa H., Takeya K., Mori N., Hamanaka T., Sonobe T., Mihara K. 1984. Isolation and antitumour activity of cyclic hexapeptides isolated from *Rubia radix*. Chem Pharm Bull. 32: 284–290.

Jagetia G.C., Baliga M.S., Venkatesh P. 2003. Effect of *Sapthaparna* (*Alstonia scholaris* Linn) in modulating the benzo(a) pyrene-induced forestomach carcinogenesis in mice. Toxicol Lett, 144: 183–193.

Jagetia G.C., Baliga M.S. 2003. Treatment with *Alstonia scholaris* enhances radiosensitivity *in vitro* and *in vivo. Cancer Biother Radiopharm.* 18: 917–929.

Jagetia G.C., Baliga M.S. 2003. Modulation of antineoplastic activity of cyclophosphamide by *Alstonia scholaris* in the Ehrlich ascites carcinoma-bearing mice. *J Exp Ther Oncol.* 3: 272–282.

Jang M. *et al.*, 1997. Cancer chemopreventive activity of Resveratrol, a natural product derived from Grapes. *Science* 10, Jan 1997; 275 No. 5297: 218–220.

Jeong J.B., Jeong H.J., Park J.H., Lee S.H., Lee J.R. *et al.*, 2007. Cancer-preventive peptide lunasin from *Solanum nigrum* L. inhibits acetylation of core histones H3 and H4 and phosphorylation of retinoblastoma protein (Rb). *J Agric Food Chem.* 55: 10707-13.

Karnick C.R. 1996. Pharmacology of Ayurvedic medicinal plants. Newdelhi, India; Sri Satguru Publications.

Khan N., Sultana S. 2005. Anticarcinogenic effect of *Nymphaea alba* against oxidative damage, hyperproliferative response and renal carcinogenesis in Wistar rats. *Mol and Cell Biochem.* 271(1-2): 1-11.

Kumar M.H., Ramesh C. 2014. Anticancer activity of Nano encapsulated formulation from the extracts of *Picrorhiza kurroa* against human cancer cell lines. *J Pharmacog and Phytochem.* 2(5): 182-185.

Lang I. 2002. Possibilities and results with Herceptin-Taxol combination in the treatment of breast cancer. *Magy Onkol.* 46: 195–196.

Lissoni P., Giani L., Zerbini S., Trabattoni P., Rovelli F. 1998. Biotherapy with the pineal immunomodulating hormone melatonin versus melatonin plus aloe vera in untreatable advanced solid neoplasms. *Nat Immun.* 16: 27–33.

Maiti B. 1994. Antineoplastic effects of the root extract of *Hygrophila spinosa*. In: Proceedings of the International Conference on Current Progress in Medicinal and Aromatic Plant Research. Calcutta, India; p. 135–140.

Matthew S., Kuttan G. 1999. Immunomodulatory and antitumour activities of Tinospora cordifolia. *Fitoterapia,* 70: 35-43.

Mazumdar U.K., Gupta M., Maiti S., Mukherjee D. 1997. Antitumour activity of *Hygrophila spinosa* on Ehrlich ascites carcinoma and sarcoma-180 induced mice. *Indian J Exp Biol.* 35: 473–477.

Mellado W., Magri N.F., Kingston D.G., Garcia-Arenas R., Orr G.A., Horwitz S.B. 1984. Preparation of biological activity of taxol acetates. *Biochem Biophys Res Commun.* 124: 329–336.

Mukund sabnis V.D. 2006. Chemistry and Pharmacology of Ayurvedic Medicinal Plants. Varanasi, India; Chaukhamba Amarabhati Publications.

Murthy K.R.S. 2001. Bhavaprakasa of bhavamisra. Vol. II. Madhya and Uttara Khanda. Varanasi: Krishnadas Academy.

Murugesa mudhaliar K.S. 2006. Siddha materia medicamedicinal plant division. Chennai, India: Directorate of Indian medicine and Homeopathy Publications.

Narah Merina, Kalita Jogen Chandra *et al.*, 2012. Medicinal plants with potent anticancer activity -A review. International Research Journal of Pharmacy, 3(6): 26.

Naveena, Bharath B.K. 2011. Selvasubramanian. Antitumor activity of *Aloe vera* against Ehrlich Ascites Carcinoma (EAC) in Swiss albino mice. *Int. J. Pharma and Bio. Sci.* 2(2): 400-409.

Nema R., Khare S., Jain P., Pradhan A. 2013. Anticancer Activity of *Withania somnifera* (Leaves) Flavonoids Compound. *Int J Pharm Sci Rev Res.* 19(1): 103-106.

Ohnuma T., Sridhar K.S., Ratner L.H., Holland J.F. 1982. Phase I study of indicine N-oxide in patients with advanced cancer. *Cancer Treat Rep.* 66: 1509–1515.

Olsen D.L., Bradley C., Johnson M., Macias J.L., Love V., Markoe A. 2001. The effect of *Aloe vera* gel/mild soap versus mild soap alone in preventing skin reactions in patients undergoing radiation therapy. *Oncol Nurs Forum.* 28: 543–547.

Oskoueian E., Abdullah N., Saad W.Z., Omar A.R., Ahmad S., Kuan W.B., Zolkifli N.A., Hendra R., Ho Y.W. 2011. Antioxidant, anti-inflammatory and anticancer activities of methanolic extracts from *Jatropha curcas* Linn. *J Med Plant Res.* 5(1): 49-57.

Pai K.S., Srilatha P., Suryakant K., Setty M.M., Nayak P.G., Rao C.M., Baliga M.S. 2012. Anticancer activity of *Berberis aristata* in Ehrlich ascites carcinoma-bearing mice: a preliminary study. *Pharm Biol.* 50(3): 270-277.

Pal S., Chakraborti S.K., Banerjee A., Mukerji B. 1968. Search for anticancer drugs from Indian medicinal plants. *Indian J Med Res.* 56: 445–455.

Parimala R., Sachdanandam P. 1993. Effect of Plumbagin on some glucose metabolising enzymes studied in rats in experimental hepatoma. *Mol Cell Biochem.* 125: 59–63.

Pecere T., Gazzola M.V., Mucignat C., Parolin C., Vecchia F.D., Cavaggioni A. *et al.*, 2000. Aloe-emodin is a new type of anticancer agent with selective activity against neuroectodermal tumours. *Cancer Res.* 60: 2800–2804.

Prasad G.C. 1987. Studies on cancer in Ayurveda and its management. *JRAS*. 3: 147–167.

Prasad G.C. 1968. Deshpande PJ. Effect of Rohitaka (*Amura rohitaka*) on leukaemia. *J Res Indian Med*. 3: 36.

Praveenkumar V., Kuttan R., Kuttan G. 1994. Chemoprotective action of Rasayanas against cyclophosphamide toxicity. *Tumouri*. 80: 306–368.

Premalatha Balachandran, Rajgopal Govindarajan. 2005. Cancer - An Ayurvedic Perspective. *Pharmacological research,* 51: 19-30.

Premalatha B., Sachdanandam P. 1999. *Semecarpus anacardium* L. nut extract administration induces the in vivo antioxidant defense system in aflatoxin B1 mediated hepatocellular carcinoma, *J Ethnopharmacol,* 66: 131–139.

Rabi T., Gupta R.C. 1995. Antitumour and cytotoxic investigations of *Amura rohitaka*. *Int J Pharmacogn*. 33: 359–364.

Rahman S., Sultana S.2007. Glycyrrhizin exhibits potential chemopreventive activity on 12-Otetradecanoyl phorbol-13-acetate-induced cutaneous oxidative stress and tumor promotion in Swiss albino mice. *J Enzyme Inhib Med Chem*. 22: 363-369.

Rao A.V., Devi P.U., Kamath R. *In vivo* radioprotective effect of *Moringa oleifera* leaves. Indian J Exp Biol 39: 858–863.

Rastogi R.P., Merhotra B.N., editors. 1993. Compendium on Indian medicinal plants. New Delhi: CSIR publications.

Rave-Frank M., Meden H., Jaschke A., Tanzer A., Boghun O., Fietkau R. 1997. The effect of paclitaxel on the radiosensitivity of gynecological tumour cells. *Strahlenther Onkol*. 173: 281–286.

Ruskin S.R., Aruna S.R. 2014. *In-vitro* and *In-vivo* antitumor activity of *Catharanthus roseus*. *Int Res J Pharm App Sci*. 4(6): 1-4.

Sakarkar D.M., Deshmukh V.N. 2011. Ethnopharmacological review of traditional medicinal plants for anti cancer activity. *International journal of Pharm Tech research, Jan–Mar*. 3(1): 298-308.

Salim E.I., Fukushima S. 2003. Chemopreventive potential of volatile oil from black cumin (*Nigella sativa* L.) seeds against rat colon carcinogenesis) seeds against rat colon carcinogenesis. *Nutr Cancer*. 45: 195–202.

Seetharam Prasath J. 1997. Anuboga vaithya dheva ragasiyam. Chennai, India; Rathna Nayakkar & Sons.

Shanmugavelu. M. 2006. Siddha maruthuva Noi naadal Noi mudhal naadal thiratu. Chennai, India; Directorate of Indian Medicine and Homeopathy Publications.

Shirin H. *et al.*, 2001. Antiproliferative effects of Sallylmercaptocysteine on colon cancer cells when tested alone or in combination with sulindac sulfide. *Cancer Res,* 61: 725–731.

Shafi G., Hasan T.N., Syed N.A., Al-Hazzani A.A., Alshatwi A.A., *et al.* 2012. *Artemisia absinthium* (AA): a novel potential complementary and alternative medicine for breast cancer. *Mol Biol Rep.* 39(7): 7373-9.

Sharada A.C., Solomon F.E., Devi P.U., Udupa N., Srinivasan K.K. 1996. Antitumour and radiosensitising effects of withaferin A on mouse Ehrlich ascites carcinoma in vivo. *Acta Oncol.* 35: 95–100.

Shirin H., Pinto J.T., Kawabata Y., Soh J.W., Delohery T., Moss S.F. *et al.* 2001. Antiproliferative effects of S-allylmercaptocysteine on colon cancer cells when tested alone or in combination with sulindac sulfide. *Cancer Res.* 61: 725–731.

Sonata S. 1986. The efficacy of Ayurveda drugs on Cancer (*Arbuda*). Workshop on cancer souvenir. Chennai: Central Research Institute for Siddha.

Sreejaya S.B., Santhy K.S. 2013. Cytotoxic properties of *Acorus calamus* in MCF-7 breast cancer cells. *Int J Curr Res Aca Rev.* 1(1): 106-111.

Subbarayan P.R., Sarkar M., Impellizzeri S., Raymo F., Lokeshwar B.L., Kumar P., Agarwal R.P., Ardalan B. 2010. Anti-proliferative and anti-cancer properties of *Achyranthes aspera*: specific inhibitory activity against pancreatic cancer cells. *J Ethnopharmacol.* 131(1): 78-82.

Sugapriya D., Shanthi P., Sachdanandam P. 2008. Restoration of energy metabolism in leukemic mice treated by *Semecarpus anacardium* Linn. nut milk extract. *Chem Biol Interact.* 173: 43-58.

Subbareddy V.V., Sirsi M. 1969. Effect of *Abrus precatorius* Linn. On experimental tumours. *Cancer Res.* 29: 1447–1451.

Sultana S., Ahmed S., Jahangir T. 2008. *Emblica officinalis* and hepatocarcinogenesis: A chemopreventive study in Wistar rats. *J Ethnopharmacol.* 118: 1-6.

Suresh V., Sruthi V., Padmaja B., Asha V.V. 2011. *In vitro* anti-inflammatory and anti-cancer activities of *Cuscuta reflexa* Roxb. *J Ethnopharmacol.* 134(3): 872-877.

Thakore *et al.* 2001. A brief review of plants having anti cancer property. *International Journal of Pharmaceutical Research and Development,* 3(11): Jan 2012: 129-136.

Trivedi N.P., Rawal U.M. 2001. Hepatoprotective and antioxidant property of *Andrographis paniculata* in BHC induced liver damage in mice. *Indian J Exp Biol,* 39: 41–46.

Utthamarayan K.S. 2005. Siddhar Aruvai Maruthuvam. Chennai, India; Directorate of Indian Medicine and Homeopathy Publications.

Verma R., Chaudhary H.S., Agrawal R.C. 2011. Evaluation of Anticarcinogenic and Antimutagenic Effect of *Tinospora cordifolia* in Experimental Animals. *J Chem Pharm Res.* 3(6): 877-881.

Vrinda B., Uma Devi P. 2001. Radiation protection of human lymphocyte chromosomes *in vitro* by orientin and vicenin. *Mutat Res.* 498: 39–46.

Yun Ruchen *et al.* 2009. Identification of an Alkyl hyroquinone from Rhus succedanea as an inhibitor of Tyrosinase and melanogenesis. *Journal of Agricultural and Food Chemistry,*7: 2200-2205.

Zaniboni A., Ardizzoni A., De Marinis F., Portalone L., Boni C., Meriggi F., *et al.* 2003. Phase II study of Taxol combined with ifosfamide and carboplatin in the treatment of stage IIIb-IV non-small-cell lung cancer. *Am J Clin Oncol.* 26: 84–88.

❑❑❑

Chapter 6

Ethno-medicinal Plants of Meghalaya: An Account on Sustainable Utilization

Ramesh Kumar*, C.S. Purohit and Vinod Maina

Botanical Survey of India, Arid Zone Regional Centre, Jodhpur-342014, *Corresponding author email: rkpaliwalbsi@yahoo.com

ABSTRACT

Medicinal plants play an important role, especially in modern times, when the damaging effects of food processing and over-medication have assumed alarming proportions. They are now being increasingly used in cosmetics, food and brewages, as well as alternative medicines in different ways. This paper is compilation of indigenous knowledge on ethno-medicinal uses and deals with the 8 endemic plant species and 88 other plant species which have been used for treatment of 63 different ailments by Khasi, Garo and Jaintia tribes in Meghalaya state.

Keywords: Ethno-medicinal, tribes, Khasi, Garo, Jintia, Meghalaya.

introduction

Medicinal plants have been known for millennia and are highly esteemed all over the world as a rich source of therapeutic agents for the prevention of diseases and ailments. Herbal medicine is still the mainstay of about 80% of the world population mainly in the developing countries for primary health care because of better body and lesser side effects (Kamboj, 2000); countries with ancient civilizations such as China, India, South America, Egypt, etc. are still using several plant remedies for various ailments. There is a urgent need to documentation the valuable information before disappearing over time.

In this regard, India has been referred to as the medicinal garden of the world, where a number of recognized indigenous systems of medicine viz., Ayurveda, Sidhha, Unani, Homeopathy, Yoga and Naturopathy are being utilized for the health care of people. All the natural herbal products are safe so these are popular among rural and urban community of India. For their non-toxicity, less side effects and easy availability, their demands are increasing in developing and developed countries (Kalia, 2005). Traditional system of medicines take an approach that since disease or prone to disease is due to imbalance in the equilibrium eliminates the disease. The treatment aims not only curing the disease but also enhancing the body vitality to combat the disease and minimize the chance of relapse. Compilation of ethnobotanical data was started by Kirtikar and Basu (1935), Anonymous (1948-1976) which was followed by several ethnobotanists Chopra, Nayar and Chopra (1956), Jain (1968, 1991), Chopra, Nayar and Verma (1969), Anonymous (1986), Caius (1986), Haridasan (2006) in the recent past.

As per Haridasan (2006) reports a Large numbers of medicinal plants have been used in different systems of medicine as (a) Indigenous systems of medicine, viz., Ayurveda (1769 plant spp.), Unani (751 plant spp.) and Sidha (1121 plant spp.); Tibetan system of medicine (279 plant spp.); (b) Modern system or allopathy of medicine (38 plant spp.) and (c) Homeopathy system of medicine (482 plant spp.). Apart from these, many plants are used in folk medicines (4671 plant spp.) by various tribal and ethnic groups of India. About 44% of the total Indian flora (Anonymous, 2007) has been reported to be used as medicinal plants. The corresponding figures for North-east India (Tiwari, 2002) and Meghalaya are 25% and 31% respectively (Lakadong & Barik, 2006).

Study area: Meghalaya is one of the eight sister states of the North Eastern region of India and one of the species rich and mega-biodiversity centre. It covers an area of 22,429 sq.km., comprising of the Khasi, Garo and Jaintia hills lying between 25°.47'-26°10' N latitude and 89°45'-92°47' E longitude. It is bounded on the north and east by the state of Assam; on the west by Assam and Bangladesh and on the south by Bangladesh. Three major tribes viz. the Khasis, Jaintias and Garos are predominantly here. Some other tribes like the Mikir, Koch, Bodo, Hmar, Hajong, Rabha, Miri etc. are also present in some pockets of the state. The non-tribal are also residing in Meghalaya mainly includes Bengalees, Assamese, Marwaris, Nepalis and some others. The tribal population is about 85% of the population.

The vegetation of the state is very rich and diverse due to favourable climatic conditions, good rainfall, temperature and relative humidity. The forest is tropical, subtropical and temperate type. Total forests area is 9,496 km^2 which is 42.34% of the state. There are one Biosphere reserve (Nokrek), 2 National Parks and 3 Wildlife sanctuaries in the state.

Methodology: Entire Meghalaya state has been selected as the study area for conducting detailed ethno-botanical explorations and gathering information on use of plant species by local people. The total number of 12 long and about 40 short duration field survey have been conducted in last 8 years for this study.

During the survey, these medicinal plants have identified by their local names used by different tribal and with the help of various regional and state floras like Flora of British India Vol 1-7 (Hooker, 1872-1897), Flora of Assam Vol 1-5 (Kanjilal *et al.*, 1934-1940), Forest flora of Meghalaya Vol 1-2 (Haridasan and Rao, 1985-1987), Flora of Jowai Vol 1-2 (Balakrishnan, 1981-1983), Flora of Nangpoh (Joseph, 1980) and Flora of India series.

Medicinal plants in Meghalaya Scenario: The traditional knowledge about indigenous medicinal plants in Meghalaya is much higher than many other states. Witnessed from the numerous ethno-botanical research papers published. Meghalaya has a number of village doctors. The traditional healers are known as *Nong Ai Dawai Kynbat* in Khasi Hills, *Uwa a Dawai* in Jaintia Hills and *Kaviraj* in Garro Hills. Use of medicinal plants to meet the family's primary health care and nutritional need is traditional and embedded in the culture of Khasis, Garos and Jaintias.

Medicinal plants are very rich in Meghalaya due to its wide range of altitudinal variation, geographical location and favourable environmental conditions for plant growth and richness is also result of people's traditional knowledge (Anonymous 2003, 2005 and 2007). An over eight hundred plant species have been reported, out of them the under mentioned eight species are endemic to Meghalaya (Table-I) are used frequently for curing different body ailments by the Khasis, Jaintias and Garo people.

Table 1. List of endemic plants used as medicine in Meghalaya

SN	Botanical name	Family	Local name	Parts used	Ailments treated
1.	*Camellia kissii* Wall.	Theaceae	Dieng Tymem (K)	Root	Skin diseases
2.	*Citrus latipes* (Swingle) Yu.Tanaka	Rutaceae	Soh Heh (J)	Fruits, leaves	Body ache, vomiting, cold, fever
3.	*Goniothalamus simon sii* Hook.f. & Thomson	Annonaceae	-	Fruits	Throat irritation
4.	*Nepenthes khasiana* Hook.f.	Nepenthaceae	Tiew Rakot (K)	Pitcher	Urinary troubles, stomach disorders, night blindness, skin diseases, leprosy
5.	*Ophiorrhiza subcapitata* Wall. ex Hook.f.	Rubiaceae	Samachik (G)	Roots, leaves	Fever, sore throat, tonsils, facial blemishes
6.	*Osbeckia capitata* Benth. ex Naudin	Melastomataceae	Soh Pythem (K)	Whole plant	Snake-bite, muscle swellings
7.	*Sophora benthamii* Steenis	Fabaceae	Palwang (G)	Bark	Pregnancy (blood purifier after delivery)
8.	*Xylosma longifolia* Clos	Salicaceae	Dieng Kani (K); Phulwal (G)	Bark	Stomach ache

Khasis and Jaintia tribes of Meghalaya used about 100 wild plants as medicines, subsidiary food, making implements, musical instruments, religious ceremonies etc. (Joseph & Kharkongor, 1981). Garos tribe used 25 plant species for food, 24 species for medicine, 5 species for fish poison, 7 species for fibres, 3 species for dyes, 4 species for magico-religious beliefs and 10 species for miscellaneous purposes (Shanpru & Vasudeva, 1981).

Several workers (Kumar *et al.,* 1987; Rao *et al.,* 1989; Maikhuri & Gangwar, 1993; Jain *et al.,* 1977; Choudhary & Neogi, 2003; Samati, 2004; Murugkar & Subbulakshmi, 2005; Ahmed & Borthakur, 2005; Kayang, 2007) have done extensive works on ethno-medicinal as well ethno-botanical aspects. A number of plant species are used for the treatment of their various diseases and various parts in different ways and different parts of the same species used to cure different ailments of the body by people of Meghalaya (Table - II). The consumption of medicinal plants is more popular in Khasis and Garo people in comparison to the Jaintias. A list of 88 medicinal plant species used very commonly in traditional system of medicine in Meghalaya is provided along with their botanical name, local names popular in Khasi, Garo and Jaintia, parts they used and the ailment treated (Table-II).

Table 2. List of plants used in Traditional system of medicine in Meghalaya

SN	Botanical name	Local name	Parts used	Ailments treated
1.	*Acorus calamus* L.	Kynbat ksuid (K)	Whole plant	Mental disease, pregnancy
2.	*Aegle marmelos* (L.) Correa	Selpri, Belethi (G)	Leaves, roots, bark, fruit	Weakness, ulcers, fever, dropsy
3.	*Ageratina adenophora* (Spreng.) King & Rob	Latnaiong (K)	Leaves	Burns, cuts, injuries
4.	*Ageratum conyzoides* (L.) L.	Saphlang ke (K)	Leaves and fruits	Prevent bleeding
5.	*Allium sativum* L.	Rasin (G); Rynsun syntiewwher (K)	Whole plant	Burning pain in the feet
6.	*Aloe vera* (L.) Burm.f.	Sal komla (G), Syntiew shylluit (K)	Whole plant	Weakness, fever, fractures, asthma, cold, gastritis, burns, anti cancer
7.	*Alpinia galanga* (L.) Willd.	Lakud (K)	Leaves	Burns
8.	*Anaphalis adnata* DC.	Tiew riem (K, J)	Leaves	Haemotoma
9.	*Aporosa octandra* (Buch.-Ham. ex D. Don) Vickery	Chamolja (G)	Bark	Leprosy
10.	*Artemisia vulgaris* L.	Jaiaw (K)	Whole plant	Antifungal, wounds
		Syingblei (K)	Tuber	Stomachache, fractures
		Syingkhmoh (K)	Tuber	Chronic cough
		Bet (K)	Tuber	Chronic cough
		Jyrwiing (K)	Tuber	Dysentery
11.	*Asparagus filicinus* Buch.-Ham. ex D. Don	Shiahjaker (K)	Tuber	Cyst, piles
12.	*Asparagus racemosus* Willd.	Memang tamatchi (G)	Roots	Diabetes, stomach disorder, blood pressure
13.	*Averrhoa carambola* L.	Amlenga (G); Sohkhloo (K)	Fruit	Jaundice, dysentery
		Bolanchi (G)	Bark	Fractures
		Dokrikra (G)	Whole plant	Fractures
14.	*Azadirachta indica* A.Juss	Neem (G, K)	Leaves	Dysentery, diarrhea (along with bark of sal)
15.	*Begonia roxburghii* A.DC.	Jagaw (K)	Leaves	Bile, dysentery, chronic cough
16.	*Bidens pilosa* L. var. *minor* (Blume) Sherff.	Soh byrthit (K, J)	Leaves	Antidote for snakebite
17.	*Colocasia esculenta* (L.) Schott.	Mania (G)	Tuber	Fractures
18.	*Calotropis procera* (Aiton) Dryand	Ka ang (K)	Whole plant	Malaria
19.	*Cannabis sativa* L.	Bhang (K)	Leaves	Dysentery, diarrhea, skin disease
20.	*Cassia alata* L.	Dopawant (G)	Leaves	Fractures
		Matasam (G)	Leaves	Fractures
		Golbira (G)	Leaves	Asthma
21.	*Cassia fistula* L.	Snaru (G)	Bark, roots	Purgative, tonic, febrifuge, jaundice

22.	*Centella asiatica* (L.) Urb.	Mana muni (G); Krah poh (K)	Leaves	Fever, cough & cold, blood pressure, dysentery, cholera and stomachache
23.	*Cinnamomum verum* Presl.	Dieng syiem (K, J)	Leaves	Cough
24.	*Cissus quadrangularis* L.	Samritchu (G)	Stem	Fractures
25.	*Clematis gouriana* Roxb. ex DC.	Bat bteng doh (K)	Leaves	Cuts and injuries
26.	*Clerodendron colebrookianum* Walp.	Jarem (K, J)	Leaves	Rheumatism
27.	*Crossocephalum crepidiodes* (Benth.) Moore	Jathymmai (K)	Leaves	Constipation, stomach disorders
28.	*Curcuma amada* Roxb.	Dike tegatchu (G)	Rhizome	Goldstone
29.	*Curcuma caesia* Roxb.	Dike beholi (G)	Rhizome	Rheumatism, fracture
30.	*Curcuma zedoaria* (Christm.) Roscoe	Holdiasok (G); Syieng khloo (K)	Tuber	Jaundice, dysentery
31.	*Cuscuta reflexa* Roxb.	Shaw (K)	Whole plant	Bodyache, rheumatism
		Jitrai (J)	Whole plant	Jaundice
32.	*Cyperus rotundus* L.	Satuinke (K)	Leaves	Dysentery and jaundice
33.	*Datura stramonium* L.	Tiew shulim (K)	Whole plant	Paralysis, rheumatism, stroke
34.	*Drymaria cordata* (L.) Willd. ex Schult.	Nai ke (K)	Leaves	Antidote for Snake bite
35.	*Dysphania ambrosioides* (L.) Mosyakin & Clemants	Iada (K)	Leaves	Typhoid, fever, pneumonia
36.	*Emblica officinalis* Gaertn.	Ambare (G), Sohmylleng (K)	Bark, fruits, leaves	Menstrual problem, blood pressure, skin disease, cough & cold, gastritis
37.	*Emilia sonchifolia* (L.) DC.	Jhurmawria (G)	Leaves	Boils
38.	*Erythrina variegata* L.	Diengsong (K)	Leaves	Burns
39.	*Fagopyrum acutatum* (Lehm.) Mansf. ex K.Hammer	Jarain (K)	Leaves, fruit	Medicinal salad, burns, dysentery, cuts and injuries
40.	*Ficus benghalensis* L.	Prap (G); Dienggri (K)	Resin extract	Ringworm, ulcers, skin disease
41.	*Gallinsoga parviflora* Cav.	Batnongrim (K, J)	Leaves	Antidote for snakebite
42.	*Gynocardia odorata* R.Br.	Bolribu (G)		Asthma
43.	*Hedychium coronarium* J.Koenig	Gong (G)	Rhizome	Vomiting, stomach pain, epilepsy
44.	*Hedyotis uncinella* Hook. & Arn.	Bat iong (K, J)	Leaves	Skin disease
45.	*Hedyotis scandens* D.Don	Bat iong (K, G)	Leaves	Eye trouble
			Root	Dysentery
46.	*Holarrhena pubescens* Wall. ex G. Don	Golmatra (G)	Bark, seeds, leaves	Dysentery, epilepsy, dizziness
47.	*Houttuynia cordata* Thunb.	Jamyrdoh (K)	Leaves	Purification of blood
		Krah lemen (K)	Leaves	Toothache
48.	*Justicia adhatoda* L.	Dieng khloo (K)	Leaves	Rheumatic pain
49.	*Justicia gendarussa* Burm.f.	Dojagipe (G)	Leaves	Fractures, injuries
		Gimbil (G)	Bark	Fractures, injuries

50.	*Kaempferia galanga* L.	Dike walsari (G)	Bark	Bleeding
		Wakprata (G)	Leaves, tuber	Body ache
		Dike chisik (G)	Leaves, tuber	Body ache
		Boga jachong (G)	Leaves	Jaundice
		Dike ban (G)	Tuber	Pneumonia
		Mongnal (G)	Flowers, fruit	Piles
51.	*Litsea monopetala* (Roxb.) Pers.	Boldokakki; laham bol (G)	Bark, seeds	Leprosy, fractures, abortion, injuries
52.	*Luffa acutangula* (L.) Roxb.	Sohpru (K)	Whole plant	Stomachache
		Bat sohmynken (K)	Leaves	Boils, cuts and injuries
		Jatangneng (K)	Fruit, flowers	Skin disease
		Tiew lamungor (K)	Whole plant	Fractures
		Bat weisi (K)	Whole plant	Fractures
		Bat shinia (K)	Whole plant	Fractures
53.	*Mikania micrantha* Kunth	Jyrmi bapur (G)	Whole plant	Stomach trouble
54.	*Molineria capitulata* (Lour.) Herb.	Rekoksi (G)	Tuber, leaves	Diarrhea, dysentery
		Dokime (G)	Leaves	Fractures
55.	*Morinda angustifolia* Roxb.	Chenong (G)	Whole plant	Jaundice, eye disease
		Matria bol (G)	Leaves	Epilepsy
56.	*Morus australis* Poir	Sohlyngolkhur (J)	Leaves	Ulcers
57.	*Mussaenda frondosa* L.	Syntu slaieh (K)	Leaves	Body swelling
58.	*Myrica esculenta* Buch.-Ham. ex D.Don	Sohphie (K)	Fruit	Headache
		Soh kwit (K)	Fruit	Headache
		Bat sohpdok (K)	Fruit, leaves	Cuts and injuries
59.	*Nepenthes khasiana* Hook.f.	Memang koksi (G); Tiew rakot (K)	Juices	Asthma, indigestion, kidney problem, night blindness and skin diseases
60.	*Nicotiana tabacum* L.	Sada (G); Duma (K)	Leaves	Toothache
61.	*Oenanthe javanica* (Bl.) DC.	Jatira (K, J)	Leaves	Stomach pain
62.	*Oroxylum indicum* (L.) Kurz.	Kiring (G)	Bark, leaves	Jaundice, fever
63.	*Osbeckia capitata* Benth. ex Naudin	Soh lakthut (K)	Whole plant	Swelling of muscles
64.	*Paederia foetida* L.	Pasim(G), Dieng jyrmi (K)	Whole plant	Injuries, fractures, dysentery, stomach disorder, gastritis, diarrhea
65.	*Passiflora nepalensis* Walp.	Sla sohbrap (K)	Leaves, fruit	Dysentery, jaundice
66.	*Phlogacanthus tubiflorus* Nees	Alot (G)	Leaves, flowers	Blood pressure
67.	*Piper longum* L.	Bat sohmarit (K)	Leaves	Dysentery
68.	*Piper peepuloides* Wall.	Sohmrit khlaw (K)	Seeds	Chronic cough
69.	*Potentilla fulgens* Diels	Lanieang kynthei (K)	Whole plant	Blood pressure

70.	*Psidium guajava* L.	Saprian (K)	Leaves	Dysentery
71.	*Rauwolfia serpentina*(L.) Benth. ex Kurz	Dogrikme (G)	Roots	Malaria, chicken pox, fever, tuberculosis, high blood pressure, snake poison, stomach pain
72.	*Rhus chinensis* Mill.	Sama (K)	Seeds	Neutralisation of mushroom poisoning
73.	*Ricinus communis* L.	Dieng kaston (K)	Leaves	Rheumatism and body swelling
74.	*Schefflera venulosa* (Wight & Arn.) Harms	Jengjil (G)	Bark, leaves	Pneumonia
		Dorimit (G)	Bark	Jaundice
		Memang wardo (G)	Stem	Fracture
		Derasteng (G)	Leaves	Menstrual problems
		Darechik budu (G)	Bark	Menstrual problems
		Sampret (G)	Leaves	Cataract
75.	*Schima wallichii* Choisy	Boldak (G); Dieng phuhlieh (K)	Leaves	Cuts, wounds, worms
76.	*Smilax guianensis* Vitman	Swarthit (K)	Leaves	Snake bite
77.	*Swertia chirayita* (Roxb.) Buch.-Ham. ex C.B. Clarke	Chirota (G), Sharita (K)	Whole plant	Malaria, fever, tuberculosis, cholera, diabetes
78.	*Stephania hernandiifolia* (Willd.) Walp.	Samkusim (G)	Tuber	Asthma
		Boga salgro (G)	Leaves	Weakness, jaundice
		Bolrasin (G)	Bark	Cholera, epilepsy
		Samkatong (G)	Rhizome	Gastritis
		Dike bisi (G)	Rhizome	Rheumatism
79.	*Tamarindus indica* L.	Cheeng (G)	Leaves	Fractures
		Dochengcheng (G)	Leaves	Fractures
80.	*Terminalia bellirica* (Gaertn.) Roxb.	Churi bol, Chiirore (G)	Bark, Fruit	Gastritis, anemia, cough & cold, menstrual disorders
81.	*Terminalia chebula* Retz.	Artak (G)	Bark, fruits	Asthma, dysentery, cough & cold, spleen disorder, gastritis, diarrhea
82.	*Tinospora cordifolia* (Willd.) Miers	Jyrmi khlaw (K)	Whole plant	Low blood pressure
83.	*Toxicodendron acuminatum* (DC.) Wu & Ming	Bolmicheng (G)	Bark	Urine infection
		Amango (G)	Bark, stem	Fractures
		Ganping (G)	Bark	Fractures
		Tasrak (G)	Bark	Fractures
		Chongi bita (G)	Tuber	Fractures
		Tasundu (G)	Tuber	Fractures
		Mecheng (G)	Roots, leaves	Fractures
		Mebitchi (G)	Roots, leaves	Fractures
84.	*Tylophora indica* (Burm.f.) Merr.	Mamloki (G)	Leaves	Cancer
		Dagal (G)	Leaves	Weakness
85.	*Viburnum colebrookeanum* Wall. ex DC.	Syntuiap bru (K)	Leaves	Sores
		Bat kynjri (K)	Leaves	Injuries, cuts and wounds

		Bat iong (K)	Leaves	Allergic
		Tiew lili (K)	Stem	Fractures
		Rynneng (K)	Tuber	Gastric, diabetes
		Shyndo (K)	Stem	Poisoning, stomachache
		Tiew lasir (K)	Rhizome	Anti cancer, tuberculosis, pregnancy
86.	*Vitex quinata* (Lour.) F.N. Williams	Rangre (G)	Bark	Anemia
87.	*Zanthoxylum armatum* DC.	Jaiur (G)	Bark	Small pox
89.	*Zingiber officinale* Roscoe	Eching (G); Sying (K)	Tuber	Cough, fever

Khasi (K); Garo (G) and Jaintia (J).

Discussion

A total number of 88 plant species have been ethno-medicinally used by tribes Khasi, Garo and Jintia in the Meghalaya state. (Table-III). The 88 plant species documented were recorded to serve as remedy for treatment of 63 different diseases and ailments with the highest number of 23 species reported for use as dysentery followed by fracture (22 species), stomachache (14 species), jaundice (13 species), fever (12 species) and the other diseases ranging from 1-11 species each.

Khasi tribes have used ethno-medicinally 56 plant species as remedy for treatment of 50 different disease. Garo tribes have used ethno-medicinal 46 plant species as remedy for treatment of 49 different disease, where as tribe Jintia used 9 plant species as remedy for treatment of 7 different disease (Table-III).

The data shows Khasi people are more familiar to the plants (56 species used in 50 different diseases) regarding medicinal uses of plants in comparison to Garo (46 species in 49 different diseases) and Jaintia (9 species used in 7 different diseases) respectively. It is interesting to know that, the same plant is used for different diseases in traditional therapy and Ayurvedic system of medicine. The multiple uses of a species show that inter-relationship between them and the tribal and there is an urgent need of conservation of their traditional knowledge. This is urgency and need to protect the fast disappearing medicinal plants in nature on the basis of their traditional knowledge.

Table 3. Uses of plants in different diseases by tribes- Khasi, Garo and Jintia

SN	Disease	Number of plants used by tribes in disease			Total
		Khasi	Garo	Jintia	
1	Mental disease	1	0	0	1
2	Pregnancy	2	0	0	2
3	Weakness	1	3	0	4
4	Ulcers	1	2	1	4
5	Fever	5	7	0	12
6	Dropsy	0	1	0	1
7	Burns	6	2	0	8
8	Cuts	7	1	0	8
9	Injuries	9	2	0	11
10	Prevent bleeding	1	1	0	2
11	Fractures	6	16	0	22
12	Asthma	2	6	0	8
13	Gastritis	4	6	0	10
14	Anticancer	2	2	0	4
15	Haemotoma	1	0	1	2
16	Leprosy	0	2	0	2
17	Antifungal	1	0	0	1
18	Wounds	3	1	0	4
19	Stomachache	7	7	0	14
20	Chronic cough	4	0	0	4
21	Dysentery	14	9	0	23
22	Cyst	1	0	0	1
23	Piles	1	1	0	2
24	Diabetes	2	2	0	4
25	Blood Pressure	4	5	0	9
26	Jaundice	5	7	1	13
27	Diarrhoea	3	4	0	7
28	Bile	1	0	0	1
29	Antidote for snakebite	4	1	2	7
30	Malaria	2	2	0	4
31	Skin disease	6	3	1	10
32	Purgative	0	1	0	1

33	Tonic	0	1	0	1
34	Febrifuge	0	1	0	1
35	Cough & Cold	5	5	1	11
36	Cholera	2	2	0	4
37	Rheumatism	5	3	1	9
38	Constipation	1	0	0	1
39	Goldstone	0	1	0	1
40	Bodyache	1	1	0	2
41	Paralysis	1	0	0	1
42	Stroke	1	0	0	1
43	Typhoid	1	0	0	1
44	Pneumonia	1	2	0	3
45	Menstrual problem	1	3	0	4
46	Boils	1	1	0	2
47	Ringworm	2	2	0	4
48	Epilepsy	0	4	0	4
49	Vomiting	0	1	0	1
50	Eye trouble	1	2	0	3
51	Dizziness	0	1	0	1
52	Toothache	2	1	0	3
53	Body swelling	3	0	0	3
54	Headache	2	0	0	2
55	Indigestion	1	1	0	2
56	Kidney problem	1	1	0	2
57	Night blindness	1	1	0	2
58	Chicken pox	0	2	0	2
59	Tuberculosis	1	2	0	3
60	Cataract	0	1	0	1
61	Anemia	0	2	0	2
62	Urine infection	0	1	0	1
63	Allergic	1	0	0	1

References

Ahmed A.A. and Borthakur S.K. 2005. Ethnobotanical wisdom of Khasis (Hynniewtreps) of Meghalaya. *Bishen Singh Mahendra Pal Singh*, Dehradun.

Anonymous 1948-1976. The Wealth of India, Raw Materials. CSIR, New Delhi.

Anonymous, 1986. The useful Plants of India. CSIR, New Delhi.

Anonymous, 2003. State of Forest Report, Forest Survey of India. Ministry of Environment and Forests, Government of India, Dehradun.

Anonymous, 2005. State of the Environment Report, Department of Environment and Forests, Government of Meghalaya, Meghalaya.

Anonymous, 2007. Foundation for revitalization of Local Health Traditions - A Report. FRLHT. Bangalore.

Balakrishnan N.P. 1981-1983. Flora of Jowai, Meghalaya. Vol. I-II. Howrah, Botanical Survey of India.

Caius J.F. 1986. The Medicinal and Poisonous Plants of India. Scientific Publishers, Jodhpur.

Chopra R.N., Nayar S.L. and Chopra I.C. 1956. Glossary of Indian Medicinal Plants. CSIR, New Delhi.

Chopra R.N., Nayar S.L. and Verma B.S. 1969. Supplement to Glossary of Indian Medicinal Plants. CSIR, New Delhi.

Choudhary D. and Neogi B. 2003. Ethnobotany of Khasi and Chakma tribes of North-east Inda. In JK Maheshwari (Eds.) *Ethnobotany and Medicinal plants of Indian subcontinent*. 583-584. Scientific Publishers.

Haridasan K. and Rao R.R., 1985 & 1987. Forest flora of Meghalaya Vol.I-II, Bishen Singh Mahendra Pal Singh, Dehradun.

Haridasan K. 2006. Diversity, Conservation and Sustainable use of Medicinal Plants in India with special reference to North-east. In: Pandey, H.N. and Barik, S.K. (eds.) *'Ecology, Diversity and Conservation of Plants and Ecosystems in India'*. Regency Publications, New Delhi, pp. 377 - 388.

Hooker J.D. 1872-1897. The Flora of British India Vol. I-VII, London.

Jain S.K. 1968. Medicinal Plants. National Book Trust, New Delhi.

Jain S.K. 1991. Dictionary of Indian Folk Medicine and ethnobotany. Deep Publications, New Delhi.

Jain S.K. Hajra P.K. and Shanpru R. 1977. A survey of edible plants in bazaars of Meghalaya. *Bull. Meg. Sci. Soc.* 2: 29-34.

Joseph J. and Kharkongor P. 1981. A preliminary ethno-botanical survey in the Khasi and Jaintia Hills, Meghalaya. In: Jain, S.K. (ed.). Glimpse of Indian Ethnobotany. Oxford and IBH Publishing Co., New Delhi.

Joseph J., 1980. Flora of Nongpoh and its vicinity. Howrah, Botanical Survey of India.

Kalia A.N. 2005. Text Book of Industrial Pharmacognosy. Oscar publication.

Kamboj V.P. 2000. Herbal Medicine. *Curr. Sci.* 78(1): 35-39.

Kanjilal U.N., Kanjilal P.C., Das A., De R.N. and Bor N.L. 1934-1940. Flora of Assam Vol. I-V, Shillong Govt. Press.

Kayang H., 2007. Tribal knowledge on wild edible plants of Meghalaya, North-east India. *Indian J. Trad. Know.* 6(1): 177-181.

Kirtikar K.R. and Basu B.D. 1935. Indian Medicinal Plants, 2nd ed. Vol. 1-4. Bishen Singh Mahendra Pal singh, Dehradun and Periodical Experts, New Delhi.

Kumar Y., Fancy S. and Rao R.R. 1987. Further contribution to the ethnobotany of Meghalaya plants used by 'War Jaintia' of Jaintia hills district. *J. Econ. Tax. Bot.* 11(1): 65-70.

Lakadong N.J. and Barik S.K. 2006. Diversity and Distribution of Endemic plants in Meghalaya. In: Pandey, H.N. and Barik, S.K. (eds.) *'Ecology, Diversity and Conservation of Plants and Ecosystems in India'.* Regency Publications, New Delhi, pp. 274-311.

Maikhuri R.K. and Gangwar A.K. 1993. Ethnobotanical notes on the Khasi and Garo tribes of Meghalaya, North-east India. *Econ. Bot.* 47(4): 345-357.

Murugkar D.A. and Subbulakshmi G. 2005. Nutritive values of wild edible fruits, berries, nuts, roots and spices consumed by the Khasi tribes of India. *Ecology and Food and Nutrition,* 44(3): 207-223.

Rao R.R., Neogi B. and Prasad M.N.V. 1989. Ethnobotany of some weeds of Khasi and Garo hills Meghalaya and North-east India. *Econ. Bot.* 43(4): 471-480.

Samati H. 2004. Kitchen garden plants of Pnar tribe in Jaintia hills district, Meghalaya. *Ethnobotany* 16: 125-130.

Shanpru R. and Vasudeva M.K. 1981. Some plants in the life of the Garos of Meghalaya. In SK Jain (Eds.) *Glimpses of India Ethnobotany* 149-152, Oxford and IBH, New Delhi.

Tiwari P.C. (ed.) 2002. Ethnobotany. Aavishkar Publishers. Xvi, pp. 455.

❑❑❑

Chapter 7

Ethnobotany of Orchidaceae

Saranjeet Kaur

Faculty, Department of Science, Cordia College, Sanghol, Punjab, India

ABSTRACT

This paper deals with ethnobotany of family Orchidaceae.The paper lucidly explains a number of Orchids alongwith various phytochemicals present in them Various uses of orchids have been dealt with.

Keywords: Orchid, Peninsula, Vedic, Charaka, Aphrodisiac.

Introduction

The orchids are beautiful creation of nature. They belong to family orchidaceae, which is one of the largest and highly evolved angiospermic family, comprising of 25,000-35,000 species (Hossain, 2011). They are scattered in all continents except Antarctica but their greatest diversity occurs in tropical and sub-tropical climates. Orchids, have naturalized within a vast geographic expanse and climatic zones, ranging from tropical to temperate regions. The Himalayan, the North - Eastern and the Peninsular regions are major orchid rich habitats in the country. Major speciation in orchids has taken place in the tropics and subtropics where their growth and development is distinctly influenced by thick vegetation, high humidity and existing micro-climatic conditions. Since ages, orchids are being used for their medicinal and aesthetic values. They reign supreme among floriculturally important plants due to their beauty and diversity of their long-lasting bewitchingly beautiful and attractive flowers. Their contribution to international trade in horticultural plants is unmatched. These ornamental species are cultivated for their bewitchingly beautiful blossoms. Apart from being highly floriferous, the orchids have proved themselves to be highly medicinal herbs as well. The history of medicinal orchids dates back to 3,000-4,000 years. The utility of orchids for their medicinal properties was found first mentioned in Japanese and Chinese herbal descriptions (Bulpitt, 2005).

The Meteria Medica written by Dioscorides around 1st century A.D. has also mentioned about the orchids. In the Indian vedic scriptures, the orchids find their names described. Numerous orchids have been described by 'Charaka' an Indian Physicist, in his compilation 'Charaka Samhita' which has first mentioned the use of Indian orchids for their therapeutic value. In Ayurvedic system of medicine, a group of eight herbs known as 'Ashtavarga' are used in the preparation of significant number of rejuvenating formulations and health tonics. 'Ashtavarga' also known as an important component of classical ayurvedic medicine/health tonic such as 'Chyavanprash'. One of the herbs namely, *Malaxis acuminata,* an important terrestrial species, commonly called 'Rishbhak' is known for its great therapeutic value as its dried pseudobulbs are an important ingredient of 'Ashtavarga' drugs. The species is used to cure tuberculosis and is a great aphrodisiac (cf. Chauhan, 1990). The aerial parts of orchid plant such as leaves, stem/ pseudobulbs, flowers and aerial roots and sub-terranean parts (tubers, roots, rhizomes) are used either in fresh or dried powdered form or as such for their medicinal use (Table 1). Not much work has carried out in exploring the medicinal properties of orchid species. From very few orchid species, some phytochemicals are isolated. These phytochemicals are mainly alkaloids, bibenzyl Triterpenoids, Phenethrenes and Phenenthropyran in nature (Table 2).

Orchids have been used in local system of medicines to cure nervous (*Cypripedium pubescens, Cymbidium elegans, Epipactis latifolia*), haematologic (*Habenaria edgeworthii, H. intermedia, H. pectinata*), dermal (*Dendrobium alpestre*), digestive (*Dendrobium nobile*), reproductive disorders (*Coelogyne cristata, Malaxis acuminata*) and rheumatic (*Rhynchostylis retusa, Vanda testacea*) ailments and for cooling purposes (*Coelogyne cristata, Saccolabium papillosum*) (Chauhan 1990, Lawler 1984). Their additional uses as restorative drugs, food, gums, glues, narcotics and poisons are also indicated (Arditti 1992). *Goodyera pubescens* and *Ansellia humilis* are used as antidote for snakebite and bad dreams respectively. *Vanda parviflora* and *Dendrobium chrysotoxum* have shown positive tests for its anti-cancerous properties and *Epipactis helleborine* and *Liparis ovata* as an antidote to HIV (Chen *et al.,* 2008; De clercq E. 1994). The essence 'Vanilla', obtained which is extracted from unripe green pods of *Vanilla planifolia,* is the most popular commercial produce of orchids. Curiously enough, the milch cattle in north-eastern India are fed on Dendrobes and its species to enhance their milk yield (Pempahishey

1974). The dried flowers of *D. chrysotoxum* are processed to prepare tea-coloured decoction that is found effective in treating diabetes and strengthening reproductivity (www.natureproducts.net/ Dendrobium/Dendrobium_chrysotoxum.html). Orchids have long been used for medicinal purposes. In ancient China, during 2800 B.C., there are records of some species of *Dendrobium* used for medicinal purposes (Hedge and Ingalhalli, 1988). In ancient India, orchids used for medicinal purposes were described as medicinal and ornamental plant.

The genus *Dendrobium,* is highly evolved and diverse group, represented by more than 1100 species. The dried or fresh stems of *Dendrobium* plants, known as Caulis Dendrobii (Shihu or Huang Cao in Chinese) are used in the traditional or folk medication as a Yin tonic, to nourish the stomach, promote the production of body fluid and anti-pyretic. Pharmacological studies have revealed that some of the chemical extractives of *Dendrobium* species are anti-inflammatory (Lin *et al.,* 2001), anti-platelet aggregation (Fan *et al.,* 2001), immune-regulatory (Zhao *et al.,* 2001), anti-angiogenic (Gong *et al.,* 2004) and anti-tumor in action (Morita *et al.,* 2000), which are partly responsible for the actions and indications of this herb in traditional remedies. Entire plant of *D. chrysotoxum* is used to extract cytotoxic compound i.e.1, 4, 5-Trihydroxy-7-methoxy-9H-fluoren-9-one that is highly anti-carcinogenic (Chen *et al.* 2008). Furthermore, the fresh or dried stems are used to treat the loss of appetite with nausea and impaired vision (Yang *et al.,* 2004). Likewise, *D. nobile,* which is a great ornamental due to its appealing floral displays is extremely medicinally valuable. It is one of the 50 fundamental herbs used in the traditional Chinese medicine where it is known by the name shi hu or shi hu lan. The drug shih- hu made from *D. nobile* has been valued greatly (Lawler, 1984). The drug is used as a tonic, strengthening medicine, reputed to nourish the yin system of body, to impart longevity, and serve as an aphrodisiac. The stems are useful in alleviating thirst, calm restlessness, accelerating convalescence and reducing dryness of the mouth. (Bensky *et al.,* 1986). Indigenous people of Eastern Himalayas used *D. nobile* flowers to cure eye ailments (Mandal and Datta, 2003). *Coelogyne flaccida* is another medicinally important orchid species. It is a cool growing, epiphytic orchid, being prettiest of all coelogynes, with heavily scented flowers harbours biologically active chemical compounds, such as flaccidin and oxaloflaccidin phenanthrene derivatives, have been isolated from the entire plant of this species

(Majumdar and Maiti 1989). The species is also reported to have antibacterial activity, and the paste of its pseudobulbs is applied to the forehead to treat headache; its juice is also consumed orally to treat indigestion (Pyakurel and Gurung 2008). *Vanda testacea* (Lindl.) Reichb. f. (= *V. parviflora* Lindl.) known as Banda or Rasna is an alkaloid enriched epiphytic orchid species. It is recognized for its therapeutic properties. All plant parts (roots, leaves, flowers) in powder form or as an extract is used as herbal medicine to cure rheumatism, bronchitis, nervous disorders, piles, inflammations and also as a potential anti-cancerous drug (Chauhan 1990).

Table 1. Orchid species and their phytochemicals (De and Medhi, 2015)

S.No.	Name of the orchid	Class of the Phytochemical	Name of the Phytochemical
1.	*Aerides crispum*	Pheanthropyran	Acridin
2.	*Agrostophyllum brevipes*	Triterpenoid	Agrostophyllinol
3.	*Agrostophyllum callosum*	Triterpenoid Stilbenoids	Isoagrostophyllol Orcinol, 6-methoxy-coelonin, agrostop-hyllin, callosin, callosinin, callosumin, callosuminin, callosumidin
4.	*Anoectochilus formosanus*	Glycoside	Kinsinoside
5.	*Arundina graminifolia*	Stilbenoids	Arundinan
6.	*Bulbophyllum* species	Phenenthrenes	Gymopsin
7.	*Coelogyne cristata*	Phenenthrenes	Coeloginanthridin, Coeloginanthrin
8.	*Coelogyne flaccida*	Phenenthrene	Flaccidin, oxaloflaccidin
9.	*Cypripedium calceolus*	1-4 phenenth-renequinone	Cypripedin
10.	*Orchis latifolia*	Glucoside	Loroglossin
11.	*Dendrobium moschatum*	Phenenthrenes	Rotundatin,moschatin
12.	*Dendrobium nobile*	Bibenzyl	Gigantol, Dendrobine, Nobilonine,

13.	*Epipactis helloborine*	Benzyl derivative	Oxycodone, benzyloxypropy-lindol; didehydro epoxymorphinan
14.	*Eulophia nuda*	Phenenthrene	Nudol
15.	*Eulophia ochreata*	Phenenthrene	Dimethoxyphenen-threne, Dihdrodime-thoxyphenenthrene
16.	*Vanda cristata*	Glycoside	Melianin
17.	*Vanilla planifolia*	Alkaloids, Flavonoids, Glycosides	-
18.	*Vanda roxburghii*	Glycoside	Kinenoside

Orchids as Food

The dishes prepared from orchids figure among delicacies on the menu cards of expensive world-class hotels around the world. The floral buds of *Cymbidium hookerianum* are used by Bhutanese to enhance flavour of the curry. In Malaysia, the leaves of *Dendrobium salaccence* are included as condiments for rice. In our country, tender leaves of *Cypripedium cordigerum* are cooked as vegetable by local rural communities (www.ionopsis.com/edible_orchids.htm).The seed capsule of *Leptotes bicolour* are used to flavour ice cream by Brazilians. The caned-stems and flowers of certain *Dendrobium* species are eaten in fried form. The tubers of *Gastrodia* are eaten as roasted potatoes. The flowers of *Dendrobium longicornu* are eaten as pickle. The pseudobulbs of *Coelogyne ovalis* and tubers of the *Peristylus constrictus* are consumed to quench thirst. The shoots and tender leaves of *Dactylorhiza hatagirea*, roots of *Epipactis royleana* or *Habenaria intermedia*, the leaves of the *Habenaria* species are also cooked as vegetable by the Ethnic community of Chepangs. The roots of *Malaxis cylindrostachya* and *Platanthera calvigera* are boiled and eaten (http://www.ionopsis.com).

Orchids as Ornaments

Apart from having floricultural and medicinal value, the orchids are also reported to have contributed into jewellery industry as well. The orchid jewellery is made with natural and cultivated orchid that are preserved using specific chemicals. Later, these handcrafted jewellery pieces are given finishing of gold, silver, resin or enamel (orchid-gifts-maderia.com).

Fragrances from Orchids

Mesmerizing fragrances are extracted from certain orchid species. Nearly 75% of orchid species are fragrant. These fragrances, as volatile substances, are stored in osmopheres (a kind of special cells) as essential oils in the glands present at the edges of the floral organs for instance-on petals, buds, roots and leaves. These volatile chemical compounds are highly fragrant. Since the orchids are insect (bee) pollinated; these fragrances also help flower to attract their pollinators to accomplish pollination. Table 2 shows some of the fragrant orchids with their chemically active compounds.

Table 2. Some of the fragrant orchids with their fragrances (De *et al.*, 2014)

Species	Chemical compounds
Brassavola digbiana	Cineole (citronellol rose-like), Linalool
Cycnoches ventricosum, Stanhopea tricornis, S. cirrhata,	Benzyl acetate (jasmine), Benzyl acetate 1,8-cineole
Catasetum discolor	D-carvone (rye bread)
Catasetum collare, Catasetum Gnomus, Catasetum candida	Methyl salicilate (wintergreen) Methyl BenzoateMethyl salicilate
Catasetum roseum,	Methyl cinnimate
Gongora quinquenervis	Eugenol, Linalool
Stanhopea cirrhata	1,8-cineole

References

Arditti J. 1992. Fundamentals of Orchid Biology. NewYork, John Wiley and Sons.

Bulpitt C.J. 2005. The uses and misuses of orchids in medicine. QJM: An *International Journal of Medicine.* 98:625–631.

Bensky D., Gamble X., Kaptchuk T. 1986. Chinese herbal medicine. Eastland Press, Seatle - Washington.

Chauhan N.S. 1990. Medicinal orchids of Himachal Pradesh. *The Journal of the Orchid Society of India,* 4: 99-105.

Chen Y.Y., Li Q., Chen Y.Z., Liu Y. 2008. A compound 1, 4, 5 Trihydro-7-methoxy-9H-Fluoren-9-one, a new cytotoxic compound compound from *Dendrobium chrysotoxum*. *Food Chemistry* 108: 973-976.

De, C.L., Pathak P., Rao A.N., Rajeevan, P.K. 2014. Medicinal and aromatic orchids. In: Commercial Orchids, (ed. Golachowska M., Monu E), De Gruter Open Ltd., Warsaw/Berlin. pp. 243-249.

De, L.C., Medhi R.P. 2015. Orchid-A diversified component of farming systems for profitability and livelihood security of small and marginal farmers. *Journal of global sciences* 4(2): 1393-1406.

De clercq E. 1994. Trends in drug development for the treatment of AIDS: Compounds interfering with the initial stages of HIV replicate cycle. *European Journal of Pharmaceutical Sciences* 2: 4-6.

Fan, C.Q., Wang, W., Wang, Y.P., Qin, G.W. and Zhao, W.M. 2001. Chemical constituents from *Dendrobium densiflorum*. *Phytochemistry,* 57: 1255-1258.

Gong, Y.Q., Fan, Y., Wu, D.Z., Yang, H., Hu, Z.B. and Wang, Z.T. 2004. *In vivo* and *in vitro* evaluation of erianin, a novel anti-angiogenic agent. *European Journal of Cancer.,* 40: 1554-1565.

Hossain, M.M. 2011. Therapeutic Orchids: Traditional uses and recent advances: An overview. *Fitroterapia* 82: 102-140.

Hedge S.N., Ingal halli R.S. 1988. A note on the medicinal usage of some orchids. *Arunachal Forest News* 6(1): 11-18.

Lawler L.J. 1984. Ethnobotany of the orchidaceae - A manual. In: Arditti J. (ed.). Orchid Biology, Reviews and Perspectives. 3rd edn. p. 27-149.

Lin, T.H., Chang, S.J., Chen, C.C., Wang, J.P. and Tsao, L.T. 2001. Two phenanthraquinones from *Dendrobium moniliforme*. *J. Nat. Prod.,* 64: 1084-1086.

Mandal A.K.A., Datta S.K. 2003. Ornamental orchids, Science Reporter, pp. 52-54.

Majumdar P.L. and Maiti D.C. 1989. Flaccidin and oxaloflaccidin, two phenenthrene derivatives of the orchid *Coelogyne flaccida*. Phytochemistry 28: 887-890.

Morita, H., Fujiwara, M., Yoshida, N., Kobayashi. 2000. New picrotoxinin-type and dendrobine-type sesquiterpenoids from *Dendrobium* snowflake Red Star. *Tetrahedron.,* 56: 5801-5805.

Pempahishey K.T. 1974. Orchid eaters of 'Shangri-La'. *American Orchid Society Bulletin* 43: 716-725.

Pyakurel D. and Gurung K. 2008. Uses of orchids. *In*: Enumeration of orchids and estimation of current stock of traded orchids in Rolpa district. A report submitted to district forest office, Rolpa, Nepal. pp. 6 28.

Yang H., Chou G-X, Wang Z., Xu L-S. 2004. Two New Compounds from *Dendrobium chrysotoxum*. *Helvetica Chimica Acta* 87(2): 394-399.

Zhao W., Ye Q., Tan X., Jiang H., Li X., Chen K., Kinghorn A.D. 2001. Three new sesquiterpene glycosides from *Dendrobium nobile* with immunomodulatory activity. *Journal of Natural Products* 64: 1196-2000.

❏❏❏

Chapter 8

Traditionally used Medicinal Plants of Coastal Purba Medinipur District, West Bengal

U.K. Sen and R.K. Bhakat*

Ecology and Taxonomy Laboratory, Department of Botany & Forestry, Vidyasagar University, Midnapore-721 102, West Bengal, India, E-mail: rkbhakat@rediffmail.com/ bhakatramkumar@gmail.com

ABSTRACT

The coastal plant species of Purba Medinipur district bear high medicinal and ecological values. The study includes direct interview and intense observation which is conducted among local communities. A total of 50 medicinal plants have been collected and their popular uses are listed. Due to the continuous loss of coastal vegetation, the associated indigenous knowledge is also gradually disappearing. So, it is necessary to protect and re-establish the coastal vegetation as an immediate priority.

Keywords: Coast, Medicinal plant, Traditional knowledge.

Introduction

Plants have a great role in the growth and development of the human race. The first and the most important necessity of human life is the oxygen, which is provided by the plants. Moreover, in day to day life, plants offer dyes, fibers, fodder, food, fruits, fuel wood, gums, latex, shelter, timber etc. Additionally, there are many plant species which have been continuously used by the local people for traditional medicines. Traditional medicines are the basic and alternative remedies to cure various human as well as animal ailments. The knowledge about the medicinal significance of plant species have passed from one generation to the next through oral communication. These are chiefly used by the old people. These medicines are popular because people think that these are easily available, cheaper and have no or less side effects (Rao *et al.*, 1970; Pattanaik *et al.*, 2008; Muthukumar and Samuel, 2010).

In this way, traditional ethnomedicinal knowledge can be conserved as part of living economic and ecological systems, while helping to maintain a sense in local traditional knowledge and practice and reinforcing links between local communities and the environment essential for conservation. The traditional and indigenous system of medicines persists all over the world through the ages. For primary health care, in our day to day life we use traditional herbal medicines a lot and it is 80% in the developing countries. The herbal medicines have a global demand and now-a-days the demand is gradually increasing (Ghosh, 2008; Jain *et al.*, 2009; Mitra and Mukherjee, 2009). In India, the Ayurvedic medicines are used widely in rural areas. The pharmaceutical drugs consumed in developing countries are 15% only and among these 15% are more affluent people. Medicinal plants can provide a significant source of income for rural people in developing countries, especially through the sale of wild harvested material. Coastal vegetation contains many species of specific flora and thus it is an ecological storehouse rich in biodiversity and also has high ecological values (Banerjee, 1994; Untawale, 1994). The length of coast line in India is 7,500 km with numerous lagoons, beaches, estuaries and mangrove swamps, which is rich in living and non-living resources. Local communities have developed their own traditional system of utilizing these coastal plants for medicinal purposes (Choudhury *et al.*, 1993; Franco and Narasimhan, 2009). Keeping this in view, the present study was carried out to inventorise the traditional medicinal plants of coastal Purba Medinipur in West Bengal, India.

Materials and Methods

Study Area

Purba Medinipur (latitude 22°57′10′′- 21°36′35′′ N and longitude 88° 12′40′′-86° 33′50′′E) is one of the 20 administrative districts of West Bengal with its headquarter located at Tamluk. The district was carved out of the erstwhile Medinipur district on January 1, 2002. It is bounded by the Bay of Bengal in the south, river Rupnarayana in the east and Subarnarekha in the west. The Rupnarayana is the joint flow of the river Dwarkeshwar and the river Shilai. The Bay of Bengal and these great rivers and their numerous branches created a prosperous and easy water navigational system fostering commerce, culture and early contacts with the people outside the region. At the same time, these rivers helped to widen the plant vegetation in this region. The surrounding districts are Paschim Medinipur, Howrah and South 24 Parganas in West Bengal and Balasore of Odisha. This coastal area is characterized by a mosaic of landscape consisting of a chain of sand dunes interspersed with interdunal depressions, creeks, salt marshes containing stunted growth of mangroves on the one hand and riverine alluvial plains with agricultural fields. The district is spread over on an area of 4295 km^2 with a population of 5094238 persons (population density; 1,076/ km^2) as per 2011 census. Around 91.71 % of the population lives in the rural areas of the district.

Survey methods and data analysis

Ten villages in 5 different blocks (Contai-I, Deshapran, Khejuri-II, Ramnagar-I and Ramnagar-II) are selected randomly for documentation of the traditional medicinal plants. Ten people were selected from each village irrespective of their age and sex. In this way, a total of 100 informants (80 are male and 20 are female) were contacted for present study. During the visits, the informants were chosen on the basis of a structured questionnaire. Prior consent was taken from the rural folks for documentation of their ethnic knowledge on medicinal plants. Informants were selected on the basis of their ability to identify a particular plant *in-situ* and their basic knowledge of ethnomedicine. Local herbal medical practitioners and elderly people were preferred during the interviews. Prior to the field visits, extensive literature survey was carried out on the previous ethnomedicinal and floral reports of the district. Rural areas were visited during summer, monsoon and winter to avail most of the plants in their flowering or fruiting conditions. The data were recorded on a data sheet with the names of the plants with families, vernacular names, habits, flowering and fruiting times, plant part (s) used and uses of medicinal plants for different diseases. After knowing the specific use of the medicinal plants, informants were taken to the field to identify the plants on the basis of vernacular names. Photographs of the plant habit and reproductive structures are taken and common plants were collected for herbarium preparation. Rare or endangered plants were kept untouched. Collected plants were compared with the literature and identified with the help of standard keys to the specimens (Prain, 1903 a&b; Pakrashi and Mukhopadhya, 2004; Paria, 2005).

Results and Discussion

The present study has shown that traditional medicine is still playing a significant role in meeting fundamental health care needs of inhabitants around the coastal area of Purba Medinipur district. The survey gathered information on 50 species (dicots 46 and monocots 4) belonging to 47 genera distributed in 24 families were recorded from the coastal Purba Medinipur district. Among these, 20 (40%), 18 (36%), 9 (18%), 3(6%) are herbs, shrubs, trees and climbers including lianas respectively. Of the 24 families, dicots are represented by 21 (87.5%) families; monocots consist of 3 (12.5%) families. Amongst the total dicots 46 (92%) and monocots 4 (8%); herbs, shrubs, trees and climbers represent 18, 17, 9, 2 and 2,1,0,1 species respectively, representing 36%, 34%, 18%, 4% and 4%, 2%, 0%, 2% of the total species (Table 1).

Fabaceae (6 species) was most represented family, followed by Amaranthaceae, Euphorbiaceae, Lamiaceae and Malvaceae (4 species each), Acanthaceae, Apocynaceae and Rhizophoraceae (3 species each) while Convolvulaceae, Poaceae and Verbenaceae comprise 2 species (each) respectively and the rest of the family included one species. The three well represented genera are *Avicennia, Crotalaria* and *Senna* (2 species each) respectively, and the rest of the 44 genera included single species (Table 1).

All parts of the various plants were used for the preparation of different herbal formulations; however, leaves (68%), stem (48%), root (38%) and seed (28%) were the most frequently plant parts used. Sometimes, the bark (14%), flower (14%), fruit (12%), whole plant (8%) and latex (2%) were also used (Table 1).

This paper summarizes traditional uses of medicinal plants distributed around the coastal areas of Purba Medinipur district. The area is characterized by high salinity, temperature and light intensity with high rainfall. Plants growing in such environments are mostly halophytes and xerophytes which are well adapted to survive under extreme conditions (Qasim *et al.*, 2011). They evolve multiple mechanisms, including synthesis of diverse group of distinctive compounds which help them to grow in conditions where other plants failed to survive (Khan and Qaiser, 2006). These chemicals, when utilized can serve as potent medicine against different ailments. Their range of active ingredients utilized traditionally to combat deadly ailments and these plants serves as a successful home remedy for deprived people living along the coast. A field-based observation of coastal areas revealed that most of the rural people do not have access to health care professionals. They used a wide range of locally available plant species against multiple diseases (Qasim *et al.*, 2010). The most common uses were against stomach related problems, body pain, arthritis, skin and sexual disorders, asthma etc. The most widespread problem was digestive disorder which reflected the real condition of rural settlements. Lack of basic amenities along with poor sanitation, unhygienic conditions and consumption of contaminated water are some basic reasons for digestive system disorders (Khan *et al.*, 2013).

Present investigation also revealed that the local people make good use of a wide range of plants from their surroundings. All the 50 plants had one or more medicinal uses in which every plant part used as an exclusive medicinal remedy. It is also a matter of deep concern that the transmission of traditional knowledge is decreasing through several generations because no proper way of documenting this heritage except to specific individuals interested in preserving their traditions (Tovey *et al.*, 2005). On the other hand the local medicinal plants are under threat due to anthropogenic and developmental activities such as urbanization; road buildings, growth of mega-structures etc. are examples of direct assault on nature resulting in the loss of local flora (Schmid, 1998). Therefore, it is strongly recommended that the proper documentation of traditional knowledge, utilization of natural resources and conservation of unique floral wealth should be done in a scientific way to prevent their loss.

Conclusion

The present investigation reveals that the practice of traditional plant medicines is still alive in the area under investigation. However, this indigenous knowledge is vanishing rapidly. Our younger generations are not interested of these practices because of non-availability of some important medicinal plants, unspecified doses and unknown side effects. However, in case of ethnomedicinal field further detailed investigations are needed. The study of ethnomedicine should be encouraged as to prepare basic data for economically important medicinal plants so that large scale cultivation techniques can be developed for the welfare of local poor peoples. Therefore, it is the need to conserve this indigenous and precious knowledge about the uses of medicinal plant remedies and also to pass on this to our present and future generations effectively and carry out further studies including phytochemical and pharmacological analysis.

Table 1. Medicinal plants in the coastal area of the Purba Medinipur district.

Sl. No.	Scientific Name and Family	Vernacular Name	Habit	Flowering and Fruiting Time	Part (s) Used	Ailment (s) Treated
1.	*Acalypha indica* L. **(Euphorbiaceae)**	*Muktojhuri*	H	Throughout the year	Le	Gastrointestinal irritation and skin disease
2.	*Acanthus ilicifolius* L. **(Acanthaceae)**	*Hargoja*	S	Apr.-Jul.	Le, Ro	Dyspepsia, paralyses and asthma
3.	*Achyranthes aspera* L. **(Amaranthaceae)**	*Apang*	H	Oct.- Jan.	Ro, Se	Astringent, cough, diuretic and dropsy
4.	*Alternanthera sessilis* (L.) R. Br. ex DC. **(Amaranthaceae)**	*Sanchi*	H	Throughout the year	Le, St	Galactogogue, indigestion, night blindness and snake-bite
5.	*Argemone mexicana* L. **(Papaveraceae)**	*Sialkanta*	H	Jan.-Aug.	Le, Se, St	Eye and skin diseases and parasitic infection
6.	*Asparagus racemosus* Willd. **(Asparagaceae)**	*Satamuli*	C	Sep.-Jan.	Ro	Biliary colic, dysentery, epilepsy and filaria
7.	*Avicennia marina* (Forssk.) Vierh. **(Acanthaceae)**	*Bina*	T	Apr.-Jul.	Ba, Se	Tonic and birth control
8.	*Avicennia officinalis* L. **(Acanthaceae)**	*Jatban*	T	Apr.-Jul.	Ba, Fr, Ro	Astringent, contraceptive, aphrodisiac and small pox
9.	*Bruguiera gymnorhiza* (L.) Lam. **(Rhizophoraceae)**	*Kankra*	T	Apr.-Jul.	Ba, Fl, Le	Cure diarrhea and diabetes
10.	*Calotropis gigantea* (L.) Dryand. **(Apocynaceae)**	*Akanda*	S	Jan.-Aug.	Fl, Le, St	Asthma, cold, cough, indigestion, intermittent fever and purgative

11.	*Carissa spinarum* L. **(Apocynaceae)**	*Bankaramcha*	C	Mar.-Oct.	Fr, Ro	Astringent to dysentery
12.	*Ceriops decandra* (Griff.) W.Theob. **(Rhizophoraceae)**	*Goran*	T	Apr.-Jul.	Ba	Bark extract is antihaemorrhage
13.	*Cissus quadrangularis* L. **(Vitaceae)**	*Harjora*	H	May-Sep.	Le, St	Scurvy, irregular menstruations, asthma, headache and heal bone fracture
14.	*Crotalaria retusa* L. **(Fabaceae)**	*Banatasi*	S	Jul.-Jan.	Le, Se, St	Astringent, emetic, expectorant, dyspepsia and cardiac disorders
15.	*Crotalaria verrucosa* L. **(Fabaceae)**	*Saikatatasi*	S	Throughout the year	Le, Se, St	Bites, scabies and wound
16.	*Cynodon dactylon* (L.) Pers. **(Poaceae)**	*Durba*	H	May-Dec.	Le, Ro	Astringent, cooling, burning sensation, chronic urinogenital troubles, depurative, diuretic, dropsy, epilepsy, haemostatic, hysteria, skin diseases, syphilis and tonic
17.	*Euphorbia hirta* L. **(Euphorbiaceae)**	*Barokherue*	H	Feb.-Dec.	Le, St	Asthma, colic, cough, dysentery and warts
18.	*Evolvulus alsinoides* (L.) L. **(Convolvulaceae)**	*Sankhapuspi*	H	Jul.-Dec.	Le, St	Antidote to poisoning, aphrodisiac, bitter, expectorant, hair growth stimulator, tonic and vermifuge
19.	*Excoecaria agallocha* L. **(Euphorbiaceae)**	*Genwa*	T	Nov.-Feb.	La	Carbuncle, leprosy and paralyses
20.	*Glinus oppositifolius* (L.) Aug.DC. **(Molluginaceae)**	*Gima*	H	Throughout the year	Le, St	Antiseptic, earache, skin diseases and stomachic
21.	*Glycosmis pentaphylla* (Retz.) DC. **(Rutaceae)**	*Banjamir*	S	Aug.- Feb.	Le, Ro	Cough, dyspepsia, intestinal worms, jaundice and skin disease

22.	*Halosarcia indica* (Willd.) Paul G.Wilson **(Amaranthaceae)**	*Nanajhau*	H	Nov.-Jul.	W	Skin diseases
23.	*Hemidesmus indicus* (L.) R. Br. ex Schult. **(Apocynaceae)**	*Anantamul*	C	Jul.-Dec.	Ro	Aphrodisiac, appetizing, astringent, bitter, burning sensation, carminative, diuretic, emollient, insect bite, leprosy, appetite, refrigerant, skin diseases and urinary troubles
24.	*Heritiera fomes* Buch.-Ham. **(Malvaceae)**	*Sundri*	T	Apr.-Jul.	Se	Diarrhea and anti-hypoglycemic
25.	*Hibiscus tiliaceus* L. **(Malvaceae)**	*Saikatjaba*	S	Throughout the year	Fl, Le, Ro, St	Chest congestion, dysentery, ear infections, fevers and coughs
26.	*Ipomoea pes-caprae* (L.) R.Br. **(Convolvulaceae)**	*Chhagalkhuri*	H	Mar.-Sep.	Le, St	Astringent, diarrhea, diuretic, laxative, piles, rheumatism and stomachic
27.	*Jatropha gossypifolia* L. **(Euphorbiaceae)**	*Lalbheranda*	S	Apr.-Aug.	Le, Se, St	Intermittent fever and ulcers
28.	*Leucas aspera* (Willd.) Link **(Lamiaceae)**	*Dronpuspa*	H	Aug.-Oct.	Le, Se, St	Anthelmintic, antiseptic, carminative, dysentery, dyspepsia and elephantiasis
29.	*Lippia javanica* (Burm.f.) Spreng. **(Verbenaceae)**	*Bannebu*	S	Mar.- Dec.	Le, Ro, St	Carbuncles, eczema, expectorant, rheumatism, scabies, stomachic and wounds
30.	*Melastoma malabathricum* L. **(Melastomataceae)**	*Futki*	S	Throughout the year	Fl, Le, St	Diarrhea, dysentery, hemorrhages, leucorrhoea and piles
31.	*Mimosa pudica* L. **(Mimosaceae)**	*Lajjabati*	S	Throughout the year	Le, Ro, St	Antispasmodic, constipating, diuretic, emetic, febrifuge, edema, rheumatism and uterus tumor
32.	*Ocimum tenuiflorum* L. **(Lamiaceae)**	*Radhatulsi*	S	Throughout the year	Le, Se, St	Anodyne, anti-inflammatory, antiscorbutic, carminative, cooling, digestive, liver tonic, refrigerant and stomachic

33.	*Opuntia dillenii* (Ker Gawl.) Haw. **(Cactaceae)**	*Nagphana*	S	Mar.-Jun.	Fl, Fr, St	Expectoration, ophthalmia, poultice, snake bite and whooping cough
34.	*Oxalis corniculata* L. **(Oxalidaceae)**	*Amrul*	H	Jun.-Dec.	Le, St	Anemia, fever, piles, tympanitis and warts
35.	*Pandanus odorifer* (Forssk.) Kuntze **(Pandanaceae)**	*Keya*	S	Jul.-May	Fl, Le, Ro	Anodyne, anti-inflammatory, antiscorbutic, carminative, cooling, digestive, liver tonic, refrigerant and stomachic
36.	*Phyla nodiflora* (L.) Greene **(Verbenaceae)**	*Bhuinokra*	H	Sep.-Apr.	Le, St	Alexeteric, anorexia, aphrodisiac, asthma, burning sensation, cardiopathy, digestive, diuretic, febrifuge, fever, flatulence and stomachic
37.	*Physalis angulata* L. **(Solanaceae)**	*Banteparia*	H	May-Nov.	Fr, Le, St	Appetizing, bitter, bronchitis, burning sensation, cooling, cough, diuretic, expectorant, laxative, sweet, tonic and ulcers
38.	*Porteresia coarctata* (Roxb.) Tateoka **(Poaceae)**	*Dhanighas*	H	Apr.-Jul.	W	Stomach ailments
39.	*Rhizophora mucronata* Lam. **(Rhizophoraceae)**	*Garjan*	T	Apr.-Jul.	Ba	Astringent and cure for diabetes
40.	*Rothia indica* (L.) Druce **(Fabaceae)**	*Nonakura*	H	Aug.-Dec.	W	Stomach ailments
41.	*Scoparia dulcis* L. **(Plantaginaceae)**	*Bandhane*	H	May-Dec.	Le, Se, St	Albumanaria, anemia, bronchitis, cough, diabetes, fever, ketonuria and renal troubles
42.	*Senna occidentalis* (L.) Link **(Fabaceae)**	*Kalkasunda*	S	Jul.-Sep.	Le, Ro, Se	Anodyne, anti-inflammatory, aphrodisiac, bitter, depurative, diuretic, expectorant, febrifuge, purgative, stomachic and tonic

43.	*Senna tora* (L.) Roxb. **(Fabaceae)**	*Chakunda*	H	Jul.- Jan.	Le, Ro	Anthelmintic, antiperiodic, cardiotonic, depurative, expectorant, laxative, liver tonic, ophthalmic, purgative, snake-bite and thermogenic
44.	*Sida cordifolia* L. **(Malvaceae)**	*Berela*	S	Aug.-Jan.	Le, Ro, Se	Astringent, diuretic, facial paralysis, frequent micturition and tonic
45.	*Suaeda monoica* Forssk. ex J.F.Gmel. **(Amaranthaceae)**	*Jirasag*	H	Nov.-Mar.	W	Wounds and skin diseases
46.	*Tephrosia purpurea* (L.) Pers. **(Fabaceae)**	*Bonnil*	S	Mar.-Nov.	Le, Ro, Se, St	Anemia, boils, dyspepsia, elephantiasis, enlarged liver and spleen, fever, pimples, skin diseases, stomachache and toothache
47.	*Thespesia populnea* (L.) Sol. ex Corrêa **(Malvaceae)**	*Paraspipul*	T	Throughout the year	Ba, Fl, Fr, Le, Ro	Astringent, diarrhea, tonic and skin diseases
48.	*Vitex negundo* L. **(Lamiaceae)**	*Nishindu*	S	Mar.-Oct.	Le, Ro, St	Anodyne, anti-inflammatory, aphrodisiac, bitter, depurative, digestive stomachic, diuretic, expectorant, febrifuge, purgative and tonic
49.	*Volkameria inermis* L. **(Lamiaceae)**	*Bamunhati*	S	Jan.-Jun.	Le, Ro	Febrifuge, poultice and rheumatism
50.	*Xylocarpus granatum* J.Koenig **(Meliaceae)**	*Dhundul*	T	Apr.-Jul.	Ba, Fr, Se	Astringent, anti-diarrhea and digestive

In Habit: H-Herb, S-Shrub, T-Tree, C-Climber.

In Flowering and Fruiting Time: Jan.- January, Feb.-February, Mar.-March, Apr.-April, Jun.-June, Jul.-July, Aug.-August, Sep.-September, Oct.-October, Nov.-November, Dec.-December

In Part (s) Used: Ba-Bark, Fl-Flower, Fr-Fruit, La-Latex, Le-Leaves, Ro-Root, Se-Seed, St-Stem, W-Whole plant

References

Banerjee, L.K. 1994. Conservation of coastal plant communities in India. *Bulletin of Botanical Survey of India.* 36(1-4): 160-165.

Choudhury, B.P., Biswal, A.K. and Subudhi, H.N. 1993. Enumeration of some potential medicinal plants in the district of Cuttack (Orissa). *Bioscience Research Bulletin.* 2(1): 11-16.

Franco, F.M. and Narasimhan, D. 2009. Plant names and uses as indicators of knowledge pattern. *Indian Journal of Traditional Knowledge.* 8(4): 645-648.

Ghosh, A. 2008. Ethnnomedicinal plants used in West Rarrh region of West Bengal. *Natural Product Radiance.* 7(5): 461-465.

Jain, S.C., Jain, R. and Singh, R. 2009. Ethnobotanical survey of Sariska and Siliserh regions from Alwar district of Rajasthan, India. *Ethnobotanical Leaflets.* 13: 171-188.

Khan, M.A., and Qaiser, M. 2006. Halophytes of Pakistan: Characteristics, distribution and potential economic usages. *In*: Khan, M.A. (Ed.). *Sabkha ecosystems.* Volume II: West and Central Asia. Springer Netherlands. pp. 129-153.

Khan, S.M., Page, S., Ahmad, H., Shaheen, H., Ullah, Z., Ahmad, M. and Harper, D.M. 2013. Medicinal flora and ethnoecological knowledge in the Naran Valley, Western Himalaya, Pakistan. *Journal of Ethnobiology and Ethnomedicine.* 9(1): 4.

Mitra, S. and Mukherjee, S.K. 2009. Some abortifacient plants used by the tribal people of West Bengal. *Natural Product Radiance.* 8(2): 167-171.

Muthukumar, K. and Samuel, A.S. 2010. Traditional herbal medicines of the coastal diversity in Tuticorin district, Tamil Nadu, India. *Journal of Phytology.* 2(8): 38-46.

Pakrashi, S.C. and Mukhopadhya, S. (Eds.). 2004. *Medicinal and aromatic plants of red laterite region of West Bengal (Bankura, Medinipore and Purulia).* West Bengal Academy of Science and Technology, Kolkata. pp. 1-508.

Paria, N. (Ed.). 2005. *Medicinal plant resources of south West Bengal.* Vol.-1. Research Wing, Directorate of Forests in collaboration with Department of Environment, Government of West Bengal, Kolkata. pp. 1-198.

Pattanaik, C., Reddy, C.S. and Dhal, N.K. 2008. Phytomedicinal study of coastal sand dune species of Orissa. *Indian Journal of Traditional Knowledge.* 7(2): 263-268.

Prain, D. 1903a. *Bengal plants* Vol. 1 (Ind. Repr. 1963). Botanical Survey of India, Calcutta. pp. 1-487.

Prain, D. 1903b. *Bengal plants* Vol. 2 (Ind. Repr. 1963). Botanical Survey of India, Calcutta. pp. 488-1013.

Qasim, M., Gulzar, S. and Khan, M.A. 2011. Halophytes as medicinal plants. *In*: Ozturk, M., Mermut, A.R. and Celik, A. (Eds.). *Urbanisation, land use, land degradation and environment*. Daya Publishing House, Delhi. pp. 330-343.

Qasim, M., Gulzar, S., Shinwari, Z.K., Aziz, I. and Khan, M.A. 2010. Traditional ethnobotanical uses of halophytes from Hub, Balochistan. *Pakistan Journal of Botany*. 42(3):1543-1551.

Rao, T.A., Banerjee, L.K. and Mukherjee, A.K. 1970. Some interesting plant records from Orissa coast. *Journal of Bombay Natural History Society*. 66(3): 659-660.

Schmid, C.K. 1998. Ethnobotany: Principles and application. *Biodiversity and Conservation*. 7(6): 844-845.

Tovey, P.A., Broom, A.F., Chatwin, J., Ahmad, S. and Hafeez, M. 2005. Use of traditional, complementary and allopathic medicines in Pakistan by cancer patients. *Rural and Remote Health*. 5(4): 447-457.

Untawale, A.G. 1994. *Coastal sand dune vegetation. Biodiversity in Western Ghats*, an information kit, (WWF-India, IIRR, SIDA). pp. 3-10.

□□□

Chapter 9

Research Methodology in Ethnobotany

Santosh Kumar Jha

Department of Botany, J. N. R. M. Port Blair, Email: yesbotany@gmail.com

ABSTRACT

This paper deals with quantitative and qualitative techniques. Further organoleptic, histochemical, anatomical methods, along with various extraction methodologies which are used in ethnobotanical researches are discussed. The paper also highlights methods of phytochemical screening, pharmacological studies, various separation techniques, antimicrobial studies and identification, evaluation and standardization of herbal drugs.

Keywords: Ethnobotany; organoleptic; pharmological; anthropology; pharmacopoeia.

Introduction

The recently born multidisciplinary science of ethnobotany which has in a short time of a few decades has experienced a mushrooming growth over the world. It aims at investigating and inventorying fast disappearing traditional knowledge systems pivoted on the direct and total relationship between human societies and plants. The great potential of under usage of plants used by these societies especially in mankind's problem like hunger and health has been fully realized by the results of studies carried out so far. Ethnobotanical research is much wider than the study of medicinal plants and the development of pharmaceuticals and natural products. Ethnobotany draws on multiple disciplines/fields to answer questions about how humans interact with their environment particularly plants.

In India the ethnobotanical works in an organized way were started by Botanical Survey of India in 1969. Since then uses of plants by the tribal are being recorded for a variety of purposes (Jain, 1981). Considerable number of works have been done on several aspects of plants *viz.* ethnomedicine, dyes, tans, narcotics, fiber, timber *etc,* by

Bhandari (1974), Jain (1981a,b), Singh and Pandey (1980,1981,1982), Katewa and Guria (1997), Sebastian and Bhandari (1984a), Nargas and Trivedi (1999) and various other scientists.

For many years, *Field Methods* (and its predecessor *Cultural Anthropology Methods*) has been publishing articles either directly related to ethno-biological methods on topics such as ethno-entomology (Kendall *et al.*, 1990), home gardens (Wichramasuriya and Pelto 1991; Vogl- Lukasser, and Puri 2004), and knowledge of plant use (Reyes-García *et al.*, 2004) or indirectly through innovations in research techniques that are often used by ethno-biologists (*e.g.*, freelisting, triads, pile sorts, and cultural consensus analysis). In recent years, many of the methods used by ethnobiologists have been compiled into field manuals, most notably the series titled "People and Plants Conservation Manuals," developed by the World Wildlife Fund/ UNESCO/Kew Royal Botanic Gardens as part of the People and Plants Initiative (*e.g.*, Martin 1995; Tuxill and Nabhan 1998; Cunningham 2000). Another well received methods manual was developed at the New York Botanical Garden (Alexiades 1996).

The pharmacological treatment of diseases began long ago with the use of herbs (Schulz *et al.*, 2001). Methods of folk healing throughout the world commonly used herbs as part of their traditions. Some of these traditions are briefly described below, provides some examples of the array of important healing practices around the world that used herbs for this purpose. Ayurveda is a medical system primarily practiced in India that has been known for nearly 5000 years includes diet and herbal remedies, while emphasizing the body, mind and spirit in disease prevention and treatment (Morgan, 2002). About 5000 traditional remedies are available in China; they account for approximately one fifth of the entire Chinese pharmaceutical market (Li, 2000). Many herbal remedies found their way from China into the Japanese systems of traditional healing. Herbs native to Japan were classified in the first pharmacopoeia of Japanese traditional medicine in the ninth century (Saito, 2000). The desire to capture the wisdom of traditional healing systems has led to a resurgence of interest in herbal medicines (Tyler, 2000), particularly in Europe and North America, where herbal products have been incorporated into so-called 'alternative', 'complementary', 'holistic' or 'integrative' medical systems. This paper deals with quantitative and qualitative techniques in ethnobotany, organoleptic, histochemical, anatomical methods, as well as various extraction methodologies. Further the paper discusses methods of phytochemical screening, pharmacological studies, various separation techniques, antimicrobial studies and identification, evaluation and standardization of herbal drugs. The ethnobotanical studies require all or some steps which are described under following heads.

Collection of Ethnobotanical Data

Collection and compilation of socioeconomic data of the study area is essentially required for ethnobotanical studies. For socio-economic study following parameters may be considered.

1. Profile of the households (Type of family).
2. Religion and caste profile (ST, SC, OBC, others).
3. Agricultural land, Crops cultivated and livestock's.
4. Irrigation and transport facilities.
5. Drinking water and toilet facilities.
6. Availability of electricity and use of fuel.
7. Awareness about local health tradition among the tribes.
8. Knowledge about medicinal plants for home remedies.
9. Use of home remedies
10. Assessment of treatment other than home remedies.
11. Comparison of cost of home remedies and other treatment method adopted by tribal.

Ethnobotanical data may be collected in different manners i.e. by enquiry, observation, interview and participation. The methodology used for collecting the ethno botanical information was put into following three major categories:

1. **Direct approach:** This included the intensive field surveys among tribal and remote areas of a region.
2. **Indirect approach:** It included collection of information from literature, museums, herbarium etc.
3. **Miscellaneous:** Some information may be also collected after discussion with the non-tribal e.g. village headman, spiritual leader, *ozha, vaidhya, hakims,* priests, teacher, physicians,

veterinary, doctor, social worker, postal authorities and Ayurvedic doctors etc.

The ethnobotanical survey includes plant collections, informal interviews with the inhabitants and the implementation of the ethnographic method of free listing and semi-structured interviews (Weller and Romney, 1998) or it may be done by following the methodology of Cotton (1996).The survey may involve individuals like :

1. Village head man and old experienced person/village resource persons.
2. Middle men or interpreter.
3. Men and women working in the field
4. Men and women in weekly markets and in other common places.
5. Local forest officers.
6. Baidyas, Sirahas, Guniyas, Gaytas and knowledgeable women.
7. Traditional birth attendant (Dai).
8. Traditional bone setter (haddi baidh) and Maatipujari of village.

Documentation of the Information

The documentation of the information obtained from folk healers and baidyas may be documented under following heads:

1. Name of the disease.
2. Name of the plant
3. Vernacular name.
4. Family name
5. Plant part used.
6. Habit of the plant
7. Other ingredients used.
8. Drug preparation.
9. Drug application and its effectiveness.
10. Cost of treatment/episode.

Information may be documented disease wise use of plants/plant parts for the preparation of drug. The cost of the treatment per episode/ per disease should be noted. The information may be cross checked by the other informants in the same locality for verification of medicinal claims by the first informants. This practice resulted credibility of data regarding the use of plants and statements of informants.

Quantitative Techniques in Ethnobotany

Quantitative techniques have been used in ethnobotany to compare the uses and the cultural importance of different plant taxa. These analyses are of great scientific interest as they reflect cultural value systems, and they may also aid in the conservation of biodiversity (Byg & Balslev 2001). It is expected that people will be motivated to con-serve resources that are most important to them, in con-trast to resources perceived as less useful (Byg & Balslev 2001, Garibaldi & Turner 2004).

The use of quantitative techniques to evaluate the relative importance of plants in a given culture is common in ethnobotanical literature. Ever since the publication of the Use-Value index proposed by Phillips and Gentry (1993a, 1993b) (modified from Prance *et al.* 1987), similar approaches have been widely used by many authors (Albuquerque *et al.*, 2005a, Cunha & Albu-querque 2006, Galeano 2000, Gomez-Beloz 2002, Kris-tensen & Balslev 2003, Kvist *et al.*, 2001, Torre-Cuadros & Islebe 2003). The most popular techniques (indices) are based on "informant consensus" - the degree of agree-ment among the different people interviewed concerning the use of a given resource (*e.g.*, Byg & Balslev 2001). Numerous authors have applied these techniques to investigate the impact of exploitation of locally important resources, based on the supposition that more important a resource is, the greater will be the exploitation pressure placed upon it. Although these interpretations have sometimes been questioned (Albuquerque & Lucena 2005, Silva & Albuquerque 2004), neither their use as tools for evaluating the importance of a given re-source, nor their limitations or scope, have been critically examined. It is necessary now for "studies that assess the reliability of the different indices queue presumably proxy for the same phenomena" (Reyes-García *et al.*, 2006). Informants' Consensus Factor, Fidelity level, Use value and the technique of Relative Importance are some of the important quantitative techniques which are being used in ethnobotanical studies.

Informants' Consensus Factor

To estimate use variability of the medicinal plants and to determine which plants are particularly interesting in the search for bioactive compounds, the Informant Consensus Factor (*F*ic) (Heinrich et al., 1998a) is calculated. This factor estimates the relationship between the "number of use-reports in each category (*n*ur) minus the number of taxa used (*n*t)" divided by the "number of use-reports in each category minus 1".

Ethnobotanical techniques, pointed out that proce-dures based on "informant consensus" tend be more ob-jective as they are designed to eliminate investigator bias in attributing relative importance to a given plant (Phillips 1996). The level of homogeneity among information provided by different informants was calculated by the Informants' Consensus Factor, FIC (Trotter and Logan, 1986) using the following formula:

$$FIC = Nur - Nt / (Nur - 1)$$

Where, Nur = number of use reports from informants for a particular plant-use category; Nt = number of taxa or species that are used for that plant use category for all informants. The product of this factor ranges from 0 to 1. A high value (close to 1) indicates that relatively few taxa (usually species) are used by a large proportion of people, while a low value indicates that the informants disagree on the taxa to be used in the treatment within a category of illness.

Fidelity level

The fidelity level (FL) is the percentage of informants claiming the use of a certain plant species for the same major purpose, is calculated for the most frequently reported diseases or ailments as: $FL (\%) = (Np / N) \times 100$.

Where, Np = number of informants that claim a use of a plant species to treat a particular disease; N = number of informants that use the plants as a medicine to treat any given disease (Alexiades, 1996).

Use value

The use value (UV) of the plant species is calculated following an adaptation of the Phillips and Gentry method, as created by Rossato et al. and Silva et al. using the following formula:

$$UV_{is} = \Sigma U_{is}/n_{is}$$

Where UV_{is} = the use value of the species s mentioned by the informant i; ΣU_{is} = the number of uses of species s mentioned in each event by the informant i; n_{is} = the number of events in which the informant i cited species.

The Technique of Relative Importance

The technique of Relative Importance (RI), proposed by Bennett and Prance (2000), was developed primarily for measuring the usefulness of medicinal plants. The RI value is derived from the number of indications (of pharmacological properties) for that species and from the number of ailments that it is used to treat. As such, the importance of a species increases if it is used to treat more infirmities. As this technique was conceived, it would be possible to calculate the Relative Importance of a medicinal plant based only on secondary sources.

The various parameters for identification, evaluation and standardization of medicine obtained from ethnobotanical sources require following all or some steps mentioned hereunder.

Pharmacological Evaluation

Preparation of Powder and Reagents: The whole plant/parts is collected and is dried under shade. These dried materials are mechanically powdered after keeping them in an oven at 35^{0}C for 1-7days. These powdered materials are used for further physico-chemical, fluorescent and phyto-chemical analyses. Usually the procedure recommended in Indian Pharmacopoeia (Anonymous, 1966; 1985; 1996) are followed for the determination of loss on drying at 110°C, total ash, acid insoluble ash, sulphated ash, water soluble ash , Na^+ and K^+ content. Separately the plant materials are grind to course powder and pass through 80 µm mesh sieve. These powdered materials are used for pharmacological evaluation.

Loss on drying: The powdered samples are dried in an electrical oven at 105^0 -110^{0}C until reaches a constant weight.

Determination of pH of Aqueous Solution: The powdered materials (80 ìm mesh) are suspended in glass filled with distilled water. After 2hrs, filtered and the clear solution is measured for pH.

Total Ash Value: 5 gm of plant powder is ignited in an electric furnace at 500-550^{0}C in silica crucible until the sample reaches a constant weight.

Sulphated Value: The ash powder is moistened with 1 ml of H_2So_4 and ignited to 800°C until reaches a constant weight.

Water Soluble Ash Value: The water insoluble matter is collected in an ash less filter paper and ignited in an electric furnace at 450^{0}C in silica crucible until reaches a constant value. The weight of insoluble matter is subtracted from the weight of the total ash to indicate the weight of water soluble ash.

Acid Insoluble Ash Value: Total ash obtained is heated with addition of 25 ml of dilute HCl for 10 min. It is filtered in an ash less filter paper (Whatman No. 41) and the residue is ignited in the furnace to get a constant weight.

Extraction Methodologies

Extraction is one of the crucial steps in the analysis of medicinal plants, because it is necessary to extract the desired chemical components from the plant materials for further separation and characterization. The basic operation included steps, such as pre-washing, drying of plant materials or freeze drying, grinding to obtain a homogenous sample and often improving the kinetics of analytic extraction and also increasing the contact of sample surface with the solvent system. Proper actions must be taken to assure that potential active constituents are not lost, distorted or destroyed during the preparation of the extract from plant samples. The selection of solvent system largely depends on the specific nature of the bioactive compound being targeted. Different solvent systems are available to extract the bioactive compound from natural products. The extraction of hydrophilic compounds uses polar solvents such as methanol, ethanol or ethyl-acetate. For extraction of more lipophilic compounds, dichloromethane or a mixture of dichloromethane/methanol in ratio of 1:1 are used. In some instances, extraction with hexane is used to remove chlorophyll (Cos *et al.,* 2006).

As the target compounds may be non-polar to polar and thermally labile, the suitability of the methods of extraction must be considered. Various methods, such as sonification, heating under reflux, soxhlet extraction and others are commonly used (United States

Pharmacopeia and National Formulary, 2002; Pharmacopoeia of the People's Republic of China, 2000; The Japanese Pharmacopeia, 2001) for the plant samples extraction. In addition, plant extracts are also prepared by maceration or percolation of fresh green plants or dried powdered plant material in water and/or organic solvent systems.

The other modern extraction techniques include solid-phase micro-extraction, supercritical-fluid extraction, pressurized-liquid extraction, microwave-assisted extraction, solid-phase extraction, and surfactant-mediated techniques, which possess certain advantages, such as the reduction in organic solvent consumption and in sample degradation, elimination of additional sample clean-up and concentration steps before chromatographic analysis, improvement in extraction efficiency, selectivity, and/kinetics of extraction. The ease of automation for these techniques also favours their usage for the extraction of plants materials (Huie, 2002).

Extractive Values

The material is weighed in a thimble and it is placed in a Soxhlet apparatus. Hexane is taken in the round bottom flask and hot extraction is carried out for 24 hrs. The extract in the round bottom flask is concentrated by distillation and the dry extract is weighed to get the hexane soluble fraction. The mare was used for successive extraction with chloroform and methanol. The percentage solubility in each case is calculated with reference to the powdered plant material (stem bark, root bark and leaves) taken initially. 80 μm mesh powders of plants are extracted sequentially (Soxlet) with alcohol. The water extracts are purified separately. The extracts are evaporated under vacuum. Extractive value (%W/W) of plant materials in water and alcohol are calculated separately.

Fluorescence Analysis

Fluorescent characteristics of the plant powders as such and after treating them with chemical reagents are observed in day light as well as under UV radiation. Fluorescent analysis of all the plant powders are carried out according to the methods of Chase and Pratt (1949) and Kokoshi *et al.* (1958). Behaviour of powdered plant materials with different chemical reagents is carried out as described by Kay (1938) and Johansen (1940).

Histochemical and Anatomical Methods

Fixation: The required samples of different organs are cut and removed from the plant and fixed in FAA [Formalin (5 ml) + Acetic acid (5 ml) + 70 %Ethyl Alcohol (90 ml)] as per the schedule given by Sass (1940). Infiltration of the specimens is carried out by gradual addition of paraffin wax (melting point 58-60°C) Tertiary Butyl Alcohol (TBA) until TBA solution attained super saturation. The specimens were cast into paraffin blocks.

Sectioning: The paraffin embedded specimens are sectioned with the help of Rotary Microtome. Usually the thickness of the sections should be 10-12 ìm. De-waxing of the sections is done by customary procedure (Johansen, 1940). The sections are stained with Toluidine blue as per method published by O'Brian *et al* (1964). Since Toluidine blue is a polychromatic stain, the staining results remain remarkably good and some cytochemical reactions are also obtained. The dye rendered pink colour to the cellulose walls, blue to the lignified cells, dark green to suberin, violet to the mucilage, blue to the protein bodies etc. Wherever, necessary sections are also stained with safranin and Fast-green and KI (for starch).

For studying stomatal morphology, venation pattern and trichome distribution, paradermal sections (sections taken parallel to the surface of leaf) as well as clearing of the leaf with 5 % sodium hydroxide or epidermal peeling by partial maceration employing Jeffrey's maceration fluid (Sass, 1940) are employed. Glycerin mounted temporary preparations are made for macerated materials. Powdered materials of different parts are cleared with NaoH and mounted in glycerine medium after staining. Different cell components are studied and measured.

Photomicrographs: Microscopic descriptions of tissues are supplemented with micrographs wherever necessary. For normal observations bright field is used. For the study of crystals, starch grains and lignified cells, polarized light is used as they appear bright against dark background. Magnification of the figures is indicated by the scale bars.

Phytochemical Screening

The phytochemical investigation of a plant involves extraction of the plant material, separation and isolation of the constituents of interest, characterization of the isolated compounds and their quantitative evaluations.

Table 1. Priliminary phytochemical screening method

Tests	Reagents used	Colour formed	Sign
1.Tests for alkaloids			
a.Mayer's tests	Potassium mercuric iodide solution	Cream precipitate	+
b.Wagner's tests	Iodine potassium solution	Brown precipitate	+
c.Hager's tests	Saturated solution of picric acid	Yellow colour	+
d.Dragendarff's tests	Potassium bismuth iodide solution	Raddish brown precipitate	+
2.Tests for amino acids			
a.Millon's tests	Millon reagents	White precipitate	+
b.Ninhydrine tests	Ninhydrin solution	Violet colour	+
3.Tests for carbohydrates			
a.Molisch's tests	Alcoholic a naphthol+sulphuric acid	Purple to violet colour rings	+
b.Barfoed's tests	Barfoed reagents	Red colour(monosaccharide) after 10 min.colour form(disaccharide)	+
c.Selivanoff's tests	Selivanoff's reagents	Rose colour(ketone)	+
d.Tests for pentoses	Hydrochloric acids + phloroglucinol	Red colour	+
4. Test forAnthraquinone			
Borntrager's test	1ml 10% ferric chloride solution + 1ml of Concentrated hydrochloric acid.	Pink or deep red coloration of aqueous layer	+
5. Cardiac glycosides			
Kellar – Kiliani test	1ml of glacial acetic acid,+1ml ferric chloride +1ml concentrated sulphuric acid.	Green-blue coloration of solution	+
6.Flavonoid			
Shinoda test	A piece of magnesium ribbon +1ml ofconcentrated hydrochloric acid.	Pink red or red coloration	+
7. Steroid			
Liebermann-Burchardt test	1ml of chloroform+ 2 -3ml of aceticAnhydride+ 1 to 2 drops of concentrated sulphuric acid	Dark green coloration	+
8.Tannin			
Braemer's test	10% alcoholic ferric chloride will be added to 2 - 3ml of methanolic extract(1:1)	Dark blue or greenish grey coloration of the solution	+
9. Terpenoid			
Liebermann-Burchardt test	1ml of chloroform, 2-3ml of acetic anhydride, 1 to 2 drops of concentrated sulphuric acid	Pink or red coloration	+
Salkowski test	2 ml of chloroform + 3 ml of concentrated sulphuric acid H_2SO_4	Reddish brown color of interface	+

Separation Techniques

Qualitative chromatography: Chromatography is a method of separation based on adsorption, repartition and ion exchange. It brings supplementary information about chemical composition. Different types of chromatographic techniques like TLC, HPLC, and GC are used to know about various phytochemicals.

Thin Layer Chromatography (TLC)

TLC is a simple, quick, and inexpensive procedure that gives the researcher a quick answer as to how many components are in a mixture. TLC is also used to support the identity of a compound in a mixture when the Rf of a compound is compared with the Rf of a known compound. Additional tests involve the spraying of phytochemical screening reagents, which cause color changes according to the phytochemicals existing in a plants extract; or by viewing the plate under the UV light. This has also been used for confirmation of purity and identity of isolated compounds.

Bio-autography is a useful technique to determine bioactive compound with antimicrobial activity from plant extract. TLC bioautographic methods combine chromatographic separation and *in situ* activity determination facilitating the localization and target-directed isolation of active constituents in a mixture. Traditionally, bioautographic technique has used the growth inhibition of microorganisms to detect anti-microbial components of extracts chromatographed on a TLC layer. This methodology has been considered as the most efficacious assay for the detection of anti-microbial compounds (Shahverdi, 2007).

High Pressure Liquid Chromatography (HPLC)

This technique is used both for identification and for assay. For example; HPLC technique is used as an identification test in the case of the herbal drugs extracted from *Echinaceae angustifoliae radix, E. pallidae radix, E. purpureae folium*. High performance liquid chromatography (HPLC) is a versatile, robust, and widely used technique for the isolation of natural products (Cannell, 1998). Currently, this technique is gaining popularity among various analytical techniques as the main choice for fingerprinting study for the quality control of herbal plants (Fan *et al.,* 2006). Natural products are frequently isolated following the evaluation of a relatively crude extract in a biological assay in order to fully characterize the active entity. The biologically active entity is often present only as minor component in the extract and the resolving power of HPLC is ideally suited to the rapid processing of such multi-component samples on both analytical and preparative scale.

Gas Chromatography-Mass Spectroscopy

Gas chromatography equipment can be directly interfaced with rapid scan mass spectrometer of various types. The flow rate from capillary column is generally low enough that the column output can be fed directly into ionization chamber of MS. The simplest mass detector in GC is the Ion Trap Detector (ITD). In this instrument, ions are created from the eluted sample by electron impact or chemical ionization and stored in a radio frequency field; the trapped ions are then ejected from the storage area to an electron multiplier detector. The ejection is controlled so that scanning on the basis of mass-to-charge ratio is possible. The ions trap detector is remarkably compact and less expensive than quadrapole instruments. GC-MS instruments have been used for identification of hundreds of components that are present in natural and biological system (Oleszek & Marston, 2000; Philipson, 2007; Daffre *et al.,* 2008).

Other techniques

NMR studies: A nuclear magnetic resonance spectrum gives the largest amount of information about the structure of a compound. In NMR Spectroscopic method, a substance is placed in a strong magnetic field that affects the spin of the atomic nuclei. A radio wave passes through the substance, and reorients these nuclei. When the wave is turned off, the nuclei release a pulse of energy that provides data on the molecular structure of the substance and that can be transformed into an image by computer techniques.

Immunoassay: Immunoassays, which use monoclonal antibodies against drugs and low molecular weight natural bioactive compounds, are becoming important tools in bioactive compound analyses. They show high specificity and sensitivity for receptor binding analyses, enzyme assays and qualitative as well as quantitative analytical techniques. Enzyme-linked immunosorbent assay (ELISA) based on MAbs are in many cases more sensitive than conventional HPLC methods. Monoclonal antibodies can be produced in specialized cells through a technique known as hybridoma technology (Shoyama *et al.*, 2006).

Fourier-transform infrared spectroscopy (FTIR)

FTIR has proven to be a valuable tool for the characterization and identification of compounds or functional groups (chemical bonds) present in an unknown mixture of plants extract (Eberhardt *et al.*, 2007; Hazra *et al.*, 2007). In addition, FTIR spectra of pure compounds are usually so unique that they are like a molecular "fingerprint". For most common plant compounds, the spectrum of an unknown compound can be identified by comparison to a library of known compounds. Samples for FTIR can be prepared in a number of ways. For liquid samples, the easiest is to place one drop of sample between two plates of sodium chloride. The drop forms a thin film between the plates. Solid samples can be milled with potassium bromide (KBr) to and then compressed into a thin pellet which can be analyzed. Otherwise, solid samples can be dissolved in a solvent such as methylene chloride, and the solution then placed onto a single salt plate. The solvent is then evaporated off, leaving a thin film of the original material on the plate.

Efficient standardisation and quality control of multiherbal formulations can be achieved by adopting a six point protocol consisting of:

a) Standardisation and authentication of ingredients.

b) Sop Development for Manufacturing Process of Formulations.

c) Triple 'P' Based Screening Coupled With Residual Analysis of Formulations.

d) Stability (Shelf Life) Studies

e) Efficacy

f) Safety

Antimicrobial Studies

The concentrate of all the extracts and isolated compounds are tested for antimicrobial activities against human pathogens. For antimicrobial studies a media is required depending upon the type of organism.

Bacterial Media

Muller Hinton Media (Hi-Media) is mixed with distilled water and then sterilized in autoclave at 15 lb pressure for 15 minutes. The sterilized media are poured in to petridishes. The solidified plates are bored with 5 mm diameter cork borer. The plates with wells are used for antibacterial studies.

Fungal Media (PDA)

200g of potato slices are boiled with distilled water. The potato infusion is used as water solute of media preparation. 29g of dextrose is mixed with potato infusion. 20g of agar is added as solidifying agent. These constituents are mixed and autoclaved. The solidified plates are bored with 6mm diameter cork borer.

Bacterial Strains

The bacterial and fungal pathogenic strains are obtained from any Institute which provides bacterial strain or these strains may be cultured.

Anti bacterial / antifungal activity of the plant extract

The aqueous extract or extracts in any other organic solvents of whole plants or any specific part may be taken. Different concentrations of these extracts are tested against different bacterial/ fungal pathogens for their antimicrobial activity. It is demonstrated by well diffusion assay.

Well Diffusion Method

Antibacterial and antifungal activity of plant extract are tested using well diffusion method (Banel *et al.,* 1996). The prepared culture plates are inoculated with different strains of bacteria and fungi using streak plate method. Wells are made on the agar surface with 5mm cork borer. The extracts are pound into the well using sterile syringe. The plates are incubated at 37+2°C for 48 hrs for fungal activity and for 24 hrs for bacterial activity. The plates are observed for zone formation around the wells.

The zone of inhibition is calculated by measuring the diameter of the inhibition zone around the well (in mm) including the well diameter. The readings are taken in three different fixed directions in all three replicates and the average values are tabulated.

Identification, evaluation and standardization of herbal drugs: The various parameters used for identification and evaluation are described in Table 2.

Table 2. Various parameters used for identification and evaluation.

Methods	Evalution Parameters
1. Authentication	A. Parts of plants collected like leaf, flower, root, stolen
	B. Regional status
	C. Family
	D. Biological source
	E. Chemical constituents
2. Morphology or Organoleptic evaluation	A. Odour
	B. Taste
	C. Size
	D. Shape
	E. Special feature
3. Microscopic evaluation	A. Leaf content
	B. Trichomes
	C. Stomata
	D. Quantitative microscopy
4. Chemical evaluation	A. Chemical test
	B. Chemical assay
	C. Phytochemical screening
5. Physical evaluation	A. Moisture content
	B. Viscosity
	C. Melting point
	D. Solubility
	E. Optical rotation
	F. Refractive index
	G. Ash value
	H. Extractive value
	I. Volatile oil content
	J. Foreign matter etc
6. Biological evaluation	A. Microbial contamination
	B. Pesticides contamination
	C. Pharmacological activity of drugs

Identification of herb is based on macroscopical and microscopical features. Macroscopical feature involves odour, taste, color, size, shape and special feature of plant and microscopical involves leaf content, trichome, stomata *etc.* Certain microscopical features and chemical test comes under evaluation and standardization of herbal drug. The three major ways followed for the standardization of herbal drugs are pharmacogonostical, physicochemical and phytochemical. Evaluation of drugs means confirmation of its identity and determination of its quality and purity and detection of adulteration.

Standard expression is used to describe all measures which are taken during the manufacturing process and quality control leading to a reproducible quality. It also involves the study from birth of plant to its clinical application. It also includes the herbal drugs preparation to a define content of a constituent or a group of substance with known therapeutic activity respectively by addition of excipients or by mixing herbal drugs preparation. In other words it ensures that every packet of medicine has correct ingredient in correct amount and will induce intended therapeutic effect.

Authentication

In India, two government organizations i.e. Central Council for Research in Unani Medicine (CCRUM) and Central Council for Research in Ayurvedic Medicine(CCRAM) are working for quality control in authentication the plant material collected from an appropriate region of the country at an appropriate stage of its growth. Authentication is done by details taxonomical study.To ensure and enhance the quality of herbal medicines, the Government of India(GOI) has notified Good Manufacturing Practices (GMP) under Schedule 'T' of the Drugs and Cosmetics Act 1940 which also ensures raw materials used in the manufacture of drugs are authentic, of prescribed quality and are free from contamination. The guidelines for Good Agricultural Practices (GAP) seek to lay down a cultivation programme designed to ensure optimal yield in terms of both quality and quantity of any crop intended for health purposes. In recent time, there has been great demand for plant derived products in developed countries.

These products are increasingly being sought out as medicinal products, nutraceuticals and cosmetics. There are around 6000 herbal manufacturers in India. More than 4000 units are producing ayurvedic medicines. World Health Organization (WHO) gives guidelines for the herbal standardization and evaluation.

Evaluation Parameters of Herbal Drugs

Macroscopic evaluation

In this methods, description, general condition of the drug size, shape outer surface inner surface are referred. A sensory or organoleptic character describes colour, odour taste, and consistency e.g. the fractured surface in cinchona, quillia, cascara barks and quassia wood are important characteristics. The ovoid tears of gum acacia, ribbon shaped characterizes of tragacanth, disc shaped structure of nux vomica, concial shape of aconite and quills of cinchona, are evaluated by this method.

Microscopic evalution

The inner pseudoparenchyma cells are oval or rounded, they contain fixed oil & protein. The whole tissue is devoid of cellulose and lignin. Various parameters included in microscopy are as follows :

a) Leaf content

b) Trichome

c) Stomata

Determination of leaf content

This includes parameter like stomatal number, stomatal index, vein islet number, vein termination number etc.

Stomatal number and stomatal index

There are several types of stomata, distinguished by the forms and arrangement of the surrounding cells.

Anomocytic (irregular - cell)

Anisoytic (unequal - cell)

Diacytic (cross- cell)

Paracytic (parallel cell)

Determination of stomatal index

The stomatal index is the percentage of the number of stomata formed by the total number of epidermal cells including the stoma being counted as one cell. This is done by placing leaf fragment of about 5x5mm in size in a test tube containing about 5ml chloral hydrate.Number of stomata and epidermal cell are counted.

Stomatal index= S / E+S×100

Where: S = Total number of stomata in a given area of leaf E= Number of epidermal cells (including trichomes) in the same area of leaf.

Physical Evaluation

Determination of foreign matter: Drugs should be free from moulds, insects, animal, faecal matter and other contamination such as earth stones and extraneous matters.

Percentage of foreign organic matter = n × W × 94, 100 × 100/S × M × P

Where: n = number of chart particles in 25 fields. S = number of spores in the same 25 field. W = weight in mg of *Lycopodium* taken.

M = weight in mg of the sample (calculation on the sample dried at 105^{0}C.

P = number of characteristics particles per mg of the pure foreign matter.

94,000 = number of spores per mg of *Lycopodium*.

Determination of total ash

The residue remaining after incineration is the ash content of drugs, which simply represents inorganic salts, naturally occurring in drugs or adhering added to it as form adulteration.

Two types of ash are determined :

i) Acid insoluble ash value.

ii) Determination of water soluble ash.

Determination of extractive value

i) Determination of alcohol soluble extractive

ii) Determination of water soluble extractive.

Determination of moisture content

10 gm of drug is taken in a evaporating dish. Then it is dried at 105°C for 3 hours and again weighed. Drying and weighing are continued at one hour interval until difference between two successive weighing corresponds to not more than 0.25 percent. The reading is taken after a constant weight is reached and the moisture content is determined.

Determination of specific optical rotation

Formula for obtaining Specific rotation D25 = 100 × Ψlc Where: Ψ = corrected observed rotation in drug at-25°

D = d line of sodium light

l = length of the polarimeter tube in dome.

c = concentration of substance in percent w/v.

Determination of pH: The pH value of an aqueous liquid may be defined as the common-logarithm of the hydrogen ion concentration expressed in grams. Potentiometrically pH value is determined by a glass electrode and a suitable pH meter.

Solubility: The presence of adulterant in a drug could be indicated by solubility studies which is identified by various solvents.

Conclusion

The above mentioned various techniques and methodologies are used frequently by ethnobiologist across the world for ethnobotanical studies. However it is impossible for the author to bring all the techniques and methodology together in a single paper.

References

A.P. Cox, J.M. Balick. 1996. Ethnobotanical Research and traditional Health Care in Developing Countries: Plants, People and Culture. W.H. Freeman and Co.

Biochemistry Assay for Bioactivity, 6. Academic Press, London, pp. 47–69.

C′aceres, A., Gir′on, L.M., Alvarado, S.R., Torres, M.F. 1987. Screening of antimicrobial activity of plants populary used in Guatemala for the treatment of dermatomucosal diseases. *Journal of Ethnopharmacology* 20: 223–237.

Canalesa, M., Hernandez, T., Caballero, J., Vivor, A.R.D., Avila, G., Duran, A. and Lira, R. 2005. Informant consensus factor and antibacterial activity of the medicinal plants used by the people of San Rafael Coxcatlan, Puebla, Mexico. *J. Ethnopharmacol.* 97: 429-439.

Cox, P.A., Balick, M.J. 1994. The ethnobotanical approach to drug discovery. *Scientific American* 270: 60–65.

Cunha, L.V.F.C. & U.P. Albuquerque. 2006. Quantitative ethnobotany in an atlantic Forest fragment of northeast-ern Brazil-implications to conservation. *Environmental Monitoring and Assessment.* 114: 1-25.

Edeoga, H.O., Okwu, D.E. and Mbaebie, B.O. 2005. Phytochemical constituents of some Nigerian medicinal plants. *Afr.J. Biotechnol.* 4: 685-688.

Fan, X.H., Cheng, Y.Y., Ye, Z.L., Lin, R.C. and Qian, Z.Z. 2006. Multiple chromatographic fingerprinting and its application to the quality control of herbal medicines. *Anal. Chim.* Acta 555: 217-224.

Flaster, T. 1996. Ethnobotanical approach to drug discovery of bioactive compounds. Process in new crops. *Proceedings of the Third Traditional Symposium, ASHS press, Alexandria,* pp. 561-565.

Heinrich, M. 2000. Ethnobotany and its role in drug development. *Phytotherapy Research* 14: 479-488.

Huie, C.W. 2002. A review of modern sample-preparation techniques for the extraction and analysis of medicinal plants. *Anal. Bioanal. Chem.* 373: 23-30.

J. Friedman, Z. Yaniv, A. Dafni, D. Palewitch. 1986. *Journal of Ethnopharmacology.* 16: 275-287.

Katewa S.S. 2009. Indegenous people and Forests: Perspectives of an Ethnobotanical study from Rajasthan (India)-Herbal Drugs: Ethnomedicine to Modern Medicine (Springer, Berlin) 33-56.

M Ali. 2009. "Pharmacognosy and Phytochemistry" 3rd ed., CBS Publishers & Distributors, Delhi, 181-182.

Monteiro, J.M., U.P. Albuquerque, E.M.F. Lins-Neto, E.L. Araújo & E.L.C. Amorim. 2006a. Use patterns and knowl-edge of medicinal species among two rural communities in Brazil's semi-arid northeastern region. *Journal of Ethnopharmacology* 105: 173-186.

Nargas J. and Trivedi P.C. 1999. Traditional and medicinal importance of *Azadirachta indica* in India. *Journal of Economic and Taxonomic Botany* 23: 33-37.

Phillips, O., A.H. Gentry, C. Reynel, P. Wilkin & C. Gálvez-Durand. 1994. Quantitative ethnobotany and Amazonian conservation. *Conservation Biology* 8: 225-48.

Reyes-Garcia, V., V. Vadez, T. Huanca, W. Leonard & D. Wilkie. 2005. Knowledge and consumption of wild plants: a comparative study in two Tsimane' villages in the Boliv-ian Amazon. *Ethnobotany Research & Applications* 3: 201-207.

R.T. Trotter, M.H. Logan. Informant consensus. A New Approach for identifying potentially effective medicinal plants. 1986. *In:* N. L Etkin, ed., Plants in indigenous medicine and diet Rdgrave, Bedford Hills, New York.

Silva V.A.S., Nascimento V.T., Soldati G.T., Medeiros M.F.T., Albuquerque U.P. 2014. Techniques for analysis of quantitative ethnobiological data: use of indices. In Methods and techniques in Ethnobiology and Ethnoecology. Edited by Albuquerque U, da Cruz Cunha L, Lucena R, Alves R. New York: Springer; 379–395.

S. Shrikumer; U. Maheshwari; A. Sughanti; T.K. Ravi. 2006. WHO guidelines for herbal drugs standardization.

Tyler, V.E. 2000. Herbal medicine: From the past to the future. *Public Health Nutr.*, 3: 447–452 WHO (1993) *Research Guidelines for Evaluating the Safety and Efficacy of Herbal Medicines*, Manila.

Trotter, R.T., III, and M.H. Logan. 1986. Informant consensus: A new approach for identifying potentially effective medicinal plants. In *Plants in indigenous medicine and diet: Biobehavioral approaches*, ed. N. L. Etkin, 91–111. Bedford Hills, NY: Redgrave.

Vogl, C.R., B. Vogl-Lukasser, and R.K. Puri. 2004. Tools and methods for data collection in ethnobotanical studies of homegardens. *Field Methods* 16: 285–306.

Vanden Berghe, D.A., Vlietinck, A.J., 1991. Screening methods for antibacterial and antiviral agents from higher plants. In: Dey, P.M.,

Harborne, J.B., Hostettmann, K. (Eds.), Methods in plant Biochemistry Assay for Bioactivity, 6. Academic Press, London, pp. 47–69.

Weller, S.C., Romney, A.K. 1998. Systematic Data Collection. Qualitative Research Methods Series, 10. SAGE. Publications, Newbury Park, USA.

□□□

Chapter 10

Ethnobotany in Andaman and Niobar Islands Prospects and Challenges

F.H. Hashmi

Assocaite Professor, P.G. Dept. of Plant Science, Jawaharlal Nehru Rajkeeya Mahavidyalaya, Port Blair, Andaman and Nicobar Islands

ABSTRACT

Ethnobotany deals with the traditional knowledge of the indigenous people regarding the use of healing property of plants and other bioresources. It is now a multidisciplinary subject including taxonomy, pharmacognosy, phytochemistry, ecology and conservation biology. Andaman and Nicobar Islands are homeland of 6 aboriginal tribes and about 11 ethnic communities. They are repository of traditional knowledge of healing (TKH). Their knowledge if tapped properly, economic bioprospecting can generate jobs and revenue in these islands. To achieve this goal, modern scientific facilities to isolate the active compounds and to determine their mechanism of action as well as facility to validate the end product should be established in the Andaman and Nicobar islands. Legal frame works will also be required to protect the intellectual property rights(IPR) of indigenous people and to check the biopiracy. This article aims to highlight the current prospects of ethnobotany and issues related to challenges faced by ethnobotany in these islands.

Introduction

Plants have been used as a mean to cure the various ailments since the time immemorial human civilisation has started and has been acquired by trial and error over millennia. The indigenous knowledge on medicinal plants appears to have accumulating when humans started and learned how to use the traditional knowledge of medicinal plants and subsequently referred as ethnobotany which is the study or comparison of the traditional medicine practised by various ethnic groups and especially by indigenous peoples. Ethnobotany is not limited only to plant but also includes algae, lichens and fungi. It is strongly linked with taxonomy, pharmacognosy, phytochemistry, ecology and conservation biology. The Andaman and Nicobar Islands are the homeland of six aboriginal tribes and about eleven ethnic populations. Aboriginal tribes living in the Andaman Islands belong to negrito stock. They are dark in colour, short in stature with peppercorn hairs and include the Great Andamanese , the Onges, the Jarawas and the Sentinelese. The Nicobarese and the Shompens belonging to mongoloid race inhabit the Nicobar group of Islands. It is believed that the negritos in the Andamans is living since last 2,000 years but their origin cannot be explained with certainty due to the absence of sufficient archaeological evidences. It has been stated that the descendant of the first man who moved out of Africa nearly 70,000 years ago took the southern coastal route to India to Southeast Asia and Australia. This assumption is based on a genetic study of the mitochondria of indigenous tribes of the Andamans.

The Andaman and Nicobar Islands

The Andaman and Nicobar Islands are also known as Bay Islands are the Union Territory of India and is the largest archipelago system in the Bay of Bengal, comprising of about 572 emerald islands, rocks and rock outcrops (islets). They are situated between 6°45'-13°34'3" N and 92°10'-94°15' E. The total geographical area is 8,249 Km^2 with a coastline of 1,962 km. They are lying in North-South direction simulating an arc stretching over a length of about 912 km with maximum width of 57 km. The northern group of islands called the Andaman group, have a total area of 6,408 km^2 comprising a total length of 467 km and maximum width of 52 km with average width of 4 km, while Nicobar groups have an area of 1,841 km^2 comprising a length of 259 km and maximum width of 57 km. The Ten Degree channel with a width of 155 km separates the two groups of Islands. Two Islands of volcanic origin, the Barren and the Narcondum Islands are found here. This archipelago is separated at a distance of 1,200 km from the East coast of mainland India. The nearest landmass in the north is Myanmar, roughly 280 km north of Landfall Island - the northern most Island in the Andaman Group. The closest landmass to the Great Nicobar Island is Sumatra, about 145 km south. Mythologically, the name Andaman is presumed to be derived from 'Hanuman', who was known to Malays as "Handuman". These Islands are described as islands of the "Marigold Sun", were known throughout the country as "Kalapani" because of their having been a penal settlement under British Rule, a reputation fast-shedding since independence. There are 38 islands which have human population. The Andaman group has 26 inhabited islands while Nicobar group has 12 inhabited islands.

The climate of these Islands is a typical tropical, warm and humid, as they are situated in the equatorial belt. The mean temperature ranges between 18° and 35°C. They receive heavy rainfall from both Southwest monsoon from May to September and Northeast monsoons from October to December with the average annual rainfall ranging from 3000 to 3500 mm. Cyclonic winds accompanied by thunder and lightning are very frequent. January to March have fairly dry weather with scanty rainfall. The mean relative humidity usually remains between 66 and 85% throughout the year.

The tropical hot and humid climate with abundant rains supports very luxuriant and rich vegetation. According to Champion & Seth (1968), the vegetation of union territory may be broadly classified into (i) Beach forests, (ii) Mangrove forests, (iii) Wet evergreen forests, (iv) Semi-evergreen forests, (v) Moist deciduous forests and (vi) Grasslands. According to 'India State of Forest Report 2015' (FSI, 2015), the total forest cover of Andaman and Nicobar Islands is 6,751 km^2, which occupies 81.84% of the total geographical area of the Islands. The most fascinating feature of this hotspot floral bounty is the high degree of endemism and mixed type of taxa associated with aborigines.

Aborigines and Ethnic Population of Andaman and Nicobar Islands

The total population of the Islands is 3,80,581 comprising 2,02,871 males and 1,77,710 females with the growth rate of 6.86 per cent as per 2011 census. Despite the increase in the general population, the population of tribal people decreased from 29,469 to 28,530 as compared to 2001 Census, registering a decrease of 3.19%. These islands are the homeland of six aboriginal tribes, inhabiting for thousands of years in the verdant tropical rain forests and about eleven other ethnic population which were either brought by British Ruler as convicts or as a labourer. Some other had been brought by Government of India as settlers after independence. Aborigines living in the Andaman Islands are short in stature, dark skin colour with peppercorn hair and identify themselves as negrito (a sub-group of negriod people). They are the 'Great Andamanese', the 'Onges', the 'Jarawas' and the 'Sentinelese'. They are however, quite distinct from the African Tribes. Each of this tribe has a different dialect and conventional territory division. However, when these tribes did come to the islands, in general or how other Negrito tribes populated the Andaman in particular, cannot be said with certainty due to the absence of conclusive paleontological evidences. In spite of their close proximity to certain foreign countries and lying close to the main commercial routes in the Bay of Bengal, their continuing primitive ways of life remains a riddle. The Nicobar group of Islands are inhabited by the mongoloid group of tribes, the Nicobarese and the Shompen having yellowish brown complexion, flat nose, broad lips with epicanthic fold in eyes and straight type of hair. It is indubitably true that for centuries they had remained isolated from other cultural influence and had in the process acquired an absolutely distinct identity. Nicobarese are the only aboriginals in these islands who continue to sustain themselves with vigour and vitality and have been flourishing as a vibrant ethnic group. The other eleven ethnic groups are 'Bangali', 'Bhantu', 'Burmese', 'Coorgi', 'Karen', 'Madarasi (Tamil)', 'Malyali', 'Moplah', 'Ranchi', 'Telugu' and 'Valmiki'. A brief account of all these ethnic groups and aborigines is as follow:

The Bengali: They are major population group. They came after independence as settler under the Government rehabilitation scheme. The settlement started as early as 1949 and continued till 1970s.They migrated from East Pakistan now Bangladesh. They are settled in North, Middle and South Andaman. Till to-date very good number of these people are coming to Andaman from West Bengal mainly from 24 Pargana District as labourer and are settling in various places

The Bhantu: They are also known as Bhatu. Bonington (1932) recorded that Bhantu of Uttar Pradesh were a wandering community. About 300 Bhatu and their family members were sent from different jails in the mainland. They are settled in different parts of South Andaman.

The Burmese: They were brought from Burma now Myanmar by British Colonial ruler during the period of 1907-1923. They came as convicts. Many were brought as labour to clear the forest. A large number of these people returned to their native country after independence when British left these islands. Still some families are settled in South Andaman.

The Coorgi: They are immigrant from a tiny district called Kodagu in Karnataka. They were brought by the British to work in the forest.

The Karen: They are immigrants of Burma now Myanmar. The name Karen is an imperfect use of the Burmese word Kayin. They were first brought by British Government of India with the help of Christian Missionaries in March1925. In 1926-27 regular settlers started coming in search of better living. They were given land mainly at Webi in Mayabunder in North Andaman.

The Madrasi (Tamil): They are Tamil speaking people who came to Andaman from Tamil Nadu and Sri Lanka mainly in search of better livelihood. They are settled all over the Andaman and Nicobar group of islands.

The Malayali: Most of the Malayali came to Andaman in search of employment. They are mostly settled in urban area of South Andaman.

The Moplah: They are also known as Mopilla but commonly addressed by the name Kaka. They were brought by the British Government after famous Moplah Rebellion in August 1921.They are settled in Bambooflat, Panighat, Stewart Ganj, Wimberly Gunj, Mannarghat, ,Govindapuram, Malapuram,Wright Myo, Ograbraj and Calicut.

The Ranchi: The Ranchi of Andaman and Nicobar Islands are a mix of migrant people of different part of Chhota Nagpur now a part of Jharkhand. They were brought in the late 1960s as labourers for road laying, tree felling, timber cutting and other such development works in Andaman. Some others have been brought by Christian Missionaries during the pre-independence period. They are scattered in various parts of Andaman and Nicobar Islands particularly Kumra Kheti, Dairy Farm, Ranchi Tekri, MES Quarry, Jirkatang and Shoal Bay in South Andaman and Bartang, Kitchen Nallah and Rangat in Middle Andaman and Car Nicobar, Katchal, Nancowry and Great Nicobar in Nicobar District (Vidyarthi-1971).

The Telugu: They are Telugu speaking people migrated to Andaman from Srikakulum District of Andhra Pradesh during 1948-1962. They inhibit in different areas in Port Blair and other like Rangat, Mayabunder, Diglipur, Neil, Havelock, Hut Bay, CarNicobar and Kamorta.

The Valmiki: Valmiki are migrant community from Utter Pradesh, Haryana, Punjab, Madhya Pradesh and Bihar. They came after independence along with their officers who came to Andaman. Later they brought more members of their community from their ancestral place in mainland.

The Great Andamanese: They are one of the four Negrito tribes residing in Andaman and Nicobar Islands. Earlier they were good in number (625 as per census-1901), but now reduced to very few in number. They are settled in Strait Island about 60 km. north east of Port Blair.

The Jarawas: They also belong to Negrito tribe. They are confined to forest area of about 765 km^2. along the South and Middle Andaman which has been declared as Jarawa Reserve.

The Onge: The Onge are one of the four scheduled Negrito tribal group of the Andaman Islands. They are the first group of Negritos to arrive in Andaman (Lidio Cipriani -1955). Earlier they were spread all over the islands, but in 1976 to save their dwindling number the Government rehabilitated them in two settlement at South Bay and at Dugong Creek in the north eastern corner of the island.

The Sentinelese: They are one of the six scheduled tribe of Andaman and Nicobar Islands. They have typical Negrito physical traits. They are living in North Sentinel Islands 64 Km. south west of Port Blair. They are most isolated tribal community and are said to be the most ferocious and still remaining hostile thereby have no interaction of any kind with other communities.

The Shompen: They are the inhabitant of the Great Nicobar Islands. They are considered an isolated group of the Malayans which later get mixed with other race possibly the Andamanese.

The Nicobarese: Nicobarese is the generic name of all indigenous people of the Nicobar Islands and the tribal community of different islands of the Nicobar group of Islands have their own local name. The inhabitant of Car Nicobar are called 'Tarik', native of Chowra are called 'Som pai' while those belonging to Teressa are known as 'Som luroo'. The indigenous people of Katchal, Kamorta, Nancowry and Trinket are called 'Somita'. The resident of Pulo Milo are referred as 'Sop piheun' and "Sop panjang' belongs to the Little Nicobar. 'Sop la mongasho" are resident of Kondul and native of Pulo Bhabi of Great Nicobar are called Sop kanalo. The Nicobarese tribes are offshoot of the Mongoloid race that has grown up without interference by or contact by outsider for quite long time (Whitehead,1924). However there are reports that Car Nicobar, Teressa, Katchal, Nancowry, Kondul and Great Nicobar etc. were frequently visited by Malayas, Burmese and other foreigners. Some of them become stranded on these islands and incorporated themselves with them (Boden Kloss, 1902).

Prospects of Ethnobotany in Andaman and Nicobar Islands

In recent years ethnobotany has become increasingly associated with the search for herbal medicines and phytonutrients which continues to expand rapidly across the world with many people now resorting to these products for treatment of various health challenges in both developing and developed countries with a belief that it will promote healthier living. It is worth mentioning that these herbal remedies become available not only in drug stores but also in food stores and supermarkets. In this context the prospect of ethnobotany in the Andaman and Nicobar Islands become very promising, if comprehensively tapped and wisely exploited, as these islands are gifted with rich biological and cultural diversity. These forest reserves evidently contain germplasm of many potentially important plants most of which do not occur elsewhere in our country. Further these islands and forests are home of six aborigine tribes and at least eleven ethnic communities who possess a treasure of traditional knowledge about various uses of plants and other resources of forest. The belly of an Onge mother after child birth is comfortably tied with the bark strip of *Mimusops littoralis* to return normalcy of the uterus and shape of the abdomen. A nicobaries boy was seen to rub some leaves (name of plant not revealed) on his body before participating in the race which he eventually won. Other ethnic communities who came here from mainland or brought by the British Ruler as convicts or as labourers were forced to retain their traditional knowledge because they faced hard time due to lack of medical facilities. Even those who were brought by Indian Government after independence and after 1971 war between India and East Pakistan(now Bangla Desh) were also deprived of any modern health facility as they were allowed to settle in places where till to date modern medical facilities is not at par with facility at Port Blair. In such circumstances their traditional knowledge retained by their old members of community came handy in the treatment of minor and common ailments and are still in use as revealed by various works reported time to time by many researchers. Bharati,P. L, Om Prakash and A. D. Jadhav (2015) reported that the Nicobarese tribes are using about 21 plants for treating various ailments. Chander,M. P, Kartick,C. and Vijayachari,P. (2017) also mentioned about the use of 23 plant for common ailments such as fever which appeared to be the most common ailment among the Nicobarese of these Islands and other common ailments including body ache, headache, snake bite, abdomen pain, ear ache, diarrhoea, vomiting and nausea, cough for short duration, paralysis, giddiness, prolonged labour, malaria, tuberculosis, skin injuries and fracture of bones. Their traditional healing knowledge is passed on from the older generation to the younger in the family through oral narrations and informal training imparted while working as helpers to the folk healers. No written texts are maintained. M.U.Sharief (2007) described about 14 plants used by the Great Andamanese, 20 plants used by the Onge and 18 plants used by the Jarawas. The other ethnic communities such as Moplah and Ranchi tribe are also using their traditional knowledge of medicinal plants for cure of common ailments.

Table 1. Medicinal Plants used by Great Andamanese Tribe

S. No.	Plant Name	Family	Parts Used	Use
1	*Anodendron manubrium Merr.*	Apocynaceae	Leaf	Abortion
2	*Argyreia hookeri* Cl.	Convolvulaceae	Leaf	Swelling & hydrocele
3	*Areca catechu* L.	Arecaceae	Fruit	Stimulant; stomach disorder
4	*Azadirachta indica* A. Juss.	Meliaceae	Leaf, bark	Dressing of cuts & wounds
5	*Citrus medica Linn.*	Rutaceae	Fruit	Stomach disorder
6	*Chromolaena odorata* L.	Asteraceae	Leaf	Cuts & wounds
7	*Erythrina orientalis* (L.) Merr.	Fabaceae	Bark	Fever
8	*Hibiscus tiliaceus* L.	Malvaceae	Leaf	Stomach disorder
9	*Ocimum sanctum Linn.*	Lamiaceae	Leaf	Cold & cough
10	*Phyllanthus amarus Schum.*	Euphorbiaceae	Leaf	Antidote

11	*Pongamia pinnata* (L.) *Pierr.*	Fabaceae	Leaf, bark	Fever, back pain, chronic headache
12	*Premna serratifolia* L.	Verbenaceae	Leaf	Body pain
13	*Rhizophora apiculata* Bl.	Rhizopho-raceae	Leaf	Antidote
14	*Rinorea macrophylla* Decre.	Violaceae	Leaf, bark	To promote lactation

**Source:* M U Sharief, Indian J. Traditional Knowledge, Vol 6, No. 3, July 2007

Table 2. Medicinal Plants used by Onges Tribe

S.N.	Plant Name	Family	Parts used	Use
1	*Ardisia solanacea* Roxb.	Myrsina-ceae	Leaves	Chest pain
2	*Canarium euphyllum* Kurz	Bursera-ceae	Resin	To repel insects
3	*Chromolaena odorata* L.	Astera-ceae	Leaf juice	Cuts & wounds
4	*Desmodium umbellatum* (Linn.) DC.	Fabaceae	Leaves	Fever
5	*Donax canaeformis* (G. Forst.) *K. Schum.*	Maranta-ceae	Root, stem, leaf	Abdominal & spinal pain, fever
6	*Dracaena angustifolia* Roxb.	Dracaen-aceae	Leaf juice applied internally	Stomach pain
7	*Drypetes assamica* (Hk.f.) Pax.& Hoffm.	Euphorbi aceae	Leaf juice applied internally/ externally	Chest pain, snakebite
8	*Glycosmis arborea* DC.	Rutaceae	Leaf juice applied	Chronic headache
9	*Horsefieldia glabra* (Bl.) Warb.	Myristic-aceae	Fruits	Abdominal pain
10	*Heritiera littoralis*	Sterculia-ceae	Leaves Preparation of Onge tea	stimulant
11	*Hibiscus tiliaceus* L.	Malva-ceae	Leaves	Stimulant

12	*Lasianthus andamanicus* Hk. f.	Rubiaceae	Fruit	Antidote
13	*Mallotus peltatus* (Geisel) Muell.	Euphorbiaceae	Green leaves placed or tied on genital parts	Menstrual pain
14	*Mucuna gigantean* DC.	Fabaceae	Gently sbeaten shoots tied around body parts	Sprain
15	*Orophea katschallica* Kurz	Annonaceae	Paste, juice, Vapours	Bee repellent
16	*Pongamia pinnata* (L.) Merr.	Fabaceae	Stem bark paste applied	Chronic headache & backache
17	*Premna serratifolia* L.	Verbenaceae	Raw fruit decoction	applied on waist & neck Body pain, cough
18	*Scaevola sericea* Vahl.	Goodeniaceae	Leaves	Rheumatic pain, fever, body ache
19	*Thespesia populnea* L. Sol.ex Corr.	Malvaceae	Leaves	Constipation
20	*Wedelia biflora* DC.	Asteraceae	Leaf paste applied	Cuts, headache

**Source:* M U Sharief, Indian J. Traditional Knowledge, Vol 6, No. 3, July 2007

Table 3. Medicinal Plants used by Jarawas Tribe

S.N.	Plant name	Family	Parts used	Use
1	*Ageratum conyzoides* Linn.	Asteraceae	Tender leaves	Fever, throat pain
2	*Amomum aculeatum* Roxb.	Zingiberaceae	Leaves, Stem	Cough, fever, curing scar
3	*Angiopteris lygodiifolia* Roscust	Angiopteridaceae	Young leaves	Cough, cold
4	*Areca triandra* Roxb.	Arecaceae	Fruits, Stem	Fruits chewed; stem used to sit during menstruation and also for post parturition.

5	*Canarium commune* Linn.	Burseraceae	Heated leaves	Headache
6	*Carica papaya* L.	Caricaceae	Raw fruit	Prophylactic
7	*Caryota mitis* Lour.	Arecaceae	Young shoots	Vomiting, stomach ache
8	*Chromolaena odorata* L.	Asteraceae	Leaves	Leech bite
9	*Curcuma zeodaria* (Christin) Rosc.	Zingiber-aceae	Rhizome, leaves	Cold, cough
10	*Dracaena angustifolia* Roxb.	Agavaceae	Twigs	To clean genitals during menstrua-tion period.
11	*Knema andamanica* (Warb.) deWilde	Myristic-aceae	Leaves, bark	Throat pain, cough
12	*Myristica andamanica* Hook. f.	Myristic-aceae	Leaves & twigs	Sickness, to stop bleeding
13	*Pandanus andamanensis* Kurz	Pandana-ceae	Leaves	Body pain
14	*Piper betle* Linn.	Piperaceae	Stem, leaves	Body pain
15	*Pseuduvaria prainii* (King) Merr.	Annonaceae	Leaves	Headache & abdominal pain
16	*Syzygium samarengense* (Bl.) Merr. & Perry	Myrtaceae	Leaves	Fever, headache
17	*Thottea tomentosa* (Bl.) Ding Hou.	Aristoloc-hiaceae	Whole plant	Chest pain, headache, cough, cold
18	*Trichosanthes bracteata* (Lamk.) Voigt.	Cucurbit-aceae	Whole plant	Throat infection

**Source:* M U Sharief, Indian J. Traditional Knowledge, Vol 6, No. 3, July 2007

Table 4. Medicinal Plants used by Nicobarese Tribes

S.N.	Plant Name	Family	Parts Used	Use
1	*Ficus microcarpa* L.f.	Moraceae	Stem & leaf	Fracture and Joint pain
2	*Sterculia rubiginosa Vent.*	Stericuli-aceae	Leaf	Fever
3	*Calophyllum inophyllum* L.	Clusiaceae	Leaf	Pain in the eye
4	*Datura metel* L.	Solanaceae	Leaf	Breathing difficulty
5	*Macaranga tanarius* (L.) Muell.	Euphorbi-aceae	Leaf	Indigestion and Abdominal pain
6	*Cassia alata* L.	Caesalpini-aceae	Leaf	Ringworm
7	*Morinda citrifolia* L.	Rubiaceae	Leaf, Root and Fruit	Fracture
8	*Cassia sophera* L.	Caesalpini-aceae	Leaf	Hypertension and Giddiness
9	*Paederia foetida* L.	Rubiaceae	Leaf	Fracture
10	*Ganophyllum falcatum Blume*	Sapindaceae	Leaf	Abdominal Pain
11	*Ricinus communis* L.	Euphorbi-aceae	Leaf	Burning Micturition
12	*Claoxylon indicum* (Reinw, e x Blume)	Euphorbi-aceae	Leaf	Eye injury and Cough
13	*Rauvolfia sumatrana* Jack	Apocyn-aceae	Leaf	Injury
14	*Ficus tinctoria* G.forst.	Moraceae	Stem & Leaf	Fracture and Joint pain
15	*Pyrossia lanceolata* (L.) Ferw	Polypodi-aceae	Stem & Leaf	Fracture and Joint pain
16	*Punica granatum* L.	Punicaceae	Leaf	Chest pain
17	*Gossypium herbeccum* L.	Malvaceae	Leaf	Burning Micturition
18	*Tabernaemontana crispa* Roxb.	Apocyn-aceae	Leaf	Injury
19	*Sterculia parviflora* Roxb.	Sterculi-aceae	Leaf	Injury

20	*Cyanthilium cinereum* (L.) H.Rob.	Asteraceae	Leaf	Pain in the Eye
21	*Oroxylum indicum* Vent.	Bignoniaceae	Leaf	Injury

Source: Bharati P.L., Om Prakash and A.D. Jadhav, International Journal of Bioassays 4.12 (2015) 4650-4652

Table 5. Medicinal Plants used by Ranchi Tribe

S.N.	Plant Name	Family	Parts Used	Use
1	*Acalypha indica* L.	Euphorbiaceae	Leaves	Gastro intestinal disorder
2	*Achyranthes aspera* Linn.	Amaranthaceae	Root	Jaundice
3	*Adhatoda vasica*	Acanthaceae	Leaves	Leucorrhoea
4	*Ageratum conizoides* Linn.	Asteraceae	Leaves	Antiseptic
5	*Aloe vera* Burm.	Liliaceae	Latex	Pain killer
6	*Ammannia baccifera* Linn.	Lythraceae	Leaves	Swelling on feet
7	*Avicennia officinalis* Linn.	Verbenaceae	Stem bark	Eosinophilia, cold and cough
8	*Azadirachta indica*	Meliaceae	Leaves and fruits	Mosquito repellent
9	*Calotropis gigantia*	Asclepiadaceae	Root	Any pain
10	*Cassia alata*	Caesalpiniaceae	Tender leaves	Ring worm
11	*Cassia fistula* L.	Caesalpiniaceae	Root	Poisonous insect bite
12	*Cassia occidentalis* Linn.	Caesalpiniaceae	Leaves	Bone fracture
13	*Centella asiatica* (Linn.) Urban	Apiaceae	Leaves	Jaundice and appetiser
14	*Chromolaena odorata* Linn.	Asteraceae	Leaves	Wound healer
15	*Cleome viscosa* Linn.	Capparidaceae		Leaves Piles
16	*Clitoria ternatea.* L.	Fabaceae	Root	Asthma

17	*Cocos nucifera*	Palmae	Fruit (Liquid endos perm)	Jaundice
18	*Colocasia esculenta*	Araceae	Flowering bud	Diabetes
19	*Crotalaria trifolia*	Fabaceae	Root	Fever
20	*Curcuma aromatica* Salisb.	Zingiberaceae	Rhizome	Stomach ache
21	*Cycas circinalis* Linn.	Convolvulaceae	Male cone	Hydrocoel
22	*Cynodon dactylon* (Linn.) Pers.	Poaceae	Leaves	Increase milk in pregnant lady
23	*Eclipta prostrata* Linn.	Asteraceae	Whole plant	High blood pressure and diabetes
24	*Euphorbia hirta* Linn.	Euphorbiaceae	Whole plant	Hemorrhagic disorder
25	*Gliricidia sepium* (Jacq.) Kunth ex.walp.	Fabaceae	Leaves	Insect bite pain and swelling
26	*Heliotropium indicum* Linn.	Boraginaceae	Leaves	Baldness
27	*Hibiscus rosa-sinensis* L.	Malvaceae	Leaves	Pain killer
28	*Ipomoea gracilis*	Convolvulaceae	Leaves	Itching
29	*Jatropha gossypifolia* L.	Euphorbiaceae	Stem	Tooth and gum infection
30	*Mangifera indica*	Anacardiaceae	Bark	Dysentery and Diarrhoea
31	*Marsilea quadrifolia* L.	Marsileaceae	Leaves	Indigestion
32	*Mimosa pudica* Linn.	Mimosaceae	Whole plant	Pain killer
33	*Morinda angustifolia*	Rubiaceae	Latex	Blood dysentery and diarrhoea
34	*Moringa olifera*	Moringaceae	Leaves	Hydrocoel
35	*Murraya koenijii* (Linn.)	*Spreng.*	Rutaceae	Leaves Blood purifier and herbal facial

36	*Musa sapiantum*	Musaceae	Unripe fruit	Dysentery
37	*Mussaenda frondosa*	Rubiaceae	Anterior sepal	Dandruff
38	*Ocimum sanctum* Linn.	Lamiaceae	Leaves	Chronic cough and cold
39	*Phyllanthus amarus* Schum. & Thom.	Euphorbi-aceae	Leaves	Jaundice and increase appetite
40	*Piper betle* Linn.	Piperaceae	Leaves	Gastrointestinal disorders
41	*Plumeria acutifolia* Poir.	Apocyn-aceae	Bark	White discharge in women
42	*Pongamia pinnata* Linn.	Fabaceae	Seed end-osperm	Malarial and normal fever
43	*Psidium guajava* Linn.	Myrtaceae	Leaves	Diarrhoea
44	*Ricinus communis*	Euphorbi-aceae	Fruit without epicarp	Mouth and stomach ulcer
45	*Scoparia dulcis* Linn.	Scrophular-iaceae	Whole plant	Urinary disorder
46	*Smilax medica* L.	Liliaceae	Stem	Tooth ache and gum infection
47	*Solanum xanthocarpum*	Solanaceae	Ripen fruit	Gum infection and tooth ache
48	*Tabernaemontana crispa* Roxb.	Apocyn-aceae	Root	Stomach pain
49	*Thevetia peruviana*	Apocyn-aceae	Latex	Rheumatic joint pain
50	*Torenia violacea* (Azaola ex Blaneo) Pennell	Scrophula-riaceae	Whole plant	Stomach ache
51	*Vitex trifolia* Linn.	Verbenaceae	Leaves	Fever and body pain
52	*Ziziphus jujuba*. L.	Rhamnaceae	Tender leaves	Stomach pain

**Source:* E.A. Johnson,(2008), dissertation entitled “Ethnomedicinal Flora used by Ranchi Tribe of Andamans” for M.Phil.Degree.

Table 6. Medicinal Plants used by Moplah Community

S.N.	Plant Name	Family	Parts Used	Use
1	*Abrus precatorius*	Fabaceae	Leaves	Fever, Cough,Cold
2	*Achras zapota*	Sapotaceae	Bark, Fruit	Fever, Diarrhoea Biliousness
3	*Acorus calamus*	Acoraceae	Rhizome Leaves	Bronchitis Sinusitis Skin eruptions Rheumatism
4	*Allamanda cathartica*	Apocynaceae	Leaves	wounds
5	*Alocasia macrorrhizos*	Araceae	Stem	Itching
6	*Aloe vera*	Liliaceae	Leaves	Headache, Body pain, Sun burn Pimples, Jaundice Constipation
7	*Anacardium occidentale*	Anacardi-aceae	Leaves & bark	Diarrhoea
8	*Annona squamosa*	Annonaceae	Seed, Fruit Root,	Headache, Anaemia Burning sensation, Cough
9	*Areca catechu*	Arecaceae	Nuts Leaves Roots	Diarrhoea, Low blood pressure, Lumbago, Bronchitis Liver disease
10	*Azadirachta indica*	Meliaceae	Bark Leaves	Fever,Boils, Chicken pox, Skin disease, Malaria, Pimples
11	*Bambusa arundinacea*	Poaceae	Root Leaves	Joint pain, Period pain Osteoarthritis
12	*Calotropis procera*	Ascelpiad-aceae	Stem bark Root bark	Diarrhoea, Dysentery Jaundice, Cough
13	*Cardiospermum halicacabum*	Sapindacea	Leaves Seeds Roots	Ear pain Arthritis Fever
14	*Carica papaya*	Caricaceae	Leaves	Malaria
15	*Cassia alata*	Caesalpini-aceae	Leaves	Ringworm
16	*Cassia tora*	Caesalpini-aceae	Leaves	Leprosy, Ringworm Itching, Arthritis

17	*Casuarina equisetifolia*	Casuarin-aceae	Leaves Twigs	Dysentery, Diarrhoea, Swelling
18	*Catharanthus roseus*	Apocyn-aceae	Whole plant	Diabetes, Malaria
19	*Centella asiatica*	Apiaceae	Leaves	Skin burns
20	*Chromolaena odorata*	Asteraceae	Young leaves	Skin wounds
21	*Cinnamomum tamala*	Lauraceae	Leaves	Acidity
22	*Citrus limon*	Rutaceaea	Fruit	Diarrhoea, Low B.P
23	*Clitoria ternatea*	Fabaceae	Whole plant	Stress, Depression
24	*Cocus nucifera*	Aracaceae	Water	Colitis, Kidney stones Acidity
25	*Colocasia esculenta*	Araceae	Petiole Leaf	Earache, Otorrhea Haemorrhage, Adenitis
26	*Eryngium foetidum*	Apiaceae	Whole plant	Diabetes, Acidity
27	*Costus speciosus*	Costoaceae	Rhizome Leaves Stem	Headache, Fever, Dysentery
28	*Curcuma longa*	Zingiber-aceae	Rhizome	Indigestion, Common Colds Throat Infection, Wounds
29	*Curcuma zedoaria*	Zingiber-aceae	Rhizome	Skin ailments, Ulcers Menstrual irregularities
30	*Cymbopogan citratus*	Poacaea	Leaves	Cough & colds
31	*Euphorbia hirta*	Euphorbi-aceae	Whole plant	Skin disease, Thrush
32	*Euphorbia tirucalli*	Euphorbi-acea	Branches	Rheumatism
33	*Gardenia gummifera*	Rubiaceae	Gum resin	Tooth ache, Wound
34	*Gliricidia sepium*	Fabaceae	Leaves	Dysentery
35	*Glycosmis pentaphylla*	Rutaceae	Leaf bark Stembark Root ark	Damaged liver, Diarrhoea
36	*Hibiscus rosasinensis*	Malvaceae	Leaf	Overweight, LDL cholesterol, Wounds

37	*Indigofera tinctoria*	Fabaceae	Leaf	Wounds, Insect bite Skin disorders, Scorpion bite
38	*Ixora coccinea*	Rubiaceae	Roots Flower	Dysentery Skin disease
39	*Jatropha gossypiifolia*	Euphorbi-aceae	Leaves Root Bark Stem Latex	Body pain Wounds Cuts Tooth problem Ringworm
40	*Kyllinga nemoralis*	Cyperaceae	Tuber	Diarrhoea, Cough Diabetes
41	*Lantana camara*	Verbenaceae	Leaves	Insect bite, Fungal infection Microbial infection
42	*Laurentia longiflora*	Campunal-aceae	Leaves	Wounds
43	*Lawsonia inermis*	Lyhraceae	Leaves	Sun burn, Rashes, Dandruff Cracking nails
44	*Mangifera indica*	Anacardi-aceae	Bark Seed Kernel Leaves	Dysentery, Diarrhoea Heart burn, Wounds Gastropathy, Helminthiasis
45	*Melastoma malabathricum*	Melastom-ataceae	Flowers, seeds, leaves	Vaginal discharge Diarrhoea Cuts & wounds
46	*Mimosa pudica*	Mimosaceae	Leaves	Wounds, Piles
47	*Moringa oleifera*	Moringaceae	Leaves Flower	Stomach pain High blood pressure, Anaemia
48	*Murraya koenigii*	Rutaceae	Leaves	Premature hair greying
49	*Musa paradisiaca*	Musaceae	Flowers young leaves Roots	Bronchitis, Dysentery Ulcers, Skin burns, Digestive disorder
50	*Ocimum sanctum*	Lamiaceae	Leaves	Cough
51	*Phyllanthus amarus*	Euphorbi-aceae	Whole plant	Jaundice, Diabetes Frequent menstruation Swelling , Itchiness
52	*Piper betle*	Piperaceae	Leaves	Ulcers ,Bruise, Dry cough Stomach disorder

53	*Piper nigrum*	Piperaceae	Stem Leaves Fruit	Body pain Fever cough
54	*Psidium guajava*	Myrtaceae	Leaves Leaves Fruit	Dysentery Bleeding gum & bad breath Hypertension, Diabetes
55	*Ricinus communis*	Euphorbiaceae	Root	Asthma
56	*Rosa indica*	Rosaceae	Rose petals	Headaches, Indigestion
57	*Solanum torvum*	Solanaceae	Fruit	Enlarged spleen Thread worms
58	*Tamarindus indica*	Caesalpiniaceae	Seed Leaves Root	Centipede bite, Diarrhoea Sprain, Sore joint Dry cough, Ringworm Boils, Fever
59	*Zingiber officinale*	Zingiberacae	Rhizome	Nausea, Cough
60	*Ziziphus mauritiana*	Rhamnaceae	Leaves, roots Root bark	Rheumatism Vomiting Eye disease

**Source:* Ms. Naziya Parveen, (2017) dissertation entitled "Plant Folk Medicine among Moplah community of Andaman and Nicobar islands." for M.Sc.Degree.

Plants produce economically important organic compounds such as oils, resins, tannins, rubber, gums, waxes, dyes, flavours and fragrances, pesticides and pharmaceuticals. Many medicines used and/ or currently in use such as: Aspirin, codeine, ipeacac, pilocarpine, pseudoephedrine, quinine, reserpine, scopolamine, theophylline and vinblastine. etc., have been derived from medicinal plants based on ethnobotanical research programs (Liu, H. *et al.*, 2013). Subramoniam, A., (2014) stated that all chlorophyll-a bearing plants have pharmaceutical potential. This statement was made based on the recently reported anti-inflammatory activity in chlorophyll-a and its related products. A number of useful drugs have been developed based on ethnobotanical approaches. Chinese traditional medicine has been a novel basis of drug development in China. Up to 2007, China has collected 3563 extracts, 64715 compositions and 5000 single compounds from 3000 Chinese herbs together with about 130 kinds of chemical drugs obtained. It is estimated that among the total flowering plants occuring in the tropical regions of the world, only few in percentages were studied for their pharmaceutical potentials.

The Andaman and Nicobar Islands harbors a total of 2649 plant taxa, comprising 2508 species, 32 subspecies, 103 varieties and 6 forma under 1109 genera in 238 families belonging to 4 different plant groups, namely bryophytes (mosses), pteridophytes, gymnosperms and angiosperms. Of which, angiosperms are the predominant group, comprising of 2314 species, 31 subspecies, 89 varieties and 6 forma under 1011 genera in 181 families, constituting 92% of entire flora of the Andaman and Nicobar Islands. Gymnosperms are represented by only 7 species and 2 varieties under 4 genera and 3 families. There are 129 species, 1 subspecies and 9 varieties of pteridophytes under 62 genera belonging to 38 families and 58 species and 3 varieties of mosses (bryophytes), under 32 genera and 16 families in the islands. (C. Murugan, S. Prabhu, R. Sathiyaseelan and R.P. Pandey- 2016). Very little has been done to explore the bioprospect potential of these flora resources of Andaman and Nicobar islands. As the global use of herbal medicinal products continues to grow and many more new products are introduced into the market, the importance of ethnobotany, wisely tapping and exploitation of traditional knowledge of aborigine tribes and other ethnic communities of Andaman and Nicobar Islands assume more significance. The prospect seems more bright and promising.

Present Challenges to Ethnobotany in Andaman and Nicobar Islands

In recent years, the discipline of ethnobotany has become increasingly associated with the search for medicines which has posed many challenges too. Oral information is gathered and disseminated without proper verification and clinical trial, sometime putting life at risk. Uncontrolled flow of information results in wild harvesting of economic high valued medicinal plants and their over exploitation, pushing them at the brink of extinction, if not conserved. The community people from whom information is gathered, are not compensated or rewarded for their traditional knowledge. Their intellectual property rights (IPR) are violated. The situation in Andaman and Nicobar Islands is also not better and challenges are enormous as the issue is not at public domain due to lack of awareness in these islands. Some of these major issues are discussed below:

Dwindling Population of aborigine tribes

The aborigine tribes are repository of vast knowledge of the flora of their surrounding environment and the uses of these resources. But their population decreased from 29,469 to 28,530 as compared to 2001 Census, registering a decrease of 3.19%. Among the four tribes of Andaman and two tribes of Nicobar Islands, the Great Andamanese paid the heaviest toll of their lives on account of their contact with the early settlers of these islands. In 1858, their number was conservatively estimated near about 3500. In 1901 their strength depleted to 625. In 1971 they were found to be only 24 in all. Their numbers have increased steadily to 26 during 1981 and 1991 and during 2001 Census, 43 Andamanese were enumerated who have largely been concentrated in Strait Island now. At present the tribe-wise population of aborigine tribes in Andaman and Nicobar Islands as per Census 2011 is as follow:

Tribes	Population
Great Andamanes	44
Jarawas	380
Sentinelese	15
Onges	101
Shom Pens	229
Nicobarese	27168

The present trend of decline in the population of tribes is very alarming. If corrective measures and steps are not taken, all their traditional knowledge of healing will be lost with them. It will be a huge loss to compensate.

Lack of Proper and Authentic Documentation

The traditional healing knowledge among the tribes and other ethnic communities is passed on from the older generation to the younger through oral narrations and informal training imparted while working as helpers to the folk healers. No written texts are maintained. Even the Nicobaries who have had a natural growth of population and have made headway on the road to civilization are not maintaining any record and are not willing to share their knowledge. The traditional healer among the Moplah and the Ranchi also do not have any written record. Whatever documents are available, that is the result of efforts done by the various researchers who documented the information obtained through the interviews with the tribal people and members of other ethnic communities.

Biodiversity and Cultural Diluation

The climatic change may affect the biodiversity of these islands also which may result in the loss of plants which may have the medicinal properties. For example loss of medicinal plants through deforestation and associated migration of traditional healer from village to town can result in villages not only loosing access to their plants but also the knowledge of medicinal value of those plants to their community. Sooner or later the fragile ecosystem of these islands will be affected by the climatic change and also by the developmental work done without proper impact assessment. As a result the biodiversity will be diluted and many organism will be lost which may have medicinal property. Therefore ethnobotanical project should be designed that can evaluate the cultural importance of plants in an area. It can play important role on plant conservation as these projects can provide quantitative and qualitative data about diversity of species used in the community.

Proximity to Civilised World

Last decade has seen a change in the approach towards the aborigine tribes of these islands. Except the Sentinelese other five tribes have come closer to the civilised world. Exclusive medical wards in Primary Health Centres (PHCs) located near the Jarawa Reserve at Tushnabad, Kadamtala and GB Pant Hospital, Port Blair have been set up for treatment of Jarawas. Medical Sub Centres have been established by the A&N Administration at Dugong Creek in Little Andaman & Strait Island for providing health care facilities to the Onges & Andamanese tribes. Health card of each individual of Particularly Vulnerable Tribal Groups (PVTGs) is maintained by Tribal Welfare Officers of Andaman Adim Janjati Vikas Samitie (AAJVS)/ Medical Sub Centres and are regularly updated. Nutritional supplements are being provided to the pregnant women, lactating mothers and infants belonging to Onge & Andamanse tribes through AAJVS. The proximity to the modern world is resulting in loss of the traditional knowledge of tribal people. Though the traditional healing practices is still in use among the old Nicobarese.

Lack of scientific and technological approaches on the use of medicinal plants

Information regarding use of various plants for healing purpose is obtained through interviews. Their statements are not clinically tested. It has been observed that within same community different people are using same plant differently for the treatment of the same disease. Some species are used to treat specific medical condition in one community but have other uses by different community. For example *Plectranthus amboinicus* (Lour). Spring, is used for cure of skin and digestive infection in Africa but the same plant is used for treatment of epilepsy and convulsion in South Am toxicity caused by herbal products (Eldeen *et al.*, 2005). In Andaman and Nicobar Islands the general awareness about the use of medicinal plants has increased in recent time and people are resorting to old methods of healing but it has been observed that some people try their own method without any experience, which may be harmful.

Lack of information on mechanism of actions of herbal drugs

There is a complete lack of scientific and technological approaches to indicate pharmacodynamic properties and mechanism of actions of ethnobotanical based-herbal drugs. The complexity of herbal ingredients, their targets in human body and the mechanism of action underlying the herbal medicine efficiencies is required to be understood. There is no such facility in Andaman where the isolation of active compound can be done and may be tested or trialled.

Intellectual property rights and biopiracy

Legislation associated with ownership and access to biodiversity in Convention on Biological Biodiversity (CBD) requires an agreement before ethnobotanists start their studies and where some information can be obtained. Information about the use of specific species of a plant in a community is usually obtained through interview of individuals or group of people. There are ethical, economic and legal issues associated with collection and dissemination of these information. Information gathered for a purely social and conservation reason can result in unforeseen consequences if disseminated without due care and agreement. For example a survey of medicinal plants in a village or in forest area might result in data being collected about endangered medicinal plants of high economic value. This information if disseminated can be used by people to target species. Similarly information about medicinal plants of an area may result into wild harvesting of those species. Therefore it is critical to find out at the start of a project what use can be made of any information gathered specially traditional knowledge. Further indigenous people can be compensated for sharing their ethnobotanical knowledge. This situation raises important questions about the contribution of ethnobotanical research to the local communities as being the primary sources of these knowledge. Invasion of unprotected indigenous resources including traditional knowledge which has been termed biopiracy, represents major challenges to ethnobotany more in Andaman and Nicobar Islands due to lack of awareness about the legislation of conservation of biodiversity in CBD and Intellectual Property Rights (IPR). There is no active Government agency also to implement them.

Conclusion

Ethnobotanical based medicine has gained popularity in recent time and has been further propelled by the alarming rate of harmful side effects of modern medicine. It has now assumed high economic value and have thus brought many issues together, leading to emergence of a new sub-discipline socio-ethnobotany dealing with the indigenous people, their traditional and cultural knowledge and resources and economic sharing for imparting their knowledge for bioprospecting. Though modern healthcare system and traditional healthcare system rarely go together there is a need to increase the economic value of bioprospecting for which both system should integrate with modern science and biotechnology. This will create good number of job opportunity in Andaman and Nicobar Islands and will bring revenue also, if the knowledge gathered form ethnobotany is used for production of various health care products - phytomedicines, nutraceuticals and food supplements. Local legal frame work will be also required to regulate bioprospecting and to check the biopiracy. But there are challenges, as these islands do not have any modern facility to determine the active compounds of medicinal plants, their mechanism of action and validation of end product which all are very much required for a patent and to protect the intellectual property rights.

References

Bharati, P.L., Om Prakash and A.D. Jadhav. 2015. "Plants used as traditional medicine by the Nicobari tribes of Andaman & Nicobar Islands". *International Journal of Bioassays,* ISSN:2278-778X CODEN:IJBNHY

Bonington, M.C.C. 1932. "Census of India, 1931. The Andaman and Nicobar Islands", Vol. II, Part I & II.

Champion, H.G. & Seth, J.K. 1968. "*A Revised Survey of Forest Types of India.*" Manager of Publications, Delhi.

Chander, M.P., Kartick C., Vijayachari P. 2017. " Herbal medicine & healthcare practices among Nicobarese of Nancowry group of Islands - an indigenous tribe of Andaman & Nicobar Islands". Indian J Med Res [serial online] 2015; 141: 720-44.

Cipriani,Lidio. 1955. "Census of India 1951; On the origin of Andamanese" Vol. XVII.

Eldeen, I.M.S., E.E. Elgorashi and J. van Staden, 2005. "Antibacterial, anti-inflammatory, anti-cholinesterase and mutagenic effects of extracts obtained from some trees used in South African traditional medicine". *J. Ethnopharmacol.,* 102: 457-464.

Kloss, C.B. 1902. "The Andaman and Nicobar", London, John Murray.

Liu, H., J.Wang, W. Zhou, Y. Wang and L. Yang, 2013. "Systems approaches and polypharmacology for drug discovery from herbal medicines: An example using licorice." *J. Ethnopharmacol.,* 146: 773-793.

Murugan, C., S. Prabhu, R. Sathiyaseelan & R.P. Pandey, 2016. "A Checklist of Plants of Andaman and Nicobar Islands" Envis Centre on Floral Diversity, Botanical Survey of India.

Sharief, M.U. 2007. "Plants folk medicine of Negrito tribes of Bay Island" *Indian J Traditional Knowledge,* Vol. 6, No. 3, July 2007.

Singh, K.S. 1994. "People of India - Andaman and Nicobar Island", Vol. XII. Anthropological Survey of India, Affiliated East-West Press Pvt. Ltd.

Subramoniam, A. 2014. "Present scenario, challenges and future perspectives in plant based medicine development". *Ann. Phytomed.,* 3: 31-36.

Vidyarthi, L.P. 1971. "Ranchi tribals in Andaman and Nicobar Islands", *Journal of Social Research,* 14(2): 50-59.

Whitehead, George 1924. "In the Nicobar Island" London, Seeley Service.

❑❑❑

Chapter 11

Occurrence and Distribution of Arbuscular Mycorrhizal Fungi in Some Selected Medicinal Plants Growing in Natural Forest of Port Blair, Andaman and Nicobar Islands

Dipon Sharmah

Department of Plant Science, Jawaharlal Nehru Rajkeeya Mahavidyalaya, Port Blair-744104, Andaman and Nicobar Islands, India. E-mail- dsharmah@yahoo.com

ABSTRACT

The present study is an attempt of its first kind to understand the biodiversity of arbuscular mycorrhizal fungi (AMF) in some important medicinal plants viz. *Melastoma affine, Ardisia elliptica, Mimosa pudica* and *Cinnamomum verum* growing in natural undisturbed forest of Port Blair, Andaman and Nicobar Islands. It was found that all the four medicinally important plants used by the local populace were colonized by AMF, and these fungi must be playing an important role in the production of useful plant metabolites. The present study provides a basis for further understanding the enormous significance of AMF diversity and efficient utilization of arbuscular mycorrhizal technology in growth and sustainable utilization of medicinal plants.

Introduction

Microorganisms are important components of earth's biota and many of these are found to be linked with the roots of higher plants. The association of the plants with soil fungi started in the Paleozoic era and it was an instrumental association because it paved the way for colonization of land by ancient plants. This symbiotic association of plants and fungi, termed mycorrhiza by Frank in 1885, is one of the most important terrestrial symbiosis on planet earth. The term mycorrhiza, literally means 'fungus-root', is a non-pathogenic symbiotic associations between plant roots and fungi. According to Brundrett (1991) mycorrhizas are symbiotic associations essential for one or both partners, between a fungus (specialized for life in soils and plants) and a root (or other substrate-contacting organ) of a living plant, that is primarily responsible for nutrient transfer. Of the seven different types of mycorrhizae recognized in plant kingdom viz., ectomycorrhizae, arbuscular, ericoid, arbutoid, monotropoid, orchid and E-strain mycorrhizas, the best studied and widespread type is the arbuscular mycorrhiza. Arbuscular mycorrhizal fungi (AMF) are a key group of plant beneficial symbionts and are an important component of rhizospheric soils. They form mutualistic associations with approximately 90% of terrestrial plants and act as a living connection between soil and plants, influencing soil fertility and plant nutrition, diversity and productivity (Smith and Read 2008; van der Heijden *et al.,* 1998). The symbiosis is characterized by reciprocal interaction wherein the host plant provides the fungus with soluble carbon sources, and at the same time the fungus provides multiple benefits to the host plant. Some of the benefits provided by the fungus includes- enhanced uptake of certain immobile nutrients (phosphorus), defends plants against pathogens, alleviates environmental stresses on plants, improves plant tolerance to drought and polluted environments and accelerates plant establishment (Smith and Read 2008). Moreover, AMF contribute to the formation and maintenance of soil structure by the production of a protein, glomalin, which improves soil aggregates stability (Bedini *et al.,* 2009). Thus, AMF improves plant and soil health by virtue of their various nutritional and non-nutritional activities. At present, AMF are considered as an important component in the restoration and reestablishment of the vegetation in fragile or degraded ecosystems, and in the maintenance of plant biodiversity and ecosystem functioning (van der Heijden *et al.,* 1998; Grime *et al.,* 1987). Because of these multiple beneficial AM fungal activities and functions, it is quite important to conserve these key groups of microorganisms in both natural and agricultural ecosystems.

Andaman and Nicobar archipelago is unique as it is covered by 86 % forest and rich in floristic and faunistic composition. However, till date no study has been carried out in this region to understand the general pattern of AMF occurrence and diversity. Although the region is very rich in medicinal plants, there is no information about the occurrence and distribution of AMF in medicinally important plants of the region. The potential use of AMF for promoting growth and disease resistance in medicinal plants is now well established

(Chanda *et al.*, 2014). Knowledge about the presence and diversity of AMF can strongly contribute to the sustainable management of nutrition for medicinal plants in a specific area Therefore, a detail survey of the presence and diversity of indigenous AMF of this biodiversity hotspot region was urgently felt as this would provide a basis for future action. This study is an attempt of its first kind to understand the biodiversity of AMF in some important medicinal plants growing in natural undisturbed forest of Port Blair, Andaman and Nicobar Islands. Four medicinally important plant species- *Melastoma affine* D. Don. (Melastomataceae), *Ardisia elliptica* Thunb. (Myrsinaceae), *Mimosa pudica* Linn. (Mimosaceae) and *Cinnamomum verum* J. Prestl. (Lauraceae) were selected for this study. The leaves and flowers of *Melastoma affine* is used by the local communities as remedies for cholera, diarrhea, stomachache, toothache and clean wounds and cuts (Dagar and Singh 1999). *Ardisia elliptica* is traditionally used by the locals for alleviating chest pains, treatment of fever and diarrhea (Dagar and Singh 1999). Recently, it has been shown that one its bioactive component (Beta-amyrin) is more potent than aspirin in inhibiting collagen-induced platelet aggregation (Ching *et al.*, 2010). The herb *Mimosa pudica* has been used traditionally for ages, in the treatment of urogenital disorders, piles, dysentery, sinus, and also applied on wounds (Dagar and Singh 1999). Apart from its use as spice, the bark of *Cinnamomum verum* is traditionally used by the locals of the region as a remedy against colds, diarrhea, various problems of the digestive system such as loss of appetite and also to lower blood sugar and cholesterol levels (Dagar and Singh 1999).

Material and Methods

Study site

The study site is located at undisturbed natural forest in Port Blair (92p 432 27.40 east longitude to 11p 362 44.62 north latitude), Andaman and Nicobar Islands, India. The climate of the region is moderate and tropical monsoon type with temperature varying from 22 p C to 32 p C with heavy or light rainfall almost throughout the year. The mean average temperature was 27.6 p C and total average rainfall was 3397.4 mm (data average of five years period) (Fig. 1). The mean relative humidity was 81.79 %. The vegetation of the study site consists of tropical evergreen and semi-evergreen forest. The top canopy includes plants like *Syzygium cumini* (L.) Skeels., *Cocus nucifera*

Linn., *Cinnamomum verum* J.Presl., *Areca catechu* Linn., *Acasia auriculiformis* A.Cunn.exBenth. Middle canopy consist of medium sized trees such as *Tabernaemontana divericata* R.Br. ex Roem. & Schult., *Melastoma caffine, Costus speciosus* (Koenig ex Retz.) j.E.Sm, *Zingiber zerumbet* Sm., *Cromolina* odorata (L.) King & H.E. Robins, *Ixora coccinia* Linn. Understory plants include *Vernonia cinerea* Less., *Cyanodon dactylon* (L.) Pers, *Tridax procumbens* Linn., *Hyptis capitata* Jacq., *Themeda triandra* Forssk., *Centella asiatica* (L) Urban.

Sampling procedure

The rhizophere soil samples associated with plant roots were collected randomly between the months of January to February 2014 and samples were collected from a depth of 5-10 cm. Each soil sample of approximately 100 g was collected from the following four medicinally important plant species: *Melastoma affine, Ardisia elliptica, Mimosa pudica* and *Cinnamomum verum.* Soil samples were stored in polythene bags at 4 ^{0}C in the laboratory until analysis.

Determination of soil character

Organic Carbon (O/C) in the soil sample was estimated by the method of Okalebo (1993). pH and Electrical conductivity (EC) was determined by suspending soil in water in the ratio of 1:10 (w/v) and kept overnight at room temperature to allow the solid to settle down. Phosphorus (P) was estimated by phosphomolylodic blue color matter using Shimadzu UV-160A spectrophotometer. Potassium (K) was estimated using the flame photometer instrument as described by Okalebo (1993).

Spore extraction

Spores from the rhizophere soil samples were isolated by wet sieving and decanting method described by Gerdemann and Nicolson (1963). 100 g of soil from each plant was suspended in a bucket of water and stirred with the help of a hand-held stirrer and kept aside for 15-20 minutes. The suspension was sieved through a series of sieves arranged in decreasing order viz., 300μm, 212μm, 106μm and 40μm. Now the sieves where washed thoroughly and the residues were collected in various beakers of 50 ml each and filtered onto a filter paper and placed in a 9 cm petridish for examination under a binocular Labomed stereo-zoom microscope. With the help of this microscope

AMF spores were counted. Permanent slides were prepared by mounting spores in polyvinylalcohol-lactophenol. Photographs were taken with the help of Digital Microscope (Labomed Vision 2000) having an external Digital Eye Camera. Spore identification was based on color, size, surface ornamentation and wall structure with reference to current online descriptions, www.amf-phylogeny.com.

Statistical Analysis

AMF composition in field samples was evaluated based on spore density (SD) and relative abundance (RA). SD was calculated from direct counts of spores or sporocarps per 100 g soil. RA was calculated as the number of given species spore divided by total number of spores.

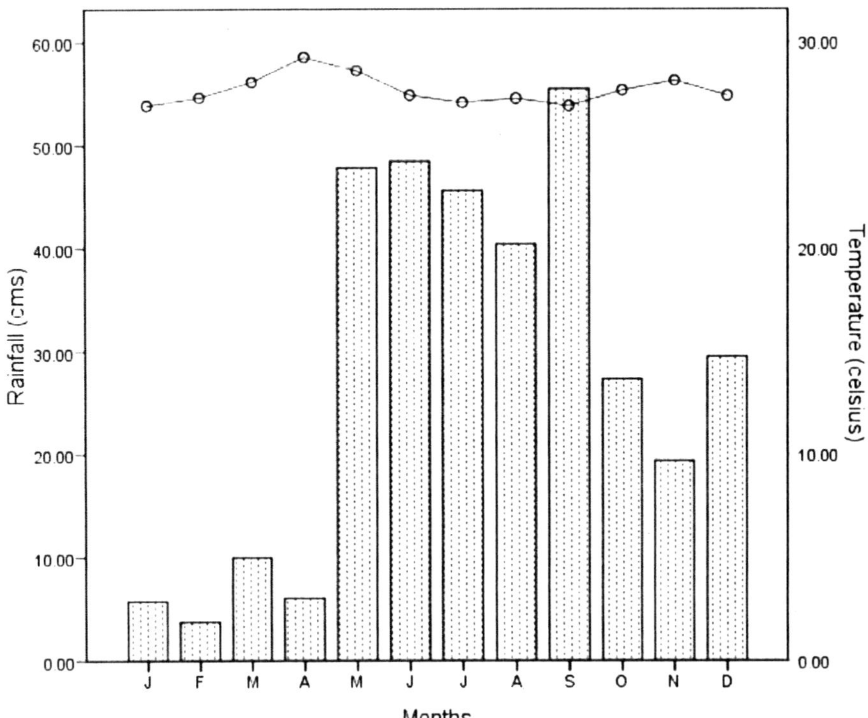

Fig. 1. Average monthly means of rainfall (columns) and temperature (continuous line) from January 2010 to December 2014. Records for rainfall and temperature courtesy of Meteorological station, Port Blair.

Result

Soil physico-chemical properties

The chemical properties of the rhizospheric soil were analyzed in soil testing laboratory, Port Blair (Table 1). Soil was described as alluvial type. Soil pH was normal and electrical conductivity (EC) was normal. Organic carbon (O/C) content was high. The amount of extractable phosphorus (P_2O_5) was low and potassium (K_2O) was completely absent.

Table 1. Chemical properties of soil

Soil Sample	pH	EC	O/C	P_2O_5	K_2O
	6.6 (N)	0.90 (N)	1.34 (H)	7.2 (L)	-

N-Normal, H-High, L-Low

AMF diversity and richness

Arbuscular mycorrhizal status of medicinal plants growing in natural forest area was studied. A total of fourteen AMF taxa (Fig. 2) were identified morphologically from the rhizophere soil samples collected from the four plant species namely *Melastoma affine* D.Don., *Ardisia elliptica* Thunb., *Mimosa pudica* Linn. and *Cinnamomum verum* J. Prestl (Table 2). Some of the AMF morphotytes could not be identified up to species level because of low number of individuals. The unidentified morphotypes were designated as sp. and were distinguished on the basis of morphological characters of spore wall, hyphal attachment and spore size. Out of the total fourteen AMF taxa, six (42.8 %) were identified up to species level, and eight (57.2%) up to genus. Ten of them belonged to genus *Glomus*, two each to *Rhizophagus* and *Acaulospora*. Of the four plant species studied, nine AMF taxa were found each in *Mimosa pudica* and *Ardisia elliptica*, ten in *Cinnamomum verum* and seven in *Melastoma affine*.

Altogether, three genera viz., *Glomus*, *Rhizophagus* and *Acaulospora* were identified during the sampling period. *Glomus* was the dominant genus in the rhizospheric soil samples of all four plant species studied followed by *Rhizophagus*. The two species, *Rhizophagus fasciculatus* and *Glomus* sp. 4 were common in all the four plants. The relative abundance of dominant genus *Glomus* was highest in *Melastoma affine* (77.2 %) than in *Mimosa pudica* (66.9 %),

Table 2. The spore density (SD) and relative abundance (RA) of AMF species isolated from the four medicinal plants species.

AMF species	*Melastoma affine*		*Ardisia elliptica*		*Mimosa pudica*		*Cinnamomum verum*	
	SD	RA	SD	RA	SD	RA	SD	RA
Glomus tortuosum	165	22.54	156	23.17	0	0	103	13.02
Glomus glomerulatum	92	12.56	43	6.38	23	3.69	0	0
Glomus fuegianum	0	0	0	0	70	11.23	0	0
Glomus sp. 1	136	18.57	0	0	106	17.01	92	11.63
Glomus sp. 2	0	0	25	3.71	53	8.50	41	5.18
Glomus sp. 3	0	0	64	9.50	0	0	89	11.25
Glomus sp. 4	46	6.28	31	4.60	58	9.30	72	9.10
Glomus sp. 5	74	10.10	0	0	36	5.77	68	8.59
Glomus sp. 6	52	7.10	49	7.28	71	11.39	0	0
Glomus sp. 7	0	0	0	0	0	0	52	6.57
Rhizophagus fasciculatus	167	22.81	108	16.04	94	15.08	212	26.80
Rhizophagus manihotis	0	0	136	20.20	112	17.97	0	0
Acaulospora foveata	0	0	0	0	0	0	22	2.78
Acaulospora sp. 1	0	0	61	9.06	0	0	40	5.05
Total	732	100	673	100	623	100	791	100

Cinnamomum verum (65.4%) and *Ardisia elliptica* (54.7%). *Rhizophagus*, the second most frequented genus showed the highest relative abundance in *Ardisia elliptica* (36.2%) than in *Mimosa pudica* (33%), *Cinnamomum verum* (26.8%) and *Melastoma affine* (22.8%). The dominant and generalist species *Rhizophagus fasciculatus*, which occurred in all the four plants studied, had the highest relative abundance in *Cinnamomum verum* (26.8%) than in *Melastoma affine* (22.8%), *Ardisia elliptica* (16%) and *Mimosa pudica* (15%).

There was a difference in the total spore density of AMF among the plants studied- *Cinnamomum verum* (791 spores per 100g soil), *Melastoma affine* (732 spores per 100g soil).

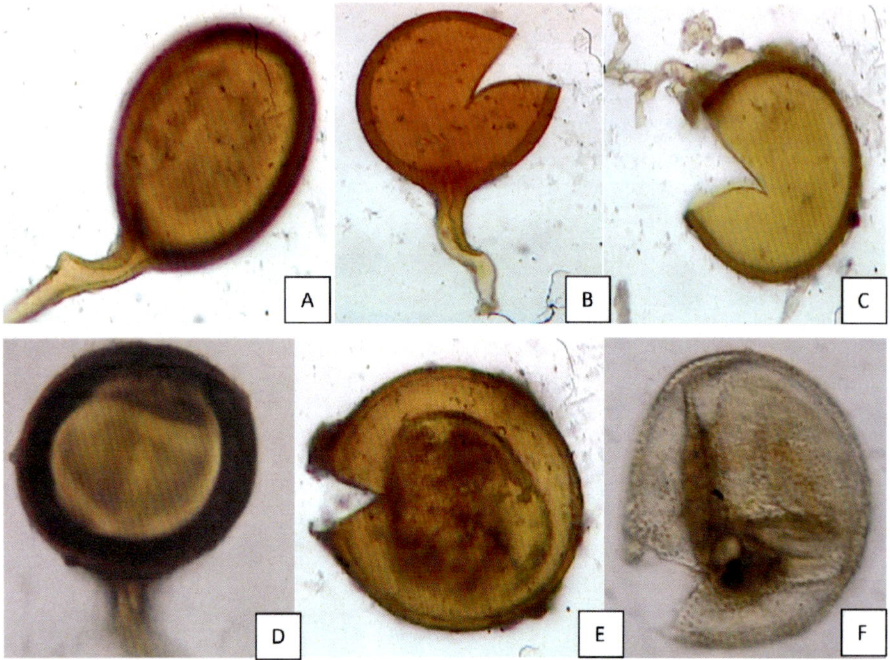

Fig. 2. Some AMF morphotypes detected from the four medicinal plants studied. A *Rhizophagus manihotis*, B *Glomus fuegianum*, C *Glomus glomerulatum*, D *Glomus* sp. 1, E *Acaulospora foveata*, F *Acaulospora* sp. 1.

Ardisia elliptica (673 spores per 100g soil) and *Mimosa pudica* (623 spores per 100g soil). The dominant genus *Glomus* had the highest spore density in *Melastoma affine* (565 spores per 100g soil) than in *Cinnamomum verum* (517spores per 100g soil), *Mimosa pudica* (417 spores per 100g soil) and *Ardisia elliptica* (368 spores per 100g soil), and followed by second dominant genus *Rhizophagus,* which had 244 spores per 100g soil in *Ardisia elliptica,* 212 spores per 100g soil in *Cinnamomum verum,* 206 spores per 100g soil in *Mimosa pudica* and 167 spores per 100g soil in *Melastoma affine.*

Discussion and Conclusion

In the present study carried out in the rich tropical evergreen forest of Port Blair, AMF diversity and community composition was described in four medicinally important plants based on spore morphological identification. Since soil sampling was carried out in between January and February 2014, we are not sure that all the AMF spores present in the study sites were detected. This is because AMF sporulation is seasonal and is controlled by a number of factors. However, many researchers still believe that AMF community estimation using spores and species richness is a valid approach for diversity related studies (Oehl *et al.*, 2005; Brundrett and Ashwath 2013).

All the four plant species studied (*Melastoma affine, Ardisia elliptica, Mimosa pudica, and Cinnamomum verum)* were mycorrhizal, as evidenced by the presence of AMF spores. In this study, a total of 14 AMF morphotype were isolated. This is in agreement to Sharmah and Jha (2011), who found twelve AMF morphotype in natural undisturbed forest of Assam. However, a host of other workers have reported higher AM fungal richness in other ecosystem (Tehabi *et al.*, 2008; Zhang *et al.*, 2003). The higher AM fungal species richness is probably due to large area studied and diversity study carried out over an extended period of time.

There are a number of factors that affect AMF sporulation. These are seasonality, edaphic factors, host-dependence, age of the host plants and the sporulation abilities of AMF (Bever *et al.*, 2001; Zhao *et al.*, 2001). In this study, the spore density found in four different plant species was variable ranging between 623 and 791 spores per 100g soil. Similarly, mean spore density ranging from 113 to 470 spores (20 g soil)$^{-1}$ was obtained in a hot and arid ecosystem (Li *et al.*, 2007). In the present study, the highest spore density was found in *Cinnamomum verum* (791 spores per 100g soil). Up to 15531 spores (25 g soil)$^{-1}$ in the dry tropical ecosystems from Costa Rica were reported by Johnson and Wedin (1997). Guadarrama and Alvarez-Sanchez (1999) reported that highest spore density is observed if sampling is done in the dry season of the year. Therefore, higher spore density observed in this study may be due to soil sampling carried out in dry season of the year.

In this study it was observed that the genera *Glomus* and *Rhizophagus* were dominant followed by *Acaulospora*. This is in agreement with a number of other workers (Zhao *et al.*, 2001; Guaderrama and Alveraz-Sanchez 1999), who also found *Glomus* and *Acaulospora* to be dominant in a number of ecosystems all over the world. The widespread occurrence of *Glomus* and *Acaulospora* can be attributed to the fact that these AM morphotype are more adapted to tolerate diverse environmental conditions and are also not specific to particular plant host species, showing wide host range.

AMF diversity determines plant diversity, productivity and ecosystem stability (vander Heijden *et al.*, 1998). In this study, which was carried out in four important medicinal plant species growing in natural undisturbed forest, considerable AMF diversity was found to be present. Further, the study was carried out over a period of short time which limits the isolation and discovery of non-sporulating AM fungal species. Considering this, it can be concluded that there is a wealth of AMF diversity in these forest ecosystem of Port Blair, Andaman and Nicobar Islands. The AMF diversity must be playing an important role in maintaining forest diversity and its stability in fragile ecosystem of this region. As phosphorus was quite low in the soil and surprisingly potassium level was nil (Table 1), therefore, it can be inferred that AMF is playing an important role in making these nutrients available plants. Panwar and Tarafdar (2006) observed that AMF can play an important role in the re-establishment and conservation endangered medicinal plants in Indian Thar Desert. Also, plants inoculated with AMF showed significant increase in growth and production of active secondary metabolites in some traditional medicinal plants (Chanda *et al.*, 2014). In this study, it was found that all the four medicinally important plants used by the local populace were colonized by arbuscular mycorrhizal fungi, and these fungi must be playing an important role in the production of useful plant metabolites. Therefore, we believe that further studies in this field should focus on larger area, covering medicinally important plant species endemic to this region, and over a considerable period of time, so that all AM fungal species sporulating at the time of sampling could be isolated. It would be interesting to understand in near future the enormous significance of AMF diversity in this biodiversity hotspot region and efficient utilization of AM technology in medicinal plants, agriculture and maintenance of sustainable forestry.

References

Bedini, S., Pellegrino, E., Avio, L., Pellegrini, S., Bazzoffi, P., Argese, E. & Giovannetti, M. 2009. Changes in soil aggregation and glomalin related soil protein content as affected by the arbuscular mycorrhizal fungal species *Glomus mosseae* and *Glomus intraradices*, *Soil Biol Biochem*, 41: 1491–1496.

Bever, J.D., Schultz, P.A., Prinkle, A. & Morton, J.B. 2001. Arbuscular Mycorrhizal Fungi: more diverse than meets the eye, and the ecological tale of why, *BioScience,* 51: 923-931.

Brundrett, M.C. & Ashwath, N. 2013. Glomeromycotan mycorrhizal fungi from tropical Australia III. Measuring diversity in natural and disturbed habitats, *Plant Soil,* DOI: 10.1007/s11104-013-1613-4.

Brundrett, M. 1991. Mycorrhizas in natural ecosystems, *Advances in Ecological Research,* 21: 171-313.

Chanda, D., Sharma, G.D., Jha, D.K. & Hijri, M. 2014. The potential use of arbuscular mycorrhiza in the cultivation of medicinal plants in Barak Valley, Assam: a review, *Current World Environment,* 9: 544-551.

Ching J. *et al.,* 2010. Beta-amyrin from *Ardisia elliptica* Thunb. is more potent than aspirin in inhibiting collagen-induced platelet aggregation, *Indian J of Exp Biology,* 48: 275-279.

Dagar, J.C. & Singh, N.T. 1999. Plant resources of the Andaman & Nicobar Islands, Vol I & II. Bishen Singh Mahendra Pal Singh Publication, New Delhi.

Gerdemann, J.W. & Nicolson, T.H. 1963. Spores of mycorrhizal Endogyne species extracted from soil by Wet Sieving and Decanting, *Trans. Br Mycol. Soc.,* 46: 235-244.

Grime J.P., Mackey, J.M.L., Hiller, S.H. & Read, D.J. 1987. Floristic diversity in a model system using experimental microcosms, *Nature,* 328: 420-422.

Guadarrama, P. & Álvarez-Sánchez, F.J. 1999. Abundance of arbuscular mycorrhizal fungi spores in different environments in a tropical rain forest, Veracruz, Mexico, *Mycorrhiza,* 8: 267-270.

Johnson, N.C. & Wedin, D. 1997. Soil carbon, nutrients, and mycorrhizae during conversion of dry tropical forest to grassland, *Ecol Appl,* 7: 171-182.

Li, L.F., Li, T. & Zhao, Z.W. 2007. Differences of arbuscular mycorrhizal fungal diversity and community between a cultivated land and an old field and a never cultivated field in a hot and arid ecosystem of South West China, *Mycorrhiza,* 17: 655-665.

Oehl, F., Sieverding, E., Ineichen, K., Ris, E.A., Boller, T. & Wiemken, A. 2005. Community structure of arbuscular mycorrhizal fungi at different soil depths in extensivelymanaged agroecosystems, *New Phytol,* 165: 273-283.

Okalebo, J.R., Gathua, K.W. & Woomer, P.L. 1993. Laboratory methods of plant and soil analysis ; A working manual. *Technical Bulletin No. 1, Soil Science Society,* East Africa.

Panwar, J. & Tarafdar, J.C. 2006. Distribution of three endangered medicinal plant species and their colonization with arbuscular mycorrhizal fungi, *Journal of Arid Environments,* 65: 337-350.

Sharmah, D. & Jha, D.K. 2011. Diversity of arbuscular mycorrhizal fungi in disturbed and undisturbed forests of Karbi Anglong hill district of Assam, *Current World Environment,* 6: 253-258.

Smith, S.E. & Read, D.J. 2008. Mycorrhizal Symbiosis, 2nd edn, Academic Press.

Tehabi, A., Coyne, D., Hountondji, F., Lawouin, L., Wiemken, A. & Oehl, F. 2008. Arbuscular mycorrhizal fungal communities in sub-Saharan savannas of Benin, west Africa, as affected by agricultural land use intensity and ecological zone, *Mycorrhiza,* 18: 181-195.

van der Heijden, M.G.A., Klironomos, J.N., Ursic, M., Moutoglis, P., Streitwolf-Engel, R., Boller, T., Wiemken, A. & Sanders, I.R. 1998. Mycorrhizal fungal diversity determines plant biodiversity, ecosystem variability and productivity, *Nature,* 396: 69-72.

Zhang, Y., Guo, L.D. & Liu, R.J. 2003. Diversity and ecology of arbuscular mycorrhizal fungi in Dujiangyan, *Acta Phytoecol Sinica,* 27: 537-544.

Zhao, Z.W., Xia, Y.M., Qin, X.Z., Li, X.W., Cheng, L.Z., Sha, T. & Wang, G.H. 2001. Arbuscular mycorrhizal status of plants and the spore density of arbuscular mycorrhizal fungi in the tropical rain forest of Xishuangbanna, southwest China, *Mycorrhiza,* 11: 159-162.

❑❑❑

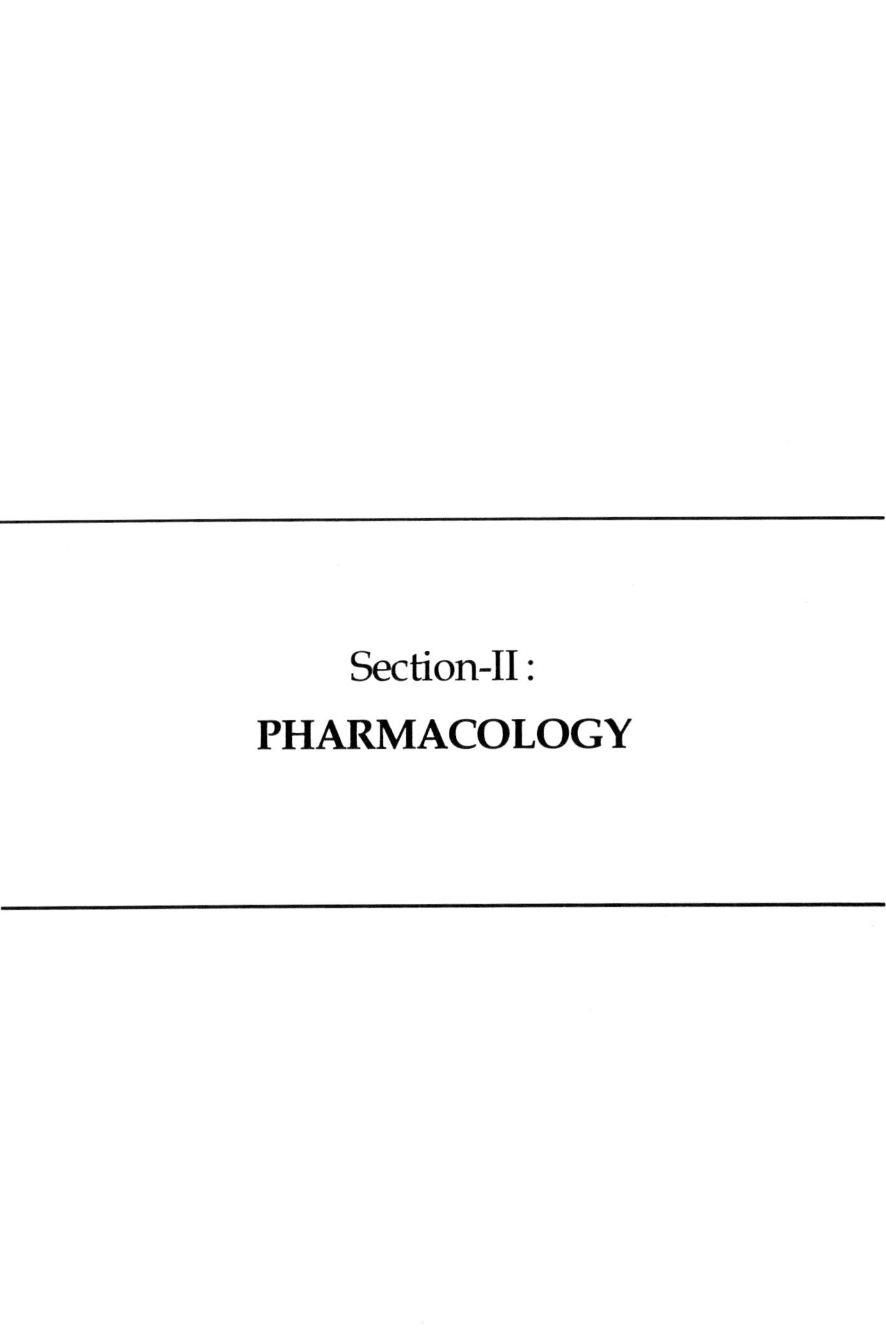

Section-II :

PHARMACOLOGY

Chapter 12

Medicinal Uses, Phytochemistry and Pharmacological Activities of Motha (*Cyperus rotundus* Linn.): A Potential Herb

Vikas Kumar*

Department of Silviculture and Agroforestry, College of Forestry, Kerala Agricultural University, KAU, Thrissur, Kerala-680656 (India), E-mail: vkskumar49@gmail.com

ABSTRACT

Motha (*Cyperus rotundus*) is a perennial plant and is one of the most invasive weeds known, which is spread in tropical, subtropical and temperate regions. People from India and other developing country have relied on traditional herbal preparations to treat themselves. Therefore, it is useful to investigate the potential herbs against various clinical disorders as well as development of new therapeutic agent in human subjects. Thorough screening of literature available on motha depicted the fact that it is a popular remedy among the various ethnic groups, Ayurvedic and traditional practitioners for treatment of ailments such as antibacterial, antimalarial, antiinflammatory, antidiarrhoeal, antimutagenic, antimicrobial, antioxidant, cytotoxic and various other activities. The paper summarizes available information that will aid in future medicine preparation by identifying active ingredients and their mode of action for a specific therapeutic activity using the latest technology. Collected information suggested that motha has limited activity against different forms of infectious diarrhoea due to its selective activity against diarrheal pathogens.

Keywords: *Cyperus rotundus*, Motha, control management, traditional medicine.

Introduction

Biodiversity of natural resources like plants, animals, microbes, minerals and marine sources has served not only the primary human needs but also health care since time immemorial (Patwardhan *et al.*,

2004). Since beginning, there are different theories of disease and different systems employed in medicine, but botanical remedies are universal. The widespread use of herbal remedies and healthcare preparations, as those described in ancient texts such as the Veda and the Bible, has been traced to the occurrence of natural products with medicinal properties. Civilized societies have bequeathed myths and compendiums of healing herbs and the herbal remedies from people of preliterate societies continue to surprise us with their extensive green pharmacy (Balick and Cox, 1996). The biogeographic position of India is unique which makes India rich in all the three levels of biodiversity such as species diversity, genetic diversity and habitat diversity (Krishnaraju *et al.,* 2005). A survey conducted by the All India Coordinated Research Project on Ethnobiology (AICRPE) during the last decade recorded over 8000 species of wild plants used by the tribals and other traditional communities in India for treating various health problems (Laloo *et al.,* 2006). One of the ancient classics, "Charak Samhita" (Chandra and Sharma, 1986) is the oldest text available on the complete treatment of diseases which specifies the use of hundreds of herbs in the complete treatment of diseases. The Ayurveda, whose history goes back to 500 B.C., is one of the ancient health care systems, which is a potential source of indigenous drugs. A large number of such herbs are mentioned in "Bhavprakash" (Vaishya, 1835) as well as "Aryavaidhya Kalanidhi" (Kavade Krishnamurthi, 1986). "Indian Materia Medica" also gives a large number of medicinal plants for the treatment of various diseases (Nandkarni, 1976). In India, over 2600 plant species have been considered useful in the traditional system of medicine like Ayurveda, Unani, Siddha and Home remedies (Khandelwal, 1999). Number of herbal drugs and their compositions are recommended for combating human ailments in the ancient texts as well as in modern medicine (Sastri, 1962; Siddique *et al.,* 1995; Katewa *et al.,* 2004).

Weeds have been reckoned as a major threat globally and act as dampener for crop production despite the farmers putting enormous effort in their removal to get better yield. Despite decades of research and development and advances in management practices aimed at their management, weeds remain as a constant threat to productivity and sustainability of soil and environment (Kumar, 2016a).*Cyperus rotundus* Linn. (Family: Cyperaceae) is a multivalent plant widely used in traditional medicine around the world for treatment of various diseases (Singh *et al.,* 2012; Kumar and Desai, 2014; Kumar, 2015; Kumar, 2016b; Kumar, 2016c). This plant is also known as purple nutsedge or nutgrass (English), Arabic: Soad, Soadekufi; Bangali: Nagarmotha; Burma: Vomonniu; Hindi: Motha; Malaya: Mushkezamin; Gujarat: Nagaramothaya; English: Nut grass; Sanskrit: Chakranksha, Charukesara; Urdu: Saad kufi (www.plantzafrica.com; Parotta, 2001; Imama, 2014; Kumar, 2016d). The genus name *Cyperus* is derived from *Cypeiros,* which was the ancient Greek kupeiros, meaning sedge and *rotundus* is Latin word for round and refers to the tuber (David *et al.*, 2000). The family comprises about 104 genera and more than 5000 species world wide, although number vary greatly due to differing taxonomic concepts of individual researchers. The largest genus is *Carex* with about 2000 species worldwide, followed by *Cyperus* with about 550 species (www.plantzafrica.com). It was formally described by De Jussieu in 1789. In India, it is common in open, disturbed habitats to an elevation of about 1800 m (Parotta, 2001). It's indigenous to India, but now it is found in tropical, subtropical and temperate regions from Asia, South Africa, South America, etc. (Qshaughnessy, 1841; Dymock *et al.*, 1893; Watt, 1908; Wazing, 1968; Kurup *et al.*, 1979; Kirtikar and Basu, 1994; Gordon-Gray, 1995; Uddin *et al.*, 2006; Qasim *et al.*, 2014; Kumar, 2016c), causing large losses in cop yield (Holm *et al.*, 1977). It has become a major weed of vegetable, row and plantation crops and it's very difficult to manage with either organic or conventional weed control strategies (William, 1976~ Bangarwa *et al.*, 2008~ Wang *et al.*, 2008), and has been called the world's worst weed (Holm *et al.*, 1991; Holm *et al.*, 1977; Kadir *et al.*, 2000; Terry, 2001). Motha is one of the most extensively researched non cultivated plant species on the planet, yet the complexities of its life cycle, and its multiple adaptations to environmental extremes and weed control tactics are as yet incompletely understood.

Taxonomy, Botanical Morphology

Cyperus rotundus is a grass like weed dark green glabrous culms, arising from underground tubers (Nadkarni and Nadkarni, 1996; Kirtikar and Basu, 2001; Riemens *et al.*, 2008). Its top growth 4-30 inches tall, an extensive underground network of basal bulbs, fibrous roots, below ground and rosettes of leaves, scrapes and umbels above ground (Isman *et al.*, 2014) and thin wiry rhizomes, and tubers borne in chains of 2-6 or more on rhizomes, with tubers spaced 210 inches apart and widely through rhizome and tubers (Khalida and Siddiqui, 2014).

Fig. 1. Phenology of *Cyperus rotundus* (Motha): **A.** Natural cultivated in agricultural land; **B.** Uprooted motha plant; **C.** Motha flower spikes; **D.** Motha seeds and **E.** Motha roots

The leaves are mostly basal, dark green, 0.1-0.25 inches wide with a prominent midrib, and abruptly tapered at the tips. Leaves long, often overlapping stem. Flowers borne in compound umbel, spikes loosely spicate of 3 8 spixelets (Chatterjee and Pakrashi, 2009) (Fig. 1). The purplish to red brown inflorescence is borne on a culm (stem) that is triangular in cross section and usually taller than the foliage (Bryson and DeFelice, 2009) as well as glabrous culm of unequal length (Bryson and Defelice, 2009). The inflorescence itself consists of an umbel of spikes, some of which are sessile, and others are borne on stalks of unequal length. The subtending leaves like bracts are usually shorter than the longest spikes. Seeds in the form of trigonous nuts, flowers and fruits almost throughout the year, but chiefly during rainy season (Riemens *et al.,* 2008; Chatterjee and Pakrashi, 2009) but the bulk of the seeds will most likely be non-viable (Tayyar *et al.,* 2003 and Riemens *et al.,* 2008). The plant forms a subterranean basal bulb, which contains the shoot meristem (site of cell division and formation of new leaves). Basal bulbs form mostly within 3 inches of the soil surface, although bulbs have been observed at 4-8 inches (Hauser, 1962~ William and Warren, 1975; William, 1976~ Holm *et al.,* 1991; Kadir *et al.,* 2000). Bulbs develop fibrous root systems that may extend 4 feet deep in the soil profile (Holm *et al.,* 1991~ California Department of Food and Agriculture). Because the shoot growing point remains in the basal bulb, leaves regrow readily if severed at the soil surface (William and Warren, 1975). Tubers are usually oblong ovate in shape and blackish to brown in color, with an internally reddish white form and a specific fragrance (Oladipupo and Oyedeji, 2009).

Each tuber has multiple buds, most of which remain dormant and act as a reserve in the event that the active shoot is destroyed. Tuber chains show apical dominance, so that the terminal tuber initiates active growth while many or all of the others on the chain remain dormant unless the terminal tuber is destroyed or the chain is broken (Kawabata and Nishimoto, 2003). Dormant tubers commonly persist in the soil for 3-4 years, and can remain viable for as long as 10 years in some conditions (California Department of Food and Agriculture). The tuber sprout consists of a sharp pointed rhizome, which grows toward the soil surface, then differentiates into shoot and leaves in response to light (Chase *et al.,* 1998). Motha have a remarkable ability to survive adverse conditions and then grow explosively under favourable conditions (Tumbleson and Kommedahl, 1961). Cyperus impact varies with location, probably due to differences in climate, crop rotation or biotype (Stoller *et al.,* 1975; Costa and Appleby, 1976; Keeley *et al.,* 1983; Horak *et al.,* 1987). Losses can result when Cyperus competes with crops to decrease yield or directly damage below ground plant parts, such as onion bulbs (Keeley, 1982).

Phytochemistry: Phytochemical studies have shown that the major chemical components of Motha herb are essential oils, flavonoids, terpenoids, and mono sesquiterpenes. The plant contains the following chemical constituents; cyprotene, acopaene, cyperene, aselinene, rotundene, valencene, cyperol, gurjunene, trans calamenene, dcadinene, gcalacorene, cadalene, amuurolene, gmuurolene, cyperotundone, mustakone, isocyperol, acyperone (Meena *et al.,* 2010). 4, 11 selinnadien 3 one and 1, 8 cineole (Visetson *et al.,* 2001). The oil of C. rotundus was mainly composed of cyperol, α cyperene, rotundine, α cyperone, α copaene, valerenal, myrtenol, α pinene, α pinene and α Selinene, sesquiterpene hydrocarbons (Caryophyllene) (Neema et al., 2008; Bisht *et al.,* 2011) (Fig. 2).

Traditional and Contemporary Activity / Scientific Reports

Motha is a traditional medicinal plant appearing among Indian, Chinese and Japanese natural drugs used against spasms, stomach and bowel disorders and menstrual irregularities (Sharif Ali *et al.,* 2008; Ahmad *et al.,* 2012; Hemanth *et al.,* 2013; Srivastava *et al.,* 2013). Based on the literature available on traditional as well as potential uses, many researchers have extensively investigated the phytochemical, physicochemical, and pharmacological properties of this plant to identify the compounds responsible for these actions (Thebtaranonth *et al.*, 1995; Sonwa and König, 2001). Motha with a large number of biologically active phytochemicals has diverse variety of pharmacological properties, as described below, has been found effective in the treatment of chronic disorders (Singh *et al.*, 2012). The essential oil of motha has reported posses antibacterial activity (Ghedira *et al.*, 2008; Nima *et al.*, 2008), antioxidant activity (Kilani, *et al.*, 2005; Pal and Dutta, 2006; Kilani-Jaziri *et al.*, 2009), antimutagenic activity (Kilani, *et al.*, 2007), antidiabetic activity (Raut and Gaikwad, 2006), antidiarrhoeal (Uddin *et al.*, 2006), antimalarial activity (Weenen *et al.*, 2009), antiinflammatory, antipyretic and analgesic activities (Gupta *et al.*, 1971), hepatoprotective (Kumar and Mishra, 2005), woundhealing properties (Shivakumar *et al.*, 2009; Dang *et al.*, 2011), protective effects against free radical-mediated protein oxidation (Bahramikia *et al.*, 2009) and inhibitory effects against superoxide and nitric oxide production (Seo *et al.*, 2001). However, allelopathic effects and competing for growth resources with crops in farmland are claimed as most important of disadvantages of motha, that causes heavy loss to crops and reduce the yield (Gilreath and Santos, 2004). The rhizome part of motha is one of the oldest known medicinal plants used for treatment of dysmenorrheal and menstrual irregularities (Bhattarai, 1993; El-Kaream, 2012; Ghannadi *et al.*, 2012). Dangwali *et al.* (2010) had reported that decoction prepared by burning and adding the ash of fresh leaves of *Ammannia baccifera* L. (10 g) and *C. rotundus* roots (10 g) and fresh ginger (5 g) in sesame oil is used to cure skin eruptions.

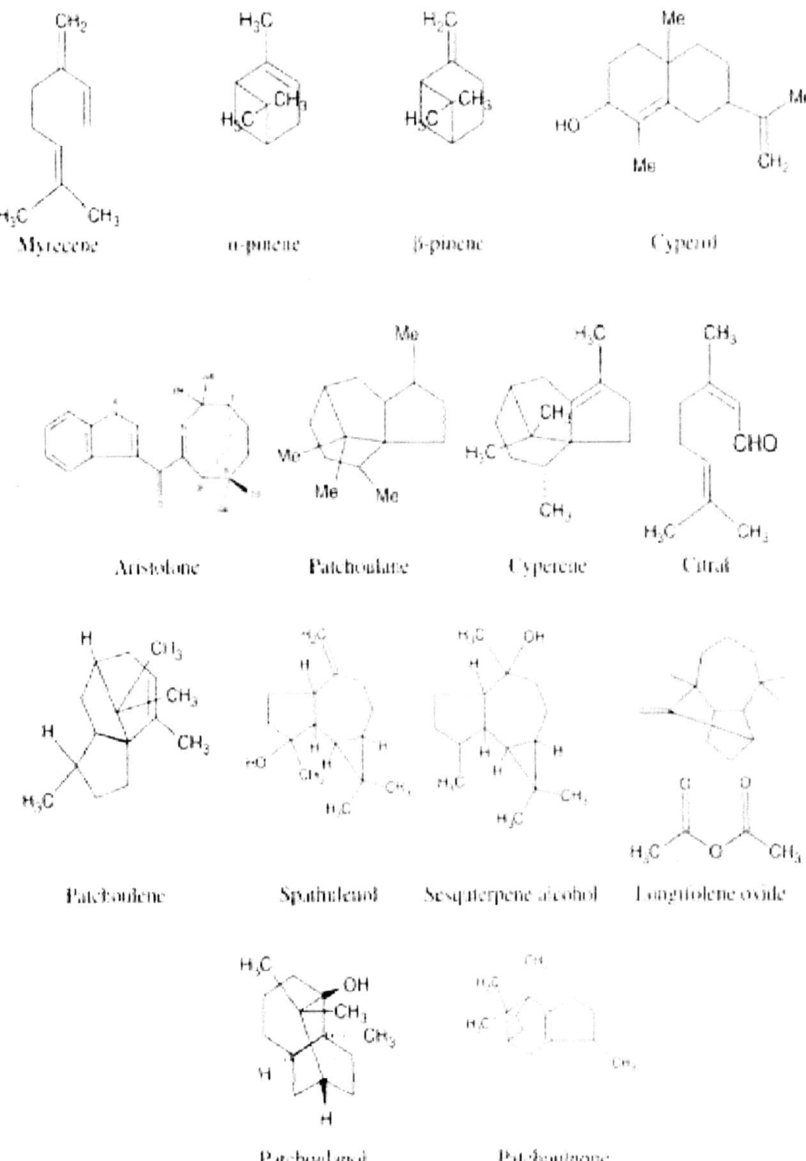

Fig. 2. Some of the chemical constituents of Motha (*Cyperus rotundus*)

Das and Mishra (1988) reported that stem bark of *Holarrhena pubescens*, fruits and leaves of *Punica granatum*, bulbous root of *C. rotundus*, dried young fruits of *Aegle marmelos*, and flowers of *Woodfordia fruticosa*, are orally used to treated dysentery. *C. rotundus* bark (200 g), *C. speciosus* rhizomes (200 g), and *Azadirachta indica* bark (200 g) are boiled in 1 L of water to reduce the volume to one-fourth. Two to four spoons of decoction are prescribed after every meal for 15 days (Paul et al., 2013). *Lawsonia inermis* and *Azadirachta indica* leaves are macerated with the plant of *C. rotundus*, which are topically applied on the infected area and also suggested that a cup of juice obtained from the macerated flowers of *C. rotundus* with 10-15 drops of honey is used to treat dysentery (Jahan *et al.*, 2011).

Pharmacological activities

Analgesic activity: Mohsin *et al.* (1989) revealed that ethanolic and hot water extracts of *C. rotundus* at 500 mg/ kg and 12.7 g/kg were inactive in the hot plate method and acetic acid writhing test, in addition to being non-analgesic. Later on, the analgesic activity of the crude extract of *C. rotundus* was evaluated by the tail-flick method on mice (Mansoor *et al.*, 2012). When administered orally, the crude extract at 300 mg/kg body weight (b.w.) (dissolved in 0.9% saline solution) showed highly significant reaction time (570.45 s) as compared with control and standard drugs. Imam and Sumi (2014) concluded that the chemical constituents of the plant contain active ingredients that could be extracted to develop promising antinociceptive medicines.

Antibacterial activity: *Proteus vulgaris* was sensitive to acetone extract (6 mg/disc) of motha, while other bacteria (*e.g.* Escherichia coli, Pseudomonas aeruginosa) were resistant to motha with 10 mg/ disc (maximum dose tested). When compared with other herbs, motha have mild antibacterial activity in high doses (Tambekar *et al.*, 2009). A mark inhibitory effect of motha was observed against *Salmonella enteritidis*, *Staphylococcus aureus* and *Enterococcus faecalis* with total oligomers flavonoides (TOFs) and ethyl acetate extracts (Kilani *et al.*, 2008). In light of the traditional claim of motha in the treatment of diabetes, investigations were carried out to evaluate its effect on alloxan induced hyperglycemia in rats. Oral daily administration of 500 mg/kg of the extract (once a day for seven consecutive days) significantly lowered the blood glucose levels. This antihyperglycemic activity can be attributed to its antioxidant activity as it showed the strong DPPH radical scavenging action in vitro (Raut and Gaikwad, 2006).

Antimalarial activity: Activity guided investigation of motha tubers led to the isolation of patchoulenone, caryophyllene or oxide,

10, 12 peroxycalamenene and 4, 7 dimethyl l tetralone. The antimalarial activities of these compounds are in the range of ECso 10 4-10 6 M, with the novel ndoperoxide sesquiterpene, 10, 12 peroxycalamenene, exhibiting the strongest effect at ECso 2.33 × 106 M (Thebtaranonth *et al.*, 1995).

Antidiarrheal activity: *C. rotundus* tubers have been traditionally used in several Ayurvedic formulations to treat diarrhea (Agarwal *et al.*, 2005). The methanol extract of motha rhizome, given orally at the doses of 250 to 500 mg/kg between, showed significant antidiarrheal activity in castor oil induced diarrhoea in mice. Among the fractions, tested at 250 mg/kg, the petroleum ether fraction (PEF) and residual methanol fraction (RMF) were found to retain the activity, the latter being more active as compared to the control. The ethyl acetate fraction (EAF) did not show any antidiarrhoeal activity (Uddin *et al.*, 2006). The antidiarrheal potential of an aqueous extract of *C. rotundus* against castor oil-induced diarrhea in mice. They concluded that pretreatment of mice with aqueous extract dependently decreases the purging frequency (47.45%) at 500 mg/kg p.o. through an antisecretory mechanism (Shamkuwar *et al.*, 2012).

Anti-inflammatory activity: Alkaloids, betacyanins, flavonoids, quinones, terpenoids, and phenols account for the antioxidant potential of *C. rotundus* (Liu, 2004). The alcoholic extract (70% alcohol) possessed anti inflammatory activity against carrageenan induced oedema and also found effective against formaldehyde induced arthritis in albino rats (Sundaram *et al.*, 2008). In another study the petroleum ether extract of the rhizomes showed anti-inflammatory activity against carrageenan induced oedema in albino rats. The triterpenoid obtained by chromatographic separation from petroleum ether extract revealed a high potent anti-inflammatory activity. This terpenoid was also found to possess significant antipyretic and analgesic effects similar to acetyl salicylic acid. Motha have also reported as protective in inflammatory bowel disease. In addition, the extract suppressed the production of O_2- by phorbol ester stimulated RAW 264.7 cells in dose and time dependent manners. Collectively, these results suggest that the methanol extract of rhizomes of motha could be developed as anti-inflammatory candidate for the treatment of inflammatory diseases mediated by overproduction of NO and $O--_2$ (Soe *et al.*, 2001). Another study on alcoholic extract of motha showed highly significant ($P<0.001$) anti-inflammatory activity against the exudative and proliferative phases of inflammation in two animal models (carrageenan induced oedema and formaldehyde induced arthritis in rats). Its anti-inflammatory relative effect was higher than that of hydrocortisone (75.9% versus 47.3% in carrageenan-induced oedema model; 55.1% versus 35.6% in formaldehyde induced arthritis model (Kapadia *et al.*, 1967; Singh *et al.*, 1969; Singh *et al.*, 1970; Singh and Kulshrestha, 2010). In general, bioactive compounds present in *C. rotundus*, including nootkatone, α-cyperone, β-selinene, and valencene, contribute to its anti-inflammatory activity through their action on hemeoxygenase-1 pathway (Tsoyi *et al.*, 2011; Khan *et al.*, 2011).

Analgestic and sedative activity: The total decocts of rhizomes of motha showed analgesic effects in the acetic acid writhing test (Vu and Mai, 1994). The petroleum ether extract and essential oil of the plant are also reported to possess analgesic activity (Gupta *et al.*, 1971; Birdar *et al.*, 2010). Further, the study conducted on isocurcumenol, a sesquiterpene isolated from motha have demonstrated that this compound acts as a benzodiazepine receptor agonist and thus, positively modulates- GABAergic neurotransmission via enhancement of interaction of aminobutyric acid (GABA) with its receptor in animals. The benzodiazepine receptor complex. GABA being an inhibitory neurotransmitter, these finding provide a pharmacological explanation for the empirical use of motha as sedative (Ha *et al.*, 2002).

Antimutagenic and radical scavenging activity: This study evaluates mutagenic and antimutagenic effects of aqueous, total oligomers flavonoids (TOF), ethyl acetate and methanol extracts from aerial parts of motha with the *Salmonella typhimurium* assay system. The different extracts showed no mutagenicity when tested with *Salmonella typhimurium* strains TA98, TA100, TA1535 and TA1538 either with or without the S9 mix. On the other hand, our results showed that all extracts have antimutagenic activity against aflatoxin B1 (AFB1) in TA100 and TA98 assay system and against sodium azide in TA100 and TA1535 assay system. TOF, ethyl acetate and methanol extracts exhibited the highest inhibition level of the Ames response induced by the indirect mutagen AFB1.Whereas, ethyl acetate and methanol extracts exhibited the highest level of protection towards the direct mutagen, sodium azide, induced response. In addition to antimutagenic activity, these extracts showed an important free radical scavenging activity toward the 1,1 diphenyl 2 picrylhydrazyl free radical. TOF, ethyl acetate and methanol extracts showed IC50 value of 15, 14 and 20 g/ml, respectively (Kairani *et al.*, 2005).

Antioxidant activity: Antioxidant activity of motha rhizomes extract (CRRE) was evaluveted in a series of in vitro assay involving free radicals and reactive oxygen species and IC50 values were determined. CRRE exhibited its scavenging effect in concentration dependent manner on superoxide anion radicals, hydroxyl radicals, nitric oxide radical, hydrogen peroxide, and property of metal chelating and reducing power. The extract was also studied for lipid peroxidation assay by thiobarbituric acid-reactive substances (TBARS) using young and aged rat brain mitochondria. The extract was also effective in preventing mitochondrial lipid peroxidation induced by $FeSO_4$ ascorbate in concentration dependent manner. The results obtained in the present study indicate that motha rhizomes extract can be a potential source of natural antioxidant (Nagulendran *et al.*, 2007).

Antipyretic activity: The alcoholic extract of motha showed highly significant ($P<0.001$) antipyretic activity against pyrexia produced in albino rats by the subcutaneous injection of suspension of dried Brewer's yeast in gum acacia in normal saline. A specific fraction obtained by chromatographic method from the petroleum ether extract was found to possess a significant anti-pyretic effect similar to acetyl salicylic acid when used on the same animal model (Gupta *et al.*, 1971).

Antispasmodic activity: An aqueous extract of rhizomes of motha (ACR) was tested for its anti diarrheal and anti spasmodic activity. Anti diarrheal effect of ACR was evaluated in castor oil induced diarrhea in mice and antispasmodic effect was evaluated by charcoal meal test in mice at a dose of 125, 250, 500 mg/kg. The % inhibition of diarrhea was 30.36%, 37.90%, 45.45% and 92.45% for ACR 125, 250, 500 mg/kg (po) and loperamide 2 mg/kg dose (po) respectively. ACR 125, 250, 500 mg/kg (po) and atropine sulfate 2 mg/kg dose (po) produced 24.35%, 31.48%, 36.75% and 55.94% inhibition of intestinal transit respectively. These results were indicated that ACR produces its anti diarrheal effect through decreasing intestinal secretions and anti spasmodic effect by inhibiting the intestinal motility (Shamkuwar *et al.*, 2012).

Antiplatelet activity: Seo *et al.* (2011) examined the antiplatelet activities of *C. rotundus* extract and its eight constituents against collagen-, thrombin-, and arachidonic acid (AA)-induced rat platelet aggregation. Among all the tested materials, (þ)-nootkatone along with the *C. rotundus* extract significantly inhibits platelet aggregation inrat.

Anti-emetic activity: The ethanolic extract of motha in the dose of 128.1± 11.6 mg/kg was found to protect 50% dogs against apomorphine induced vomiting (Singh *et al.,* 1970).

Insect repellency activity: Hexane extract of tuber of motha was screened under laboratory conditions for repellent activity against mosquito vector Anopheles culicifacies, Anopheles stephensi and Culex quinquefasciatus. The motha tuber extract was used to determine their effect on mosquito vector and comparison with the N, N diethyl 3 methylbenzamide. Result obtained from the laboratory experiment showed that the tuber extracts are more effective for repellency of the entire mosquito vector even at a low dose (Singh *et al.,* 2009).

Antispatic activity: Ethanolic extract of motha produced relaxation of rabbit ileum and spasmolytic effect against contractions induced by acetylcholine, barium chloride and 5-hydroxitriptamine, showing a direct relaxant action on the smooth muscle (Singh *et al.,* 1970).

Inhibition of gastric motility activity: The rhizome of motha was assessed for its cytoprotective effects against ethanol induced gastric damage. Decoctions of Rhizoma Cyperi were given orally to rats 30 min. before ethanol was administered. It suggested that the protective action of motha is related to its inhibition of gastric motility and endogeneous prostaglandins may play an important role (Zhu *et al.,* 1997).

Lactogenic activity: Badgujar and Bandivdekar (2015) had reported that it is worth noting that *C. rotundus* rhizomes were widely used to develop mammary glands and increase the quantity of breast milk in humans (Ahmad, 2007). Recently, the lactogenic activity of *C. rotundus* rhizomes was studied in rats.

Ovicidal and larvicidal activity: The ovicidal and larvicidal effect of essential oils extracted from the tubers of motha was studied on eggs and fourth instar larvae of Aedes albopictus. The eggs and larvae were exposed to serial concentration of the oils ranging from 5 to 150 ppm and kept under observation for 24 h. Both the oils showed remarkable ovicidal and larvicidal activities indicated by EC50 values of<5 ppm and LC50 and LC90 values of <20 ppm. The results obtained suggested that the essential oils of these Cyperus species can serve as a potential source of natural mosquitocidal agents (Kempraj and Bhat, 2008).

Wound healing activity: The present study was aimed to evaluate the wound healing activity of extract of tuber of motha. On the basis of traditional use and literature references, this plant was selected for evaluation of wound healing potential. An alcoholic extract of tuber of motha was examined for wound healing activity in the form of ointment in three types of wound models on rats: the excision, the incision and dead space wound model. The extract ointments showed considerable difference in response in all the above said wound models as comparable to those of a standard drug nitrofurazone ointment (0.2 % w/w NFZ) in terms of wound contracting ability, wound closure time and tensile strength (uratchikody *et al.,* 2006).

Cultural Control Methods

The manual removal of weeds is the predominant method of weed management practices in tropics. It is highly laborious, inefficient and often uneconomical. Weed competition during the early stages of crop growth and also competing for light, moisture, nutrients and space which will be ultimate adverse effect on growth of the major plant and yield. Ideal, weeding is done 3-4 weeks after the planting (Kumar, 2017).

Hand-weeding: When hand-weeding, the wiry connections between tubers make them easier to remove than if the connections have been severed by soil cultivation. The tuber, or basal bulb, of an emerged shoot must be removed to control motha. Simply removing the top growth is ineffective, and an inch or so of new growth will emerge the following day. Hence, dig up and remove tubers from the soil to greatly reduce the populations of motha (Kumar, 2016c).

Mulching ornamental areas: Mechanical control of an invasive perennial weed infestation begins with an initial vigorous tillage to fragment the weed, followed by additional cultivations whenever fragments have regenerated new shoots with 3-4 leaves, at which time the weed's underground reserves have been drawn down to their lowest point (Mohler and DiTommaso, 2012). Motha will not be controlled with organic mulches. Sheet mulches also are ineffective against this species. Black plastic sheet mulch is not effective because the sharp, pointed shoot tip of the motha plant can easily penetrate it. Weed cloth, or woven black polypropylene weed mat, can be effective in suppressing motha when used properly. It is porous to air and water and can be an effective tool for reducing underground tubers without the use of chemicals or tedious hand-weeding.

Chemical Control Methods

Non-selective herbicides

One of the most effective herbicides for non-selective control of motha is glyphosate (several products with this active ingredient are commercially available). Glyphosate is translocated throughout the entire plant, allowing it to eradicate the underground tubers connected to the leaves. Application of timing play important role to control motha. To obtain maximum control, motha flowers should be present before herbicide application.

Selective herbicides

There are many selective herbicides that can be used effectively to control motha in warm-season turfgrasses. Herbicide applications are most effective when applied to motha plants having at least three to eight leaves. Suggested application rates for these herbicides are Sulfosulfuron (1.25 oz per acre), Trifloxysulfuron sodium (20 grams per hectare), Monosodium methyl arsenate (2.0 lb *a.i* per acre) and Halosulfuron (1.333 oz per acre (0.062 lb *a.i.* per acre) (Kumar, 2016c). It indicated that Motha density could be reduced with glyphosate applied in soybean (Reddy and Bryson, 2009). Glyphosate has considerable activity on Motha and has effectively reduced populations in various cropping systems (Bryson *et al.*, 2003; William and Hirase, 2004; Edenfield *et al.*, 2005; Webster *et al.*, 2008).

To control an existing stand of motha in an actively growing warm-season, at least two applications of most herbicides will usually be required. Four to six weeks after making an initial herbicide application, new motha shoots will emerge. The plants arise from dormant tubers that were not connected to motha shoots when the initial herbicide application was made. This new stand of motha must be treated to prevent new tubers from repopulating the soil. If possible, wait 2-3 months between herbicide applications to allow the entire population of motha shoots to emerge before reapplication.

Positive economic impact: There are no known modern positive economic impacts of motha.

Negative economic impact: A variety of crops other than cotton have seen loss of yields ranging up to 75% in sugar cane (Skinner, 2007). Approximate quantities of fertilizer that may be mobilized and stored in motha equal 815 kilograms of ammonium sulfate, 320 kilograms of potash, and 200 kilograms of phosphate per hectare (Skinner, 2007).

Ecological impacts: Many noxious annual and perennial weeds have been regarded as species with allelopathic potential and can severely affect crop survival and productivity (Qasem 1994; Qasem, 2001). Phytotoxicity of allelochemicals present in the weed extracts might be caused synergistic activity on the germination and growth of rice seedlings rather than single chemical. The statistically observed significances are evident for the inhibitory effects of *C. ortundus* and *C. dactylon* on the growth of rice cultivars (Geethambigai and Prabhakaran, 2014). Motha contains allelopathic chemicals that inhibit the growth of surrounding vegetation, giving it a competitive edge that can crucially impact native ecosystems. The inhibitory compounds of motha are released through root exudation, volatilization and decaying of plant residues (Alsaadawi and Salih, 2009). The release of volatile and non-volatile compounds into the environment inhibits the growth of surrounding plant species, particularly impacting the development of seedlings. The impact of phytotoxicity resulting from the allelopathic depends on the receptor plant species (Alsaadawi and Salih, 2009). Motha compete for moisture, nutrient and sunlight resources, and can produce up to 40,000 kilograms of subterranean plant material per hectare (Skinner, 2007).

References

Agarwal, S.S., Tamrakar, B.P. and Paridhavi, M. 2005. Clinically Useful Herbal Drugs. Ahuja Book Company, Darya Ganj, Delhi, India.

Ahmad, M., Mahayrookh, M., Rehman, A.B. and Jahan, N. 2012. Analgesic, antimicrobial and cytotoxic effect of *Cyperus rotundus* ethanol extract. *Pak. J. Pharmacol.* 29: 7-13.

Alsaadawi, I.S. and Salih, N.M.H. 2009. Allelopathic potential of *Cyperus rotundus* L. II. Isolation and identification of Phytotoxins. *Allelopathy Journal* 23(2): 85-90.

Badgujar, S.B. and Bandivdekar, A.H. 2015. Evaluation of a lactogenic activity of an aqueous extract of *Cyperus rotundus* Linn. *J. Ethnopharmacol.* 163: 39-42.

Bahramikia, S., Ardestani, A. and Yazdanparast, R. 2009. Protective effects of four Iranian medicinal plants against free radical-mediated protein oxidation. *Food Chem.* 115: 37-42.

Balick, M.J. and Cox, P.A. 1996. Plants, People and Culture: The Science of Ethnobotany. New York: Scientific American Library.

Bangarwa, S.K., Norsworthy, J.K., Jha, P. and Malik, M. 2008. Purple nutsedge (*Cyperus rotundus*) management in an organic production system. *Weed Science* 56: 606-613.

Bhattarai, N.K. 1993. Folk herbal remedies for diarrhea and dysentery in central Nepal. *Fitoterapia* 64: 243-250.

Birdar, S., Kangralkar, V.A., Mandavkar, Y., Thakur, M. and Chougule, N. 2010. Anti-inflammatory, anti-arthritic, analgesic and anticonvulsant activity of Cyperus essential oils. *Int J Pharm Pharmaceut Sci* 2(4): 112-115.

Bisht, A., Bisht, G.R., Singh, M., Gupta, R. and Singh, V. 2011. Chemical composition and antimicrobial activity of essential oil of tubers of *Cyperus rotundus* Linn. collected from Dehradun (Uttarakhand). Int *J Res Pharm Biomed Sci* 2: 661 665.

Brosnan, J.T. and DeFrank, J. 2008. Purple Nutsedge Control in Turf and Ornamentals. Cooperative Extension Service. Revises Nutgrass control in the lawn, landscape, and garden by Nishi. K., DeFrank, *J. and Hensley*, D.L. 1998.

Bryson, C.T. and DeFelice, M.S. 2009. Weeds of the South. University of Georgia Press, Athens, GA.

Bryson, C.T., Reddy, K.N. and Molin, W.T. 2003. Purple nutsedge (Cyperus rotundus) population dynamics in narrow row transgenic cotton (*Gossypium hirsutum*) and soybean (*Glycine max*) rotation. *Weed Technol* 17: 805-810.

Chandra, K.A. and Sharma, P. 1986. Charak Samhita. Vol. I to IV., Sri Satguru Publications , Delhi.

Chase, C.A., Sinclair, T.R., Shilling, D.G., Gilreath, J.P. and Locascio, S.J. 1998. Light effects on rhizome morphogenesis in nutsedges (Cyperus spp): Implications for control by soil solarisation. *Weed Science* 46: 575-580.

Chatterjee, A. and Pakrashi, S.C. 2009. The Treatise on Indian Medicinal Plants. Vol.VI. New Delhi: National Institute of Science Communication (CSIR), p. 809 1250.

Dangwal, L.R., Sharma, A., Kumar, N., Rana, C.S. and Sharma, U. 2010. Ethno-medico botany of some aquatic Angiospermae from North-West Himalaya. *Researcher* 2(4): 49-54.

Das, P.K. and Misra, M.K. 1988. Some ethnomedicinal plants of Koraput district Orissa. *Anc. Sci. Life* 8(1): 60-67.

Daswani, P.G., Brijesh, S., Tetali, P. and Birdi, T.J. 2011. Studies on the activity of *Cyperus rotundus* Linn. tubers against infectious diarrhea. *Indian J. Pharmacol.* 43(3): 340.

David, W.H., Vernon, V.V. and Jason, A.F. 2012. Purple nutsedge (*Cyperus rotundus* L.). Florida (U.S.A): Institute of Food and Agricultural Sciences, University of Florida, p. 02 15.

Dymock, W., Warden, C.J.H. and Hooper, D. 1893. In: Pharmacographia Indica, Kegan Paul, Trench, Trubner & Co. Ltd London.

Edenfield, M.W., Brecke, B.J., Colvin, D.L., Dusky, J.A. and Shilling, D.G. 2005. Purple nutsedge (*Cyperus rotundus*) control with glyphosate in soybean and cotton. *Weed Technol* 19: 947-953.

El-Kaream, G.F.A. 2012. Role of *Cyperus rotundus* oil in decreasing hair growth. *J. Intercult. Ethnopharmacol.* 1: 111-118.

Family: Cyperaceae. Available from: http://www.plantzafrica.com/plantcd/cyperaceae.htm. [Last cited on 2016 Apr 20].

Geethambigai, C.S and Prabhakaran, J. 2014. Allelopathic Potential of *Cyperus Rotundus* L. and *Cynodan Dactylon* L. on Germination and Growth Responses of Some Rice Cultivars, *Int J Curr Biotechnol* 2(12):41-45.

Ghannadi, A., Rabbani, M., Ghaemmaghami, L. and Malekian, N. 2012. Phytochemical screening and essential oil analysis of one of the Persian sedges, *Cyperus rotundus* L. *Int. J. Pharm. Sci.* 3: 424-427.

Ghedira, K., Sghaier, M.B., Kilani, S., Limem, I., Bouhlel, I., Boubaker, J., Bhouri, W., Skandrani, I., Neffatti, A., Ammar, R.B., Dijoux-Franca, M.G. and Chekir-Ghedira, L. 2008. In vitro evaluation of antibacterial, antioxidant, cytotoxic and apoptotic activities of the tubers infusion and extracts of *Cyperus rotundus*. *Bioresour Technol* 99: 9004-9008.

Gilreath, J.P. and Santos, B.M. 2004. Herbicide dose and incorporation depth in combination with 1, 3-dichloropropene plus chloropicrin for *Cyperus rotundus* control in tomato and pepper. *Crop Prot* 23: 205-210.

Gordon-Gray, K.D. 1995. Cyperaceae in Nepal, National Botanical Institute, Pretoria, South Africa, pp. 45-76.

Gupta, M.B., Palit, T.K., Singh, N. and Bhargava, K.P. 1971. Pharmacological study to isolate the active constitutents of *Cyperus rotundus* responsible for anti-flammatory, antipyretic and analgestic activity. *Indian J Med Res* 59: 76-82

Ha, J.H., Lee, K.Y., Choi, H.C. and Cho, J. 2002. Modulation of radioligand binding to the GABA A-benzodiazepine receptor complex by a new component from *Cyperus rotunudus. Bio Pharm Bull* 25(1): 128-130.

Hauser, E.W. 1962. Development of purple nutsedge under field conditions. *Weeds* 10: 315-321.

Hemanth, K.K., Tamatam, A., Pal, A. and Khanum, F. 2013. Neuroprotective effects of *Cyperus rotundus* on SIN-1 induced nitric oxide generation and protein nitration: ameliorative effect against apoptosis mediated neuronal cell damage. *Neurotoxicology* 34: 150-159.

Holm, L.G., Plucknett, D.L., Pancho, J.V. and Herberger, J.P. 1977. The World's Worst Weeds, In: Distribution and Biology, University Press of Hawaii, Honolulu.

Holm, L.G., Plucknett, D.L., Pancho, J.V. and Herberger, J.P. 1991. The world's worst weeds. Kriegar Publishing Company, Malabar, FL.

Imam, H., Zarnigar, Sofi, G., Aziz, S. and Lone, A. 2014. The incredible benefits of Nagarmotha (*Cyperus rotundus*). International Journal of Nutrition, Pharmacology, *Neurological Diseases* 4(1): 23-27.

Imam, M.Z. and Sumi, C.D. 2014. Evaluation of antinociceptive activity of hydromethanol extract of *Cyperus rotundus* in mice. *BMC Complement Altern. Med.* 14(1): 83-87.

Jahan, F.I., Hasan, M.R.U., Jahan, R., Seraj, S., Chowdhury, A.R., Islam, M.T., Khatun, Z. and Rahmatullah, M, 2011. A comparison of medicinal plant usage by folk medicinal practitioners of two adjoining villages in Lalmonirhat district, Bangladesh. American Eurasian. *J. Sustain. Agric.* 5(1): 46-66.

Kadir, J. and R. Charudattan, R. 2000. Dactylaria higginsii, a fungal bioherbicide agent for purple nutsedge (*Cyperus rotundus*). Biological control: theory and applications in pest management 17: 113-124.

Kapadia, V.H., Naik, V.G., Wadia, M.S. and Dev, S. 1967. Sesquiterpenoids from Essential oil of *Cyperus rotundus. Tetrahedron Lett* 46-61.

Katewa, S.S., Choudhari, B.L. and Jain, A. 2004. Folk herbal medicines from tribal areas of Rajasthan, *India. J Ethnopharmacol* 92: 41-46.

Kavade Krishnamurthi, G.G. 1986. Aryavaidyakalanidhi, Shri Venkateshwar (steam) Mudranalaya, Mumbai.

Kawabata, O. and Nishimoto, R.K. 2003. Temperature and rhizome chain effect on sprouting of purple nutsedge (*Cyperus rotundus*) ecotypes. *Weed Science* 51: 348-355.

Kempraj, V. and Bhat, S.K. 2008. Ovicidal and larvicidal activities of Cyperus giganteus Vahl and *Cyperus rotundus* Linn. essential oils against *Aedes albopictus* (Skuse). *Nat Prod Radiance* 7: 416 419.

Khalida, S. and Siddiqui, S. 2014. Weeds of Pakistan: Cyperaceae. *Pak. J. Weed Sci. Res.* 20: 233-263.

Khan, S.M., Choi, R.J., Lee, D.U. and Kim, Y.S. 2011. Sesquiterpene derivatives isolated from *Cyperus rotundus* L. inhibit inflammatory signaling mediated by NF-êB. *Nat. Prod. Sci.* 17(3): 250-255.

Khandelwal, N.K. 1999. Newsletter of AIDOCOC. 2: 11.

Kilani, S., Ben Sghaier, M., Limen, I., Bouhlel, I., Boubaker, J., Bhouri, W., Skandrani, I., Neffatti, A., Ben Ammar, R., Dijoux-Franca, M.G., Ghedira, K. and Chekir-Ghedira, L. 2008. *In Vitro* evaluation of antibacterial, antioxidant, cytotoxic and apoptotic activities of the tubers infusion and extracts of *Cyperus rotundus. Bioresour Technol* 99(18): 9004-9008.

Kilani, S., Ben, A.R., Bouhlel, I., Abdelwahed, A., Hayder, N., Mahmoud, A. and et al. 2005. Investigation of extracts from (Tunisian) *Cyperus rotundus* as antimutagens and radical scavengers. *Environ Toxicol Pharmacol* 20: 478 484.

Kilani, S., Bouhlel, I., Ammar, R.B., Sghair, M.B., Skandrani, I., Boubaker, J., Mahmoud, A., Dijoux-Franca, M.G., Ghedira, K. and Chekir-Ghedira, L. 2007. Chemical investigation of different extracts and essential oil from the tubers of (Tunisian) *Cyperus rotundus.* Correlation with their antiradical and antimutagenic properties. *Ann Microbiol* 57: 657-664.

Kilani-Jaziri, S., Neffati, A., Limem, I., Boubaker, J., Skandrani, I., Sghair, M.B., Bouhlel, I., Bhouri, W., Mariotte, A.M., Ghedira, K., Dijoux Franca, M.G. and Chekir-Ghedira, L. 2009. Relationship correlation of antioxidant and antiproliferative capacity of *Cyperus rotundus* products towards K562 erythroleukemia cells. *Chem-Biolo Interact* 181: 85-94.

Kirtikar, K.R. and Basu, B.D. 1944. Indian Medicinal Plants, Vol. IV, Basu, L.N., 49, Leader Road, Allahabad, India, p. 2638.

Kirtikar, K.R. and Basu, B.D. 2001. Indian Medicinal Plants, 2nd Edition. Oriental Enterprise, Dehradun, pp. 3633-3635.

Krishnaraju, A.V., Rao, T.V.N., Sundararaju, D., Vanisree, M., Tsay, H.S. and Subbaraju, G.V. 2005. Assessment of bioactivity of Indian medicinal plants using brine shrimp (*Artemia salina*) lethality assay. *Int J App Sci Engg* 2: 125-134.

Kumar, S.V.S. and Mishra, H.S. 2005. Hepatoprotective activity of rhizomes of *Cyperus rotundus* linn. against carban tetrachloride induced hepatotoxicity. *Ind J pharm Sci* 671: 84 88.

Kumar, V. 2015. Ethno-Medicinal plants in five forest ranges in Dang's district, South Gujarat. *Indian J Trop Biodiv* 23(2): 148-156.

Kumar, V. 2016a. Weeds in tropics: problems and prospects. *Van Sangyan* 3(2): 1-6.

Kumar, V. 2016b. Utilization of Plant Biodiversity for Treatment and Cure of Diseases of Human Beings in Thalamalai hills, Namakkal District, Tamil Nadu, India. In: Jha, S.K. (eds), Status and advancement in ethnobotany (In press).

Kumar, V. 2016c. Diversity of ethno medicinal plants in Thalamalai hills, Namakkal district, Tamil Nadu, India. Annals of Pharmacy and Pharmaceutical Sciences (In press).

Kumar, V. 2016d. Importance of Motha (*Cyperus rotundus* Linn.) and its control management. *Van Sangyan* 3(5): 37-43.

Kumar, V. and Desai, B.S. 2014. Indigenous Knowledge of wild plant species of South Gujarat. In: Sanjeev Kumar, Ethnobotanical Studies in India. Deep Publication, p. 303-310.

Kumar, V. 2017. Advances forest plantation nursery management. Scientific Publication. New Delhi (In press).

Kurup, P.N.V., Ramdas, V.N.K. and Joshi, P. 1979. Musta, Handbook of Medicinal Plants, Revised and enlarged, CCRAS, New Delhi, pp. 146-147.

Laloo, R.C., Kharlukhi, L., Jeeva, S. and Mishra, B.P. 2006. Status of medicinal plants in the disturbed and the undisturbed sacred forests of Meghalaya, northeast India: population structure and regeneration efficacy of some important species. *Curr Sci* 90: 225-232.

Li, Y. 2014. Determination of ferulic acid content in *Cyperus rotundus* by HPLC. *J. Chem. Pharm. Res.* 6(3): 1496-1500.

Mansoor, A., Mehjabeen, M., Rehman, A.B. and Jahan, N. 2012. Analgesic, antimicrobial and cytotoxic effect of *Cyperus rotundus* ethanol extract. *Pak. J. Pharmacol.* 29: 7-13.

Meena, A.K., Yadav, A.K., Niranjan, U.S., Singh, B., Nagariya, A.K. and Verma, M. 2010. Review on *Cyperus rotundus* A potential herb. *Int J Pharm Clin Res* 2: 20 22.

Mohler, C.A. and DiTommaso, A. 2012. Unpublished. Manage weeds on your farm: A guide to ecological strategies. Department of Crop and Soil Sciences, Cornell University. Prepublication draft, version 5.1 (2008). Publication through SARE expected in 2012.

Mohsin, A., Shah, A.H., Al-Yahya, M.A., Tariq, M., Tanira, M.O.M. and Ageel, A.M. 1989. Analgesic, antipyretic activity and phytochemical screening of some plants used in traditional Arab system of medicine. *Fitoterapia* 60(2): 174-177.

Nadkarni, A.K. and Nadkarni, K.M. 1996. Indian Materia Medica. Popular Prakashan Private Ltd, Bombay, pp. 428-429.

Nagulendran, kr., Velavan, S., Mahesh, R. 2007. *In vitro* Antioxidant Activity and Total Polyphenolic Content of *Cyperus rotundus* Rhizomes, *E-Journal of Chemistry* 4(3): 440-449.

Nandkarni, A.K. 1976. Indian Materia Medica. Vol. I to II, Popular Prakashan Pvt. Ltd., Mumbai.

Nima, Z.A., Jabier, M.S., Wagi, R.I. and Hussain, H.A. 2008. Extraction, identification and antibacterial activity of Cyperus oil from Iraqi C. *rotundus. Eng. Technol.* 26(10): 1156-1160.

Oladipupo, A.L. and Oyedeji, A.O. 2009. Chemical composition of the essential oils of *Cyperus rotundus* l. from South Africa. *Molecules* 14: 2909-2917.

Pal, D.K. and Dutta, S. 2006. Evaluation of the Antioxidant activity of the roots and Rhizomes of *Cyperus rotundus* L. *Indian J Pharm Sci* 68: 256-258.

Parotta, J.A. 2001. Healing Plants of Peninsular India. New York: CABI Publishing, pp. 02 66.

Patwardhan, B., Vaidya, A.D.B. and Chorghade, M. 2004. Ayurveda and natural product drug discovery. *Curr Sci* 86: 789-799.

Paul, S., Devi, N. and Sarma, G.C. 2013. Ethnobotanical utilization of some medicinal plants by Bodo People of manas biosphere reserve in the treatment of malaria. *Int. Res. J. Pharm.* 4(6): 102-105.

Puratchikody, A., Devi Nithya, C. and Nagalakshmi, G. 2006. Wound healing activity of *cyperus rotundus* linn. *Indian Journal of Pharmaceutical Sciences* 68: 97-101.

Qasem, J.R. 1994. Allelopathic effect of white top (*Lepidium draba*) on wheat and barley. *Allelopathy J* 1(1): 29-40.

Qasem, J.R. 2001. Allelopathic potential of white top and Syrian sage on vegetable crops. *Agron J* 93(1): 64-71.

Qasim, M., Abideen, Z., Adnan, M.Y., Ansari, R., Gul, B. and Khan, M.A. 2014. Traditional ethnobotanical uses of medicinal plants from coastal areas. *J. Coat. Life Med.* 2(1): 22-30.

Qshaughnessy, W.B. 1841. The Bengal Dispensatory, Thacker Spink and company, Calcutta, pp. 627-628.

Raut, N.A. and Gaikwad, N.J. 2006. Antidiabetic activity of hydro-ethanolic extract of Cyperus rotundus in alloxan induced diabetes in rats. *Fitoterapia* 77: 585-588.

Reddy, K.N. and Bryson, C.T. 2009. In-Crop and Autumn-Applied Glyphosate Reduced Purple Nutsedge (*Cyperus rotundus*) Density in No-Till Glyphosate-Resistant Corn and Soybean. *Weed Technol* 23: 384-390.

Riemens, M.M., van der Weide, R.Y. and Runia, W.T. 2008. Biology and Control of *Cyperus rotundus* and *Cyperus esculentus*, review of a literature survey. Plant Research International 1-18.

Sastri, B.N. 1962. The Wealth of India, Raw Materials. Vol. VI, New Delhi, India. Council of Scientific and Industrial Research, pp. 380-381.

Seo, E.J., Lee, D.U., Kwak, J.H., Lee, S.M., Kim, Y.S. and Jung, Y.S. 2011. Antiplatelet effects of *Cyperus rotundus* and its component (+)-nootkatone. *J. Ethnopharmacol.* 135(1): 48-54.

Seo, W.G., Pae, H.O., Oh, G.S., Chai, K.Y., Kwon, T.O., Yun, Y.G., Kim, N.Y. and Chung, H.T. 2001. Inhibitory effects of methanol extract of *Cyperus rotundus* rhizomes on nitric oxide and superoxide productions by murine macrophage cell line, RAW 264.7 cells. *J Ethnopharmacology* 76: 59-64.

Shamkuwar, P.B., Hoshamani, A.H. and Gonjari, I.D. 2012. Antispasmodic effect of *Cyperus rotundus* L. (Cyperaceae) in diarrhea. *Der Pharm. Lett.* 4: 522-524.

Sharif Ali, S., Kasoju, N., Luthra, A., Singh, A., Sharanabasava, H., Sahu A., and Bora, U. 2008. Indian medicinal herbs as sources of antioxidants. *Food Res Int* 41: 1-15.

Shivakumar, S.I., Suresh, H.M., Hallikeri, C.S., Hatapakki, B.C., Handiganur, J.S., Sankh, K. and Shivakumar, B. 2009. Anticonvulsant effect of *Cyperus rotundus* Linn. rhizomes in rats. *J. Nat. Rem.* 9: 192-196.

Siddhiqui, M.A.A., John, A.Q. and Paul, T.M. 1995. Status of some important medicinal and aromatic plants of Kashmir Himalaya. *Advances in Plant Sciences.* 8: 134-139.

Singh, N. and Gilca, M. 2010. Herbal Medicine – Science embraces tradition - A new insight into the ancient Ayurveda, Lambert Academic Publishing, Germany, pp. 139-148.

Singh, N., Kulshrestha, V.K., Gupta, M.B. and Bhargava, K.P. 1969. Pharmacological studies on *Cyperus rotundus. Indian J Pharm* 1(2): 9.

Singh, N., Kulshrestha, V.K., Gupta, M.B. and Bhargava, K.P. 1970. A pharmacological study of *Cyperus rotundus. Indian J Med Res* 58: 103-109.

Singh, N., Pandey, B.R., Verma, P., Bhalla, M. and Gilca, M. 2012. Pytopharmacotherapeutics of *Cyperus rotundus* Linn. (Motha): An overview. *Indian Journal of Natural Products and Resources* 3(4): 467-476.

Singh, S.P., Raghavendra, K. and Dash, A.P. 2009. Evaluation of hexane extract of tuber of root of *Cyperus rotundus* Linn (Cyperaceae) for repellency against mosquito vectors. *J Parasitol Res* 1: 1 5.

Skinner, M. 2007. United States of America. Plant Guide: Purple Nutsedge. USDA.

Sonwa, M.M. and König, W.A. 2001. Chemical study of the essential oil of *Cyperus rotundus. Phytochemistry* 58: 799-810.

Srivastava, R.K., Singh, A. and Shukla, S.V. 2013. Chemical investigation and pharmaceutical action of *Cyperus rotundus* - A Review. *J. Biol. Active Prod. Nat.* 3:166-172.

Sundaram, M.S., Sivakumar, T. and Balamurugan, G. 2008. Anti-inflammatory effect of *Cyperus rotundus* Linn. Leaves on acute and subacute inflammation in experimental rat models. *Biomedicine* 28: 302-304.

Tambekar, D.H., Khante, B.S., Chandak, B.R., Titare, A.S., Boralkar, S.S. and Aghadte, S.N. 2009. Screening of antibacterial potentials of some medicinal plants from Melghat forest in India. *Afr J Trad Compl Altern Med* 6(3): 228-232.

Tayyar, Rana, and Jodie Holt. 2003. Genetic and morphological analysis of two novel nutsedge biotypes from California. *Weed Science* 51: 731-739.

Terry, P.J. 2001. The Cyperaceae-Still the world's worst weed? *Proc BCPC Symp,* 77.

Thebtaranonth, C., Thebtaranonth, Y., Wanauppathamkul, S. and Yuthavong, Y. 1995. Antimalarial sesquiterpenes from tubers of *Cyperus rotundus*: Structure of 10, 12 peroxycalamenene, a sesquiterpene endoperoxide. *Phytochemistry* 40: 125 128.

Tsoyi, K., Jang, H.J., Lee, Y.S., Kim, Y.M., Kim, H.J., Seo, H.G., Lee, J.H., Kwak, J.H., Lee, D. and Chang, K.C., 2011. (+)-Nootkatone and (+)-valencene from rhizomes of *Cyperus rotundus* increase survival rates in septic mice due to heme oxygenase- 1 induction. *J. Ethnopharmacol.* 137: 1311-1317.

Uddin, S.J., Mondal, K., Shilpi, J.A. and Rahnan, M.T. 2006. Antidiarrhoeal activity of *Cyperus rotundus*. *Fitoterapia* 77(2): 134-138.

Vaishya, L. 1835. Bhavprakash, Shri Venkateshwar (steam) Mudranalaya, Mumbai.

Visetson, S., Milne, M. and Milne, J. 2001. Toxicity of 4, 11 Selinnadien 3 one from nutsedge (*Cyperus rotundus* L.) tuber extracts to diamondback moth larvae (*Plutella xylostella* L.), detoxification mechanisms and toxicity to non target species. *Kasetsart J Nat Sci* 35: 284 292.

Vu, V.D. and Mai, T.T. 1994. Study on analgesic effects of *Cyperus stoloniferus*. *Retz Tap Chi Duoc Hoc* 1: 16-17.

Waazing, E.J. 1968. The Pharmacopeia of India. Allen, W.H. and Company, London, p. 250.

Wang, G., McGiffen Jr. M.E. and Ogbuchiekwe, E.J. 2008. Crop rotation effects on *Cyperus rotundus* and *C. esculentus* population dynamics in southern California vegetable production. *Weed Research* 48: 420-428.

Watt, G. 1908. A dictionary of the Economic Products of India. 2nd reprint edn, Periodical Experts, Delhi, Vol. 2, pp. 686-687.

Webster, T.M., Grey, T.L., Davis, J.W. and Culpepper, A.S. 2008. Glyphosate hinders purple nutsedge (*Cyperus rotundus*) and yellow nutsedge (*Cyperus esculentus*) tuber production. *Weed Sci* 56: 735-742.

Weenen, H., Nkunya, M.H.H., Bray, D.H., Mwasumbi, L.B., Kinabo, L.S., Kilimali, V.A.E.B. 1990. Antimalarial activity of Tanzanian medicinal plants. *Planta Med* 56: 368–370.

William T.M. and Hirase, K. 2004. Comparison of commercial glyphosate formulations for control of prickly sida, purple nutsedge, morningglory and sicklepod. *Weed Biol Manag* 4(3): 136-141.

William, R. D. 1976. Purple nutsedge: Tropical scourge. *Hort Science* 11: 357-364.

William, R.D. and Warren, G.F. 1975. Competition between purple nutsedge and vegetables. *Weed Science* 23: 317-323.

Zhu, M., Luk, H.H., Fung, H.S. and Luk, C.T. 1997. Cytoprotective effects of *Cyperus rotundus* Linn. against ethanol induced gastric ulceration in rats. *Phytother Res* 11(5): 392-94.

□□□

Chapter 13

Bioactive Metabolites from Plant Associated Endophytic Fungi

Ramesha Alurappa[1], Radhakrishnan Narayanaswamy[2], Rudramurthy Gudepalya Renukaiah[3] and Mallappa Kumara Swamy[4*]

[1] Padmashree Institute of Management and Sciences, Kommagatta, Kengeri, Bangalore 560060, India. [2] Department of Nutrition and Dietetics, Faculty of Medicine and Health Sciences, Universiti Putra, Malaysia (UPM), Serdang 43400, Selangor, Malaysia. [3] Department of Biotechnology, East-West College of Science, Bangalore-560091, India. [4] Department of Crop Science, Faculty of Agriculture, Universiti Putra Malaysia (UPM), Serdang 43400, Selangor, Malaysia

ABSTRACT

Endophytic fungi are unique microbes that reside in the plant tissues and cause no harm or any symptoms of diseases. Although plants are the major source of modern drugs, there is a continuous search for new sources to obtain new lead molecules with higher biological properties for treating deadly diseases.Many plants are associated with several kinds of endophytic fungi capable of producing bioactive secondary metabolites.Thus, endophytic fungi can act as a reservoir of bioactive principles which are yet to be explored in detail. In addition, plant-endophytic fungal association stimulates plant growth, increase resistance towards disease causing pathogens, suppress the weed, increase tolerance to abiotic and biotic stresses. In this chapter, various aspects of endophytic fungi including their symbiosis with plants, biological implications and important secondary metabolites production are discussed in detail. This information would certainly help to improve the pace of modern drug discovery.

Keywords: Endophytes, symbiosis, bioactive compounds, metabolites, anticancer, antimicrobial.

Introduction

Alexander Fleming invented world's first antibiotic penicillin on 28th September 1928, which revolutionized the class of bacteria-killer drugs. From then onwards, more number of antibiotics were discovered towards the treatment of infectious diseases. But, there is no end for new discoveries due to the drug resistance in bacteria (Xing *et al.,* 2011). Provision of drugs has always been a challenge in medicinal field, seeking therapeutic drugs from natural products. Research in natural products for drug discovery methods are competitive with other synthetic drugs, due to lesser toxicity and broad spectrum activities in less quantity of compound administration. Research on plant based products requires continuous improvements in the screening process, extraction, isolation and structure interpretation. Moreover, various issues related to large-scale supply of novel compounds should be addressed in order to evaluate their bioactivities so that novel drug molecules can be discovered based on the available traditional knowledge (Butler, 2004; Kumara Swamy *et al.,* 2011; Swamy *et al.,* 2015a; Greetha *et al.,* 2016). Many of the plant based products are having several biological properties and hence, they are widely used in therapeutics to treat many diseases. Interestingly, most of the modern drugs are either plant based compounds or their synthetic derivatives. Many of these bioactive compounds are derived from medicinal plants based on the available traditional practices and knowledge (Mohanty *et al.,* 2014; Swamy *et al.,* 2015a).

In addition, medicinal plants have become the source of natural compounds for the discovery of novel drug molecules with improved biological activities and less toxicity. Plant extracts and their purified compounds provide immense possibilities in the discovery of new lead molecules with higher effectiveness. Plants produce enormous secondary metabolites with high chemical diversity (Sasidharan *et al.,* 2011; Kumara *et al.,* 2012; Swamy *et al.,* 2015a; Swamy *et al.,* 2015c; Greetha *et al.,* 2016). The utilization of medicinal plantmaterials may result in species deterioration and cause loss of biodiversity. Moreover, obtaining natural products from plants is pretty expensive. Therefore, microorganisms residing inside the medicinal plants are suitable alternatives in drug discovery process as they minimize such problems and can act as important sources of natural products with vast potential for the discovery of new molecules for pharmaceutical use (Nisa *et al.,* 2015). These microorganisms, so called as endophytes, can be isolated and grown in the laboratory and tested for bioactivity. Endophytes include mainly bacteria and fungi and among them, fungi are most frequently isolated endophytes. However, occurrence and significance of endophytes associated with many number of plant species are yet to be screened and studied in detail. Hence, there is a large scope for the discovery of novel bioactive metabolites from these microbes associated with therapeutically valued medicinal plants (Nisa *et al.*, 2015). Conversely, endophytic fungi remain under explored because of their inability to produce higher quantities of bioactive compounds required for drug discovery. Endophytic fungi may be at least 1 million species distributed in plants worldwide suggesting a rich source of novel and undescribed diverse species (Chandra, 2012). Previous reports showed that endophytic fungi live inside the plant tissues causing no harm or any disease. In addition, endophytic fungi promote plant growth and produce secondary metabolites which are required for plant defense mechanisms (Tan and Zou, 2001); Chandra *et al.*, 2012). These metabolites synthesized or derived from the endophytic fungi are biologically active compounds. Endophytic fungal species are reported to produce vast number of antimicrobial, antioxidant and anticancer compounds (Chandra 2012; Nisa *et al.*, 2015; Demain 2014). In this chapter, we highlight the importance of endophytic fungi and their symbiosis with the host plants. In addition, several biological implications of endophytic fungi with special reference to the production of various classes of bioactive compounds are discussed in detail.

About endophytic fungi

Endophytic fungi are generally noticed in most of the plant species on earth. They reside in the host plant tissues established either through symbiotic or pathogenic relationships. These microbes are reported to produce a plethora of biomolecules involved in promoting plant growth or providingplant protection (Strobel *et al.*, 2004). Endophytes are defined by many researchers in many ways which is mainly based on the viewpointsof endophytes isolated, examined and identified from host plants. Bacon and White (2000) confer a comprehensive and ordinarily accepted explanation of endophytes as "Microbes that colonize and live in the internal tissues of plants without causing any immediate, adverse effects". While, Tan and Zou (2001) explains endophytes as "an endophyte is a bacterial (including Actinomycete) or fungal microorganism, which spends the whole or part of its life cycle colonizing inter- and/or intra-cellularly inside the healthy tissues of the host plant, typically causing no apparent symptoms of disease". Schulz and Boyle, (2005) explained "Endophytic fungi, which colonize plants internally without apparent adverse effects, occur mutualistically, ubiquitously in plants and they produce number of compounds which can inhibit pathogens". Although the asymptomatic nature of endophytes living within plant tissues has incited focus on mutualistic or symbiotic relationships among host plants and endophyes, the biological diversity of endophytic microbes suggest that they can also be aggressive opportunistic or saprophytic pathogens (Strobel and Daisy, 2003).

Origin and evolution of fungal endophytes

The occurrence of endophytic fungi associated with plants has been revealed in the fossilized plant stems and leaf tissues (Taylor and Taylor, 2000). Many harmless endophytic fungi are inactive pathogens, and can result in plant diseaseswhen the plant is under stress or aged. Gene mutation could have occurred leading to an endophytic mutant during their evolution in association with its host plant. *In vitro* conditions, both the host and the endophytes are able to produce secondary metabolites which are toxic to each other (Peters *et al.*, 1998). It has been suggested that endophytic fungi achieve asymptomatic colonization by a balanced antagonism between fungal virulence and response of plant defense mechanisms (Rodriguez *et al.*, 2009). The multilocus phylogenetic framework of ancestral-state reconstructions suggests that endophytes have arisen from ancestors of insect-parasites and later got differentiated through a successions of inter-kingdom host jumps (Torres *et al.*, 2007). Endophytic fungi such as *Neotyphodium* and *Balansia*are reported to be evolved from insect parasites before progressing to become epibiotic plants. The microbes initially obtained nutrients by infecting insects and whiteflies. Later, they entered grass hosts directly and aroused as different forms of endophytes (Rodriguez *et al.*, 2008; Rodriguez *et al.*, 2009; Nisa *et al.*, 2015).

The interactions between host-endophyte and host-pathogen involve perpetual mutual antagonisms that are mediated by the secondary metabolites secreted by the symbionts. Any imbalance in these host-pathogen interactions results in diseases whereas, host-endophyte interactions will always have a balanced antagonism.Thus, it can be speculated that these long-held associations have resulted in the possible transfer of genetic information between themselves and the host plants and viceversa. Interaction of endophytes is specific to host which confers the evolution of endophytes when they particularly infect the plant. Among numerous endophytes, at least one host specific species can be observed. The endophytic fungi, *Acremonium* species inhabiting*Lolium perenne* L. and a lot of grasses are anticipated to be host specific to grasses. Interestingly, *Epichloe* species are known to colonizewith the intercellular space of host tissues systematically and can remain asymptomatic during their lifecycle and when the infected plants produce seeds, they get transmitted vertically. Likewise, some of the other *Epichloe* species can cause choke disease, where inflorescence maturation stops to produce no seeds (Clay and Schardl, 2002). A naturally occurring non-pathogenic endophyte, *Colletotrichum magna* was developed from the mutation of wild-type strain causing anthracnose diseases in cucurbit plants resulting in dark, sunken lesions on stems, leaves, flowers as well as fruits. Also, this mutant has the ability to propagate systematically within the host plant showing no symptoms of diseases. However, retains the characteristics such as host specific infection, wild-type levels of *in vitro* sporulation, appressoria formation and spore adhesion (Nisa *et al.,* 2015).

Symbiosis between plant and endophytic fungi

Endophytes are a representative symbiotic association of the fungi and their host plants. The description of symbiosis refers to the living together of dissimilar organisms where both hosts and symbionts are benefited (Akthar *et al.,* 2016; Swamy *et al.,* 2016a, Swamy *et al.,* 2016b). These symbiotic fungal associations relatively influence onrhizospheric soil nutritionand plant fitness (Brundrett, 2006; Akthar *et al.,* 2016). Research efforts have witnessed that most of the plants in nature have a symbiotic relationship with endophytic fungi. Endophytic fungi live inside the plant, analogous to an epiphytes living on the plant surface. For example, some plant pathogens like smuts can exist internally and asymptomatically within host plant for many years before they finally become evident. Nevertheless, a large variety of heterotrophic organisms exist within the plant tissues during their life cycle and produce no disease symptoms. Asymptomatic endophytic fungi may be ever-present in the plant kingdom, rivaling insects in their species diversity (Arnold *et al.,* 2000). Most attention has been emphasized on endophytic fungi existing in plant leaves, roots and stems (Clay and Schardl, 2002; Park and Eom, 2007). Many researchers have showed that there exists a intra and inter specific competition and mutualistic symbiosis of mycorrhizae and endophytes. These interactions are further reported to affect the variety of the plant community and evolution of the species. This symbiosis between host plants and fungi is distributed widely affecting the diversity and structure of ecosystem (Park and Eom, 2007).

The mode of penetration and colonization of endophytic *Fusarium* species and pathogen *Drechslera* sp. in bean and barley plants proved that the endophytic fungi are different from pathogenic fungi (Barbara *et al.,* 2002). The endophytic fungi penetrated through the stomata along the anticlinal epidermal cells and colonization was limited, localized and intercellular in the shoots of bean and barley; in contrast, the pathogen was penetrated directly through the cell wall and colonized extracellularly. These differences of endophytes with respect to pathogenic fungi may not be able to cause disease to the host plant. Endophytic fungi are reported to promote the host plant growth and developmentas well as plant physiology (Muhammad *et al.,* 2010; Swamy *et al.,* 2016a). Endophytic fungi receive nutrition, protection and multiplication opportunities from their associated hostplants. Likewise, host plants are also benefited from this symbiosis (Clay and Schardl, 2002). The endophytes produce cytokinins, indole acetic acid and many other plant growth promoting elementswhich help to control or regulate germination, growth, metabolism, or other physiological activities (Tan and Zou, 2001). Partial involvement of these fungi can enhancethe uptake of soil nutrients such as nitrogen and phosphorus by host plants (Reis *et al.,* 2000). Inoculation of an endophyte, *Colletotrichum gloeosporioides* toin vitro cultures of *Artemisia annua* L. have shown to promote their growth effectively (Tan and Zou, 2001). The presence of these endophytes enhances the fitness of host plants. These endophytic fungi release a wide range of secondary metabolitesfor instance alkaloids and theseact as metabolic inhibitors of insects, anti-feedants, promotes drought tolerance of host plant, and reduces microbial infections. The herbs infected with endophytic fungishows better resistance towardshigher temperature (Rodriguez *et al.,* 2009). Few endophytic fungi, like *Acremonium lolli* of perennial Ryegrass produce poisonous secondary metabolic products (Ergopeptine alkaloids) that are involved in protecting the plants from pathogens (Philippe, 2016). Some of the endophytic *Neotyphodium* species can only spread by infecting seeds of the host plants (Schardl and Phillips, 1997). Plant-endophytic fungal association also provides defensive mutualism by producingseveral classes of bioactive metabolites such as ergopeptine, lolitrems, loline (alkaloids) and pyrrolopyrazine alkaloids and (Bush *et al.*, 1997; 76. Saikkonen et al., 2010; Philippe, 2016). Ergovaline produced from endophytic fungi is expected to be the key agent of fescue toxicosis observed in livestock (Philippe, 2016).

Endophytic Fungi Versus Mycorrhizae

Mycorrhizae and endophytic fungi in plants represent symbiotic association. There is no significant difference between their association and mechanism with the host plants, but few differences are present in their appearance and structure (Jumpponen, 2001). Mycorrhizal fungi associates with plants by colonizing roots and nurture inthe surrounding rhizospheric soil, while endophytes live exclusively inside the plant tissues and emerge to sporulate when the host plant tissue or whole plant is senesced (Rodriguez *et al.*, 2009). Both mycorrizae and endophytic fungi are beneficial to the fungi and the host plants (Smith and Read, 1997). However, endophytic fungi appear ubiquitously in roots and shoot of monocotyledonous and dicotyledonous of the plant and have been found in all the species of plants. Whereas, Mycorrizae of most groups appear in higher plants of the root parts (Hamayun *et al.*, 2009). Mycorrhizal association can be internal, where the fungus colonizes the host plant's roots (Endomycorrhizae) or as observed in Vesicular Arbuscular Mycorrhizal fungi (VAM), or extracellular (Ectomycorrhizae) (Akthar *et al.*, 2015; Swamy *et al.*, 2016a). VAM association closely attaches plants through their hyphal networks which are usually found in excessof 100 meters of hyphae per cubic centimeter of soil (Parniske, 2008). VAM significantly enhances the nutrient and water uptake ability of plants through roots from soil. Moreover, VAM also increases the plant ability to toleratevarious harsh environmental situations including mineral deficiencies, water stress, soil toxicities, and soil erosion (Parniske, 2008; Akthar *et al.*, 2015).Endophytes are a diverse cluster of fungi having superficial implications on plants through plant ecology, fitness, and provides biotic and abiotic stress tolerance, decreases water consumption and increases plant biomass (Brundrett, 2006). Therefore, the mycorrhizae of plants are grouped under root endophytes (Schulz *et al.*, 2002; Lima *et al.*, 2012).

Biological implications of endophytic fungi

Endophytic fungican stimulate plant growth, increase resistance towards disease causing pathogens, suppress the weed, increase tolerance to abiotic and biotic stresses (Sturz *et al.*, 2000). In addition, they also have potential to produce vast bioactive secondary metabolites with pharmaceutical importance (Tan and Zou, 2001; Demain, 2014). Many endophytic fungi provide protection to the host by inducing defense mechanisms in plants against a broad range of pathogens. The endophytes are known to produce an antibiotic substance which inhibits the pathogen growth, or may compete with pathogen for space and nutrition. For example, Barley plants associated with endophyte *Piriformospora indica* have exhibited antagonistic to vascular pathogen, *Fusarium culmorum* and leaf pathogen, *Blumeria graminis* (Johnson *et al.*, 2014). Endophytic fungi,namely *Cordana* sp. and *Nodulisporium* sp. showed effective activity against Anthracnose disease cause fungi *Colletotrichum musae* in wild Banana (Nuangmek *et al.*, 2008). Endophytic fungi also produce herbicidal, nematicidal and pesticidal compounds and have the ability to trigger plant defense mechanism. Endophytic fungi like *Acremonium, Paecilomyces, Trichoderma, Fusarium, Chaetomium,* and *Phyllosticta* species have exhibited nematicidal activity against*Radopholus similis* and *Meloidogyne incognita* (Vu *et al.*, 2006; Kalele *et al.*, 2007; Goswamia *et al.*, 2008; Yan *et al.*, 2011). The endophytic fungus *Hypoxylon pulicicidum* sp., produce novel indole diterpenes, Nodulisporic acids (NAs) which exhibited insecticidal activity against wide range of insects such as mosquitoes, fruit flies and dog flea (Bills *et al.*, 2012). *Neotyphodium coenophialum* and *Neotyphodium lolii* are the grass (*Lolium arundinaceum*) endophytes which produced ergot alkaloids, toxic to herbivores (Simons *et al.*, 2008). The metabolites of endophytic fungi can also be successfully used in weed management. For example, Brefeldin A produced by *Cladosporium*sp. suppress the pollen tube maturity of weed *Picea meyeri* (Wang *et al.*, 2007) and Ascotoxin isolated from *Paraconiothyrium* sp. affect the seed germination of weed *Lactuca sativa* and *Echinochloa crus-galli* (Khan *et al.*, 2012).

Endophytic fungi provide defense mechanism and also significantly influence the development of plant community. The interaction between the endophytic fungi with host plants increases plant productivity as well as the species diversity (Park and Eom, 2007). Endophytes can direct or indirectly promote the plant growth through different mechanisms. Endophytes produce some kind of growth hormones or induces the host plants to secrete hormones and thus stimulatesthe host growth by improving nutrient metabolism (Schulz and Boyle, 2005). For example, an endophytic fungus *Sebacina vermifera* significantly promoted the growth of *Nicotiana attenuata* by inhibited the ethylene signaling leads decrease in ethylene production (Barazani *et al.*, 2007) and seed-borne fungal endophyte, *Stagonospora* species isolated from *Phragmites australis* enhanced reed biomass formation in axenic microcosms by providing a competitive advantage to germlings (Ernst *et al.*, 2003).

Endophytic fungi enhancing drought tolerance in host plant is a great benefit under dry environmental conditions; the mechanisms vary from plant to plant. For example, tall fescue grass growth increased with an endophyte *Neotyphodium* association and its growth more than uninfected tall fescue due to water stressed (West *et al.*, 1993). Drought stress resulted in altered stomatal behavior and osmotic regulation in plants (West, 1994). *Piriformospora indica* provides tolerance to drought stress in Chinese cabbage leaves through stimulating the expression of drought linked genes and plastid-localized Ca^{2+} sensing receptor protein. Endophytic fungi significantly alter the antioxidant activity of its host which enhances tolerance from abiotic stress. The accumulation of loline alkaloids produced by endophytes affected osmotic potential and increased tolerance to other environmental stresses like heat, low light and low soil fertility (Siegel and Bush, 1997).

Endophytic fungi are involved in biodegradation of ecological wastes. The fungal endophytes have the ability to degrade several kinds of organic compounds including carbon and nitrogen sources by producing enzymes. Endophytic fungi extensively metabolize benzoxazinones to less toxic metabolites by the oxido-reductases (Zikmundova *et al.*, 2002). In addition, endophytes have been reported to to improve the soil micro-environment by decomposing environmental pollutants. For example, 89.51% of phenanthrene was removed by *Ceratobasidum stevensii* isolated from the *Euphorbiaceae* (Dai *et al.*, 2010). Endophytic fungus *Pestalotiopsis microspora* degraded the synthetic polymer polyester polyurethane, polycyclic aromatic hydrocarbons and phenanthrene with bioremediation applications (Russel *et al.*, 2004; Wang and Dai, 2011).

The secondary metabolites from endophytic fungi play a wide range of biological activities. In the recent years, abundance of secondary metabolites extracted from endophytic fungi are found to have applications in medicinal drug as antimicrobial agents, anticancer and antitumor agents, antidiabetic agents, cholesterol inhibitors and immunosuppressive agents (Strobel, 2003). In the search of new bioactive molecules, endophytic microorganisms have been demonstrated to produce many metabolites that are reported to possess antimicrobial, antioxidant, anti-inflammatory, insecticidal, anticancer and other biological properties (Sun *et al.,* 2004; Luo and Fang, 2008; Demain 2014; Gubiani *et al.,* 2014; de Felício *et al.,* 2015). Apart from biological activity, they have been successfully screened for application in diverse areas, in agriculture as pesticides and herbicides (Tétard Jones *et al.,* 2015).

Endophytic fungi and their secondary metabolites

The secondary metabolites of fungal origin display a wide array of structural types, and their study has certainly improved the field of organic chemistry during the past several decades. Endophytic fungi are reported to produce a huge number of low molecular weight compounds, also grouped as secondary metabolites. "The metabolites which are produced after the active growth phase and are not directly involved in growth and development of the organism are collectively known as secondary metabolites". In comparison to the primary metabolites which are essentially same for all living systems, the secondary metabolites are usually species specific and are derived from some of the intermediates of primary metabolism. Secondary metabolites from endophytic fungi comprises a wide variety of structures including terpenoids, alkaloids, quinones, xanthones, peptides, steroids, flavonoids, phenols and phenolic compounds. The secondary metabolites obtained from grass endophytic fungi includes alkaloids like peramine, ergovaline, ergotamine, lolitrem etc (Wang *et al.,* 2004; Kunkel *i* 2004; Nisa *et al.,* 2015; Demain 2014). As stated by Strbel, (2002), fungal endophytes occurring within the plants have the capability to secrete metabolites which are either similar to, or with higher activity than that of their respective hosts. The secondary metabolites of fungal origin exhibit a very wide range or structural types, and their study has significantly improved the modern discovery. However, recovery of the fungal metabolites still remains as the major challenge in drug discovery (Demain, 2014). Apart from anti-microbial activity, several secondary metabolites have been successfully screened for application in diverse areas. These include their application in agriculture as pesticides and herbicides, in pharmaceuticals as cholesterol inhibitors, immunosuppressive agents, anticancer and antitumor agents. In the last few years, many secondary metabolites obtained from endophytic fungi are reported to have many applications in the field of biomedicine and agrochemistry (Strobel, 2003; Huang *et al.*, 2007; Nisa *et al.*, 2015). Plants are potential reservoir of many endophytic fungi and may be the treasure for many useful bioactive metabolites. There is a need and increasing demand to hunt for the bioactive metabolites turned in recent years towards hidden metabolites in the endophytic fungi which are used as antibacterial, antifungal and anticancer drugs. Endophytic fungi from medicinal plants have played an important role as an alternative source of novel bioactive metabolites (Nisa *et al.*, 2015; Demain, 2014). Some of these bioactive secondary metabolites produced from endophytic fungi are discussed below.

Antibacterial compounds

There are a number of antibacterial metabolites reported from endophytic fungi and few of them have been reviewed in this section. Krohn *et al.* (1999) reported antibacterial compound Ketodivinyllac-tonic steroid, herbarulide from *Pleospora herbarum,* an endophytic fungus. Lu *et al.,*(2000) determined bactericidal propertyof 10 steroid compounds isolated from *Colletotrichum* sp. isolated from the medicinal plant, *Artemisia annua.* Among the tested compounds 3-oxo-ergosta-4,6,8(14),22-tetraene, 3b-hydroxy-ergosta-5-ene, 3b-hydroxy-5a,8a-epidioxy-ergosta-6,3b,5a-dihydroxy-6b-acetoxy-ergosta-7,22-diene, 22-diene 6-isoprenylindole-3-carboxylic acid and 3b,5a-dihydroxy-6b-phenylacetyloxy- ergosta-7,22-diene showed effective inhibition against *Staphylococcus aureus, Bacillus subtilis, Pseudomonas* sp. and *Sarcina lutea.*Wagenaar andClardy (2001) isolated antibacterial compound, dicerandrols A-C from the endophytic fungus, *Phomopsis longicolla* of *Dicerandra frutescens.* Strobel *et al.* (2001) reported antibacterial activity of5 classes of volatile compounds including esters, alcohols, ketones, lipids and acids from a novel endophytic fungus,*Muscodor albus* associated with *Cinnamomum zeylanicum.* The compounds showed effective inhibition for *E. coli, S. aureus, Micrococcus luteus,* and *B. subtilis.* Out of them, esters of 1-butanol 3-methyl- acetate were found to the most effective class of inhibitory constituents. A polyketide Citrinin compound isolated from *Penicillium janthinellum* associated with *Melia azedarach,* exhibited antibacterial activity against *Leishmania* sp. (Marinho *et al.,* 2005).

Tsuchinari *et al.* (2007) found the antibacterial activity of 3,4,5-trisubstituted N-methyl-2-pyridone alkaloid, fusapyridons A characterized from the extract of *Fusarium* sp. YG-45 isolated from *Maackia chinensis.* The compound showed activity against *S. aureus* and *P. aeruginosa*with the minimum inhibitory concentration (MIC) at 50 and 6.25µg/mL respectively. Huang *et al.* (2007) evaluated the efficiency of metabolites extracted from endophytic fungi of *Nerium oleander* against the growth of *E. coli, B. cereus, S. annatum, S. aureus,* L. *monocytogenes* and *C. krusei.* They found that the extract of 7 predominant fungal isolates exhibited antimicrobial activity than the host plant metabolites. Zhuang *et al.* (2008) characterized anti-bacterial compounds from *Penicillium* sp. isolated from mangrove plant *Cerbera manghas.* They identified compounds 4-(3-hydroxybutan-2-yl)-3,6-dimethylbenzene-1,2-diol, and 3,4,5-trimethyl-1,2-benzenediol which possessed the inhibitory effects on Methicillin Resistant *Staphylococcus aureus* (MRSA) while 4-(3-hydroxybutan-2-yl)-3-acetyl-6-methylbenzene-1,2-diol had no effect on MRSA. Kjer *et al.,* (2009) conducted antibiotic activity of the compounds from *Alternaria* sp., against multidrug-resistant bacteria. Two new compounds xanalteric acids I and II exhibited less bactericidal activity against *S. aureus* while, another compound altenusin exhibited broad spectrum activity against several bacteria like *Enterococcus faecium, Enterococcus cloacae, Streptococcus pneumonia* and *Pseudomonas aeruginosa* with an MIC between 31.5 to 125µg/mL.

Fernandes *et al.* (2009) determined the MIC and minimum bactericidal concentration (MBC) for crude extract of *Alternaria alternata* endophyte obtained from *Coffea arabica* L. The ranges of MIC and MBC values were found to be 50-100ìg/mL for *S. aureus* and 400-800ìg/mL for *E. coli.* Kusari *et al.* (2009) conducted the antibacterial activity for Deoxypodophyllotoxin, isolated from endophytic fungus *Aspergillus fumigatus,* which revealed effective inhibition towards the pathogenic bacteria *S. aureus, K. pneumonia* and *P. aeruginosa.* Lima *et al.* (2011) documented the 15 culture filtrates of endophytic fungi isolated from *Piper aduncum* L. inhibited the growth of *M. tuberculosis* by more than 90% as determined by qualitative and quantitative bioassays. Lv *et al.* (2010) studied the antimicrobial potential of endophytes isolated from *Saussurea involucrate.* Extracts isolated from the fermentation broth of 49 endophytic fungi were tested for their antimicrobial properties against pathogenic microorganisms *S. aureus, E. coli, B. subtilis, A. fumigatus, C. albicans* and *C. neoformans.* The study showed thatat least 1 test microbe was inhibited by 12 fungi isolated. Among them, 5 strains were reported to possess broad spectrum antimicrobial activity. However, only 4 strains were effective against the tested fungal pathogens.

Nithya and Muthumary (2010) reported antibacterial compound from endophytic fungus *Phomopsis* sp. associated with *Plumeria acutifolia*. The compound was extracted with ethylacetate and identified as terpenoid by thin layer chromatography (TLC), UV, FT-IR spectroscopic analysis. The compound was found to be active against the growth of the bacterial pathogens *E. coli*, *Pseudomonas* sp., *Klebsilla* sp., *B. subtilis*, *S. aureus* and *S. typhi* whereas there was no significant effect on *C. albicans*. Similarly, antibacterial terpene compound was found in *Phomopsis* sp. isolated from the medicinal plant, *Allamanda cathartica*. Twenty microlitres of ethylacetate fraction of terpene compound revealed effective inhibitory activity against *E. coli*, *Pseudomonas* sp, *Klebsilla* sp., *S. aureus*,*S. typhi* and *B. subtilis*with the zone of inhibition ranging from 15±0.30 to 25±0.50mm (Nithya and Muthumary, 2011). Tayung *et al.* (2011) isolated *Fusarium solani* from the bark of *Taxus baccata* which produced antibacterial compounds 1-tetradecene, 8-pentadecanone, 8-octadecanone, 10-nonadecanone and octylcyclohexane displaying considerable bactericidalactivity against *S. epidermidis*, *S. aureus*,*S. flexneri*, *B. subtilis*, *E. coli* and*K. pneumonia*. Ho *et al.* (2012) examined the endophyes (*Lasmenia* sp, *Ophioceras tenuisporum*, *Xylaria cubensis* and *Cyanodermella* sp.)obtained from *Citrus* and *Zanthoxylum* of *Rutaceae* and *Cinnamomum* of *Lauraceae* against three phytopathogenic bacteria like *Erwinia carotovora*, *Xanthomonas campestris*and *Ralstonia solanacearum*. Amongtested isolates, *Cyanodermella* sp. showed better inhibition of all the pathogens. In an another study, Nathe *et al.* (2012) showed that the crude extracts of *Xylaria* sp. and *Diaporthe* sp. with superior antimicrobial activity against clinical pathogens namely,*Salmonella paratyphi* and *Enteroccocus faecalis* (Nath *et al.*, 2012). Subbulakshmi *et al.* (2012) tested antibacterial activity of endophytic fungi *Alternaria* spp., *Colletotrichumgloeosporioides*, *Pestalotiopsis* sp., *Fusarium* sp., *Pestalotiopsis* sp isolated from the leaf samples of *Biota orientalis*, *Pinus excels* and *Thujaoccidentalis*. Methanol extract of *Pestalotiopsis* sp. isolated from *Biota orientalis* had significant inhibition of growth of the bacteria *Streptococcus faecalis* and *Salmonella typhi*.

Senthilmurugan *et al.* (2013) reported the antibacterial activity of crude extract of endophytic fungus *Botrytis* sp. isolated from *Ficus benghalensis* which inhibited the growth of *Escherichia coli* and *Klebsiella* sp.Pinheiroa *et al.* (2013) conducted the Antibacterial activity for two alkaloids, fumigaclavine C and pseurotin A produced by endophytic fungus *Aspergillus* sp. EJC08 associated with medicinal plant *Bauhinia guianensis.* The compound fumigaclavine C inhibited *E. coli, P. aeruginosa, S. aureus* and *B. subtilis* with MIC at 62.50, 31.25, 15.62 and 7.81µg/mL respectively. Whereas, pseurotin A inhibited*E. coli, P. aeruginosa, S. aureus* and *B. subtilis* with MIC at 31.25, 31.25, 15.62 and 15.62µg/mL respectively. Subban *et al.* (2013) reported antibacterial activity of the compound 4-(2,4,7-trioxa-bicyclo[4.1.0]-heptan-3-yl) extracted from *Pestalotiopsis mangiferae,* an endophyte of *Mangifera indica* Linn. The MIC of this compound for *B. subtilis* and *K. pneumoniae* was at 0.039mg/mL, whereas for *E. coli* and *M. luteus* it was 1.25mg/mL followed by *P. aeruginosa* at 5mg/mL. de Felício *et al.* (2015) isolated 45 endophytic fungi from *Bostrychia tenella* (Ceramiales). The crude extracts of these fungi were evaluated for antifungal, antibacterial and cytotoxic activities and the results revealed that endophytes Penicillium decaturense and P. waksmanii had the highest bioactivities and for the first time they isolated cytochalasin D, a well-known antibiotic and antitumor compound. More recently, Mani *et al.* (2015) have isolated 169 endophytes from the medicinal plant, Aegle marmelos. The strains belonged to *Curvularia australiensis, Alternaria alternate, Alternaria citrimacularis,* Aspergillus *niger* and *Cladosporium cladosporioides.* The extracts of these fungi were effective against the clinical pathogens such as clinical pathogens *S. epidermidis, S. aureus, Shigella sp., P. aeruginosa, E. faecalis, E. coli, K. pnuemoniae, P. mirabilis and S. typhi.*

Antifungal compounds

Endophytic fungi exhibit antagonism towards pathogenic fungal organisms. Krohn *et al.* (1999) reported herbarulide, a novel ketodivinyllactonic steroid from *Pleospora herbarum,* an endophyte. Lu *et al.* (2000) conducted antifungal activity for secondary metabolites isolated from the endophytic fungus *Colletotrichum* sp. of *Artemisia annua.* The metabolites 3b,5a-dihydroxy-6b-phenylacetyloxy-ergosta-7,22-diene, 3-oxo-ergosta- 4,6,8(14),22-tetraene, 3b,5a-dihydroxy-6b-acetoxy-ergosta-7,22-diene and 3b-hydroxy-ergosta-5-ene effectively inhibited *Aspergillus niger* and *Candida albicans* with theMICs between 50 to 100µg/mL. Another compound, 6-isoprenylindole-3-corboxylicacid showed inhibitory activity againstfungal pathogens such as*Phytophthora capsici, Gaeumannomyces graminis* var. *tritici*and *Rhizoctonia cerealis*. Likewise, *Cryptosporiopsis* cf. *quercina* obtained from the stem explants of *Tripterygium wilfordii* producedcryptocin and this compound was found to effectively inhibit the growth of *Pyricularia oryzae* and other plant pathogens (Li *et al.*, 2000).

Strobel *et al.* (2001) reported a novel endophyte *Muscodor albus* associated with *Cinnamomum zeylanicum* and extracted 5 classes of volatile compounds including alcohols, ketones, esters, acids and lipids that exhibited antifungal activity against the pathogens namely, *Pythium ultimum, Rhizoctonia solani, Phytophthora cinnamomi, Stagnospora nodorum, Ustilago hordei, Sclerotinia sclerotiorum, Fusarium solani, Aspergillus fumigates, Verticillum dahlia, Tapesia yallundae, Cercospora beticola* and *Candida albicans*. Harper *et al.* (2003) isolated pestacin from endophytic fungus *Pestalotiopsis microspora* exihibited moderate antifungal activity for root-invading pathogen *Pythium ultimum*. Liu *et al.* (2004) isolated 12 compounds from endophytic fungus *Aspergillus fumigatus* CY018 of *Cynodon dactylon*. These compounds showed antifungal activity against human pathogens *C. albicans, A. niger* and *T. rubrum*. The compounds, asperfumoid, fumitremorgin C, fumigaclavine C, helvolic acid and physcion showed inhibitory activity against *C. albicans* with MICs of 75.0, 62.5, 31.5, 31.5 and 125µg/mL respectively and no significant activity was found against *A. niger* and *T. rubrum*. However, other compounds were observed with no antimicrobial effect.

Silva *et al.* (2006) evaluated antifungal activity of 5 compounds derived from *Phomopsis cassiae* associated with*Cassia spectabilis*. Among them, the metabolite 3,12-dihydroxycadalene exhibited activity against phytopathogens *Cladosporium cladosporioids* and *C. sphaerospermum*. Kjer *et al.* (2009) isolated altenusin metabolite from endophytic fungus *Alternaria* sp., which showed broad spectrum activity against multidrug-resistant fungi*Aspergillus faecalis* and *Candida albicans*, with an MIC 125 and 62.5µg/mL respectively. Oliveira *et al.* (2010) reported dihydroisocoumarin (3R,4R)-3,4-dihydro-4,6-dihydroxy-3-methyl-1-oxo-1H-isochromene-5-car- boxylic acid from the endophyte, *Xylaria* sp. associated with *Piper aduncum*.The compound showed moderate antifungal activity against *Cladosporium cladosporioides and C. sphaerospermum*. In addition,they also isolated 2 known compounds, (3R,4R)-4,7-dihydroxymellein and (R)-7-hydroxymellein from *Penicillium* sp. associated with *Alibertia macrophylla*. These compounds effectively inhibited *C. cladosporioides* *and C. sphaerospermum*. From the same plant, Gubiani *et al.* (2014) isolated 2 novel eremophilane-type sesquiterpenes (xylarenones F and G) which showed potent anti-inflammatory properties.

Tayung *et al.* (2011) isolated and identified antifungal compounds from *Fusarium solani* associated with the bark of *Taxus baccata*; 1-tetradecene, 8-octadecanone, 8-pentadecanone, octylcyclohexane and 10-nonadecanone, whichdisplayed activity against *C. albicans* and *C. tropicalis*. Sugijanto *et al.* (2011) determined the fungicidal activity of lecythomycin extracted and purified from the endophytic fungus *Lecythophora*sp. The results showed activity against the fungal strains such as *Aspergillus fumigatus* and *Candida kruzei* with MIC of 62.5-125mg/mL.Nath *et al.* (2012) study revealed the crude extracts of *Xylaria* sp. and *Phomopsis* sp. inhibited the growth of *C. albicans*. Subbulakshmi *et al.* (2012) tested fungicidal activity of endophytic fungi *Alternaria* spp., *Colletotrichum gloeosporioides, Pestalotiopsis* sp., *Fusarium* sp., *Pestalotiopsis* sp obtained from the leaf samples of *Biota orientalis, Pinus excels* and *Thujaoccidentalis*. Methanol extract of *Pestalotiopsis* sp. isolated from *Biota orientalis* had higher antifungal property against *C. albicans and Beauveria bassiana.*

Santiago *et al.* (2012) isolated 5-hydroxyramulosin, a polyketide compound from the culture filtrate of *Phoma* sp., an endophyte of the medicinal plant, *Cinnamomum mollissimum*. The extracted compound was reported to exhibit higher activity against *Aspergillus niger*. Ho *et al.* (2012) isolated endophytic fungi *Lasmenia* sp, *Ophioceras tenuisporum, Xylaria cubensis* and *Cyanodermella* sp. from *Citrus* and *Zanthoxylum* of *Rutaceae* and *Cinnamomum* of *Lauraceae* which showed different level of antagonistic effects against phytopathogens like *Alternaria solani, Botrytis cinerea, Colletotrichum gloeosporioides, Colletotrichum higginsianum, Cylindrocladiella lageniformis, Fusarium oxysporum, Monilinia fructicola, Penicillium digitatum, Pestalotiopsis psidii,Pythium aphanidermatum*. Among these endophytic fungi *Cyanodermella* sp. was extremely antagonistic to the fungal pathogens *Cynlidocladiella lageniformis, Fusarium oxysporum* and *Monilinia fructicola*. Gherbawy and Gashgari (2013) tested antifungal activity of 33 endophytes isolated from the leaves of *Calotropis procera*. All the isolates were tested against 4 plant pathogensnamely, *Alternaria alternata, Botrytis cinerea, Fusarium oxysporum* and *Pythium ultimum*. Among the isolates, *Chaetomium globosum* and *Myrothecium verrucaria* exhibited superior activity against the pathogenic fungal strains. Wu *et al.* (2013) evaluated the fungicidal property of the isolated steroid compounds of Phomopsis sp. associated with the plant,Aconitum carmichaeli. The identified compounds 6-ethoxy-5,15-dihydroxyergosta-7,22-dien-3-one, 9,14-dihydroxyergosta-4,7,22-triene-3,6-dione, ganodermaside D and calvasterols possessed moderate or weak antifungal activities. Subban *et al.*, (2013) reported antifungal activity of 4-(2,4,7-trioxa-bicyclo[4.1.0]-heptan-3-yl), a phenolic compound isolated from an endophytic fungus *Pestalotiopsis mangiferae* associated with *Mangifera indica* Linn. The MIC of the compound for *C. albicans* was found at 0.039mg/mL.

Anticancerous compounds

Cancer is one of the major causes of health hazard all over the world. The identification of secondary metabolites with cytotoxicity improved the development of new approach for anticancer therapy for several decades. Endophytes are able to produce novel metabolites that exhibit anticancer activities (Strobel and Daisy, 2003). For the first time, Stierle *et al.* (1993) obtained Paclitaxel, an anticancer agent from *Taxomyces andreanae* (endophytic fungus) isolated from the Yew tree (*Taxus brevifolia*). Paclitaxel, a highly functionalized diterpenoid which is found in many species of *Taxus* (Strobel *et al.*, 1993; Suffness, 1995). Zhang *et al.*, (2000) reported anticancer compound vincristine from endophytic fungus *Fusarium oxysparum* isolated from the phloem (inner bark) of *C. roseus* L. Metabolic products of cultured fungus were analysed by TLC and HPLC. Similarly, Yang *et al.* (2004) selected the endophytic fungi which produced vincristine by isolating fungi from the leaves of *Catharanthus roseus* (L.) G. Don. An endophytic fungus, *Mycelia sterilia* 97CY was isolated from the leaves of *Cantharanthus roseus* and the zymotic extracts were analyzed by TLC and HPLC found an anticancer agent vincristine. Puri *et al.* (2005) reported an anticancer compound camptothecin from an endophytic fungus *Entrophospora infrequens* obtained from *Nothapodytes foetida*. Silva *et al.* (2006) determined the antiproliferative activity of 5 compounds identified from *Phomopsis cassia*e isolated from *Cassia spectabilis* against the HeLa cervical cancer cells, the metabolite 3,12-dihydroxycadalene inhibited at an IC_{50} of 20ìM/L and 3,12-dihydroxycalamenene and 3,11,12-trihydroxy-cadalene showed weakly inhibition. Similarly, Teles *et al.*,(2006) extracted secondary metabolites from *Periconia atropurpurea*, an endophyte isolated from *Xylopia aromatica*, using ethyl acetate. Before identifying, these metabolites were tested for their biological activity and proved the cytotoxic activity of the compounds.

Phongpaichit *et al.* (2007) screened the antiproliferation and cytotoxicity of 65 crude extracts of 51 chosen endophytes of *Garcinia* species (5 species from *Garcinia atroviridis,* 23 from *G. dulcis,* 6 from *G. nigrolineata,* 16 from *G. mangostana,* and 1 from *G. scortechinii*). Among them 11.1% of extracts showed the activity against the proliferation of NCI-H187 cells and 12.7% against KB cells. Forty percent of extract showed cytotoxicity against normal Vero cell lines. Gangadevi and Muthumary (2008) isolated endophytes from different plants and screened their anticancer properties. Taxol is one of the widely studied fungal metabolite successfully used in the treatment of cancer. The fungus *Colletotrichum gloeosporioides* (strain JGC-9) isolated from *Justicia gendarussa,* produced 163.4ìg/L of taxol which showed higher cytotoxicity towards the human cancer cell lines BT 220, Int 407, H116, HLK 210 and HL 251. They also observed the dose dependent cytotoxicity of the extracted taxol suggesting the potential of this fungus in producing taxol in vitro. Ge *et al.* (2009) reported cytotoxic alkaloids from *Aspergillus fumigatus* associated with the stems of *Cynodon dactylon.* Compounds 9-deacetylfumigaclavine C and 9-deacetoxyfumigaclavine C revealed cytotoxicity against K562cells (leukemia cancer cell line) with IC_{50} values of 41.0±4.6 and 3.1±0.9μM respectively. The compound 9-deacetoxyfumigaclavine C exhibited similar activity as that of doxorubicin hydrochloride (1.2±0.2μM), anapproved drug to treatleukemia. The metabolite 14-norpseurotin considerably induced neurite outgrowth of rat pheochromocytoma cells (PC12) at a concentration of 10μM and observed to show higher activity compared to the other natural drug, pseurotin A.Secalonic acid D, a mycotoxin (ergochrome class) extracted from the mangrove endophytewas reported to have high cell toxicity against K562 and HL60 leukemia cells. It was shown to induce toxicity through apoptosis (Zhang *et al.* 2009). Nithya and Muthumary (2009) characterized an anticancer compound using UV, thin layer chromatography, and Fourier transform infrared spectroscopy (FTIR) analysis revealed the presence of taxol from culture filtrate of endophytic fungus *Colletotrichum gloeosporioides* associated with *Plumeria acutifolia.* Similarly, Srinivasan and Muthumary (2009) reported that the endophytic fungus *Pestalotiopsis* sp. associated with *Catharanthus roseus* was able to produce taxol.

Fernandes *et al.* (2009) isolated *Alternaria alternate,*an endophyte from *Coffea arabica* L. and its crude extracts was determined with antitumor activity. The cytotoxic activity against HeLa cells showed

IC_{50} at 400µg/mL. Likewise, Kjer *et al.* (2009) reported thatxanalteric acids I and II of endophytic fungus *Alternaria* sp., exhibiting higher growth inhibition against L5178Y cells at a concentration 10µg/mL. In another report by Deshmukh *et al.* (2009), a novel anticancer agent (ergoflavin)was isolated from *Claviceps purpurea* (PM0651480) associated with *Mimusops elengi* (*Sapotaceae*). Zhou *et al.* (2009) reported anticancer compound taxol from *Mucor* sp. an endophytic fungus associated with *Taxus chinensis*. Shweta *et al.* (2010) found camptothecin derivatives from endophytic fungal strains of *Fusarium solani* MTCC 9667 and MTCC 9668. The camptothecin derivatives identified as 9-methoxycamptothecin and 10-hydroxycamptothecin. Pandi *et al.* (2011) also reported Taxol from endophytic fungus *Lasiodiplodia theobromae* isolated from the medicinal plant *Morinda citrifolia* and revealed the cytotoxic effect of fungal taxol on MCF-7 cells with an IC_{50} at 300µg/mL.

An active metabolite,Sclerotiorin isolatedfrom *Cephalotheca faveolata*was shown to induce apoptosis in cancer cells (Giridharan *et al.*, 2012). Sclerotiorin was found to be a potent anti-proliferative agent against different cancer cells. It induced apoptosis in colon cancer (HCT-116) cells via BAX and inhibition of BCL-2 proteins those further activated cleaved caspase-3 enzyme causing apoptosis of cancer cells. Santiago *et al.* (2012) isolated 5-hydroxyramulosin (a polyketide compound) from the culture filtrate of *Phoma* sp. obtained from the plant,*Cinnamomum mollissimum*. The compound showed cytotoxicity against P388 murine leukemic cells with IC_{50} value 2.10µg/mL. Likewise, Lu *et al.* (2012) reported that the extracts of endophytic fungi from *Actinidia macrosperma* with cytotoxic and antitumor activities against brine shrimp and 5 types of cancer cell lines. Cytotoxic activity was found in most of the isolates and also comparatively higher toxicity against brine shrimp. From the MTT assay, it was found that about 82.4% of fungal isolatesrevealed growth inhibitory activity against cancer cells (50% inhibitory concentration IC50 <100ìg/mL). Some of the fungal isolates showed strong antitumor activity againstall cancer cell lines tested suggesting the role of endophytic fungi as a novel metabolite against cancer. Sun *et al.* (2013) evaluated cytotoxicity of 7 terpenoid compounds arisugacins B, F, G, I, J, territrem B and territrem C isolated from an endophytic fungus *Penicillium* sp. SXH-65 against Hela, HL-60 and K562 cell lines. Compound arisugacin B and F revealed the cytotoxicity at IC_{50} values ranging from 24 to 60µM.

Conclusion

Endophytic fungi are an excellent source of various bioactive natural compounds and possess considerably effective biological properties. Thus, utilization of these microbes can certainly benefit the current demand for novel lead molecules by medical, pharmaceutical and agriculture industries. However, large scale production of these metabolites in is a major challenge to the scientific world. Also, detail understanding of the plant-fungal symbiosis may reveal possible pathways involved in their relationship as well the synthesis of secondary metabolites. Hence, application of fungal biotechnology is a need of the future research to reveal the involvement of genetic control mechanisms that exists to control the production of secondary metabolites. Also, research should emphasize on detailed understanding of physiology, defensive roles and biochemical pathways involved in secondary metabolite secretion by endophytic fungi.

References

Akhtar M.S., Panwar J., Abdullah S.N.A., Siddiqui Y., Swamy M.K., Ashkani S. 2015. Biocontrol of Plant Parasitic Nematodes by Fungi: Efficacy and Control Strategies. In: Meghvansi MK and Varma A (eds), Organic Amendments and Soil Suppressiveness in Plant Disease Management. Springer International Publishing, Switzerland pp. 219-247.

Arnold A.E., Maynard Z., Gilbert G.S., Coley P.D. and Kursar T.A. 2000. Are tropical fungal endophytes hyperdiverse? *Ecol Lett* 3: 267-274.

Bacon C.W. and White J.F. 2000. Microbial endophytes. Marcel Dekker Inc., New York.

Barazani O., Von D.C.C. and Baldwin I.T. 2007. *Sebacina vermifera* promotes the growth and fitness of *Nicotiana attenuata* by inhibiting ethylene signaling. *Plant Physiol* 144(2): 1223-32.

Barbara S., Christine B., Siegfried D., Anne-Katrin R. and Karsten K. 2002. Endophytic fungi: a source of novel biologically active secondary metabolites. *Mycol Res* 106(9): 996-1004.

Bills G.F., González-Menéndez V., Martín J., Platas G., Fournier J. *et al.* 2012. *Hypoxylon pulicicidum* sp. nov. (Ascomycota, Xylariales), a Pantropical Insecticide-Producing Endophyte. *Plos One* 7(10): 466-87.

Brundrett M.C. 2006. Understanding the roles of multifunctional mycorrhizal and endophytic fungi. In: Schulz BJE, Boyle CJC, Sieber TN, eds. Microbial root endophytes. Berlin, Germany: Springer-Verlag, 281-293.

Bush L.P., Wilkinson H.H. and Schardl C.L. 1997. Bioprotective alkaloids of grass-fungal endophyte symbioses. *Plant Physiol* 114: 1-7.

Butler M.S. 2004. The role of natural product chemistry in drug discovery. *J Nat Prod* 67(12): 2141-2153.

Chandra S. 2012. Endophytic fungi: novel sources of anticancer lead molecules. *Appl Microbiol Biotechno* 195(1): 47–59.

Clay K. and Holah J. 1999 Fungal endophyte symbiosis and plant diversity in successional fields. *Science* 285: 1742-1745.

Clay K. and Schardl C. 2002. Evolutionary origins and ecological consequences of endophyte symbiosis with grasses. *Am Nat* 160: 99-127.

Dai C.C., Tian L.S., Zhao Y.T., Chen Y. and Xie H., 2010. Degradation of phenanthrene by the endophytic fungus *Ceratobasidum stevensii* found in *Bischofia polycarpa*. *Biodegradation* 21(2): 245-55.

Demain A.L. 2014. Importance of microbial natural products and the need to revitalize their discovery. *J Ind Microbiol Biotechnol* 41: 185–201.

Deshmukh S.K., Mishra P.D., Almeida A.K., Verekar S., Sahoo M.R., Periyasamy G., Goswami H., Khanna A., Balakrishnan A. and Vishwakarma R. 2009. Anti-inflammatory and anticancer activity of ergoflavin isolated from an endophytic fungus. *Chem Biodiver* 6: 784-789.

Ernst M., Mendgen K.W. and Wirsel S.G.R. 2003. Endophytic fungal mutualists: seed borne *Stagonospora* spp. enhance reed biomass production in axenic microcosms. *Mol Plant Microbe Int* 16: 580-587.

Fernandes M.R.V., Costa S.T.A., Pfenning L.H., Costa-Neto C.M., Heinrich T.A., Alencar S.M. *et al.*, 2009. Biological activities of the fermentation extract of the endophytic fungus *Alternaria alternata* isolated from *Coffea arabica* L. *Braz J Pharm Sci* 45(4): 678-685.

Gangadevi V. and Muthumary J. 2008. Isolation of *Colletotrichum gloeosporioides*, a novel endophytic taxol-producing fungus from the leaves of a medicinal plant, *Justicia gendarussa*. *Mycol Balc.* 5: 1-4.

Ge H.M., Yu Z.G., Zhang J., Wu J.H. and Tan R.X. 2009. Bioactive alkaloids from endophytic Aspergillus fumigatus. *J Nat Prod* 72: 753-755.

Gherbawy Y.A. and Gashgari R.M. 2013. Molecular characterization of fungal endophytes from *Calotropis procera* plants in Taif region (Saudi Arabia) and their antifungal activities. *Plant Biosyst* 148(6): 1085-92.

Giridharan P., Verekar S.A., Khanna A., Mishra P.D. and Deshmukh S.K. 2012. Anticancer activity of Sclerotiorin isolated from an endophytic fungus *Cephalotheca faveoleta* yaguchi, Nishim & Udagawa. *Ind J Exp Biol* 50: 464-468.

Goswamia J., Pandey R.K., Tewaria J.P. and Goswami B.K. 2008. Management of root knot nematode on tomato through application of fungal antagonists, *Acremonium strictum* and *Trichoderma harzianum*. *J Environ Sci Health* B 43(3): 237-240.

Greetha A., Swamy M.K., Sinniah U.R. 2016. *Plectranthus amboinicus* (Lour.) Spreng: Botanical, Phytochemical, Pharmacological, and Nutritional significance. *Molecules* 2016.

Gubiani J.R., Zeraik M.L., Oliveira C.M., Ximenes V.F., Nogueira C.R., Fonseca L.M., Silva D.H., Bolzani V.S. and Araujo A.R. 2014. Biologically active eremophilane-type sesquiterpenes from Camarops sp., an endophytic fungus isolated from *Alibertia macrophylla*. *J Nat Prod* 77(3): 668-672.

Hamayun M., Khan S.A., Ahmad N., Tang D., Sang-Mo K., Chae-In N., Eun-Young S., Young-Hyun H. *et al.*, 2009. *Cladosporium sphaerospermum* as a new plant growth-promoting endophyte from the roots of *Glycine max* (L.) Merr. *World J Microbiol Biotechnol* 25: 627-632.

Harper J.K., Ford E.J., Strobel G.A., Arif A., Grant D.M., Porco J., Tomer D.P. and Oneill, K., 2003. Pestacin: a 1,3,-dihydro isobenofuran from *Pestalotipsis microspora* possessing antioxidant and antimycotic activities. *Tetrahedron,* 59: 2471-2476.

Ho M.Y., Wen-Chuan C., Hung-Chang H., Wen-Hsia C. and Wen-Hsin C. 2012. Identification of endophytic fungi of medicinal herbs of Lauraceae and Rutaceae with Antimicrobial Property. *Taiwania* 57(3): 229-241.

Huang W.Y., Yi-Zhong C., Kevin D.H., Harold C. and Mei S. 2007. Endophytic fungi from *Nerium oleander* L (*Apocynaceae*): main constituents and antioxidant activity. *World J Microbiol Biotechnol* 23: 1253-1263.

Jagan M.R.A., Manas R.B., Gajendra L.M. and Parthasarathy T. 2012. Computational approach for designing and development of potent inhibitor for Hepatitis - B Virus X- associated protein through molecular docking studies. *J Chem Pharm Res* 4(1): 265-271.

Johnson J.M., Alex T. and Oelmüller R. 2014. *Piriformospora indica*: The versatile and multifunctional root endophytic fungus for enhanced yield and tolerance to biotic and abiotic stress in crop plants. *J Trop Agri* 52(2): 103-122.

Jumpponen A. 2001. Dark septate endophytes - are they mycorrhizal? *Mycorrhiza* 11(4): 207-211.

Kalele D.N., Affokpon A. and Coosemans J. 2007. Efficacy of *Paecilomyces lilacinus* strain 251 against root knot nematodes in tomato under greenhouse conditions. Communications in Agricultural and Applied Biological Sciences 72(1): 209-213.

Khan A.L., Muhammad H., Javid H., Sang-Mo K. and In-Jung L. 2012. The newly isolated endophytic fungus *Paraconiothyrium* sp. LK1 produces Ascotoxin. *Molecules* 17: 1103-1112.

Kjer J., Wray V., Edrada-Ebel R., Ebel R., Pretsch A., Lin W. and Proksch P. 2009. Xanalteric acids I and II and related phenolic compounds from an endophytic Alternaria sp. isolated from the mangrove plant *Sonneratia alba*. *J Nat Prod* 72(11): 2053-2057.

Krohn K., Carsten B., Hans-Jurgen A., Siegfried D. and Schulz B. 1999. Herbarulide, a ketodivinyllactone steroid with an unprecedented homo-6-oxaergostane skeleton from the endophytic fungus *Pleospora herbarum*. J Nat Prod 62: 629-630.

Kumara Swamy M., Pokharen N., Dahal S., Anuradha M. 2011. Phytochemical and antimicrobial studies of leaf extract of *Euphorbia nerifolia*. *J Med Plants Res* 5: 5785–5788.

Kumara S.M., Sudipta K.M., Lokesh P., Neeki A., Rashmi W., Bhaumik H., Darshil H., Vijay R., Kashyap S.S.N. 2012. Phytochemical screening and in vitro antimicrobial activity of *Bougainvillea spectabilis* flower extracts. *Int J Phytomedicine* 4: 375–379.

Kunkel B.A., Grewal P.S. and Quigley M.F. 2004. A mechanism of acquired resistance against an entomopathogenic nematode by *Agrostis ipsilon* feeding on perrennial ryegrass harboring a fungal endophyte. *Biological Control* 29: 100-108.

Kusari S., Lamshoft M. and Spiteller M. 2009. *Aspergillus fumigatus* Fresenius, an endophytic fungus from *Juniperus communis* L. Horstmann as a novel source of the anticancer pro-drug Deoxypodophyllotoxin. *J Appl Microbiol* 107: 1019-1030.

Li J.Y., Strobel G., Harper J., Lobkovsky E. and Clardy J. 2000. Cryptocin, a potent tetramic acid antimycotic from the endophytic fungus *Cryptosporiopsis* cf. *quercina*. *Org Lett* 2: 767-70.

Lima A.M.D., Julia I.S., Joao V.B.S., Ana C.A.C., Clarice M.C., Francisco C.M.C. and Valdir F.V.J. 2011. Effects of culture filtrates of endophytic fungi obtained from *Piper aduncum* L. on the growth of *Mycobacterium tuberculosis*. *Electron J Biotechnol* 14(4): 1-6.

Lima J.S., Figueiredo J.G., Gomes R.G., Stringari D., Goulin E.H., Adamoski D., Kava-Cordeiro V. *et al.*, 2012. Genetic diversity of *Colletotrichum* spp. an endophytic fungi in a medicinal plant, Brazilian Pepper tree. *ISRN Microbiology* 2012, 1-7.

Liu J.Y., Song Y.C., Zhanga Z., Wang L., Guo Z.J., Zou W.X. and Tan R.X. 2004. *Aspergillus fumigatus* CY018, an endophytic fungus in *Cynodon dactylon* as a versatile producer of new and bioactive metabolites. *J Biotechnol* 114: 279-287.

Lu H., Wen X.Z., Jun C.M., Jun H. and Ren X.T. 2000. New bioactive metabolites produced by *Colletotrichum* sp., an endophytic fungus in *Artemisia annua*. *Plant Sci* 151: 67-73.

Luiz H.R., Nurhayat T., Natascha T., David E.W., Zhiqiang P., Ulrich R.B., James J.B., Natasha M.A. *et al.*, 2012. Diversity and biological activities of endophytic fungi associated with micropropagated medicinal plant *Echinacea purpurea* (L.) Moench. *American Journal of Plant Sci* 3: 1105-1114.

Luo D. and Fang B. 2008. Structural identification of ginseng polysaccharides and testing of their antioxidant activities. *Carbohyd Poly* 72(3): 376-381.

Lv Y.L., Fu-shengm Z., Juan C., Jin-long C., Yong-mei X., Xiang-dong L., Shun-xing G. 2010. Diversity and antimicrobial activity of endophytic fungi associated with the alpine plant *Saussurea involucrata*. *Biol Pharm Bull* 33(8): 1300-1306.

Mani V.M., Soundari A.P.G., Karthiyaini D. and Preeth K. 2015. Bioprospecting Endophytic Fungi and Their Metabolites from Medicinal Tree Aegle marmelos in Western Ghats, India. *Mycobiology* 43(3): 303-310.

Marinho A.M.R., Rodrigues-Filho E., Antonio G.F., Lourivaldo S.S. 2005. C25 Steroid epimers produced by *Penicillium janthinellum*, a fungus isolated from fruits *Melia azedarach*. *J Braz Chem Soc* 6(6B): 1342-1346.

Miller K. 2003. Encyclopedia and dictionary of medicine, nursing, and allied health, seventh Edition. by Saunders, an imprint of Elsevier, Inc.

Mohanty S.K., Mallappa K.S., Godavarthi K., Subbanarasiman B., Maniyam A. 2014. Evaluation of antioxidant, *in vitro* cytotoxicity of micropropagated and naturally grown plants of *Leptadenia reticulata* (Retz.) Wight & Arn.: An endangered medicinal plant. *Asian Pac J Trop Med* 7: 267–271.

Muhammad S., Mamona N, Muhammad S., Hidayat H., Yong S.L., Naheed R., Abdul J. 2010. Antimicrobial natural products: an update on future antibiotic drug candidates. *Nat Prod Rep* 27: 238-254.

Nath A., Raghunatha P. and Joshi S.R. 2012. Diversity and biological activities of endophytic fungi of *Emblica officinalis*, an ethnomedicinal plant of India. *Mycobiology*, 40(1): 8-13.

Nisa H., Kamili A.N., Nawchoo I.A., Shafi S., Shameem N., Bandh S.A. 2015. Fungal endophytes as prolific source of phytochemicals and other bioactive natural products: A review. *Microb Pathog* 82: 50-59.

Nithya K. and Muthumary J. 2010. Secondary metabolite from *Phomopsis* sp. isolated from *Plumeria acutifolia* Poiret. *Recent Research in Science and Technology* 2(4): 99-103.

Nithya K. and Muthumary J. 2011. Bioactive metabolite produced by *Phomopsis* sp., an endophytic fungus in *Allamanda cathartica* Linn. *Recent Research in Science and Technology* 3(3): 44-48.

Nithya K. and Muthumary J. 2009. Growth studies of *Colletotrichum gloeosporioides* (Penz.) Sacc. - A taxol producing endophytic fungus from *Plumeria acutifolia*. *Indian Journal of Science and Technology* 2(11): 14-19.

Nitya K.M., Pushpalatha K.C. and Chandrashekhar G.J. 2011. Antioxidant activity and phytochemical analysis of endophytic fungi isolated from *Lobelia nicotianifolia*.*Journal of Chemical and Pharmaceutical Research*, 3(5): 218-225.

Nuangmek W., McKenzie E.H.C. and Lumyong S. 2008. Endophytic fungi from wild banana (*Musa acuminate* Colla) works against anthracnose disease caused by *Colletotrichum musae*. *Res J Microbiol* 3: 358-374.

Oliveira C.M., Luis O.R., Geraldo H.S., Ludwig H.P., Maria C.M.Y., Roberto G.S.B., Vanderlan S., *et al.*, 2010. Dihydroisocoumarins produced by *Xylaria* sp. and *Penicillium* sp., endophytic fungi associated with *Alibertia macrophylla* and *Piper aduncum*. *Phytochem Lett* 179: 1-4.

Omacini M., Chaneton E.J., Ghersa C.M. and Müller C.B. 2001. Symbiotic fungal endophytes control insect host-parasite interaction webs. *Nature* 409: 78-81.

Pandi M., Rangarajulu S.K., Yong-Keun C., Hyung J.K. and Muthumary J. 2011. Isolation and detection of taxol, an anticancer drug produced from *Lasiodiplodia theobromae,* an endophytic fungus of the medicinal plant *Morinda citrifolia. Afr J Biotechnol* 10(8): 1428-1435.

Park S.H. and Eom A.H. 2007. Effects of mycorrhizal and endophytic fungi on plant community: a Microcosm Study. *Mycobiology* 35(4): 186-190.

Parniske M. 2008. Arbuscular mycorrhiza: the mother of plant root endosymbioses. Nature reviews. *Microbiology* 6(10): 763-775.

Peters S., Draeger S., Aust H.J. and Schulz B. 1998. Interactions in dual cultures of endophytic fungi with host and nonhost plant calli. *Mycologia,* 90: 360-367.

Petrini O. 1986. Taxonomy of endophytic fungi of aerial plant tissues. *In:* Fokkema N.J., van den Huevel J., eds. Microbiology of the phyllosphere. Cambridge, UK: Cambridge University Press, pp. 175-187.

Philippe G. 2016. Lolitrem B and Indole Diterpene Alkaloids Produced by Endophytic Fungi of the Genus Epichloë and Their Toxic Effects in Livestock. *Toxins* 8(2): 47.

Phongpaichit S., Nikom J., Rungjindamai N., Sakayaroj J., Hutadilok-Towatana N., Rukachaisirikul V. and Kirtikara K. 2007. Biological activities of extracts from endophytic fungi isolated from *Garcinia* plants. *FEMS Immunology & Medical Microbiology* 51: 517-525.

Pinheiroa E.A.A., Josiwander M.C., Diellem C.P.S., Andre de O.F., Patri S.B.M., Giselle M.S.P.G. *et al.,* 2013. Antibacterial activity of alkaloids produced by endophytic fungus *Aspergillus*sp. EJC08 isolated from medical plant *Bauhinia guianensis. Nat Prod Res* 27(18): 1633-1638.

Puri S.C., Verma V., Amna T., Qazi G.N. and Spiteller M. 2005. An endophytic fungus from *Nothapodytes foetida* that produces camptothecin. *J Nat Prod* 68: 1717-1719.

Reis V.M., Baldani J.I., Baldani V.L.D. and Dobereiner J. 2000. Biological nitrogen fixation in graminae and palm trees. *Crit Rev Plant Sci* 19: 227-247.

Rodriguez R.J., Henson J., Van V.E., Hoy M., Wright L., Beckwith F., Kim Y. and Redman R.S. 2008. Stress tolerance in plants via habitat-adapted symbiosis. *Int Soc Microbial Ecol* 2: 404-416.

Rodriguez R.J., White J.F., Arnold A.E. and Redman R.S. 2009. Fungal endophytes: diversity and functional roles. *The New phytologist*, 182(2): 314-30.

Russell K.H., Richard E.F. and James J.G. 2004. Plant-assisted degradation of phenanthrene as Assessed by solid-phase microextraction (SPME). *Int J Phytorem* 6(3): 253-268.

Saikkonen K., Saari S and Helander M. 2010. Defensive mutualism between plants and endophytic fungi.*Fungal Diver* 1(1): 101-113.

Santiago C., Chris F., Murray H.G.M., Juriyati J. and Jacinta S. 2012. Cytotoxic and antifungal activities of 5-hydroxyramulosin, a compound produced by an endophytic fungus isolated from *Cinnamomum mollissimum*. *Evid-Based Compl Alt Med* 2012: 1-6.

Sasidharan S., Chen Y., Saravanan D., Sundram K.M. and Latha Y.L. 2011. Extraction, isolation and characterization of bioactive compounds from plants' extracts. *Afr J Trad Compl Alt Med* 8(1): 1-10.

Schardl C.L. 2001. *Epichloe festucae* and related mutualistic symbionts of grasses. *Fungal Gen Biol* 33: 69-82.

Schardl C.L. and Phillips T.D. 1997. Protective grass endophytes: where are they from and where are they going? *Plant Dis* 81: 430-438.

Schulz A., Römmert K., Dammann U., Aust H.J. and Strack D. 1999. The endophyte-host interaction: a balanced antagonism? *Mycol Res* 103(10): 1275-1283.

Schulz B. and Boyle C. 2005. The endophytic continuum. *Mycol Res* 109: 661-686.

Schulz B., Boyle C., Draeger S., Römmert A.K. and Krohn K. 2002. Endophytic fungi: a source of novel biologically active secondary metabolites. *Mycol Res* 106(9): 996-1004.

Scott B. 2001. *Epichloe* endophytes: fungal symbionts of grasses. *Curr Opin Microbiol* 4: 393-398.

Senthilmurugan, Viji, Sekar G.R., Kuru S. and Balamurugan S. 2013. Phytochemical screening, enzyme and antibacterial activity analysis of endophytic fungi *Botrytis* sp. isolated from *Ficus benghalensis* (L.). *Int J Pharm biosci* 2(4): 264-273

Shweta S., Zuehlke S., Ramesha B.T., Priti V., Mohana Kunar P. *et al.*, 2010. Endophytic fungal strains of *Fusarium solani*, from *Apodytes dimidiata* E. Mey. ex Arn (*Icacinaceae*) produce camptothecin, 10-hydroxycamptothecin and 9-methoxycamptothecin. *Phytochem* 71: 117-122.

Siegel M.R. and Bush L.P. 1997. Toxin production in grass/endophyte associations. In G. Carroll and P. Tudzynski, eds. The mycota V. Pt. A. Plant relationships. Springer, Berlin. Pages 185–207.

Silva G.H., Teles H.L., Zanardi L.M., Young M.C.M., Eberlin M.N., Hadad R., Pfenning L.H. *et al.*, 2006. Cadinane sesquiterpenoids of *Phomopsis cassia*, an endophytic fungus associated with *Cassia spectabilis* (*Leguminoseae*). *Phytochem* 67: 1964-1969.

Simons L., Bultman T.L., Sullivan T.J. 2008. Effects of methyl jasmonate and an endophytic fungus on plant resistance to insect herbivores. *J Chem Ecol* 34: 1511-1517.

Smith S.E. and Read D.J. 1997. Mycorrhizal Symbiosis. Academic Press, London.

Srinivasan K. and Muthumary J. 2009. Taxol production from *Pestalotiopsis* sp an endophytic fungus Isolated from *Catharanthus roseus*. *Journal of Ecobiotechnology* 1(1): 028-031.

Strobel G. and Daisy B. 2003. Bioprospecting for microbial endophytes and their natural products. *Microbial Mol Biol Rev* 67(4): 491-502.

Strobel G., Daisy B., Castillo U. and Harper J. 2004. Natural products from endophytic microorganisms. *J Nat Prod* 67(2): 257-268.

Strobel G., Stierle A., Stierle D. and Hess W.M. 1993. *Taxomyces andreanae* a proposed new taxon for a bulbilliferous hyphomycete associated with Pacific yew. *Mycotaxon* 47: 71-78.

Strobel G.A. 2003. Endophytes as source of bioactive products. Microb Infect 5: 535-544.

Strobel G.A., Emilie D., Joe S., Chris M. 2001. Volatile antimicrobials from *Muscodor albus*, a novel endophytic fungus. *Microbiology* 147: 2943-2950.

Sturz A.V., Christie B.R. and Nowak J. 2000. Bacterial endophytes: potential role in developing sustainable systems of crop production. *Crit Rev Plant Sci* 19: 1-30.

Subban K., Ramesh S. and Muthumary J. 2013. A novel antibacterial and antifungal phenolic compound from the endophytic fungus *Pestalotiopsis mangiferae.Nat Prod Res* 27(16): 1445-1449.

Subbulakshmi G.K., Thalavaipandian A., Ramesh V., Bagyalakshmi, Rajendran A. 2012. Bioactive endophytic fungal isolates of *Biota orientalis* (L) Endl., *Pinus excels* Wall. and Thuja *occidentalis* L. *Int J Adv Life Sci* 4: 9-15.

Suffness M. 1995. Taxol Science and Applications. CRC Press, Boca Raton, Florida, USA.

Sugijanto N.E., Diesel A., Rateb M., Pretsch A., Gogalic S., Zaini N.C., Ebel R., Indrayanto G. 2011. Lecythomycin, a new macrolactone glycoside from the endophytic fungus *Lecythophora* sp. *Nat Prod Comm* 6(5): 677-678.

Sun C., Wang J.W., Fang L., Gao X.D. and Tan R.X. 2004. Free radical scavenging and antioxidant activities of EPS2, an exopolysaccharide produced by amarine filamentous fungus *Keissleriella* sp. YS 4108. *Life Sciences* 75(9): 1063-1073.

Sun X., Kong X., Gao H., Zhu T., Wu G., Gu Q. and Li D. 2013. Two new meroterpenoids produced by the endophytic fungus *Penicillium* sp. SXH-65.*Arch Pharm Res* 37(8): 978-82.

Swamy M.K., Sinniah U.R. A. 2015a. Comprehensive review on the phytochemical constituents and pharmacological activities of *Pogostemon cablin* Benth.: An aromatic medicinal plant of industrial importance. *Molecules* 20: 8521–8547.

Swamy M.K., Sinniah U.R., Akhtar M.S. 2015b. *In vitro* pharmacological activities and GC-MS analysis of different solvent extracts of *Lantana camara* leaves collected from tropical region of Malaysia. Evid-Based Compl Alt Med 2015: 1–9.

Swamy M.K., Akhtar M.S., Sinniah U.R. 2016a. Root Exudates and Their Molecular Interactions with Rhizospheric Microbes. In: Hakeem KR and Akthar MS (eds), Plant, Soil and Microbes Volume 2. Springer International Publishing, Switzerland.

Swamy M.K., Akhtar M.S., Sinniah U.R. 2016b. Response of PGPR and AM Fungi Toward Growth and Secondary Metabolite Production in Medicinal and Aromatic Plants. In: Hakeem KR and Akthar MS (eds), Plant, Soil and Microbes Volume 2. Springer International Publishing, Switzerland.

Tan R.X. and Zou W.X. 2001. Endophytes: a rich source of functional metabolites. *Nat Prod Rep* 18(4): 448-459.

Taylor T.N. and Taylor E.L. 2000. The rhynie chert ecosystem: a model for understanding fungal interactions, in: Bacon C.W., White, J.F., eds. Microbial Endophytes. New York: Marcel Decker: 31-48.

Tayung K., Barik B.P., Jha D.K. and Deka D.C. 2011. Identification and characterization of antimicrobial metabolite from an endophytic fungus, *Fusarium solani* isolated from bark of Himalayan yew. *Mycosphere* 2(3): 203-213.

Teles H.L., Sordi R., Silva G.H., Castro-Gamboa I., Bolzani V.S., Pfenning L.H., Abreu L.M. *et al.* 2006. Aromatic compounds produced by *Periconia atropurpurea*, an endophytic fungus associated with *Xylopia aromatica*. *Phytochem* 67: 2686-2690.

Torres M.S., White J.F.J., Bischoff J.F. 2007. *Hypocrella panamensis* sp. nov (Clavicipitaceae Hypocreales): evaluation on the basis of morphological and molecular characters. *Mycol Res* 111: 317-323.

Tétard Jones C. and Edwards R. 2016. Potential roles for microbial endophytes in herbicide tolerance in plants. *Pest Manag Sci*72(2): 203-209.

Tsuchinari M., Keiko S., Fuminori H., Tetsuya M., Takuya K. and Yoshihito S. 2007. Fusapyridons A and B, novel pyridone alkaloids from an endophytic fungus, *Fusarium* sp. YG-45. *Zeitschrift für Naturforschung*, 62: 1203-1207.

Vu T.T., Hauschild R. and Sikora R.A. 2006. *Fusarium oxysporum* endophytes induced systemic resistance against *Radopholus similis* on banana. *Nematology* 8(6): 847-852.

Wagenaar M.M. and Clardy J. 2001. Dicerandrols, new antibiotic and cytotoxic dimers produced by the fungus *Phomopsis longicolla* isolated from an endangered mint. *J Nat Prod* 64(8): 1006-1009.

Wang F.W., Jiao R.H., Cheng A.B., Tan S.H. and Song Y.C. 2007. Antimicrobial potentials of endophytic fungi residing in *Quercus variabilis* and brefeldin A obtained from *Cladosporium* sp. *World J Microbiol Biotechnol* 23: 79-83.

Wang J., Machado C., Panaccione D.G., Tsai H.F. and Schardl C.L. 2004. The determinant step in ergot alkaloid biosynthesis by an endophyte of perennial ryegrass. *Fungal Gen Biol* 41: 189-198.

Wang Y. and Dai C.C. 2011. Endophytes: a potential resource for biosynthesis, biotransformation, and biodegradation. *Ann Microbiol* 61: 207-215.

West C.P. 1994. Physiology and drought tolerance of Endophyte-infected grasses. *In:* CW Bacon, JF Jr. White, eds; Biotechnology of Endophytic Fungi of Grasses), CRC Press, Boca Raton, FL. 87-98

West C.P., Izekor E., Turner K.E. and Elmi A.A. 1993. Endophyte effects on growth and persistence of tall fescue along a water supply gradient. *Agronomy Journal* 85: 264-270.

Wu H., Hong-Yan Y., Xiang-Lin Y. and Yu-Hua L. 2013. Diversity of endophytic fungi from roots of *Panax ginseng* and their saponin yield capacities. *Springer Plus*, 2(107): 1-9.

Wu S.H., Rong H., Cui-Ping M. and You-Wei C. 2013. Two new steroids from an endophytic fungus *Phomopsis* sp. Chem. Biodivers 10(7): 1276-1283.

Xing Y.M., Chen J., Cui J.L., Chen X.M. and Guo S.X. 2011. Antimicrobial activity and biodiversity of endophytic fungi in *Dendrobium devonianum* and *Dendrobium thyrsiflorum*. *Curr Microbiol* 62(4): 1218-24.

Yan X., Sikora R.A. and Zheng J. 2011. Potential use of cucumber (*Cucumis sativus* L.) endophytic fungi as seed treatment agents against root-knot nematode *Meloidogyne incognita. Journal of Zhejiang University Science* B 12(3): 219-225.

Yang X., Zhang L., Guo B. and Guo S. 2004. Preliminary study of a vincristine- producing endophytic fungus isolated from leaves of *Catharanthus roseus. Chinese Traditional and Herbal Drugs* 35(1): 79-81.

Zhang L., Guo B., Li H., Zeng S., Shao H., Gu S. and Wei R. 2000. Preliminary study on the isolation of endophytic fungus of *Catharanthus roseus* and its fermentation to produce products of therapeutic value. *Chinese Traditional and Herbal Drugs* 31(11): 805–807.

Zhang W., Draeger S., Schulz B. and Krohn K. 2009. Ring B aromatic steroids from an endophytic fungus, *Colletotrichum* sp. *Nat Prod Comm* 4(11): 1449-54.

Zhou X., Zheng W., Zhu H. and Tang K. 2009. Identification of a taxol-producing endophytic fungus EFY-36. *Afr J Biotechnol* 8 (11): 2623-2625.

Zhuang H., Wen-Li M., Hai-Bin C., Yan-Bo Z., Hai-Peng L. and Kui H. 2008. Antibacterial constituents from the endophytic fungus *Penicillium* sp of mangrove plant *Cerbera manghas. Chemical journal of Chinese Universities-Chinese* 29(4): 749-752.

Zikmundova M., Drandarov K., Bigler L., Hesse M. and Werner C. 2002. Biotransformation of 2-Benzoxazolinone and 2-Hydroxy-1, 4-Benzoxazin-3-one by endophytic fungi isolated from *Aphelandra tetragona. Appl Environ Microb* 10: 4863-4870.

❑❑❑

Chapter 14

Berries: Source of Natural Antioxidant

Rojita Mishra*[1] and Satpal Singh Bisht[2]

[1] Lecturer, Department of Botany, Govt. Science College Chatarapur, Odisha, E-email:rojitamishra@gmail.com, [2] Professor, Department of Zoology, Kumaun University, Nainital, Uttarakhand, India, E-email: sps.bisht@gmail.com

ABSTRACT

Berries are well-known for their nutrition and health-promotion values. This is mainly due to the phytochemical composition and the bioactive compounds present in the berries.In recent years, the antioxidant activity of berry fruits and their roles in the prevention and treatment of various human chronic and degenerative diseases have attracted extensive attention from all stake holders. Berries are suggested to be a good source of dietary antioxidants. To have a better understanding of the mechanism underlying the antioxidant activity of berries the present chapter is focused on various aspects of berries as powerful antioxidant.

Keywords: Berries, antioxidant activity, phytochemicals and bioactive compounds.

Introduction

In the last few decades researchers are interested more towards all kinds of research related to fruits. Particularly fruit berries are well studied, as the main dietary sources of bioactive compounds more abundantly in highly-coloured berries. Bioactive compounds from berries are obtained from plants belong to families i.e., Rosaceae (strawberry, raspberry, blackberry), and Ericaceae (blueberry, cranberry). Grape berries and their products are great sources of bioactive compounds. The group of berries having low bioactive compounds but they play significant role in nutraceuticals includes bilberries (*Vaccinuim myrtillus*), elderberries (*Sambucus* spp.), gooseberies (*Ribes uva-crispa*), cape gooseberies (*Physalis peruviana*), chokecheries(*Prunus virginiana*), arctic brambles (*Rubus articus*), cloudberries (*Rubus chamaemorus*), crowberries (*Empetrum nigrum, E.*

hermaphroditum), lingonberries (*Vaccinium vitis-idaea*,loganberies (*Rubus loganobaccus*), marionberries (*Rubus* spp.), honeyberies (*Lonicera caeulea*), Saskatoon berries (*Amelanchier alnifolia*), Rowan berries (*Sorbus* spp.), maqui, and seabuckthorn (*Hippophae rhamnoides*). Bioactive compounds that are found mainly in berries can be grouped as antioxidants like phenolic compounds and fruit colorants (anthocyanins and carotenoids). Berries' phenolics represent a diverse group of compounds including phenolic acids, such as hydroxybenzoic and hydroxycinnamic acid conjugates; flavonoids, such as flavonols, flavanols, and anthocyanins. In addition, tannins, divided into condensed tannins (proanthocyanidins) and hydrolysable tannins. Apart from the above antioxidant groups' berries are the rich source of vitamins (ascorbic acid) and minerals with antioxidant properties. These compounds are of great interest for nutritionists and food technologists as it can be used as functional foods ingredients. Due to the market for functional foods in the EU having grown, the years 1999 to 2006 saw the market increase from about $1.8 billion to $8 billion[1].

Antioxidant activity denotes the ability of a bioactive compound to maintain cell structure and function by effectively clearing free radicals, inhibiting lipid peroxidation reactions, and preventing other oxidative damage. According to Sies, 1986 in a normal healthy human, the generation of pro-oxidants in the form of ROS and RNS are effectively kept in check by the various levels of antioxidant defence. When it gets exposed to adverse physicochemical, environmental or pathological agents such as atmospheric pollutants, cigarette smoking, ultraviolet rays, radiation, toxic chemicals this maintained balance is shifted in favour of pro-oxidants resulting in 'oxidative stress'. The antioxidants may mediate their effect by directly reacting with ROS, quenching them and chelating the catalytic metal ions. Several synthetic antioxidants, *e.g.*, BHA and butylated hydroxytoluene are commercially available but are quite unsafe and their toxicity is a problem of concern. Natural antioxidants, especially phenolics and flavonoids are safe and bioactive. Therefore, in the recent past, considerable attention has been directed towards identification of plants with antioxidant ability that may be used for human consumption[2]. It is also a foundation of many other biological functions, such as anti-cancers, anti-inflammation and anti-aging. More importantly, the prevention of many chronic diseases, such as cancer, diabetes and cardiovascular disease, has been suggested to be associated with the antioxidant activity. Therefore, a deep study of natural antioxidants, such as those from fruits and vegetables, is of great importance to human health[3].

Phytochemicals Present in Berries

The biologically active constituents of medicinal, commercial and poisonous plants have been studied throughout the development of organic chemistry. Many of these compounds are secondary metabolites. These are mainly considered as phytochemicals or natural products. They often have ecological role in regulating interaction between plants, microbes insects and animals. Phytochemical survey can reveal natural products that are markers for botanicals and evolutionary relationship *i.e,* chemotaxonomy. The classes of phytochemicals are :

- Polyketides and fatty acids
- Terpenoids and steroids
- Phenylpropanoids
- Alkaloids
- Specialised amino acids and peptides
- Specialized carbohydrates

Main phytochemicals that are present in berries includes phenolic compounds, flavons, flavonols, anthocyanins, phenolic acids and hydrolysable tannins. Few phytochemical compositions of important berries and their health benefits are given in the Table 1.

Table 1. Chemical composition and health benefits of few berries.

Berries	Phytochemicals	Health benefits
Strawberries	Phenolic compounds	Antioxidants,
	Flavonols	**Immunity boosters,**
	Kaempferol glycosides	Prevents the
	(Kaempferol-3-glucoside,	clouding
	Kaempferol-glucuronide,	over the eye lenses,
	Kaempferol-3-malonylglucoside,	Prevention of cancer,
	Kaempferol-coumaroyl-glucoside	Wrinkle fighter,
	Quercetin glycosides	Lowering of blood

	Quercetin-3-glucuronide, Quercetin-3-malonylglucoside, Quercetin-3-rutinoside = rutin, Quercetin-3-glucoside **Anthocyanins** **Cyanidin glycosides** Cyanidin-3-glucoside, Cyanidin-3-rutinoside, Cyanidin-3-galactoside, Cyanidin-3-malonylglucoside) **Pelargonidin glycosides** Pelargonidin-3-glucoside, Pelargonidin-3-rutinoside, Pelargonidin-3-galactoside, Pelargonidin-3-arabinoside, Pelargonidin-3-malonylglucoside, Pelargonidin-3-malylglucoside) **Peonidin glycosides** Peonidin-3-glucoside) **Phenolic acids and Hydrolyzable tannins** Ellagic acid and its glycosides Ellagitannins Gallic acid Gallotannins Caffeic acid p-coumaric acid and coumaroyl glycosides	cholesterol, Reduces inflammation, **Regulate blood pressure,** Prevention of constipation and diverculitis, Helps in weight management.
Black berries	**Phenolic compounds** **Flavonols and Flavons** **Kaempferol glycosides** Kaempferol-gacetylgalactoside, Kaempferol-glucoside) **Quercetin glycosides** Quercetin-3-galactoside, Quercetin-3-glucuronide, Quercetin-3-glucoside, Quercetin-3-rutinoside = rutin, Quercetin-3-rhamnoside) **Myricetin glycosides**	Better digestive health, strengthened immune defense Healthy functioning of heart, prevention of cancer and relief from endothelial **dysfunction. Blackberry provides cognitive benefits and aids in enhancing memory**

	Myricetin-3-galactoside, Myricetin-3-glucoside) **Anthocyanins** **Cyanidin glycosides** (Cyanidin-3-glucoside, Cyanidin-3-rutinoside, Cyanidin-3-arabinoside **Pelargonidin glycosides** (Pelargonidin-3-glucoside) **Peonidin glycosides** (Peonidin-3-glucoside) **Phenolic acids and Hydrolyzable tannins** **Ellagic acid and its glycosides** Ellagitannins (sanguiin H-6 and lambertianin C) Gallic acid and galloyl esters p-coumaric acid and coumaroyl glycosides	**weight management, keeping the bones strong, healthy skin, improved vision and disease-free eyes, normal blood clottingserve as a valuable food during pregnancy owing to an impressive gamut of healthful nutrients**
Blue berries	**Phenolic compounds** **Flavonols** Myricetin glycosides (Myricetin-3-glucoside, Myricetin-3-rhamnoside) Quercetin glycosides (Quercetin-3-galactoside, Quercetin-3-glucoside, Quercetin-3-rutinoside) **Anthocyanins** Cyanidin glycosides (Cyanidin-3-galactoside, Cyanidin-3-glucoside, Cyanidin-3-arabinoside) Delphinidin glycosides (Delphinidin-3-galactoside, Delphinidin-3-arabinoside, Delphinidin-3-glucoside) Malvidin glycosides (Malvidin-3-galactoside, Malvidin-3-arabinoside, Malvidin-3-glucoside)	AntioxidantsUsed in weight management Protects DNA damage so help in curing cancer and ageing Controls blood cholesterol Prevention of heart disease Lower blood pressure Brain function and improvement of memory Antidiabetic effects Controlling urinary tract infections. Helps in preventing muscle damage

	Petunidin glycosides (Petunidin-3-galactoside, Petunidin-3-arabinoside, Petunidin-3-acetylglucoside) Peonidin glycosides (Peonidin-3-galactoside, Peonidin-3-arabinoside)	
Cranberries	**Phenolic compounds** **Flavonols** Kaempferol glycosides (Kaempferol-3-glucoside) Quercetin glycosides (Quercetin-3-galactoside, Quercetin-3-arabinoside, Quercetin-3-rhamnoside **Anthocyanins** Cyanidin glycosides (Cyanidin-3-glucoside, Cyanidin-3-galactoside, Cyanidin-3-arabinoside) Peonidin glycosides (Peonidin-3-glucoside, Peonidin-3-galactoside, Peonidin-3-arabinoside) Pelargonidin glycosides (Pelargonidin-3-galactoside, Pelargonidin-3-arabinoside) Malvidin glycosides (Malvidin-3-galactoside, Malvidin-3-arabinoside) Delphinidin glycosides (Delphinidin-3-arabinoside) Petunidin glycosides (Petunidin-3-galactoside) **Phenolic acids** p-coumaric acid	They cure urinary tract infections Cardiovascular disease cure Colon cancer treatment Treatment of gum disease

Methods for The Antioxidant Activity Evaluation of Berries

These assays include perhaps the most popular, but often misunderstood by its name, total phenols assay by Folin Ciocalteu reagent (FCR). In addition, also grouped into this category are the Trolox equivalent antioxidant capacity (TEAC) assay, the ferric ion reducing antioxidant power (FRAP) assay, the N, N-dimethyl-p-phenylenediamine (DMPD) assay, and the Cu(II) reduction capacity assay. These methods involve two components in the reaction mixture, antioxidants and oxidant (also the probe). The probe itself is an oxidant that abstracts an electron from the antioxidant, causing colour changes of the probe. The degree of the colour change is proportional to the antioxidant concentrations. The reaction end point is reached when colour change stops. The change of absorbance (¢A) is plotted against the antioxidant concentration to give a linear curve. The slope of the curve reflects the antioxidant's reducing capacity, which is expressed as Trolox equivalence (TE) or gallic acid equivalent (GAE).

Total Phenols Assay by Folin-Ciocalteu Reagent

FCR was initially intended for the analysis of proteins taking advantage of the reagent's activity toward protein tyrosine (containing a phenol group) residue[4]. Many years later, Singleton and co-workers, 1999 extended this assay to the analysis of total phenols in wine; since then the assay has found many applications [5]. The FCR-based assay gained popularity and is commonly known as the total phenols (or phenolic) assay. The FCR actually measures a sample's reducing capacity, but this is not reflected in the name "total phenolic assay". Numerous publications applied the total phenols assay by FCR and an ET-based antioxidant capacity assay (*e.g.*, FRAP, TEAC, etc.) and often found excellent linear correlations between the "total phenolic profiles" and "the antioxidant activity". This is not surprising if one considers the similarity of chemistry between the two assays. One of the assays may just be redundant. A recent report of using polyphenol oxidase for assaying total phenols in tea may be more specific to phenolic compounds. The FCR is typically made by first boiling (for 10 h) the mixture of sodium tungstate (100 g), sodium molybdate (25 g), concentrated hydrochloric acid (100 mL), 85% phosphoric acid (50 mL), and water (700 mL). After boiling, lithium sulfate (150 g) is added to the mixture to give an intense yellow solution the FC reagent. Contamination of reductants leads to a green color, and the addition of oxidants such as bromine can restore the desired yellow color. The exact chemical nature of the FC reagent is not known, but it is believed to contain heteropolyphosphotunstates-molybdates. Sequences of reversible one- or two-electron reduction reactions lead to blue species. In essence, it is believed that the molybdenum is easier to be reduced in the complex and electron-transfer reaction occurs between reductants and Mo(VI): Obviously, the FC reagent is nonspecific to phenolic compounds as it can be reduced by many nonphenolic compounds [*e.g.*, vitamin C, Cu(I), *etc.*]. Phenolic compounds react with FCR only under basic conditions (adjusted by a sodium carbonate solution to pH _10). Dissociation of a phenolic proton leads to a phenolate anion, which is capable of reducing FCR. This supports the notion that the reaction occurs through electrontransfer mechanism. The blue compounds formed between phenolate and FCR are independent of the structure of phenolic compounds, therefore ruling out the possibility of coordination complexes formed between the metal centre and the phenolic compounds.

Trolox Equivalent Antioxidant Capacity Assay

The TEAC assay was first reported by Miller et al., in 1993 and later improved Re et al.,1999)[6-7]. In the improved version the oxidant, was generated by persulfate oxidation of 2,2'-azinobis (3-ethylbenzothiazoline-6-sulfonic acid) (ABTS). 7 mmol of ABTS ammonium was dissolved in water and treated with 2.45 mmol of potassium persulfate, and the mixture was then allowed to stand at room temperature for 12-16 h to give a dark blue solution.

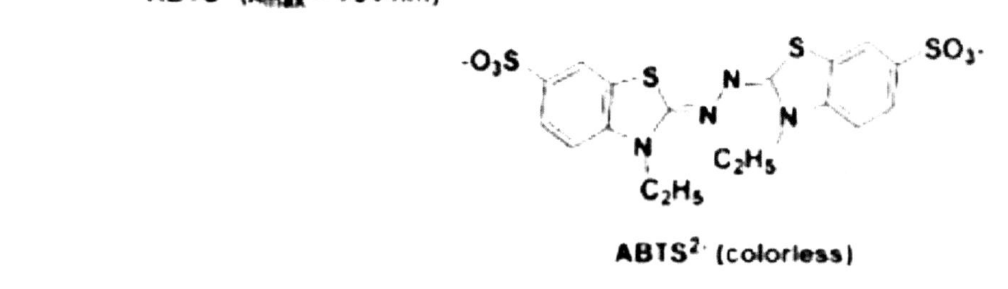

+ antioxidant

- K_2SO_5

ABTS˙ (λ_{max} = 734 nm)

ABTS^{2-} (colorless)

This solution was diluted with ethanol or buffer (pH 7.4) until the absorbance reached 0.7 at 734 nm. One millilitre of the resulting solution was mixed with 10 μL of sample. The absorbance was read at 30 °C, 1, 4, and 6 min after mixing at 30 °C. The difference of the absorbance reading is plotted versus the antioxidant concentrations to give a straight line. The concentration of antioxidants giving the same percentage change of absorbance of the ABTS as that of 1 mM Trolox was regarded as TEAC[7].

Ferric Ion Reducing Antioxidant Power Assay

The FRAP assay also takes advantage of electron-transfer reactions. Here in a ferric salt, Fe(III)(TPTZ) (TPTZ :2,4,6-tripyridyls-triazine), is used as an oxidant by Benzie and Strain ,1999) [8]. The redox potential of Fe(III) salt (0.70 V) is comparable to that of ABTS anion radical (0.68 V).

+ antioxidant

- e

$[Fe(III)(TPTZ)_2]^{3+}$ $[Fe(II)(TPTZ)_2]^{2+}$, λ_{max} = 593 nm

Therefore, essentially, there is not much difference between TEAC assay and the FRAP assay except TEAC is carried out at neutral pH and FRAP assay under acidic (pH 3.6) conditions.

The FRAP assay involves the following procedures: The oxidant in the FRAP assay is prepared by mixing TPTZ (2.5 mL, 10 mM in 40 mM HCl), 25 mL of acetate buffer, and 2.5

mL of ferric chloride (20 mM). The conglomerate is referred to as "FRAP reagent". The final solution has Fe(III) of 1.67 mM and TPTZ of 0.83 mM. Therefore, the TPTZ is deficient as the ideal reaction stoichiometry between Fe (III) and TPTZ is 1 to 2. The oxidant is not just ferric ion of TPTZ it also contains other Fe(III)(0.68) species which can lead to potential problems as many metal chelators in food extract could bind Fe(III) and form complexes that are also capable of reacting with antioxidants. To measure FRAP value, 300 µL of freshly prepared FRAP reagent is warmed to 37 °C and a reagent blank reading is taken at 593 nm; then 10 µL of sample and 30 µL of water are added. Absorbance readings are taken after 0.5 s and every 15 s until 4 min. The change of absorbance (¢A) (A 4min – A)is calculated and related to ¢A of an Fe(II) standard solution. ¢A is linearly proportional to the concentration of antioxidant. One FRAP unit is arbitrarily defined as the reduction of 1 mol of Fe(III) to Fe(II).

Total Antioxidant Potential Assay Using Cu(II) as an Oxidant

The method is based on reduction of Cu(II) to Cu(I) by reductants (antioxi-ants) present in a sample. A chromogenic reagent, bathocuproine (2, 9-dimethyl-4, 7-diphenyl-1, 10-phenanthroline), forms a 2:1 complex with Cu(I), which has a maximum absorbance at 490 nm Schiltz, 1996(9). It was found that 1 mol of R-tocopherol can reduce 2 mol of Cu(II) to Cu(I) (10)Yamashita *et al.*, 1998. More recently, Zaporozhets *et al.*, 2004 (11) reported a method for measuring the antioxidant power of herbal products based on solid-phase spectrophotometry using tetrabenzo [b,f,j,n] [1,5,9,13] tetraazacyclohexadecine-Cu(II) complex immobilized on silica gel. The absorbance of the modified sorbent (712 nm) increases when the Cu (II) is reduced.

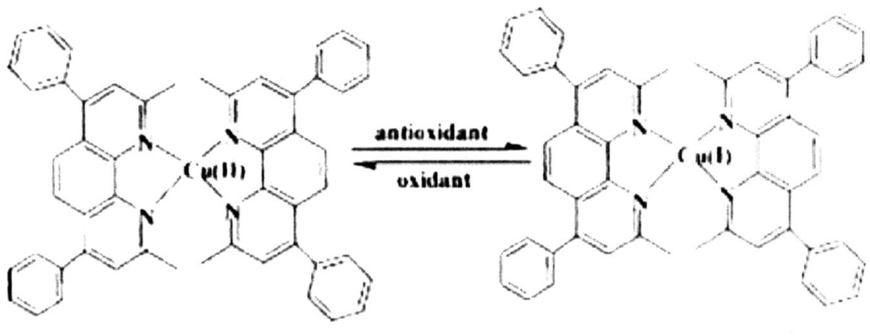

2,2-Diphenyl-1-picrylhydrazyl Radical ScavengingCapacity Assay

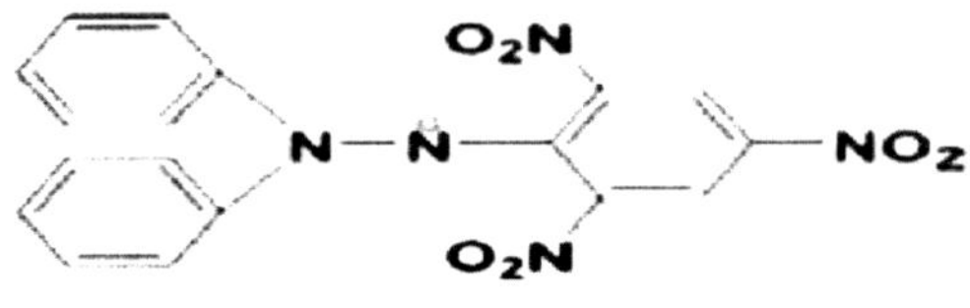

DPPH

DPPH is one of a few stable and commercially available organic nitrogen radicals and has a UV-vis absorption maximum at 515 nm. Upon reduction, the solution colour fades; the reaction progress is conveniently monitored by a spectrophotometer. The DPPH assay is typically run by the following procedure: DPPH solution (3.9 mL, 25 mg/L) in methanol is mixed with sample solution (0.1 mL). The reaction progress absorbance of the mixture is monitored at 515 nm for 30 min or until the absorbance is stable. Upon reduction, the colour of the solution fades. The percentage of the DPPH remaining is calculated as :

$$\%DPPH_{rem} = 100 \times [DPPH]_{rem}/[DPPH]_{T=0}$$

The percentage of DPPH remaining is proportional to the antioxidant concentrations, and the concentration that causes a decrease in the initial DPPH concentration by 50% is defined as EC50. The time needed to reach the steady state with EC50, is calculated from the kinetic curve and defined as T_{EC50}. Sanchez-Moreno and co-workers classified the kinetic behaviour of the antioxidant compound as follows: <5 min (rapid), 5-30 min (intermediate), and >30 min (slow). They further proposed a parameter, called "antiradical efficiency (AE)" (12), to express the antioxidant capacity of a certain antioxidant. AE is calculated as :

$$AE = (1/EC_{50})T_{EC50}$$

The DPPH assay is technically simple, but some disadvantages limit its applications. Besides the mechanistic difference from the HAT reaction that normally occurs between antioxidants and peroxyl radicals, DPPH is long-lived nitrogen radical, which bears no similarity to the highly reactive and transient peroxyl radicals involved in lipid peroxidation. Many antioxidants that react quickly with peroxyl radicals may react slowly or may even be inert to DPPH. This is evident from the T values ranging from 1.15 min (ascorbic acid) to 103 min (rutin). Consequently, the antioxidant capacity is not properly rated. The reaction kinetics between DPPH and antioxidants are not linear to DPPH concentrations.

Assays measuring other ROS scavenging capacity

Experimental evidence has directly or indirectly suggested that there are six major reactive oxygen species causing oxidative damage in the human body. These species are superoxide anion, hydrogen peroxide, peroxylradicals, hydroxyl radical, singlet oxygen, and peroxynitrite . To counteract the assault of these ROS, living cells have a biological defense system composed of enzymatic antioxidants that convert ROS/RNS to harmless species. For example, superoxide anion is converted to oxygen and hydrogen peroxide by superoxide dismutase (SOD) or reacts with nitric oxide to form peroxynitrite. Hydrogen peroxide can be converted to water and oxygen by catalase. In contrast, no enzymatic action is known to scavenge peroxyl radicals, and Peroxynitrite. The burden of defense relies on a variety of nonenzymatic antioxidants such as vitamins C and E and many phytochemicals that have the property of scavenging oxidants and free radicals. To comprehensively evaluate the oxidant-scavenging capacity of a food sample, assays have to be designed to include these ROS. However, so far the majority of assays are designed to measure a sample's capacity to react with one oxidant (either organic radical or redox active compounds). The peroxyl radical has been the most frequently used ROS in the assays because it is a key radical in autoxidation and it can be generated conveniently from the thermal decomposition of azo compounds.

In vivo antioxidant activity measurements

Antioxidant activity of natural bioactive compounds present in berries is measured with the help of following tests and parameters [13]

- Assay of superoxide dismutase (SOD) activity
- Assay of catalase activity
- Assay of peroxidase activity
- Glutathione peroxidase assay
- Assay of ascorbate oxidase activity
- Assay of glucose-6-phosphate
- Estimation of reduced glutathione
- Quantification of vitamins(C,E and A)
- Assay of malondialdehyde (MDA)
- Assay of diene conjugates.

Factors Influencing The Antioxidant Capacity of Berries and Products

Cultivar, genotype, variety, Growing location, Cultivation techniques, (conventional, organic) Cultivation condition, (greenhouse, plastic tunnel, open-field, light), Growing season, ripening Processing, Storage (time, temperature). Following studies substantiates these predictions.

Viskelis *et al.*, 2016, the chemical composition and antioxidant capacity of berry fruits and suggested that antioxidant activity mainly controlled by gentic factors[14]. Based on phenolic compounds concentration the investigated berry species may be ranked as follows: *Aronia melanocarpa*>*Sambucus nigra* L.>*Ribes nigrum* L.>*Rubus occidentalis* L.>*Rubus idaeus* L.>*Hippophae rhamnoides* L. Antioxidant capacities varied significantly among berries and cultivars and were highly correlated with phenolic compounds content. Among all berry species tested, fruits of *Aronia melanocarpa* were found to have the highest amounts of anthocyanins, followed by *Sambucus nigra* L., *Ribes nigrum* L., and black raspberry *Rubus occidentalis* L. The highest ascorbic acid content among investigated berry species was found in black currant fruits, followed by the sea buckthorn fruits[14].

Many studies suggest that antioxidant activity is influenced by the polarity of phenolic compounds. The present work indicated that the cranberry phenols are good free radical-scavengers, but they were less efficient at inhibiting the lipid peroxidation. The antioxidant effect was dose dependent at the concentration levels used; in general, fractions were good antioxidant only at the higher concentration of 313 μg/mL. Also, the technological process to manufacture cranberry juice has negatively affected the antioxidant and antiradical activities of all fractions regardless of polarity. However, phenols present in most fractions of fruit extracts have shown very interesting antiradical properties while phenols in the most hydrobobic fractions of extract rich in watersoluble phenolic compounds and the anthocyanin-rich cranberry extract had a remarkable action against lipid oxidation[15].

Many researchers developed techniques and methodologies to among many strategies two important strategies to enrich phenolic antioxidants are: (i) Genetic improvement of fruits and vegetables to produce plants that will yield fruits and vegetables with higher phenolic concentration and (ii) Bioprocessing of botanicals using solid-state systems and synergies to generate phytochemical profiles with improved health benefits. Currently in terms of genetic improvement breeding strategy coupled with micropropagation using tissue culture is being developed. A second exciting strategy that can be used is the bioprocessing of botanicals using solid-state systems using food grade microorganisms common in Asian food systems and synergies to generate phytochemical profiles with enhanced functionality and health benefits[16].

Mechanism of Antioxidant Action of Berries

Oxidative stress and inflammation play a pivotal role in the initiation and progression of atherosclerosis and cardiovascular disease (CVD). There are evidences that berry anthocyanins plays significant role in decreasing oxidative damage and inflammation in cellular and animal models of CVD [17, 18, 19]. Elderberry anthocyanins significantly decreased cytotoxicity caused by chemical inducers of oxidative stress. Anthocyanins from blackberry extract were shown to protect against peroxynitrite-induced oxidative damage in human umbilical vein endothelial cells [18]. Mulberry anthocyanins have also exhibited antioxidative and antiatherogenic affects, by inhibiting oxidation of LDL and formation of foam cells, respectively, in an in vitro model of atherosclerosis[19]. Anthocyanins from berries commonly consumed in the United States, such as blueberries and cranberries, have been reported to reduce TNF-α induced up regulation of inflammatory mediators in human microvascular endothelial cells[20]. De Furia et al.,2009 showed the protective effects of blueberries against insulin resistance and hyperglycemia, thus reducing the risk factors for CVD[21]. Nitric oxide (NO), when formed through activation of inducible nitric oxide synthase (iNOS) has proinflammatory effects, leading to increased vascular permeability, induction of inflammatory cytokines, and the formation of peroxynitrite, a strong oxidizing agent. Pergola *et al.*, 2006 have reported inhibitory effects of the anthocyanin fraction of blackberry extract on NO biosynthesis in the murine monocyte/ macrophage J774 cell line stimulated with lipopolysaccharide[22]. The study also reported that blackberry anthocyanin extract inhibited inducible iNOS protein expression, thereby decreasing the inflammatory response in macrophages and inhibiting the formation of foam cells. While increased iNOS expression leads to the proinflammatory effects of NO, generation of NO by endothelial nitric oxide synthase (eNOS) plays a crucial role in maintaining cardiovascular homeostasis by favorably modulating blood pressure and reducing endothelial dysfunction. Xu *et al.*, 2004 and Lazze *et al.*,2006 (23;24) have reported the upregulation of eNOS by cyanidin-3-glucoside in bovine artery endothelial cells, and increased protein levels of eNOS by anthocyanin treatment (cyanidin and delphinidin) in human umbilical vein endothelial cells. Berry anthocyanins have also been shown to affect lipid metabolism in cellular and animal models of dyslipidemia. Administration of chokeberry juice for 30 days in rats fed a standard or 4% cholesterol-containing diet, showed the anti-hyperlipidemic effects of chokeberry juice in the cholesterol-fed group Purified anthocyanins from blueberries and strawberries added to drinking water were shown to prevent the development of dyslipidemia and obesity in mice fed a high-fat diet for a period of 90 days (25,26). Anthocyanin treatment of human umbilical vein endothelial cells was further demonstrated to regulate cholesterol distribution by interfering with the recruitment of tumor necrosis factor receptor associated factors (TRAF)-2 in lipid rafts, thereby inhibiting CD40-induced proinflammatory signaling (27).

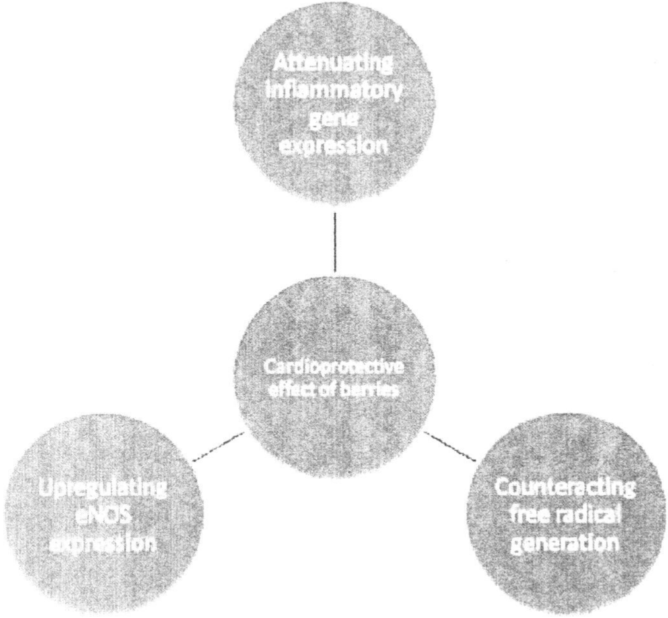

Fig. 1. Cardioprotective effect of berries (Schematic)

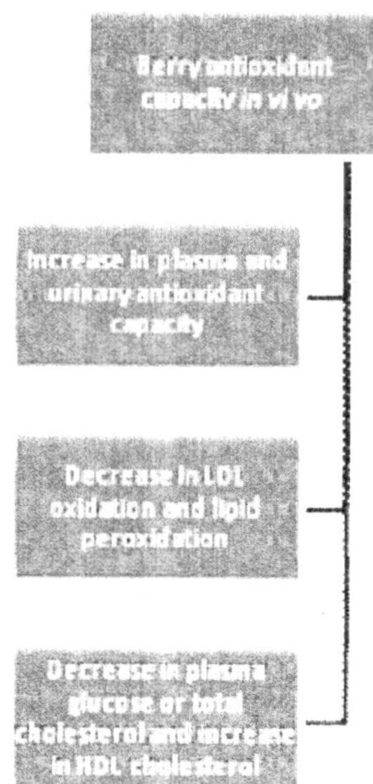

Fig. 2. Antioxidant potential of berries after intervention of berry antioxidants along with other natural antioxidants in *in vivo* conditions.

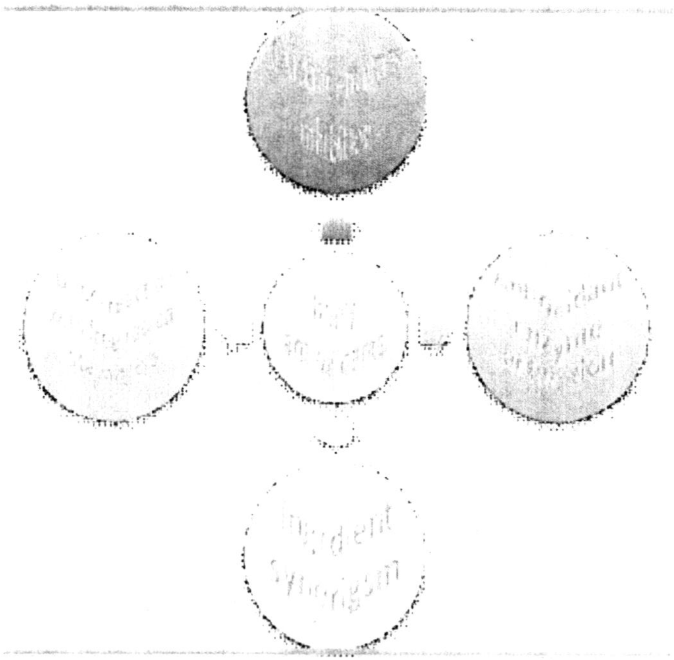

Fig. 3. Basic steps showing mechanism of action of berry antioxidants.

Table 2. Major market trade of berries throughout world

Name Berries	Family	Cultivated countries	Sequencial Market share	Reference
Strawberries	Rosaceae	Europe, North America(open fields) China (green houses)	Europe(largest) Russian Federation and USA(Second)	28,29,30
Red raspberries	Rosaceae	Europe,North America,Asia, China	Europe(first) Asia(second) North America (third)	31,28
Black berries	Rosaceae	Europe and North america	Europe(first)	32,28
Bluberries	Ericaceae	USA,Australia, Canada, Soth korea, China and turkey,Europe	USA(first) Australia and Canada (second)	33,28
Cranberry	Ericaceae	USA,Canada, Chile	USA(first) Canada (Second)	34,28

Conclusion

Berries are a rich source of natural antioxidants. Some important scientific issues are still need to be addressed. For example, current studies are mainly focused on revealing the function of individual berry components and their antioxidant activities. Along with the rapid development in purification and identification technologies of plant bioactive compounds, studies concerning berry bioactive compounds and antioxidant activities will attract more and more attention in the future. It is very important to standardize the current analytical methods and express the results as standard equivalents in order to make it possible to compare different studies from different laboratories. It is necessary to develop the simple, rapid and accurate in vivo methods for the antioxidant capacity evaluation of berries. For future study, the following fields should be taken care off (a) the molecular and cellular mechanisms by which berry antioxidants function in an in vivo system. (b) The synergism and/or antagonism between different antioxidants in berries (c) the internal and external factors which influence the content of berry antioxidants and their activity during the processing. (d) Novel unique antioxidant compounds from berries should be explored.

References

Basu, S.K., Thomas, J., Acharya, S.N. 2007. Prospects for growth in global nutraceutical and functional food markets a canadian perspective. *Aust. J. Basic Appl. Sci.* 1: 637–649.

Basu A., Rhone M. Lyons T.J. 2010. Berries: emerging impact on cardiovascular health,Nature Rev. 2010 March ; 68(3): 168–177. doi:10.1111/j.1753-4887.2010.00273.x.

Benzie, I.F.F. and Strain, J.J. 1999. Ferric reducing/antioxidant power assay: Direct measure of total antioxidant activity of biological fluids and modified version for simultaneous measurement of total antioxidant power and ascorbic acid concentration. *Methods Enzymology,* 299: 15–27.

Caillet S. *et al.,* 2012. Free Radical-Scavenging Properties and Antioxidant Activity of Fractions from Cranberry Products, Food and Nutrition Sciences, 3, 337-347 http://dx.doi.org/10.4236/fns.2012.33049 Published Online March 2012 (http://www.SciRP.org/journal/fns) 337

DeFuria J., Bennett G., Strissel K.J., *et al.,* 2009. Dietary blueberry attenuates whole-body insulin resistance in high fat-fed mice by reducing adipocyte death and its inflammatory sequelae. *J Nutr.* 139: 1510–1516.

Folin, O. and Ciocalteu, V. 1927. Tyrosine and tryptophan determinations proteins. *Journal of Biological Chemistry,* 73: 627.

Jime´nez-Escrig, A., Jime´nez-Jime´nez, I., Sa´nchez-Moreno, C. and Saura-Calixto, F. 2000. Evaluation of free radical scavenging of dietary carotenoids by the stable radical 2,2-diphenyl-1-picrylhydrazyl. *Journal of Science Food Agriculture,* 80: 1686–1690.

Fu, Y., Zhou, X., Chen, S., Sun, Y., Shen, Z., Ye, X. 2015. Chemical composition and antioxidant activity of Chinese wild raspberry (Rubus hirsutus Thunb.). *LWT Food Sci. Technol.* 60: 1262–1268.

Koca, I., Karadeniz, B, 2009. Antioxidant properties of blackberry and blueberry fruits grown in the Black Sea Region of Turkey. *Sci. Hortic.* 121: 447–450.

Lazzè M.C., Pizzala R., Perucca P., *et al.,* 2006. Anthocyanidins decrease endothelin-1 production and increase endothelial nitric oxide synthase in human endothelial cells. *Mol Nutr Food Res.* 50: 44–51.

Liu LK, Lee HJ, Shih YW, Chyau CC, Wang CJ. 2008 Mulberry anthocyanin extracts inhibit LDLoxidation and macrophage-derived foam cell formation induced by oxidative LDL. *J Food Sci.* 73:H113–H121.

Miller, N.J., Rice-Evans, C.A., Davies, M.J., Gopinathan, V. and Milner, A. 1993. A Novel method for measuring antioxidant capacity and its application to monitoring antioxidant status in premature neonates. *Clinical Science,* 84: 407–412.

Mishra R., & Bisht, S.S. S. 2015. "Herbal Antioxidants: Approaches for Isolation and Quantification" in Recent advances in natural products, Studium press, LLC,USA, ISBN : 1-62699-060-3, Chapter-8, pp. 231-278.

P. Rani., K. Meena Unni., & J. Karthikeyan 2004. Evaluation of antioxidant properties of berries. *Indian Journal of Clinical Biochemistry,* 19(2): 103-110.

Penney, B.G., McRae, K.B., Bishop, G.A. 2003. Second-crop N fertilization improves lowbush blueberry (Vaccinium angustifolium Ait.) production. *Can. J. Plant Sci.* 83: 149–155.

Pergola C., Rossi A., Dugo P., Cuzzocrea S., Sautebin L. 2006. Inhibition of nitric oxide biosynthesis by anthocyanin fraction of blackberry extract. *Nitric Oxide.* 15: 30–39.

Prior R.L., Wu X., Gu L., *et al.,* 2009. Purified berry anthocyanins but not whole berries normalize lipid parameters in mice fed an obesogenic high fat diet. *Mol Nutr Food Res.* 53: 1406–1418.

Re, R., Pellegrini, N., Proteggente, A., Pannala, A., Yang, M. and Rice-Evans, C. 1999. Antioxidant activity applying an improved ABTS radical cation decolorization assay. *Free Radical Biology and Medicine,* 26: 1231–1237.

Schilt, A.A. 1996. Analytical application of 1, 10-phenanthroline and related compounds; Pergamon Press: New York, pp. 19–66.

Singleton, V.L., Orthofer, R. and Lamuela-Raventos, R.M. 1999. Analysis of total phenols and other oxidation substrates and antioxidants by means of Folin-Ciocalteu reagent. *Methods of Enzymology,* 299: 152–178.

Skrovankova S., Sumczynski D., Mlcek J., Jurikova T., & Sochor, J. 2015. Bioactive Compounds and Antioxidant Activity in Different Types of Berries, Int. J. Mol. Sci. 16, 24673-24706. doi:10.3390/ ijms161024673.

Strik, B.C. 2007. Berry crops: Worldwide area and production systems. In *Berry Fruit: Value-Added Products for Health Promotion,* 1st ed.; Zhao, Y., Ed.; CRC Press: Boca Raton, FL, USA, pp. 3–51.

Serraino I., Dugo L., Dugo P., *et al.*, 2003. Protective effects of cyanidin-3-O-glucoside from blackberry extract against peroxynitrite-induced endothelial dysfunction and vascular failure. *Life Sci.* 73: 1097–1114.

Vattem, D.A.; Ghaedian, R.; Shetty, K. 2005. Enhancing health benefits of berries through phenolic antioxidant enrichment: focus on cranberry. *Asia Pac. J. Clin. Nutr.* 14: 120–130.

Valcheva-Kuzmanova S., Kuzmanov K., Mihova V., Krasnaliev I., Borisova P., Belcheva A. 2007. Antihyperlipidemic effect of Aronia melanocarpa fruit juice in rats fed a high-cholesterol diet. *Plant Foods Hum Nutr.* 62: 19–24.

Vattem D.A., *et al.*, 2005. Enhancing health benefits of berries through phenolic antioxidant enrichment: focus on cranberry, *Asia Pac J Clin Nutr.* 14 (2): 120-130.

Wang, Z.; Cang, T.; Qi, P.; Zhao, X.; Xu, H.; Wang, X.; Zhang, H.; Wang, X. 2015. Dissipation of four fungicides on greenhouse strawberries and an assessment of their risks. *Food Control.*, 55: 215–220.

www.intechopen.com, Viskelis et al Chemical Composition and Antioxidant Activity of Small Fruits.Accessed on 22-02-2016

Xia M., Ling W., Zhu H., *et al.*, 2007. Anthocyanin prevents CD40-activated proinflammatory signaling in endothelial cells by regulating cholesterol distribution. *Arterioscler Thromb Vasc Biol.* 27: 519–524.

Xu J.W., Ikeda K., Yamori Y. 2004. Upregulation of endothelial nitric oxide synthase by cyanidin-3-glucoside, a typical anthocyanin pigment. *Hypertension.* 44: 217–222.

Yamashita, N., Murata, M., Inoue, S., Burkitt, M.J., Milne, L. and Kawanishi, S. 1998. R-Tocopherol induces oxidative damage to DNA in the presence of copper(II) ions. *Chemical Research and Toxicolgy,* 11: 855–862.

Youdim K.A., Martin A., Joseph J.A. 2000. Incorporation of the elderberry anthocyanins by endothelial cells increases protection against oxidative stress. *Free Radic Biol Med.* 29: 51–60.

Zaporozhets, O.A., Krushynska, O.A., Lipkovska, N.A. and Barvinchenko, V.N. 2004. A new test method for the evaluation of total antioxidant activity of herbal products. *Journal of Agricultural Food Chemistry,* 52: 21–25.

Zou *et al.*, 2016. Antioxidant activity of Citrus fruits, Food Chemistry 196: 885–896.

❑❑❑

Chapter 15

Drugs from Plants

Radhakrishnan Narayanaswamy [1*], Mallappa Kumara Swamy [2] and Norhaizan Mohd Esa[1]

[1] Department of Nutrition and Dietetics, Faculty of Medicine and Health Sciences, Universiti Putra Malaysia (UPM), 43400 Serdang, Selangor, Malaysia, [2] Department of Crop Science, Faculty of Agriculture, Universiti Putra Malaysia (UPM), 43400 Serdang, Selangor, Malaysia. [*] Corresponding author E-mail: nrkishnan@gmail.com.

ABSTRACT

Plants have been an integral part of our ancient culture and they have been used as drugs for thousands of years ago by our native peoples and tribes in various parts of the world. Even today, documentations on traditional knowledge of medicinal plants and practices are still on-going throughout the world. Numerous therapeutic agents have been emerged from ethno medicine experiential base. Moreover, plants have been continuously serving as a main source of new drugs/drug leads and chemical entities (NCE). In this book chapter, we had discussed about the seven key steps involved in plant drug discovery research and have highlighted the drugs from plants.

Keywords: Neanderthal; drogue; immunomodulatory; atropine; aspirin; cocaine.

Introduction

Plants/herbs have been an integral part of the ancient culture of China, Egypt and India as medicine, and their importance even dates back to the Neanderthal period (Solecki, 1975). Now the usage of herbal medicinal products has increased tremendously across the universe, mainly for primary health care treatments. Moreover, the use of herbal medicines has been widely embraced in many developed countries with complementary and alternative medicines now becoming main stream in the Australia and North America, as well as in UK and the rest of Europe (Ekor, 2013). Throughout the world majority of the people prefer natural food, herbal medicines and natural curing practices for healthy life (Kapoor, 2005), which has been proven and stated by the World Health Organization (WHO) report that about 80% of the world population relies on traditional herbal medicines to cure various diseases (Priya *et al.*, 2002). In addition, approximately 11% of drugs in the WHO's essential medicines list have been exclusively derived from herbs/plants and another 25% of the drugs prescribed worldwide are plant-derived products (Prachayasittikul *et al.*, 2015).

Drugs from plants: A historical perspective

Plants/herbs have been playing a crucial role in sustaining human health and well being of mankind. The word "Drug" has been derived from French word "Drogue" means dry herb, which strongly suggests that earliest drugs were originated from plant sources (Wadud *et al.*, 2007). The most important plant-derived drugs were developed between 1800 and 1950 period; for instances in 1805 a young German Pharmacist (Friedrich Serturner) isolated and described "morphine" from *Papaver somniferum* (Fan *et al.*, 2006). In 1815, Vogel and Pelletier isolated "curcumin" from *Curcuma longa* rhizome (Gupta *et al.*, 2012). In 1820 Pelletier and Caventou isolated "quinine and cinchonine" from Cinchona plant bark (Kupchan, 1971). In 1885, Nagayoshi Nagai isolated "matrine" from *Sophora flavescens* and two year later same researcher again isolated "ephedrine" from *Ephedra vulgaris*. In 1888, Warden isolated "embelin" from *Embelia ribes* and in 1900, Heffter and Feurstein isolated "vilangin" from *Embelia robustra* (Rao and Venkateswarlu, 1964). In 1918, Chatterji and Ray first extracted & reported the "margosic acid" from neem oil (Chatterji and Ray, 1918). Atropine, aspirin, cocaine, codeine, digitoxin, emetine and strychnine were the few other plant-derived drugs reported between 1800 and 1950 period (Newman and Cargg, 2010). These above stated examples illustrate the rich history of plant-derived drugs. In addition to many new drug leads, recently plant-derived drugs were well represented in the top 35 worldwide selling prescription drugs in the year 2000, 2001 and 2002 (Butler, 2004).

Plants/herbs have been used as drugs for thousands of years ago by native peoples and tribes in various parts of the world. These herbal medicines were initially taken in the form of crude drugs such as powders, poultices, teas, tinctures and other herbal preparations/ formulations. The exact plants to be utilized and the procedure of practice for specific diseases were passed down through oral history/ communications (Balunas and Kinghorn, 2005). Knowledge of plants/ herbs and their medicinal properties as well as usage should be preserved, since these plants serve as renewable source for new drugs. Moreover, plants have formed the basis of well refined traditional medicine systems that have been used for many years by people in China, Egypt, India, Korea, Rome and many other countries (Newman and Cargg, 2010). A few of the previous reports on the use of herbs as medicines are documented in the Artharvaveda (Indian ayurvedic medicine, since 2000 BCE), Mesopotamia (Ancient medicine in Greek, dating back to 1700 BCE), Eber Papyrus (Ancient medicine in Egypt, dating back to 1550 BCE), Wu Shi Er Bing Fang (Traditional Chinese medicine, dating back to 1100 BCE). Each country has developed unique traditional medicine systems/practices of its own for examples Sasang constitution medicine (Traditional medicine of Korea), Kampo medicine (Traditional/Oriental medicine of Japan), Canon medicinae (Traditional medicine of Medieval Europe), Jamu (Traditional medicine of Java, Indonesia), Siddha medicine (Traditional medicine of South India, India) and Thuoc Nam (Traditional Vietnamese medicine). Few other famous literature sources on medicinal plants or practices include Thirumanthiram written by Thirumoolar around 6-7th CE (Karunamoorthi *et al.*, 2012), De Materia Medica written by Dioscorides (between 60 and 78 CE) and Shen-nung pen ts'ao ching (Divine husbandman's materia medica) written around 200 CE (Salim *et al.*, 2008). Malaysian traditional medicine practices mostly rely on old literatures such as Mujarabat Melayu, Tajul Muluk, Tajus as Salatin and Surat Tib Ubat (Jamal, 2006).

Numerous molecules/therapeutic agents have been emerged from ayurvedic experiential base; few examples include holarrhena alkaloids as antiamoebic agent, curcumin as anti inflammatory agent, phyllanthins as antiviral, piperidines as bioavailability enhancers, picrosides as hepato-protective agent, rauwolfia alkaloids for hypertension treatment, guggulsterons as hypolipidemic agent, steroidal lactones and glycosides as immunomodulatory agent, *Mucuna pruriens* for treatment of parkinson's disease and psoralens for vitiligo treatment (Patwardhan *et al.*, 2004). Similarly, arteether (antimalarial) another drug emerged from traditional Chinese medicine experiential base and galanthamine for treatment of Alzheimer's disease emerged from ethno botanical lead of Russia (Jachak and Saklani, 2007).

Documentations/databases on traditional knowledge

Even today, documentations on traditional knowledge of medicinal plants and practices are still on-going throughout the world. This knowledge, if wisely utilized could draw out promising herbal leads of the future and whole world gets benefit out of this knowledge (Narayanaswamy and Ismail, 2015). Ethno-medical information, biological evaluation of both plants extracts and their active constituents, the chemistry of natural sources and the clinical evaluation of plant extracts are still not accessible globally. Few useful online databases and registries of Traditional Knowledge cum Genetic Resources that are available include Traditional Chinese Medicine Patents Database (China), Traditional Knowledge Digital Library (TKDL, India), Korean Traditional Knowledge Portal (South Korea) and GENESYS Gateway to Genetic Resources; however these databases have limited access only to patent attorney general. In addition to this some other useful databases that are available include Indian plant anticancer compound database (InPACdb), The herb information Knowledge base (THINKherb), Traditional Chinese medicines integrated database (TCMID), Traditional Chinese medicine information database (TCHM-ID), Traditional Chinese Medicine Database @ Taiwan (TCM Database@Taiwan), Indian Medicinal Plants Database, Pandanus Database of Plants, Global Information Hub on Integrated Medicine (GLOBinMeD), Natural Products Alert database (NAPRALERT), Data Base on Antidiabetic Plants Based on Clinical/Experimental Trials (Pan et al., 2013; Farnsworth, 1988; Singh *et al.,* 2009).

Natural Product Research: Conventional Vs Advanced Method

Research and development of therapeutic materials (drugs) from plant origin using conventional way involves more money (cost) and time (patience). For instance, each new drug requires an investment of around 100-300 million US dollar. And a minimum of 10 years of hard work, with only 1 in 10,000 (Ten thousand) tested compounds being considered promising one and only 1 in 4 of these being approved as new drug (Rates, 2001). Secondly researchers should have sufficient starting material (crude plant material), according to McChesney (1995) report: 50 kg of raw material are required to provide 500 mg of pure compound for analytical analysis, bioassays, toxicology and *in vivo* (animal experiments) evaluation; and other 2 kg of pure compound (obtained from 200 tons of raw material) is needed for pre-clinical and clinical studies. The author further, summaries that the quantitative considerations of both active compound and starting plant (raw) material are mandatory for the discovery, development and launch of a new drug on the market. It also helps the researchers to plan (with regard to collection, authentication and storage of crude plant material) to have sufficient raw material in advance. In nut shell natural product drug discovery approach is multi-disciplinary one (Verpoorte, 1989), which needs different fields of expertise to work collectively towards a common goal of new drug development (Rates, 2001). Recent advancement in technology (automated high-throughput screening (HTS), DNA fingerprinting, DNA bar coding, cap-NMR, metabolomics and ayugenomics) has tremendously helped the natural product drug discovery researchers to save effort, money and time.

Key steps involved in plant drug discovery, research and development

Step-1 : Selecting plant material

According to Atanasov and co-workers (2015), five different approaches (random, ethnopharmacological, chemosystematic, ecological and computational) are followed to select plant (raw) material for natural product drug discovery and development programme.

a) **Random method**: In this method plants are randomly selected based on their availability. For example few researchers have used random method to select common weed plants for studying various biological activities. Random plant selection approach had a higher hit rate compare to that of ethnopharmacological plant selection method (Gyllenhaal *et al.*, 2012).

b) **Ethnopharmacological method**: In this method plants are selected based on traditional knowledge of medicine. For example in our previous study we had selected *Datura alba* plant for studying wound healing potential based on folklore (siddha/ ayurvedic) medicinal use of *Datura* (fresh leaf juice) in Mathan Thailam (Priya *et al.*, 2002). Ethnopharmacological plant selection approach had higher hit rates for tuberculosis compare to that of random plant selection method (Gyllenhaal *et al.*, 2012).

c) **Chemosystematic method**: In this method plants are selected based on chemotaxonomy and phylogeny, considering that plant species are known to produce certain chemicals/compounds having therapeutic potential. For example Larsen and co-workers (2010) had used phylogenic approach to select target plants against acetyl cholinesterase (AChE) activity.

d) **Ecological method**: In this method plants are selected based on the interactions between organisms and their environment, considering that plant species are known to produce certain secondary metabolites having therapeutic potential. Obbo and co-workers (2013) used ecological approach to select *Khaya anthotheca* plant based on chimpanzees self medication against protozoa.

e) **Computational method**: In this method plants are selected based on *in silico* bioactivity predicted for certain phyto-constituents. For example Temml and co-workers (2013) had used docking approach (one of the computational method) to study *Carthamus tinctorius* constitutents against indoxyl indoleamine 2, 3-dioxygenase (IDO) activity. Similarly, in our recent study (Narayanaswamy *et al.*, 2016), we had used docking approach to study eleven constituents of *Clinacanthus nutans* as inhibitors of six targeted enzymes such as xanthine oxidase, nitric oxide synthase, human neutrophil elastase, matrix metalloproteinase 2, matrix metalloproteinase 9 and squalene synthase.

According to Srikanth and co-workers (2015) report, that collection of suitable part of plant in specific season will definite possesses more active (bioactive) principles (as shown in the Table 1).

Step-2: Screening for biological activity/activities

The selection of the biological assays, mainly depends on the study objectives and it should be optimally one combined with simplicity, high selectivity, good sensitivity and reproducibility nature (Atanasov *et al.*, 2015). According to Atanasov and co-workers (2015), six different types of bioassays (*in vitro* cell based assays, *in vitro* phenotypic cell based assays, high throughput assays, *in situ*/*ex vivo* assays, *in vivo* rodent model assays and *in vivo* zebra fish & *Caenorhabditis elegans* model assays) are available for screening the biological activity/ activities of plant extracts.

a) ***In vitro* cell based assays**: In this bioassay plant extracts are tested against cultured cells to know their effects at cellular level.

b) ***In vitro* phenotypic cell based assays**: In this bioassay plant extracts are tested against cultured primary and tumor cells, in order to know their effects both in normal and tumor cells.

c) **High throughput assays**: In this bioassay plant extracts are tested against purified protein targets.

d) ***In situ/ ex vivo* assays**: In this bioassay plant extracts are tested against cultured animal tissues or organs.

e) ***In vivo* rodent model assays**: In this bioassay plant extracts are tested against experimental induced disease in animals' model. Now transgenic animal models are also available for screening plant extracts against various disease conditions.

f) ***In vivo* zebra fish & *Caenorhabditis elegans* model assays**: In this bioassay plant extracts are tested against experimental induced diseases in zebra fish & *Caenorhabditis elegans* model. Now transgenic zebra fish models are also available for screening plant extracts against various disease conditions.

In addition to the above bio assays, Shukla and co-workers (2011) had employed *Drosophila melanogaster* model for screening plant extracts against epilepsy and insects. Moreover, Jansen and co-workers (2014) had used *Drosophila melanogaster* mutant strain (PINK1) model for screening plant extracts against Parkinson's disease.

Step-3: Identification of bioactive compounds

A known routine approach is to start biological analysis using crude plant extracts and which further helps to isolate, purify and characterize the constituents responsible for the bioactive of the extract (Koehn and Carter, 2005). According to Atanasov and co-workers (2015), five different types of strategies (direct phytochemical isolation, bioactivity guided fractionation, metabolism-directed approach, metabolic profiling approach and synergy- directed fractionation) are available to identify bioactive compounds from plant extracts.

a) **Direct phytochemical isolation**: In this strategy, isolation and identification of plant constituents are carried out without immediate determination of bioactivity. It is otherwise called as conventional way of approach.

b) **Bioactivity guided fractionation**: In this strategy, isolation and identification of plant constituents are carried out with immediate evaluation of bioactivity for each fraction isolated, which ultimately results in isolate the pure active compound.

c) **Metabolism-directed approach**: In this strategy, isolation and identification of metabolites which are formed as result of metabolic transformation (biotransformation) by the human body or by microorganisms present in human intestine.

d) **Metabolic profiling approach**: In this strategy, isolation and identification of known/new metabolites, this leads to comprehensive qualitative and quantitative metabolite analysis. This approach otherwise called as metabolomic approach, which correlates well with the chemical nature and the biological activity profile of plant extracts.

e) **Synergy- directed fractionation**: In this strategy, isolation and identification of synergistically interacting phyto-constituents which could be probably missed with the conventional bioactivity guided fractionation approach.

Step-4: Structure elucidation

All proton-detecting experimental techniques such as proton Nuclear Magnetic Resonance (^{1}H NMR), two-dimensional NMR correlation spectroscopy (COSY), heteronuclear single-quantum correlation spectroscopy (HSQC), heteronuclear multiple-bond correlation spectroscopy (HMBC) and nuclear overhauser effect spectroscopy (NOESY) are important tools used for the structural elucidation of bioactive compounds (Nguta *et al.*, 2015). Advancement in probe technology (example capillary NMR probe) and emerging higher magnetic field-strength NMR has led to a significant contribution in shortening in acquisition time for NMR data and the structural elucidation of bioactive compounds now can be achieved with less than one milligram (1 mg) concentration. In addition to above, the combination of NMR spectroscopy with other analytical "hyphenated" techniques (such as LC-NMR-MS and LC-SPE-NMR) will allow for quick eliminating known active compounds which have been reported previously from the plant extracts (Salim et al., 2008). Moreover, significant progress has also been made in automated structure resolving algorithms techniques (Butler, 2004).

Step-5: Developing plant drugs in sustainable way

Plants are serving as renewable source of new drugs and therefore their conservation in sustainable environment is urgently needed which in turn helps to safeguard their demand in the global market (Dubey *et al.,* 2004). In this crucial case natural products can be revitalize using plant cell/tissue culture technique, heterologous production of plant secondary metabolites, semi-synthesis or total chemical synthesis from isolated precursors which are occurring surplus in the nature (Atanasov *et al.,* 2015). Moreover, natural products compounds are typically isolates in small quantities which are insufficient for lead optimization, development and clinical trials. Therefore, there is a significant need to develop collaborations with medicinal and synthetic chemists to explore the possibilities of its semi-synthesis or total synthesis apart from plant cell/tissue culture and heterologous production via by biotechnologist (Jachak and Saklani, 2007).

a) **Cultivation of medicinal plants**: Vriksayurveda (Indian ancient science of plant life) an unique branch of ancient science, which emphasis about the cultivation (farming), nourishment (support) and management of medicinal, edible and aromatic plants (Debnath *et al.,* 2015). In recent years, both wild and cultivated plants are used for drug developed and formulations. However, in many cases cultivated plants are better than wild plants owing to its improved quality nature. Some indigenous medicinal plants are successfully cultivated in India, in order to meet the global market demand (Dubey *et al.,* 2004).

b) **Plant cell and tissue culture approach**: High seed dormancy, low seed set, low percentage of seed germination and improper seed viability are the few problems associated with propagation of some medicinal plants. To overcome these problems, plant cell and tissue culture approaches are employed and got few successes in propagating some medicinal plants using advance biotechnology techniques (Dubey *et al.,* 2004). ROOTec (Swiss company) has developed and standardized bioreactor system for large-scale (commercial) cultivation of hair roots which results in production a few biological active compounds such as atropine, vitamin D_3 derivative, anabasine, nicotine, nor-nicotine and camptothecin (Atanasov *et al.,* 2015).

c) **Heterologous production of plant secondary metabolites**: In this approach biosynthesis of plant-derived compounds, i.e. the reconstitution of the target compounds by altering biochemical pathway in a foreign host (plant/ microbe) in order to increase bioactive compound yield. For instance, heterologous production of strictosidine (*Catharanthus roseus*) has been produced using *Nicotiana benthamiana* heterologous host system (Miettinen et al., 2014). Similarly, heterologous production of curcuminoids has been produced using *Escherichia coli* heterologous host system (Rodrigues *et al.,* 2015).

d) **Semi-synthesis approach**: Semi-synthesis is performed by the simple chemical modification of (existing) natural products, in order to improve their potency, selectivity and other properties. This approach has been employed to yield a number of bioactive compounds (Prachayasittikul et al., 2015). For example, vilangin (dimeric form of embelin) has been derived using embelin and formaldehyde by simple chemical reaction (Balachandran *et al.,* 2013).

e) **Total chemical synthesis**: Total chemical synthesis approach is used to synthesis complex natural products compounds. Moreover natural product total chemical synthesis is very crucial and well refined approach using novel synthetic strategies and explorations in chemical biology through molecular design and mechanistic studies. For example, total synthesis of vitamin B_{12} has been reported by Woodward (1973). Similarly, Ranganathan and co-workers (1980) has successful synthesized the prostaglandin F_1alpha ($PGF_1\alpha$) by re-structuring of castor oil.

Natural product compounds development can be improved by creating natural products libraries that combine the features of natural products with combinatorial chemistry (Jachak and Saklani, 2007). Thus presently, efforts has been made to deposit all the synthesized bioactive compounds by chemical means in to four different libraries, which includes synthetic compound library, combinatorial compound library, diversity-oriented synthetic (DOS) library and fragment library and further compounds from libraries can be tested for various biological activities (Singh and Culberson, 2010). However, the choice of biological targets to screen against natural compound library is essential and critical to achieve a long term success in the natural product drug discovery journey (Butler, 2004). In addition to above, Nicholas Piramal (Indian Company) has created a plant extract (PE) library containing 6000 extracts (prepared from around 2300 plant species collected from all over India) and however, it is not available in public domain (Jachak and Saklani, 2007).

Step-6: Important checkpoints

It is essential to have checkpoints in each and every above step (step 1 to 5). For example

a) First and foremost check whether you have collected correct plant (plant authentication) or plant parts and documented the same

b) Check whether you have collected sufficient plant samples and also make sure it is free from physical, chemical and biological contamination.

c) Check for the proper infrastructure and method in place (for screening, isolation and identification purpose).

d) Check for their safety nature (both in crude and pure form).

e) Check for chemical structure and renewable source material (natural/synthetic way).

f) Check for securing more research funds (including collaborative projects).

g) Check for their new synergistic/therapeutic effect.

Step-7: Available guidelines, treaties and agreements

Plants have been serving a major resource of renewable drugs and several international organizations have demonstrated against abuse of these resources. Legal enforcements are being in place to safeguard the natural resources and germplasm conservation. In addition to above preservation of both indigenous knowledge and intellectual property wealth are being actively implemented, throughout the world (Rates, 2001). Moreover, World Health Organization (WHO) had developed universal guidelines on good agricultural practices (GAP) & good collection practices (GCP) for medicinal plants, in order to promote sustainable (renewable) plant collection methods and to reduce the ecological problems associated with wild crafting of medicinal plants (Atanasov et al., 2015). The another international body, namely the International Union for the Conservation of Nature (IUCN) has published various conservation methods related to the sustainable use of medicinal plant species and also actively involved in preparing as well as regular monitoring endanger medicinal plant species/red list (Cordell, 2011). The Nairobi conference conducted on May (1992), which led to formation multinational treaty known as the Convention on Biodiversity (CBD) and it had been signed by 168 countries (Cragg and Newman, 2013) and those countries that have signed the Convention on Biodiversity (CBD) are eligible to sign or ratify the Nagoya protocol. As on December (2011), 72 countries had signed the Nagoya protocol (Cordell and Colvard, 2012). In India National Biodiversity Authority (NBA) act covers conservation, use of biological resources and associated knowledge for commercial or research purposes. And it also provides framework for accessing biological resources and benefit sharing arising out of such access and use (Jachak and Saklani, 2007). Bioprospecting is the process which enables us to make use of our rich biological heritage and also used to conserve our indigenous medicinal plants. Above all it also, helps us to check illegal exploitation by using biopiracy or gene robbing biotechnology technique (Dubey *et al.*, 2004). *Jeeva sampada* is the most comprehensive and largest bioresources database of India, which consistent in the terms of our commitment to the Convention on Biodiversity (CBD) treaty (Natesh, 2006).

Drugs from Plants: A Boon to Humans

Plants have been serving as a main source of new drugs/ drug leads and (NCE) new chemical entities (Saklani and Kutty, 2008). Farnsworth and co-workers (1985) had reported the list of more than 91 drugs obtained from plants purely based on their ethno medicine use (as shown in the table 2). According to Harvey (2008) 46 plant drugs have been reported in preclinical development stage, whereas 14 & 41 plant drugs are at Phase I and II stages respectively. And in addition to these two more drugs have been at pre-registration stage. According to Saklani and Kutty (2008) 91 plant drugs have been reported in clinical trials, mainly focused for the treatment of cancer, immunological and central nervous system (CNS) related diseases. According to Newman and Cragg (2010) 13 drugs have been approved for clinical use between years 1961-2008. Of these only three drugs (vinblastine, vincristine and taxol) are directly obtained from plants, the rest are derivatives. Similarly in 2012, they had reported 1356 new chemical entities (NCE) between years 1981-2010 (Newman and Cragg, 2012). Of these only five new chemical entities are directly obtained from plants such as solamargines (anticancer), XP-828L (antipsoriatic), curemisinin (antiallergic), sinecatechins (genital warts treatment) and oral air grasses (allergic rhinitis treatment). Furst and Zundorf (2014) had reviewed six plant-derived anti-inflammatory compounds such as capsaicin, colchicines, curcumin, epigallocatechin-3-gallate (EGCG), resveratol & quercetin and highlighted their ongoing clinical trials.

Conclusion

Even today, a majority of the world's population mainly relies on plants as a primary source of medicine, which prompted more interest among the researchers to investigate the scientific basis for their uses. And, thus natural product research has results in discovering numerous clinically used drugs (especially from plants) and every year this list keeps on increasing in great manner. Owing to tremendous growth in natural product research, researchers have responsibility to ensure accessibility (renewable raw material, infra structure to employ advance techniques), understandability (securing scientific knowledge), acceptability (economically viable in terms of methods and process) and safety (safety to humans as well as environment). Above all, research finding should benefit the end-user (patients) as well as to traditional knowledge owner/community (original owner of traditional knowledge) and developer (researcher). Moreover, sufficient funding and collaboration among the researchers, will lead to achieve more successful drugs from plants in near future.

Table 1. Collection of suitable part of plant, at specific harvesting season.

Collection of suitable part of plant (with example)	**Specific season**
Barks (*Cinchona ledgeriana*)	Early winter
Branches (*Commiphora molmol*)	Rainy season & spring
Flowers (*Papaver somniferum*)	Spring (early summer)
Fruits (*Ananas comosus*)	Spring (early summer)
Latex (*Carica papaya*)	Early winter
Leaves (*Digitalis purpurea*)	Rainy season & spring
Roots (*Rauvolfia serpentina*)	Summer (late winter)
Stems (*Chondrodendron tomentosum*)	Early winter
Harvesting in general (whole plant-*Vinca minor*)	Early winter

Table 2. List of drugs obtained from plants, based on their ethno medicine use.

Drug name (chemical class)	Plant name (common name)
Acetyldigoxin (cardiac glycoside)	*Digitalis lanata* (Grecian Foxglove)
Adoniside (cardenolide glycoside)	*Adonis vernalis* (Pheasant's Eye)
Aescin (saponin)	*Aesculus hippocastanum* (Horse Chestnut)
Aesculetin (coumarin derivative)	*Fraxinus rhynchophylla* (Ash tree)
Agrimophol (polyphenol)	*Agrimonia eupatoria* (Agrimony)
Ajmalicine (alkaloid)	*Rauvolfia serpentina* (Indian Snake Root)
Allantoin (urea derivative)	Several plants
Allyl isothiocyanate (glucosinolate)	*Brassica nigra* (Black Mustard)
Anabasine (alkaloid)	*Anabasis aphylla* (Anabasis)
Andrographolide (diterpenoid)	*Andrographis paniculata* (Creat)
Anisodamine & Anisodine (alkaloid)	*Anisodus tanguticus* (Shan Lang Dang)
Asiaticoside (triterpenoid glycoside)	*Centella asiatica* (Spadeleaf)
Atropine (alkaloid)	*Atropa belladonna* (Deadly Nightshade)
Benzyl benzoate (benzoic acid derivative)	Several plants
Berberine (alkaloid)	*Berberis vulgaris* (Barberry)
Bergenin (polyphenol)	*Ardisia japonica* (Marlberry)
Borneol (terpene)	Several plants
Bromelain (protein)	*Ananas comosus* (Pineapple)
Caffeine (methylxanthine alkaloid)	*Camellia sinensis* (Tea Plant)
Camphor (terpenoid)	*Cinnamomum camphora* (Camphor Tree)
Catechin (flavonoids)	*Potentilla fragarioides* (Potentilla)
Chymopapain (protein)	*Carica papaya* (Papaya)
Cissampeline (alkaloid)	*Cissampelos pareira* (Velvet Leaf)
Cocaine (alkaloid)	*Erythroxylum coca* (Coca Bush)
Codeine (alkaloid)	*Papaver somniferum* (Opium Poppy)
Colchicine amide & Colchicine (alkaloid)	*Colchicum autumnale* (Meadow Saffron)
Convallatoxin (glycoside)	*Convallaria majalis* (Lily of the Valley)

Curcumin (phenol)	*Curcuma longa* (Turmeric)
Cynarin (cinnamic acid derivative)	Cynara scolymus (Globe Artichoke)
Danthron (quinone)	*Cassia* species
Demecolcine (alkaloid)	*Colchicum autumnale* (Meadow Saffron)
Deserpidine (alkaloid)	*Rauvolfia canescens* (Rauvolfia)
Deslanoside (cardiac glycoside)	*Digitalis lanata* (Grecian Foxglove)
Digitalin & Digitoxin (glycoside)	*Digitalis purpurea* (Purple Foxglove)
Digoxin (glycoside)	*Digitalis lanata* (Grecian Foxglove)
L-Dopa (amino acid)	*Mucuna deeringiana* (Velvet Bean)
Emetine (alkaloid)	*Cephaelis ipecacuanha* (Ipecac)
Ephedrine (alkaloid)	*Ephedra sinica* (Ephedra)
Galantamine (alkaloid)	*Lycoris squamigera* (Magic Lily)
Gitalin (glycoside)	*Digitalis purpurea* (Purple Foxglove)
Glaucarubin (glycoside)	*Simarouba glauca* (Paradise Tree)
Glaucine (alkaloid)	*Glaucium flavum* (Yellow Hornypoppy)
Glaziovine (alkaloid)	*Ocotea glaziovii* (Sweetwood)
Glycyrrhizin (saponin glycoside)	*Glycyrrhiza glabra* (Licorice)
Gossypol (phenolic aldehyde)	*Gossypium* species
Hemsleyadin (tetracyclo triterpenoid)	*Hemsleya amabilis* (Luo Guo Di)
Hesperidin (flavanone glycoside)	*Citrus* species
Hydrastine (alkaloid)	*Hydrastis Canadensis* (Goldenseal)
Hyoscyamine (alkaloid)	*Hyoscyamus niger* (Black Henbane)
Kawain (unsaturated lactone)	*Piper methysticum* (Kava)
Khellin (coumarin derivative)	*Ammi visnaga* (Toothpick Weed)
Lanatosides A, B & C (glycosides)	*Digitalis lanata* (Grecian Foxglove)
á-Lobeline (alkaloid)	*Lobelia inflata* (Indian Tobacco)
Menthol (alcohol)	*Mentha* species
Methyl salicylate (salicyclic acid derivative)	*Gaultheria procumbens* (Wintergreen)
Monocrotaline (alkaloid)	*Crotalaria sessiliflora* (Blue Rattlesnake)
Morphine (alkaloid)	*Papaver somniferum* (Opium Poppy)

Neoandrographolide (diterpenoids)	*Andrographis paniculata* (Creat)
Nicotine (alkaloid)	*Nicotiana tabacum* (Tobacco)
Nordihydroguaiareticacid (polyphenol)	*Larrea divaricata* (Chaparral)
Norpseudoephedrine (alkaloid)	*Ephedra sinica* (Ephedra)
Noscapine (alkaloid)	*Papaver somniferum* (Opium Poppy)
Ouabain (cardiac glycoside)	*Strophanthus gratus* (Climbing Oleander)
Pachycarpine (alkaloid)	*Sophora pachycarpa* (Pea-flowered Tree)
Palmatine (alkaloid)	*Coptis japonica* (Japanese Goldthread)
Papain (protein)	*Carica papaya* (Papaya)
Papaverine (alkaloid)	*Papaver somniferum* (Opium Poppy)
Phyllodulcin (coumarin derivative)	*Hydrangea macrophylla* (Hydrangea)
Physostigmine (alkaloid)	*Physostigma venenosum* (Calabar Bean)
Picrotoxin (sesquiterpene)	*Anamirta cocculus* (Indian Berry)
Pilocarpine (alkaloid)	*Pilocarpus jaborandi* (Jaborandi)
Pinitol (cyclic polyol)	Several plants
Podophyllotoxin (lignan)	*Podophyllum peltatum* (Mayapple)
Protoveratines A & B (alkaloid)	*Veratrum album* (White Hellebore)
Pseudoephedrine (alkaloid)	*Ephedra sinica* (Ephedra)
Quinidine & Quinine (alkaloid)	*Cinchona ledgeriana* (Cinchona)
Quisqualic acid (non-protein amino acid)	*Quisqualis indica* (Rangoon Creeper)
Rescinnamine & Reserpine (alkaloid)	*Rauvolfia serpentine* (Indian Snake Root)
Rhomitoxin	*Rhododendron molle* (Chinese Azalea)
Rorifone (sulfones)	*Rorippa indica* (Indian Field Cress)
Rotenone (isoflavone)	*Lonchocarpus nicou* (Barbasco)
Rotundine (alkaloid)	*Stephania sinica* (Ru Lan)
Rutin (flavonoid glycoside)	*Citrus* species
Salicin (alcoholic glucoside)	*Salix alba* (White Willow)
Sanguinarine (alkaloid)	*Sanguinaria Canadensis* (Bloodroot)

Santonin (sesquiterpene lactone)	*Artemisia maritime* (Sea Wormwood)
Scillaren A (steroidal glycoside)	*Urginea maritime* (Sea Squill)
Scopolamine (alkaloid)	*Datura metel* (Thorn Apple)
Sennosides A & B (glycosides)	*Cassia acutifolia* & *Cassia angustifolia* (Senna Plant)
Silymarin (flavonoid)	*Silybum marianum* (Milk Thistle)
Sparteine (alkaloid)	*Cytisus scoparius* (Scotch Broom)
Stevioside (glycoside)	*Stevia rebaudiana* (Sugar Leaf)
Strychnine (alkaloid)	*Strychnos nux-vomica* (Strychnine Tree)
Tetrandrine (alkaloid)	*Stephania tetrandra* (Han Fang Ji)
Tetrahydrocannabinol (terpenoid)	*Cannabis sativa* (Cannabis)
Theobromine (alkaloid)	*Theobromo cacao* (Cocoa Tree)
Theophylline (xanthine derivative)	*Camellia sinensis* (Tea Plant)
Thymol (phenol obtained from thyme oil)	*Thymus vulgaris* (Thyme)
Trichosanthin (protein)	*Trichosanthes kirilowii* (Chinese Snake Gourd)
Tubocurarine (alkaloid)	*Chondrodendron tomentosum* (Curare)
Valepotriates (triesters of monotrepene alcohol)	*Valeriana officinalis* (Valerian)
Vincamine (alkaloid)	*Vinca minor* (Common Periwinkle)
Vinblastine & Vincristine (alkaloid)	*Catharanthus roseus* (Madagascar Periwinkle)
Xanthotoxin (coumarin derivative)	*Ammi majus* (Bishop's Weed)
Yohimbine (alkaloid)	*Pausinystalia yohimbe* (Yohimbe)
Yuanhuacine & Yuanhuadine (diterpenoids)	*Daphne genkwa* (Lilac Daphne)

References

Antiprotozoal activity of *Khaya anthotheca*,(Welv.) CDC a plant used by chimpanzees for self-medication. *Journal of Ethnopharmacology*, 147(1): 220-223.

Atanasov, A.G., Waltenberger, B., Pferschy-Wenzig, E.M., Linder, T., Wawrosch, C., Uhrin, P., Temml, V., Wang, L., Schwaiger, S., Heiss, E.H. and Rollinger, J.M., 2015. Discovery and resupply of pharmacologically active plant-derived natural products: A review. *Biotechnology Advances*, 33(8): 1582-1614.

Balachandran, C., Duraipandiyan, V., Balakrishna, K., Sundaram, R.L., Vijayakumar, A., Ignacimuthu, S. and Al-Dhabi, N.A., 2013. Synthesis and medicinal properties of plant-derived vilangin. *Environmental Chemistry Letters*, 11(3), pp.303-308.

Butler, M.S., 2004. The role of natural product chemistry in drug discovery. *Journal of Natural Products*, 67(12): 2141-2153.

Balunas, M.J. and Kinghorn, A.D., 2005. Drug discovery from medicinal plants. *Life Sciences*, 78(5): 431-441.

Cordell, G.A., 2011. Sustainable medicines and global health care. *Planta Medica*, 77(11): 1129-1138.

Cragg, G.M. and Newman, D.J., 2013. Natural products: a continuing source of novel drug leads. *Biochimica et Biophysica Acta* (BBA-General Subjects), 1830(6): 3670-3695.

Cordell, G.A. and Colvard, M.D., 2012. Natural products and traditional medicine: turning on a paradigm. *Journal of Natural Products*, 75(3): 514-525.

Chatterji, K.K. and Ray, C., 1918. Observations on Margosic Acid and its Salts and their Use in Cases oï Syphilis and Skin Diseases. *Indian Journal of Medical Research*, 5(4): 655-674.

Datura alba on burn wounds in albino rats. *Journal of Ethnopharmacology*, 83(3): 193-199.

Debnath, P.K., Banerjee, S., Debnath, P., Mitra, A., Mukherjee, P.K., 2015. Ayurveda-Opportunities for developing safe and effective treatment choices for the future. In: Evidence-based validation of herbal medicine (Ed. Pulok K. Mukherjee), Elsevier publisher, Amsterdam, Netherlands, pp.427-454.

Dubey, N.K., Rajesh Kumar, and Pramila Tripathi., 2004. Global promotion of herbal medicine: India's opportunity. *Current Science*, 86(1): 37-41.

Ekor, M., 2013. The growing use of herbal medicines: issues relating to adverse reactions and challenges in monitoring safety. *Frontiers in Pharmacology*, 4: 177.

Fan, T.P., Yeh, J.C., Leung, K.W., Yue, P.Y. and Wong, R.N., 2006. Angiogenesis: from plants to blood vessels. *Trends in Pharmacological Sciences*, 27(6): 297-309.

Farnsworth, N.R., 1988. Screening plants for new medicines. *Biodiversity*, 1: 83-97.

Fürst, R. and Zündorf, I. 2014. Plant-derived anti-inflammatory compounds: hopes and disappointments regarding the translation of preclinical knowledge into clinical progress. *Mediators of Inflammation*, 2014.

Farnsworth, N.R., Akerele, O., Bingel, A.S., Soejarto, D.D. and Guo, Z., 1985. Medicinal plants in therapy. *Bulletin of the World Health Organization*, 63(6): 965-981.

Gupta, S.C., Patchva, S., Koh, W. and Aggarwal, B.B., 2012. Discovery of curcumin- a component of golden spice, and its miraculous biological activities. *Clinical and Experimental Pharmacology and Physiology*, 39(3): 283-299.

Gyllenhaal, C., Kadushin, M.R., Southavong, B., Sydara, K., Bouamanivong, S., Xaiveu, M., Xuan, L.T., Hiep, N.T., Hung, N.V., Loc, P.K. and Dac, L.X., 2012. Ethnobotanical approach versus random approach in the search for new bioactive compounds: support of a hypothesis. *Pharmaceutical Biology*, 50(1): 30-41.

Harvey, A.L., 2008. Natural products in drug discovery. *Drug Discovery Today*, 13(19): 894-901.

Jachak, S.M. and Saklani, A., 2007. Challenges and opportunities in drug discovery from plants. *Current Science*, 92(9): 1251.

Jamal, J.A., 2006. Malay traditional medicine: An overview of scientific and technologies progress. *Asia-Pacific Tech Monitor*, 23: 37-49.

Jansen, R.L.M., Brogan, B., Whitworth, A.J. and Okello, E.J., 2014. Effects of five Ayurvedic herbs on locomotor behaviour in a Drosophila melanogaster Parkinson's disease model. *Phytotherapy Research*, 28(12): 1789-1795.

Karunamoorthi, K., Jegajeevanram, K., Xavier, J., Vijayalakshmi, J. and Melita, L., 2012. Tamil traditional medicinal system-siddha: an indigenous health practice in the international perspectives. TANG *[Humanitas Medicine]*, 2(2): 12-21.

Kapoor, V.P., 2005. Herbal cosmetics for skin and hair care. *Natural Product Radiance*, 4(4): 306-314.

Koehn, F.E. and Carter, G.T., 2005. The evolving role of natural products in drug discovery. *Nature Reviews Drug Discovery*, 4(3): 206-220.

Kupchan, S.M., 1971. Drugs from natural products – plant sources. In: *Drug Discovery* (Eds. Barry Bloom & Glenn E. Ullyot), Advances in

Chemistry Series, American Chemical Society (ACS) Publisher, Washington, USA, pp. 1-13.

Larsen, M.M., Adsersen, A., Davis, A.P., Lledó, M.D., Jäger, A.K. and Rønsted, N., 2010. Using a phylogenetic approach to selection of target plants in drug discovery of acetylcholinesterase inhibiting alkaloids in Amaryllidaceae tribe Galantheae. *Biochemical Systematics and Ecology*, 38(5): 1026-1034.

McChesney, J.D., 1995. The promise of plant-derived natural products for the development of new pharmaceuticals and agrochemicals. In: *Chemistry of the Amazon* (Eds. Peter Rudolf Seidl, Otto Richard Gottleib & Maria Auxiliadora Coelho Kaplan), American Chemical Society (ACS) Publisher, Washington, USA, pp. 66-78.

Mizoguchi, U., 1950. Historical outline of the works of pharmaceutical chemists in Japan. *Yakugaku Zasshi*, 70(7-8): 1-9.

Miettinen, K., Dong, L., Navrot, N., Schneider, T., Burlat, V., Pollier, J., Woittiez, L., van der Krol, S., Lugan, R., Ilc, T. and Verpoorte, R., 2014. The seco-iridoid pathway from *Catharanthus roseus*. *Nature communications*, 5.

Narayanaswamy, R. and Ismail, I.S., 2015. Cosmetic potential of Southeast Asian herbs: an overview. *Phytochemistry Reviews*, 14(3): 419-428.

Narayanaswamy, R., Isha, A., Wai, L.K. and Ismail, I.S., 2016. Molecular docking analysis of selected Clinacanthus nutans constituents as xanthine oxidase, nitric oxide synthase, human neutrophil elastase, matrix metalloproteinase 2, matrix metalloproteinase 9 and squalene synthase inhibitors. *Pharmacognosy Magazine*, 12(45): S21-S26.

Natesh, S., 2006. Digitized inventory of biological resources of India. *Current Science*, 91(7): 860.

Newman, D.J. and Cragg, G.M., 2010. Natural products as drugs and leads to drugs: the historical perspective. In: *Natural Product Chemistry for Drug Discovery* (Eds. Antony D. Buss & Mark S. Butler), Royal Society of Chemistry (RSC) Publisher, Cambridge, UK, pp. 1-27.

Newman, D.J. and Cragg, G.M., 2012. Natural products as sources of new drugs over the 30 years from 1981 to 2010. *Journal of Natural Products*, 75(3): 311-335.

Verpoorte, R., 1989. Some phytochemical aspects of medicinal plant research. *Journal of Ethnopharmacology*, 25(1): 43-59.

Nguta, J.M., Appiah-Opong, R., Nyarko, A.K., Yeboah-Manu, D. and Addo, P.G., 2015. Current perspectives in drug discovery against tuberculosis from natural products. *International Journal of Mycobacteriology*, 4(3): 165-183.

Obbo, C.J.D., Makanga, B., Mulholland, D.A., Coombes, P.H. and Brun, R., 2013.

Pan, S.Y., Zhou, S.F., Gao, S.H., Yu, Z.L., Zhang, S.F., Tang, M.K., Sun, J.N., Ma, D.L., Han, Y.F., Fong, W.F. and Ko, K.M., 2013. New perspectives on how to discover drugs from herbal medicines: CAM's outstanding contribution to modern therapeutics. *Evidence-Based Complementary and Alternative Medicine*, 2013.

Patwardhan, B., Vaidya, A.D. and Chorghade, M., 2004. Ayurveda and natural products drug discovery. *Current Science*, 86(6): 789-799.

Priya, K.S., Gnanamani, A., Radhakrishnan, N. and Babu, M., 2002. Healing potential of Prachayasittikul, V., Worachartcheewan, A., Shoombuatong, W., Songtawee, N., Simeon, S., Prachayasittikul, V. and Nantasenamat, C., 2015. Computer-aided drug design of bioactive natural products. *Current Topics in Medicinal Chemistry*, 15(18): 1780-1800.

Rodrigues, J.L., Prather, K.L., Kluskens, L.D. and Rodrigues, L.R., 2015. Heterologous Production of Curcuminoids. *Microbiology and Molecular Biology Reviews*, 79(1): 39-60.

Ranganathan, D., Ranganathan, S. and Mehrotra, M.M., 1980. The synthesis of PGF 1 α by re-structuring of castor oil. *Tetrahedron*, 36(12): 1869-1875.

Rao, T.P. and Venkateswarlu, V., 1964. Chemical examination of embelia ribes—IV: Some condensation reactions of embelin with primary amines. *Tetrahedron*, 20(4): 969-971.

Rates, S.M.K., 2001. Plants as source of drugs. *Toxicon*, 39(5): 603-613.

Salim, A.A., Chin, Y.W. and Kinghorn, A.D., 2008. Drug discovery from plants. In: *Bioactive Molecules and Medicinal Plants*, (Kishnan Gopal Ramawat & Jean-Michel Merillon), Springer Publisher, Heidelberg, Germany, pp. 1-24.

Solecki, R.S., 1975. Shanidar IV, a Neanderthal flower burial in northern Iraq. *Science*, 190: 880-881.

Shukla, N., Hasan, S. and Pal, A., 2011. Screening of selected medicinal plant extracts for antiepileptic and insecticidal activity in *Drosophila melanogaster* model. *Journal of Natural Science, Biology and Medicine*, 2(3): 103-104.

Singh, S., Gupta, S.K., Sabir, G., Gupta, M.K. and Seth, P.K., 2009. A database for anti-diabetic plants with clinical/experimental trials. *Bioinformation*, 4(6): 263-268.

Srikanth, N., Tewari, D. and Mangal, A.K., 2015. The science of plant life (Vriksha ayurveda) in archaic literature: An insight of botanical, agricultural and horticultural aspects of ancient India. *World Journal of Pharmacy and Pharmaceutical Sciences*, 4(6): 388-404.

Saklani, A. and Kutty, S.K., 2008. Plant-derived compounds in clinical trials. *Drug Discovery Today*, 13(3): 161-171.

Temml, V., Kuehnl, S., Schuster, D., Schwaiger, S., Stuppner, H. and Fuchs, D., 2013. Interaction of *Carthamus tinctorius* lignan arctigenin with the binding site of tryptophan-degrading enzyme indoleamine 2, 3-dioxygenase. *FEBS Open Bio*, 3: 450-452.

Wadud, A., Prasad, P.V., Rao, M.M. and Narayana, A., 2006. Evolution of drug: a historical perspective. *Bulletin of the Indian Institute of History of Medicine (Hyderabad)*, 37(1): 69-80.

Woodward, R.B., 1973. The total synthesis of vitamin B12. *Pure and Applied Chemistry*, 33(1), pp. 145-178.

□□□

Chapter 16

Screening of Phytochemical Components and Study of Antioxidant: Mediated Protective Effect of The Leaves of *Murraya koenigii*

Riti Thapar Kapoor*[1] and Nitya Rohatgi[2]

Amity Institute of Biotechnology, Noida, Uttar Pradesh, India, *Corresponding author E-mail: rkapoor@amity.edu

ABSTRACT

Murraya koenigii is widely used as herb and condiment and also used to treat various types of diseases in Indian traditional system. The present investigation was designed to assess the phytochemical components present in the leaves of *Murraya koenigii* and to evaluate their antioxidant potential against oxidative damage. The plant *Murraya koenigii* was screened for phytochemical constituents and due to its antioxidant capacity this plant has potential to act as a source of useful drug. It can improve the health status of the consumers as a result of the presence of various compounds that are vital for good health.

Keywords: Antioxidant activity, *Murraya koenigii,* phytochemical components.

Introduction

Plants have been used by human-beings in many ways to meet their basic needs such as food, clothing and shelter. India is a country with a vast reserve of natural resources and a rich history of traditional medicine. Plants supply spices, functional food ingredients, medicinal raw material, essential oil, flavouring products and in this way these are important source of income and employment to the rural and urban communities. The different system of medicinal usage practiced in India such as Ayurveda, Siddha, Unnani and other local healthcare traditions which utilize large number of plants for treatment of human diseases (Handral *et al.*, 2012). A medicinal plant is a plant that can be used for therapeutic purposes or may act as a precursor for the synthesis of useful drugs. The medicinal value of the plants lie in their phytochemicals and these phytochemicals can be used as food and medicine which may provide protection against illness (Afolabi et al. 2007). Medicinal plants contain numerous biologically active compounds which are helpful in treatment of several diseases and have been utilized by developing countries as a primary healthcare system since very long period of time. According to the World Health Organization approximately 80% of the population, mostly in developing countries relies on plant-based medicines for primary health care. The advantages claimed for therapeutic uses of medicinal plants in various ailments are their safety besides being economical, effective and their easy availability.

Murraya Koenigii belongs to the family Rutaceae, commonly known as curry-leaf tree in English and meetha neem in Hindi. It is an important medicinal plant of our country and is grown in almost every house for its aromatic leaves. It is a native of India, Sri Lanka and other south Asian countries and it is found almost everywhere in the Indian subcontinent. Curry leaf is an important leafy vegetable and its leaves are widely used in Indian cookery for flavouring the foodstuffs. The leaves have a slightly pungent, bitter and acidic taste and they retain their flavour and other qualities even after drying. Recent researches have revealed that different plant parts of *M. koenigii* contain several chemical constituents that may show antioxidative, cytotoxic, antimicrobial, antibacterial, antiulcer and cholesterol reducing activities (Gahlawat *et al.*, 2014). *Murraya koenigii* is also an important export commodity from India to foreign countries as it can generate revenue due to its medicinal properties. The bark and roots of *Murraya koenigii* can be used as stimulant and these are used externally to cure eruptions and bites of poisonous animals. The leaves of *M. koenigii* are rich source of vitamins and minerals such as calcium and iron. The green leaves can be eaten raw for curing dysentery, diarrhoea and vomiting. Leaves and roots are also useful in the treatment of leucoderma and blood related disorders (Jain *et al.*, 2012). Cataract development in eyes can be prevented by using fresh juice of curry leaves and kidney pain can be cured by using juice of roots of *Murraya koenigii*. However, a critical perusal of the literature revealed the healing properties of curry leaf in treatment of various diseases but no effort seems to have been made to study the antioxidant activity of *Murraya koenigii* by the earlier workers. In order to bridge the gap in literature, this study was designed to know the antioxidant potential of the leaves of *Murraya koenigii*.

Materials and Methods

The experiments were conducted in the Plant Physiology Laboratory, Amity Institute of Biotechnology, Amity University, Noida, India.

Collection of plant material

The extensive survey was conducted to assess the occurrence and distribution pattern of *Murraya koenigii* in Noida. The fresh and healthy leaves of *Murraya koenigii* were collected at the vegetative stage. Fresh leaves of *Murraya koenigii* were removed, washed gently with tap water only for few seconds to avoid leaching losses of water soluble components, followed by quick rinsing in distilled water and drying with clean absorbent paper.

Preparation of dry leaf powder of *Murraya koenigii*

The leaves of *Murraya koenigii* were green in colour with aromatic odour and bitter in taste. The fresh leaves of *Murraya koenigii* were kept in single layer on plastic trays under the shade for air drying for 72 hours. After air drying under the shade, leaves were powdered in a grinder and dry leaf powder was stored in sterilized polythene bags to avoid contamination.

Experimental design

In the present study, fluorescence analysis, phytochemical screening and antioxidant potential of the leaves of *Murraya koenigii* were determined by different biochemical tests.

1. Fluorescence analysis of leaf powder
2. Screening of phytochemical components
3. Analysis of antioxidant potential

1. Fluorescence Analysis of Leaf Powder

The dry leaf powder of *Murraya koenigii* was analyzed for its fluorescence with different chemical reagents such as ethanol, hydrochloric acid and sodium hydroxide (Handral *et al.*, 2010). The fluorescence analysis of leaf powder of *Murraya koenigii* is given in the Table 1.

2. Screening of Phytochemical Components

For the analysis of phytochemical components present in the leaves of *Murraya koenigii* the leaf extracts were prepared in methanol.

Preparation of methanolic leaf extract of *Murraya koenigii*

The leaf powder of *Murraya koenigii* (100 grams) was taken in a beaker and 500 ml of methanol was added. The mixture was kept on rotary shaker for 48 hours at 190-220 rpm. After 48 hours, mixture was filtered, the pellet was discarded and supernatant was evaporated to one-fourth of its original volume. After the preparation of methanolic leaf extract from *Murraya koenigii*, the leaf extract was stored at 4°C in refrigerator for further use (Vaghasiya, 2011). Different phytochemicals which were present in the methanolic leaf extracts of *Murraya koenigii* were qualitatively analyzed by the standard procedures described by Harborne (1973) and Trease and Evans (1989).

1. Test for tannin

   Approximately 0.5 g of the dried leaf powder of *Murraya koenigii* was boiled in 20 ml of distilled water in a test tube and then filtered. Few drops of 0.1% ferric chloride solution was added in the filtrate. A brownish green or blue black colour of the test solution indicates the presence of tannin in a given sample.

2. Test for saponin

   Two grams of the powdered leaf sample of *M. koenigii* was boiled with 20 ml of distilled water and filtered. The 10 ml of the filtrate was mixed with 5 ml of distilled water and shaken vigorously for a stable persistent froth. The foamy leather formation indicates the presence of saponin in the test solution.

3. Test for protein

   Qualitative test of protein was done by biuret test. In 0.5 ml of leaf extract, few drops of biuret reagent (4 drops of 40% NaOH + 2-3 drops of 1% $CuSO_4$) was added. The test solution turns into violet colour indicates the presence of proteins.

4. Test for carbohydrate

   The carbohydrates present in the leaf extracts of *Murraya koenigii* were analyzed by Fehling's test. The 0.5 ml of methanolic leaf extract was mixed with equal volume of Fehling's reagent, heated on water bath for 10 minutes. Formation of red colour indicates the presence of carbohydrate in the leaves of *Murraya koenigii.*

5. Test for resin

   In 0.5 g of leaf powder of *M. koenigii,* 5 ml of warm ethanol was added. It was filtered and filtrate was diluted with 4 ml of 15N HCl. The formation of a heavy resinous precipitate indicates the presence of resin.

6. Test for carboxylic acid

   A pinch of solid sodium bicarbonate was added in 2 ml of methanolic leaf extract. The evolution of carbon dioxide with brisk effervescence confirms the presence of carboxylic acid in the leaves of M. *koenigii.*

7. Test for quinone

   To 0.5 ml of methanolic leaf extract, 5 drops of 40% 1N NaOH was added. Blue green or red colour indicates the presence of quinone.

8. Borntrager's test for anthraquinone

   Two ml methanolic leaf extract of *Murraya koenigii* was mixed with 5 ml of 10% ammonia. Appearance of pink red or violet colour at the lower phase indicates the presence of anthraquinone.

9. Keller-Kiliani test for glycosides

   Few drops of glacial acetic acid and 2-3 drops of ferric chloride solution were added to 2 ml of methanolic leaf extract along with 1 ml of concentrated sulfuric acid. Appearance of brown ring at the interface confirms the presence of glycosides in the test sample.

10. Test for carbonyl group

    Two ml of leaf extract was mixed with 2-3 drops of 2,4 diphenyl hydrazine. Appearance of yellow coloured crystals confirms the presence of carbonyl group in the leaves of *M. koenigii.*

11. Test for phenol

    Dry leaf powder of *Murraya koenigii* (500 mg) was dissolved in 5 ml of distilled water. To this, few drops of 5% ferric chloride solution was added. A dark green colour indicates the presence of phenolic compounds in the test sample.

12. Test for flavonoid

One gram of the powered dried leaves of *Murraya koenigii* was boiled with 10 ml of distilled water for 5 minutes and filtered. Few drops of 20% NaOH solution was added to 1 ml of cooled filtrate. A change to yellow colour which on addition of acid changed to colourless solution shows the presence of flavonoids in the leaves of *Murraya koenigii.*

13. Test for terpenoid

The presence of terpenoids in the leaves of *Murraya koenigii* was analyzed by Salkowski test. Leaf extract (5 ml) was mixed with 2 ml of chloroform and 3 ml of concentrated H_2SO_4 was also added from the sides of the test tube. The reddish brown colour of test solution shows the presence of terpenoids in the leaf extract.

14. Test for steroid

Two ml of acetic anhydride was added to 0.5 ml of methanolic leaf extract of *M. koenigii.* 2 ml of concentrated H_2SO_4 was also added from the sides of the test tube The change in colour of test solution from violet to blue green colour indicates the presence of steroids in the leaves of *Murraya koenigii.*

3. Antioxidant Potential of Leaf Extract of *Murraya koenigii*

Antioxidant potential of the leaf extract of *Murraya koenigii* was analyzed by DPPH assay (Rushender et al. 2012). The free radical scavenging ability of the leaf extract against DPPH free radical was evaluated. One ml of 0.1 mM DPPH (1,1-diphenyl-2-picrylhydrazyl) in ethanol was prepared and to this solution different concentrations (50-300ìg/ìl) of leaf extract of *Murraya koenigii,* 1ml ethanol and 0.95 ml Tris HCl were added. The mixture was left for 30 minutes and absorbance was measured at 517nm. The DPPH free radical scavenging activity was calculated by the given formula:

$$\text{Radical scavenging activity (\%)} = \frac{\text{Control (absorbance) - sample (absorbance)}}{\text{Control (absorbance)}} \times 100$$

Statistical analysis

All the experiments were laid out in a complete randomized block design with three replicates (Snedecor, 1957).

Results and Discussion

The present study was conducted to assess the phytochemical constituents and antioxidant potential of the leaves of *Murraya koenigii* which is widely used by the people for various purposes. The dry leaf powder of *Murraya koenigii* was analyzed for its fluorescence with different chemical reagents (Table 1). The preliminary screening of the leaves of *Murraya koenigii* revealed that leaves contain several phytochemicals (secondary metabolites) such as tannin, saponin, terpenoid, phenolic compounds, glycosides, flavonoids and steroids which may be responsible for its medicinal properties (Table 2).

Table 1. Fluorescence analysis of the leaves of *Murraya koenigii.*

S.No.	Treatment	Day light	UV- light
1.	Leaf powder	Green	Fluorescent green
2.	Leaf powder mixed with distilled water	Bluish green	Fluorescent bluish green
3.	Leaf powder mixed with ethanol	Olive green	Fluorescent orange
4.	Leaf powder mixed with 50% HCl	Dark green	Fluorescent yellow
5.	Leaf powder mixed with 20% NaOH	Light brown	Fluorescent brown

Table 2. Analysis of phytochemical components present in the leaves of *Murraya koenigii.*

S.No.	Phytochemical components	Methanolic extract of leaves of *M. koenigii*
1.	Tannin	+
2.	Saponin	+
3.	Protein	+
4.	Carbohydrate	+
5.	Resin	+
6.	Carboxylic acid	+
7.	Quinone	+
8.	Anthraquinone	+
9.	Cardiac glycosides	+
10.	Carbonyl group	+
11.	Phenol	+
12.	Flavonoids	+
13.	Terpenoids	+
14.	Steroids	+

The oxidation reaction produces reactive oxygen species (ROS), which may start chain reactions that can damage biomolecules but antioxidants can terminate these chain reactions by removing free radical intermediates. More than 90% diseases are caused by free radicals or reactive oxygen species (ROS) which may cause oxidative damage to the cells (Sukandar *et al.*, 2015). The most common reactive oxygen species (ROS) are superoxide (O_2^-) anion, hydrogen peroxide (H_2O_2), peroxyl (ROO^-) and hydroxyl (OH^-) radicals. Oxidative stress caused by reactive oxygen species (ROS) may degrade biomolecules and increases risk of many diseases such as cardiovascular diseases, cancer, diabetes mellitus, cataract, inflammatory diseases and Alzheimer's disease (Attanayake and Jayatilaka, 2016). In the past few years research in the area of antioxidants has been expanded due to its potential benefits in disease prevention and healthcare. Antioxidants may directly react with the free radicals to destroy them by accepting or donating electrons or they may indirectly decrease the formation of free radicals. All the cells of human-beings protect themselves by multiple mechanisms such as enzymatic and non-enzymatic antioxidant systems against free radical damage but these protective mechanisms may not be enough to prevent the oxidative stress (Lu *et al.*, 2010). Several plant secondary metabolites have been reported to protect free radical induced damage in various experiments by earlier workers (Scartezzini and Speroni, 2000). The phytochemicals which can be obtained from natural resources are capable of protecting against ROS mediated damage and they have potential in preventing or curing diseases (Nagmoti *et al.*, 2012).

The antioxidant activity of the leaf extract of *Murraya koenigii* was studied by using DPPH assay. The higher antioxidant activity was observed with the higher concentration of the leaf extract of *Murraya koenigii* (Fig. 1). The basic function of antioxidant molecules is to help in preventing the oxidative stress and to help in protecting the cells by scavenging the free radicals and in this way they may play important role in the treatment of various diseases (Suganya *et al.*, 2014). Leaves of *Murraya koenigii* have high antioxidant potential and it may be due to the presence of various phytochemicals. The data of the present study clearly indicate that leaf extracts of *Murraya koenigii* have the potential to act as an antioxidant and it can be used as a source of useful drug due to the presence of various phytochemicals.

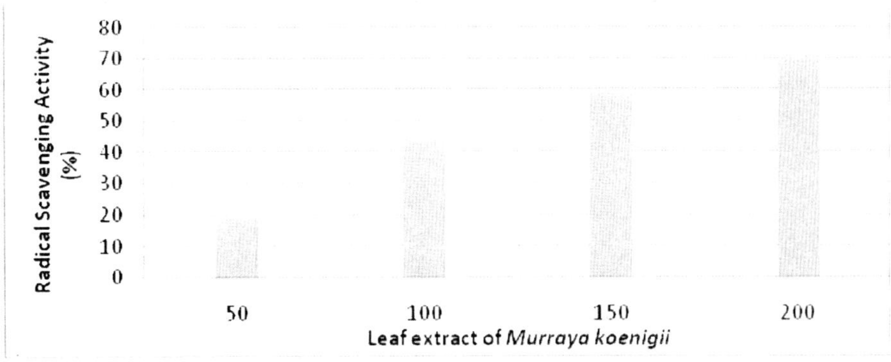

Fig. 1. Estimation of antioxidant activity of the leaf extract of *Murraya koenigii* analyzed by DPPH assay.

Conclusion

The findings of the present study clearly indicate the pharmacological activities and antioxidant potential of the leaves of *M. koenigii*. Hence *M. koenigii* can be utilized to alleviate different diseases and to boost our immune system. The wide spread availability of *M. koenigii* in India and its utilization in cure of various diseases makes it an attractive biological resource for further clinical research.

Acknowledgements

The authors are thankful to Dr. Chanderdeep Tandon, Director, Amity Institute of Biotechnology, Amity University, Noida for offering laboratory and library facilities to carry out this study.

References

Afolabi C.A., Ibukun E.O., Emmanuel A., Obuotor E.M. and Farombi E.O. 2007. Phytochemical constituent and antioxidant activity of extract from the leaves of *Ocimum gratissimum*. *Sci Res Essay*. 2(5): 163-166.

Attanayake A.P. and Jayatilaka K.A.P.W. 2016. Evaluation of antioxidant properties of 20 medicinal plant extracts traditionally used in Ayurvedic medicine in Sri Lanka. *Indian J of Tradit Know*. 15(1): 50-56.

Gahlawat D.K., Jakhar S. and Dahiya P. 2014. *Murraya koenigii* (L.) Spreng: an ethnobotanical, phytochemical and pharmacological review. *J Pharmacogn and Phytochem*. 3(3): 109-119.

Handral H.K., Jha P.K. and Shruthi S.D. 2010. Pharmacognostic and phytochemical studies on the leaves of *Murraya koenigii* L. Spreng. *Pharmacophore* 1(3): 231-238.

Handral H.K., Jha P.K. and Shruthi S.D. 2012. A review on *Murraya Koenigii*: multi potential medicinal plant. *Asian J Pharm and Clin Res* 5: 5-14.

Harborne J.B. 1984. Phytochemical methods: A guide to modern techniques in plants analysis. 2nd Edition, chapman and Hall, London. 1(10): 100-117.

Jain V., Momin M. and Laddha K. 2012. *Murraya Koenigii*: An Updated Review. *International J Ayur and Herb Med.* 2(4): 607-627.

Lu J.M., Lin P.H., Yao Q. and Chen C. 2010. Chemical and molecular mechanisms of antioxidants: Experimental approaches and model systems. *J Cell Mol Med (Berl).* 14: 840-860.

Nagmoti D.M., Khatei D.K., Juvekar P.R. and Juvekar A.R. 2012. Antioxidant activity and free radical scavenging potential of *Pithecellobium dulce* Benth seed extracts. *Free Rad Antioxid.* 2: 37-43.

Rushender R., Madhavieerike C., Madhusudhanan N. and Venugoalarao-konda 2012. In vitro antioxidant and free radical scavenging activity of *Nymphaea pubescens.* J. *Pharm. Res.* 5(7): 3804-3806.

Scartezzini P. and Speroni E. 2000. Review on some plants of Indian traditional medicine with antioxidant activity. *J Ethnopharma.* 71(1-2): 23-43.

Snedecor G.W. 1957. Statistical Methods. Ames, USA : Iowa State University Press. pp. 594.

Suganya P., Nandhini R., Jeyadoss T. and Velavan S. 2014. In vitro antioxidant activity of methanolic extract of *Shorea robusta* in hepatocytes. *International J Pharma and Pharmaceu Sci.* 6(6): 227-230.

Sukandar E.Y., Adnyana I.K. and Nurfitria R.S. 2015. Antioxidant potential of garlic and turmeric mixture- a traditional Indonesian formulation. *Indian J Tradit Know.* 14(4): 632-636.

Trease G.E. and Evans W.C. 1989. Pharmacognosy. 13th (ed). ELBS/ Bailliere Tindall, London. pp. 345-6, 535-6, 772-3.

Vaghasiya Y., Dave R. and Chanda S. 2011. Phytochemical analysis of some medicinal plants from western region of India. *Res J Med Plant.* 5: 567-576.

❑❑❑

Chapter 17

Herbal Drugs

Debdutta Bhattacharya* and Namita Mahapatra

Regional Medical Research Centre (Indian Council of Medical Research, Dept. of Health Research, Ministry of Health & Family Welfare, Govt. of India), Chandrasekharpur, Bhubaneswar-751023, Odisha

ABSTRACT

Phytochemicals from Indian system of medicine, particularly from Ayurveda, are beginning to attract interest across the world and huge efforts have been initiated to include traditional herbal medicine in modern medicine.The paper describes use of herbal drugs their present scenario and prospects in future.

Keywords: Artemisinin, Traditional knowledge, Diversity, Endemic.

Introduction

About 100 years ago, natural herbs were the main remedy for treating human diseases. It has been estimated that 25% of modern medicines are made from plants first used traditionally [Zhang *et al.*, 2015], such as aspirin, artemisinin, ephedrine, and paclitaxel. However, there is limited scientific evidence to establish the safety and efficacy of most herbal products [Bent, 2008]. Although modern medicine is well developed in most of the world, large sections of the population in developing countries still rely on the traditional practitioners, medicinal plants and herbal medicines for their primary care. Moreover during the past decades, public interest in natural therapies has increased greatly in industrialized countries, with expanding use of medicinal plants and herbal medicines. Antibiotic resistance has become a global concern (Bhattacharya *et al.*, 2014). In recent years there is increasing incidence of multiple resistances in human pathogenic microorganisms, largely due to the indiscriminate use of commercial antimicrobial drugs commonly employed in the treatment of infectious diseases (Westh *et al.*, 2004, Bhattacharya *et al.*, 2015). This has forced scientist to search for new antimicrobial substances from various sources like medicinal plants. Search for new antibacterial agents should be continued by the screening of many plant families (Parekh & Chanda, 2007).

Herbal medicines which formed the basis of health care throughout the world since the earliest days of mankind are still widely used, and have considerable importance in international trade. Recognition of their clinical, pharmaceutical and economic value is still growing, although this varies widely between countries [Jayasuriya, 1990]. Medicinal plants are important for pharmacological research and drug development, not only when plant constituents are used directly as therapeutic agents, but also as starting materials for the synthesis of drugs or as models for pharmacologically active compounds. Regulation of exploitation and exportation is therefore essential, together with international cooperation and coordination for their conservation so as to ensure their availability for the future (Jayasuriya, 1990). The United Nations Convention on Biological Diversity states that the conservation and sustainable use of biological diversity is of critical importance for meeting the food, health and other needs of the growing world population, for which purpose access to and sharing of both genetic resources and technologies are essential [WHO, 1998].

Although modern medicine is well developed in most of the world, large sections of the population in developing countries still rely on the traditional practitioners, medicinal plants and herbal medicines for their primary care. Moreover during the past decades, public interest in natural therapies has increased greatly in industrialized countries, with expanding use of medicinal plants and herbal medicines. The many and various forms of traditional medicinal products have evolved against widely different ethnological, cultural, climatic, geographical, and even philosophical backgrounds.The evaluation of these products and ensuring their safety and efficacy through registration and regulation present important challenges.

Herb: What does it mean?

An herb can be any form of a plant or its product which include leaves, stems, flowers, roots, and seeds. These plants can either be sold raw or as extracts, where the plant is macerated with water, alcohol, or other solvents to extract some of the chemicals. The resulting products contain dozens of chemicals, including fatty acids, sterols, alkaloids, flavonoids, glycosides, saponins, and others (Rotblatt &

Ziment, 2002). Because any given herb contains multiple ingredients, some manufacturers attempt to create standardized herbal products by identifying a suspected active ingredient and altering the manufacturing process to obtain a consistent amount of this chemical.

India & Herbal medicine

India has a large diversity of plant species. India consists of 2.4% of the total geographical area of the world and accounts for 8% of global biodiversity, with around 49,000 plant species, of which 5,150 are endemic (Ramakrishnappa, 2002; Singh, 2007; National Biodiversity Authority, 2012). The Himalayan range, Western Ghats, North-Eastern Indian hills (Khasi and Mizo hills), and Vindhya and Satpura ranges of the northern peninsula of India are the gold mine of higher plant species. Herbal medicine has special importance in the society, culture, and traditional medicine of India (Ramakrishnappa, 2002). Plant-based medicines are at the root of the modern health care system, and are acknowledged for their economic importance also (Sen & Chakraborty, 2015). Traditional medicinal knowledge and plants play a central role in biological research and drug development. Herbal products or constituents are not only used directly as curative agents, but also as lead molecule in the discovery of new drug. Current research and understanding suggest that the use of crude herbs or herbal products can confer real benefits on health when used long-term (Dubey *et al.*, 2004; Singh, 2007).

Increasing population and the incidence of side effects of synthetic medicines also accelerate the popularity of alternative medicines. The World Health Organization acknowledged that the goal of "Health for All" cannot be accomplished without herbal medicines. In a wider context, demand for medicinal plants, herbal medicines, health products, pharmaceuticals, food supplements, cosmetics, etc, is growing in all parts of the world, which indicates the popularity and belief of people in herbal medicines. This in turn has created great scope for India to utilize its traditional knowledge of herbal medicine and repository of medicinal plants in the service of the world population and for the economic growth of India (Singh, 2007; Sharma *et al.*, 2008; Report of the Task Force on Conservation and Sustainable use of Medicinal Plants, 2000).

Concepts and practices of different traditional medicinal systems in India are about several thousand years old. A large proportion of the Indian population still believes in and receives traditional medical care, which is based on the principles of three ancient codified Indian systems of medicine (ISMs): Ayurveda, Siddha, and Unani (Borins, 1987; Subbarayappa, 2001; AYUSH, 2010). Though different chemicals, minerals, and animal products are used in such system to prepare curative agents, but use of plants have been the basis of treatment in these system. It is estimated that Ayurveda uses 1,200 species of plant, while Siddha and Unani include 900 and 700 species of plant, respectively, in their medicinal preparations (Kannaiyan, 2008). Currently, more that 1.5 million traditional medical practitioners in India are using medicinal plants for preventive, promotional, and curative purposes (Wakdikar, 2004).

Ayurveda

It is one of the most ancient (6000 BC) of the different organized traditional medicinal systems, is native to the Indian subcontinent and has been practiced since the beginning of the Indian civilization. It can be described as the "science of life", accounting for an integrated observation of the mental, physical, spiritual, and social facets of human beings (Ravishankar & Shukla, 2007; Singla *et al.,* 2012). Preventive and curative measures are the key components of the Ayurvedic system. Major treatment approaches include the use of "aushadhi" (drugs); "anna" (diet); and "vihara", which includes exercises and a healthy mode of life (Srinivasan, 1995; Ravishankar & Shukla, 2007; Singla *et al.,* 2012).

Siddha

The Siddha system of medicine originated in the pre-Vedic period (approx. 3000 BC-2000 BC), mostly in the southern part of India (Srinivasan, 1995), "siddhas" (ancient practitioners of Siddha medicine) are believed to have developed this system which are written in Tamil and mainly practiced in Tamil Nadu (Singla *et al.,* 2012; Zysk, 2008). The Siddha medicinal system recognizes three humoural concepts: "vata" (wind), "pitta" (bile), and "kapha" (phlegm). Examination of the pulse, urine, and different anatomical features like the tongue, voice, complexion, eyes, touch (to find dry, warm, cold, sweating condition), and stools are commonly used as diagnostic criteria in Siddha medicine (Srinivasan, 1995; Zysk, 2008; Karunamoorthi *et al.,* 2012).

Unani

The basics of the Unani system of medicine were laid by Hippocrates and later by Galen. In the eleventh century, the Unani system was introduced in India by Arabs and Persians. The fundamental theory of the Unani system is "humoral theory", which presupposes the presence of four humors - blood, phlegm, yellow bile, and black bile - in the body, and for Unani practitioners diagnosis mainly depends on pulse reading, and examination of the urine and stools (Ahmad, 2008; Unani Medicine, 2014).

Folk medicine

Also known as "tribal" or "indigenous" medicine, also plays an important role in Indian society, mostly in rural/indigenous/ethnic communities. The knowledge is usually passed verbally from ancestors of the particular group of people through generations without any written script. It has been estimated that more than 8,000 species of plants are used by the tribal and ethnic communities in India as part of their health care systems (Government of India Planning Commission, 2000; Devanna *et al.*, 2014). Approximately 25,000 effective plant-based formulations are used in folk medicine and are commonly used by rural and ethnic communities in India (Wakdikar *et al.*, 2004).

Expanding Complementary and Alternative (CAM) Approaches

More than 80 percent of people in developing countries cannot afford the most basic medical procedures, drugs, and vaccines. Among wealthier populations in both developed and developing countries, complementary and alternative practices are popular although proof of their safety and effectiveness is modest. Evidence-based research in Ayurveda is receiving larger acceptance in India and abroad (Mashelkar, 2008; Cooper, 2008; Cooper, 2008; Joshi *et al.*, 2011). The National Center for Complementary and Alternative Medicine has been inaugurated as the United States Federal Government's lead agency for scientific research in this arena of medicine. Its mission is to explore complementary and alternative healing practices in the context of rigorous science, support sophisticated research, train researchers, disseminate information to the public on the modalities that work, and explain the scientific rationale underlying discoveries. The centre is committed to explore and fund all such therapies for which there is sufficient preliminary data, compelling public health need and ethical justifications (Cooper, 2005; Gavaghan, 1994).

Importance of Herbal Medicine

Herbal medicines are truly in a league of their own, and have stood the test of time until now. But unfortunately the utilization of herbal medicine for the management of diseases is less despite of its potential. A large number of ethnic and rural people use and stoically play a crucial role in protecting the ancient medicinal knowledge related to plants from fading away into oblivion. Herbal medicine is due a revival. However, incorporating herbal medicine into the true mainstream of modern health care and ensuring modern safety and efficacy standards is not an easy task.

In India, about 65% of the population mainly uses traditional medicine for their health care needs. Inequities in the accessibility, availability, and affordability of modern health care make herbal drugs more popular in rural and remote areas (Kamboj, 2000; Payyappallimana, 2010). Demand for traditional medicine in developed countries is also increasing. About 40%–50% people in Germany, 42% in the USA, 48% in Australia, and 49% in France are using traditional medicine. In the twenty-first century, herbal drugs and products from plant sources are increasingly being acknowledged in developed countries and also among the people who can afford costlier allopathic medicines, with the hope of a more eco-friendly, bio-friendly, and relatively safer treatment strategy (Kamboj, 2000; Payyappallimana, 2010).

Nearly 1 in 5 adults in the United States report taking an herbal product (Barnes *et al.*, 2002). Written records of the use of herbal medicine date back more than 5,000 years (Swerdlow, 2000). In fact, for most of history, herbal medicine was the only medicine. Even as recently as 1890, 59% of the listings in the US Pharmacopeia were from herbal products (Swerdlow, 2000), and it has been estimated that as many as one third to one half of currently used drugs were originally derived from plants (Barrett *et al.*, 1999).

Several countries (e.g., Germany, France, Sweden, and Australia) have implemented strategies for licensing herbal remedies. In Germany, such products can be registered as medicines on the basis of information in approximately 300 monographs on herbs ("positive" monographs with concise information about terminology, composition, uses, contraindications, side effects, drug interactions, dosage, mode of administration, and actions, and "negative" monographs explaining insufficient benefits or unacceptable risks) (Blumenthal, 1998). The European Commission (which governs the European Union) has recently promulgated a draft directive on the licensing of traditional herbal preparations (Wood, 2002).

A key area for the popularity of herbal products is nutritional supplements. Herbal supplements afford nutrients that are lacked or not consumed in an adequate quantity through the diet. Herbal supplements may contain vitamins, minerals, macronutrients, and antioxidants, etc, which are essential for good health. Thus, the demand of several Indian formulations - like chyawanprash, musli pak, and ashwagandhadi lehyam - is increasing (Pandey *et al.,* 2013).

Present Scenario and Future Prospects

Review of different national pharmacopeia revels that at least 120 distinct chemical products/moieties from herbal sources have been utilized as lifesaving drugs. It is predicted that among the estimated 250,000-400,000 plant species of world only 6% have been screened systematically for their biological activity and 15% have been investigated phytochemically (Patwardhan *et al.,* 2005). Between 1981 to 2002, around 119 drugs were approved and of those around 60% of anticancer and 75% anti-infective drugs could be related to natural substances (Gurib-Fakim, 2011). Several bioactive molecules from plants, especially from those used in Ayurveda, have been discovered, such as reserpine for high blood pressure, psoralens for vitiligo, alkaloids from *Holarrhena antidysenterica* (L.) against amebiasis, *Mucuna pruriens* for Parkinson's disease, piperidines as bioavailability enhancers, vasicine and vasicinone as bronchodilators, hydroxycitric acid for obesity, bacosides to treat mental retention, picrosides as hepatoprotective agents, phyllanthins for viral infections, and withanolides and steroidal lactones as immunomodulators. Current scientific knowledge, better clinical observation, and superior thoughtful consideration help to explore Indian traditional medicine in an advanced way and utilize old molecules for new therapeutic applications. For instance, forskolin is currently acknowledged as a potent adenylate cyclase activator and antimicrobial berberine alkaloids are used to cure dyslipidemia (Patwardhan & Mashelkar,

2009). A typical, systematic research on plants to find new bioactive molecules is expensive and inefficient. The high cost of and the time taken for this process are the main hurdles. One of the most important approaches in the discovery of a new drug would be recourse to the information accessible in traditional Indian medicinal systems, which are based on the proven therapeutic utility of the medicinal plants of India (Sen & Chakraborty, 2015).

Indian herbal medicine makes numerous medical claims for the treatment of many acute and chronic diseases and symptoms, the prevention of disease, and the improvement of quality of life. The advancement and success of Indian traditional medicine, especially Ayurveda, around the world is mainly due to its being a holistic approach to treatment (Tripathi, 2000). Ayurveda, a well-written medical text, describes scientific view on diseases and treatment that makes it unique. Currently, Ayurveda is included in the Indian national health care system. Several countries, like the UK, the United Arab Emirates, Sweden, Indonesia, and the USA have acknowledged the Ayurvedic health care system and more than 30 countries are on the verge of doing the same (Sheth, 2005). It has been estimated that almost two-thirds of people in the USA use one or more alternative treatment strategies, of which most are drugs from herbal sources. The demand for dietary supplements and traditional medicines is high in the US and thus more than 1,500 herbals are sold for such proposes (Patwardhan & Mashelkar, 2009). A survey in 2007 revealed that more than 200,000 American adults have used Ayurvedic medicine (NCCIH, 2005). Ayurveda is gaining momentum as a successful alternative to the conventional medicinal system through its systematic strategy of curing and preventing diseases using natural resources (Chaudhary & Singh, 2011).

Safety, Toxicity, and Side Effects

Because herbs are plants, they are often perceived as "natural" and therefore safe (Ernst, 1998). In reference to medications, safety is the likelihood of not causing harm under the proposed conditions of use, while efficacy is the capacity to induce a clinical benefit. Both safety and efficacy depend on the drug's therapeutic indication; in principle, a substance has no clinical usefulness if it is "safe" but lacks efficacy or if it is active on a relevant therapeutic target but its use is unsafe. Although these are recognized as equally essential attributes of any medicine, safety has taken precedence over proof of efficacy in drug regulation history. In the US, for instance, the Federal Food Drug and Cosmetic Act of 1938 required that safety of new drugs had to be proven by pre-marketing testing, whereas similar requirements to demonstrate drug efficacy were introduced only 25 years later by the Kefauver-Harris Amendments of 1962.

However, many different side effects to herbs have been reported and recently reviewed (Ernst, 1998; De Smet, 2002; Bent & Ko, 2004), including effects from biologically active constituents from herbs, side effects caused by contaminants, and herb–drug interactions. Case reports of nephropathy caused by the use of certain Chinese herbs are common. A particularly morbid case series describes 105 patients in Belgium who had been taking a Chinese herbal product for weight loss and developed nephropathy caused by the herb *Aristolochia fangchi*. Forty-three patients developed end-stage renal failure, and 39 had prophylactic kidney removal. Eighteen of these patients were found to have urothelial carcinoma, which was shown to be related to the formation of DNA adducts from the aristolochic acid in this herb (Nortier, 2000). Another common toxicity to herbal medicines involves pyrrolizidine alkaloids, which are complex molecules found in certain plants that may be used or inadvertently added to herbal medicines (including comfrey, which is still available in the United States). These alkaloids produce hepatotoxicity through a characteristic veno-occlusive disease that may be rapidly progressive and fatal (Stickel *et al.,* 2005). The safety of using most herbs with drugs is not well established. Some herbs are known to interact with pharmaceutical drugs, although most of this information comes from case reports rather than systematic investigations (De Smet, 2002). Because many herbs contain pharmacologically active compounds, some herbs may cause side effects through excessive biological effects. Unfortunately, the true frequency of side effects for most herbs is not known because most have not been tested in large clinical trials and because surveillance systems are much less extensive than those in place for pharmaceutical products (Bent, 2008).

According to the prevailing concept of evidence-based medicine (EBM; "...use of current best evidence in making decisions about the care of individual patients") (Sackett *et al.,* 1996), controlled and randomized clinical trials, and an unbiased systematic review with and without a meta-analysis, rank first in the hierarchy of sources of evidence for the safety and efficacy of therapeutic interventions. Clinical efficacy cannot be presumed on the basis of pharmacological actions described in animal and/or in vitro experiments, nor on physicians'/experts' opinions only. It has to be demonstrated by adequately designed and conducted phase III studies, or exceptionally by phase II trials (Davyson L. Moreira *et al.*, 2014). Many supporters of herbal medicines argue that products with a long history of popular use are generally safe when used properly at common therapeutic doses (Fong, 2002). A crucial question underlying this statement is the extent to which the absence of evidence of toxicity could be taken as evidence of the absence of toxicity or safety of herbal medicines. Whether the absence of records of adverse effects is an indication of lack of toxicity depends on the type of toxic effect and the likelihood of observing such an adverse outcome under the conditions prevailing in the traditional usage. Acute symptoms and short term toxic effects, such as gastro-intestinal disturbances and dermatological effects, are likely to be recognized and associated to herbal medicine. Therefore, the absence of such observations provides some evidence of safety in these particular endpoints (Davyson L. Moreira *et al.*, 2014).

Need of the Hour and Way Ahead

Several experts have previously suggested a number of important changes to the regulation of herbs that could improve the safety and appropriate use of these products (Lewis & Strom, 2002; Marcus & Grollman, 2002). These include: (1) requiring manufacturers to register with the concerned approving authority of the government, (2) mandating safety tests similar to those required for over-the-counter drugs, (3) requiring all health claims to be supported by data approved by the concerned approving authority of the government, and (4) ensuring that product labels provide an accurate list of all ingredients. While these changes will clearly help the safety of herbal products, additional changes are needed to improve and promote high-quality research. The most critical element will be to define specific standards for herbal products to ensure consistency between studies (Bent, 2008).

Phytochemicals from Indian system of medicine, particularly from Ayurveda, are beginning to attract interest across the world and huge efforts have been initiated to include traditional herbal medicine in modern medicine. Several government and nongovernmental organizations from different countries have actively started researching on plants and formulations described in Ayurveda. India has enormous facilities for research; the various national laboratories like Central Drug Research Institute (CDRI), Council of Scientific and Industrial

Research (CSIR), Central Institute of Medicinal and Aromatic Plants, National Botanical Research Institute, Regional Research Laboratories, Regional Medical Research Centre (ICMR), Belagavi and National Chemical Laboratory must play vital roles in this regard and should work in collaboration to study the safety and efficacy of these herbal drugs.In the present scenario of developing pan resistance among various pathogens to various synthetic drugs. Pipelines of discovery of new antibiotics are drying up as major pharmaceutical companies are losing interest in investing money in this endeavour, mainly due to the short shelf-life of the antibiotics and also due to the fast emergence of drug resistance. The present government should continue promoting and supporting research in herbal medicine to evade the impending danger of no treatment alternate for deadly pathogens in our near future.

References

Ahmad S. Unani medicine: Introduction and present status in India. *Int J Altern Med* [serial on the Internet]. 2008: 6(1). http://ispub.com/IJAM/6/1/3747.

Barnes P.M. Powell-Griner E., McFann K., Nahin R.L. Complementary and alternative medicine use among adults: United States., 2002. *Adv Data.* 2004; 343: 1–19.

Barrett B., Kiefer D., Rabago D. Assessing the risks and benefits of herbal medicine: an overview of scientific evidence. *Altern Ther Health Med.* 1999; 5: 40–49.

Bent S., Ko R. Commonly used herbal medicines in the United States: a review. *Am J Med.* 2004; 116: 478–85.

Bent S. Herbal medicine in the United States: review of efficacy, safety, and regulation: grand rounds at University of California, San Francisco Medical Centre. *J Gen Int Med* 2008; 23(6): 854–859.

Bhattacharya D., Bhattacharya H., Sayi D.S., Bharadwaj A.P., Singhania M., Sugunan A.P., Roy S. Changing patterns and widening of antibiotic resistance in *Shigella* sp. over a decade (2000-2011), Andaman Islands, India. *Epidemiol Infect* 2015; 143(3): 470-77.

Bhattacharya D., Bhattacharya H., Thamizhmani R., Sayi D.S., Reesu R., Anwesh M., Kartick C., Bharadwaj A.P., Singhania M., Sugunan A.P., Roy S. Shigellosis in Bay of Bengal Islands, India: clinical and seasonal patterns, surveillance of antibiotic susceptibility patterns and molecular characterization of multi-drug resistant Shigella strains isolated during a 6-year period, 2006 to 2011. *European J Clin Microbiol Infect Dis* 2014; 33: 157-170.

Blumenthal M. ed. The complete German Commission E monographs – therapeutic guide to herbal medicines. Austin, Tex.: American Botanical Council, 1998.

Borins M. Traditional medicine in India. *Can Fam Physician.* 1987; 33: 1061-1064.

Chaudhary A., Singh N. Contribution of world health organization in the global acceptance of Ayurveda. *J Ayurveda Integr Med.* 2011; 2(4): 179–186.

Cooper E.L. Ayurveda and eCAM: a closer connection. *Evid Based Complement Alternat Med* 2008; 5(2): 121–122.

Cooper E.L. Ayurveda is embraced by eCAM. *Evid Based Complement Alternat Med* 2008; 5(1): 1–2.

Cooper EL. CAM, eCAM, bioprospecting: the 21st century pyramid. *Evid Based Complement Alternat Med* 2005; 2(2): 125–127.

De Smet PA. Herbal remedies. *N Engl J Med.* 2002; 347: 2046–56.

Devanna N., Chakraborty R., Sen S., De B. *Tribal Medicinal Plants of Tripura, India: A Scientific Search.* New Delhi: Serial Publications; 2014.

Dubey NK., Kumar R., Tripathi P. Global promotion of herbal medicine: India's opportunity. *Curr Sci.* 2004; 86(1): 37–41.

Ernst E. Harmless herbs? A review of the recent literature. *Am J Med.* 1998; 104: 170–8.

Fong HH. Integration of herbal medicine into modern medical practices: issues and prospects. *Integr Cancer Ther.* 2002; 1: 287-293.

Gavaghan H. Koop may set up new centre for alternative medicine. *Nature* 1994; 370(6491): 591.

Government of India Planning Commission. *Report of the Task Force on Conservation and Sustainable use of Medicinal Plants.* New Delhi: Government of India Planning Commission; 2000. Available from: http://planningcommission.gov.in/aboutus/taskforce/tsk_medi.pdf.

Government of India Planning Commission. Report of the Task Force on Conservation and Sustainable use of Medicinal Plants. New Delhi: Government of India Planning Commission; 2000.

Gurib-Fakim A. Traditional roles and future prospects for medicinal plants in health care. *Asian Biotechnol Develop Rev.* 2011; 13(3): 77–83.

Jayasuriya DC. A review of legislation concerning medicinal plants. 1990 Unpublished Report.

Joshi K., Ghodke Y., Patwardhan B. Traditional medicine to modern pharmacogenomics: Ayurveda Prakriti type and CYP2C19 gene polymorphism associated with the metabolic variability. *Evid Based Complement Alternat Med* 2011; 2011: 249-528.

Kamboj V.P. Herbal medicine. *Curr Sci.* 2000; 78(1): 35–39.

Kannaiyan S. Diversity, sustainable use and conservation of medicinal plants. Inaugural address to the International Seminar on Medicinal Plants and Herbal Products, Tirupati, India, March 7–9, 2008. Available from: http://ismphpabstracts.blogspot.in/. Accessed August 20, 2014.

Karunamoorthi K., Jegajeevanram K., Xavier J., Vijayalakshmi J., Melita L. Tamil traditional medicinal system - siddha: an indigenous health practice in the international perspectives. *Int J Genuine Tradit Med.* 2012; 2(2): 1–11.

Lewis J.D., Strom B.L. Balancing safety of dietary supplements with the free market. *Ann Intern Med.* 2002; 136: 616–8.

Marcus D.M., Grollman A.P. Botanical medicines—the need for new regulations. *N Engl J Med.* 2002; 347: 2073–6.

Mashelkar R.A. Second world Ayurveda congress (theme: Ayurveda for the future)—inaugural address: part III. *Evid Based Complement Alternat Med* 2008; 5(4): 367–369.

Ministry of Ayurveda, Yoga and Naturopathy, Unani, Siddha and Homoeopathy (AYUSH). About the systems [web page on the Internet]. New Delhi: Ministry of AYUSH; nd [updated September 1, 2010]. Available from: http://indianmedicine.nic.in//index1.asp?lang=1&linkid=17&lid=40.

Moreira D.L., Teixeira S.S., Monteiro M.H.D., Oliveira A.C., Francisco J.R. Paumgartten FJR. Traditional use and safety of herbal medicines. *Rev Bras Farmacogn* 2014; 24: 248-257.

National Biodiversity Authority. Annual Report 2011–2012. Chennai: National Biodiversity Authority; 2012. Available from: http://nbaindia.org/uploaded/pdf/Annual_Report_2011_12_%20Eng.pdf.

National Centre for Complementary and Integrative Health. Ayurvedic medicine: an introduction [web page on the Internet]. NCCAM Pub No D287. Washington DC: National Center for Complementary and Integrative Health; 2005 [updated November 12, 2013]. Available from: http://nccam.nih.gov/health/ayurveda/introduction.htm.

Nortier J.L., Martinez M.C., Schmeiser H.H. *et al.*, Urothelial carcinoma associated with the use of a Chinese herb (Aristolochia fangchi). *N Engl J Med.* 2000; 342: 1686–92.

Pandey M.M., Rastogi S., Rawat A.K. Indian traditional Ayurvedic system of medicine and nutritional supplementation. *Evid Based Complement Alternat Med.* 2013; 2013: 1-12.

Patwardhan B., Mashelkar R.A. Traditional medicine-inspired approaches to drug discovery: can Ayurveda show the way forward? *Drug Discov Today.* 2009; 14(15–16): 804–811.

Patwardhan B., Warude D., Pushpangadan P., Bhatt N. Ayurveda and traditional Chinese medicine: a comparative overview. *Evid Based Complement Alternat Med.* 2005; 2(4): 465–473.

Payyappallimana U. Role of traditional medicine in primary health care: An overview of perspectives and challenges. *Yokohama J Social Sci.* 2010; 14(6): 57–77.

Ramakrishnappa K. Impact of cultivation and gathering of medicinal plants on biodiversity: Case studies from India. In: Food and Agriculture Organization of the United Nations (FAO). Biodiversity and the Ecosystem Approach in Agriculture, Forestry and Fisheries. Satellite event on the occasion of the Ninth Regular Session of the Commission on Genetic Resources for Food and Agriculture, Rome, Italy, October 12–13, 2002. Rome: FAO; 2002. Available from: http://www.fao.org/docrep/005/aa021e/AA021e00.htm.

Ravishankar B., Shukla V.J. Indian systems of medicine: a brief profile. *Afr J Tradit Complement Altern Med.* 2007; 4(3): 319–337.

Rotblatt M., Ziment I. Evidence-based herbal medicine. Philadelphia, PA: Hanley & Belfus; 2002.

Sen S. & Chakraborty R. Toward the integration and advancement of herbal medicine: a focus on traditional Indian medicine. *Botanics: Targets and Therapy* 2015; 5: 33–44.

Sen S., Chakraborty R., De B. Challenges and opportunities in the advancement of herbal medicine: India's position and role in a global context. *J Herb Med.* 2011; 1(3–4): 67–75.

Sharma A., Shanker C., Tyagi L.K., Singh M., Rao ChV. Herbal medicine for market potential in India: an overview. *Acad J Plant Sci.* 2008; 1(2): 26–36.

Sheth P.P. Global opportunities and challenges for medicinal uses of Ayurveda, herbal products, neutraceuticals and alternatives. *Health Administrator.* 2005; 19(1): 74–75.

Singh H. Prospects and challenges for harnessing opportunities in medicinal plants sector in India. *Law Environ Develop J.* 2007; 2(2): 196–211.

Singla R.K., Yadav V., Jayalakshmi S. Traditional systems of medicine – now and forever. *Webmedcentral.* 2012; 3(4): WMC003299.

Srinivasan P. National health policy for traditional medicine in India. *World Health Forum.* 1995; 16(2): 190–193.

Stickel F., Patsenker E., Schuppan D. Herbal hepatotoxicity. *J Hepatol.* 2005; 43: 901–10.

Subbarayappa B.V. The roots of ancient medicine: an historical outline. *J Biosci.* 2001; 26(2): 135–144.

Swerdlow J.L. Medicine Changes: late 19th to early 20th century. Nature's Medicine: Plants that Heal. Washington, D.C.: *National Geographic Society;* 2000: 158–91.

Swerdlow J.L. Modern Science Embraces Medicinal Plants. Nature's Medicine: Plants that Heal. Washington, DC: *National Geographic Society;* 2000: 110–57.

Tripathi Y.B. Molecular approach to Ayurveda. *Indian J Exp Biol.* 2000; 38(5): z409–z414.

Unani medicine; Modes of treatment [webpage on the Internet]. Chicago, IL: Encyclopædia Britannica, Inc.; nd [updated July 7, 2014]. Available from: http://www.britannica.com/EBchecked/topic/1899343/Unani-medicine/306595/Modes-of-treatment.

Wakdikar S. Global health care challenge: Indian experiences and new prescriptions. *Electron J Biotechnol.* 2004; 7(3): 214–220.

Wood A.J.J. Herbal remedies. *N Engl J Med.* 2002; 347(25): 2046-56.

World Health Organisation. Regulatory Situation of Herbal Medicines - A Worldwide Review. WHO 1998.

Zhang J., Onakpoya I.J., Posadzki P., Eddouks M. The Safety of Herbal Medicine: From Prejudice to Evidence. *Evid Based Complement Alternat Med* 2015; 2015: 1-3.

Zysk KG. *Siddha Medicine in Tamil Nadu*. Tranquebar Initiative Skrift-serie No 4. Copenhagen: National Museum's Tranquebar Initiative; 2008. Available from: http://natmus.dk/fileadmin/user_upload/natmus/forskning/dokumenter/Tranquebar/Skriftserie/Tranquebar_Initiativets_Skriftserie_nr_04_2008.pdf.

❏❏❏

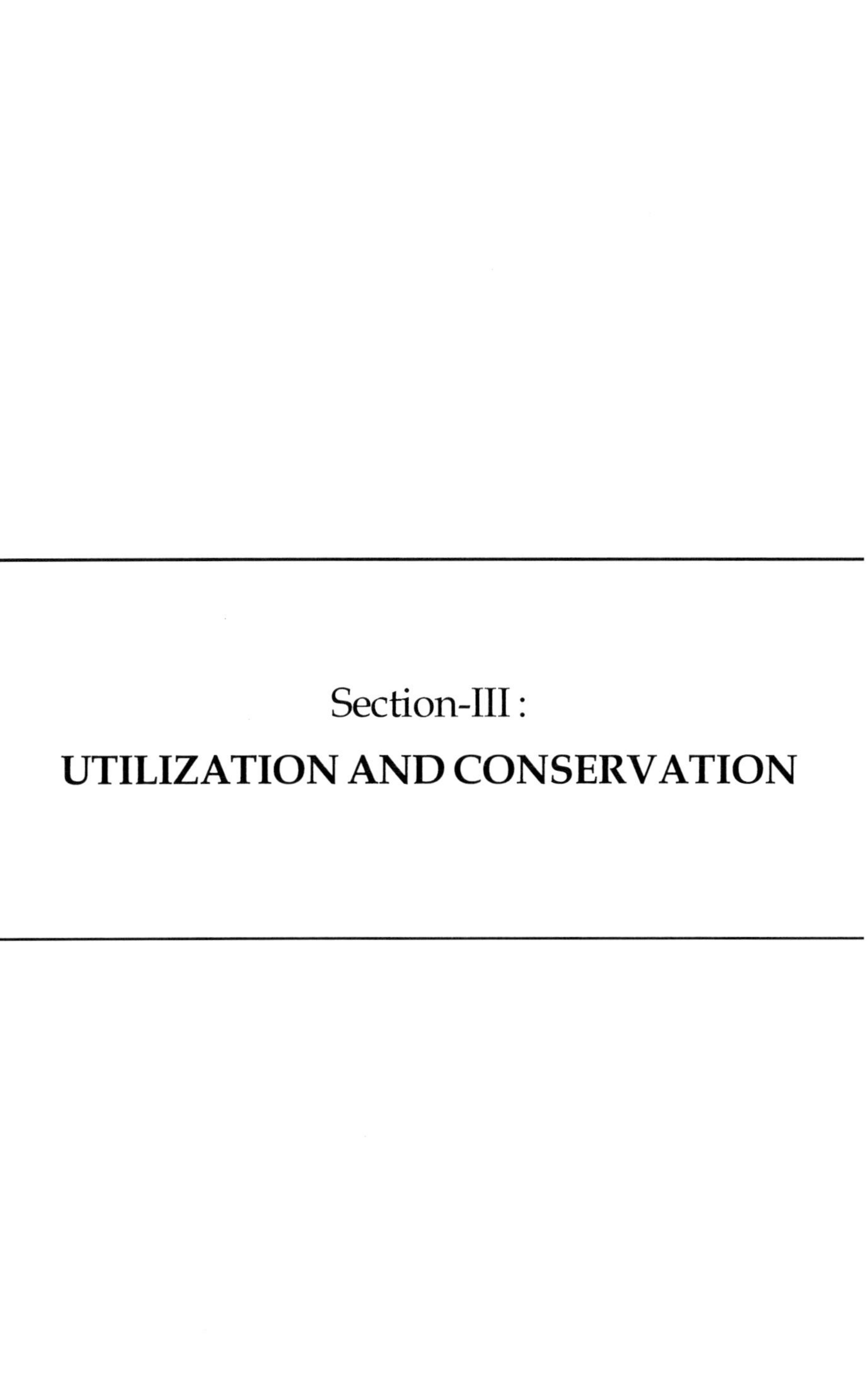

Section-III :

UTILIZATION AND CONSERVATION

Chapter 18

Medicinal and Economic Importance of *Rhododendron* in Sikkim Himalaya: An overview

C.S. Purohit*, Ramesh Kumar, Jitendra Kumar Shukla and Vinod Maina**

Botanical Survey of India, Arid Zone Regional Centre, Jodhpur-342014

** Regional Centre of Institute of Bioresources and Sustainable Development, Gangtok, Sikkim

* Corresponding author email: chandansinghpurohit@yahoo.com;

ABSTRACT

The Rhododendrons play an important role in ecological balance of surrounding ecosystem in Sikkim Himalayas. They forms a wide range of forests components in temperate, sub-alpine and alpine regions, supporting a large range of biodiversity, which provides food and shelter to birds and animals and stabilizes the mountain's soil profile by preventing soil erosion and colonize waste lands which also help in to check soil erosion and regeneration of vegetation. Ethnic group of Sikkim Himalayan region of India, such as the *Lepchas, Bhutias, Nepalese* and *Limbus* are more diverse in their food habit, life style, culture and rituals which play a major role in survival and conservation of Sikkim Himalayan rhododendrons. Amongst them the *Lepchas* are the aboriginal inhabitants of Sikkim and more dependent on forests and forest produce, Rhododendrons are also a part of their lives directly or indirectly. The present compilation reports dealt proper identification with author citation followed by local names short description, phenology, uses, habitat range in Sikkim Himalayas and specimens examined observed in Sikkim Himalayan region.

Keywords: Economic; Medicinal, Rhododendron; Sikkim-Himalaya.

Introduction

The term *Rhododendron* has been derived from two Greek words *rhodon* (rose) and *dendron* (tree) meaning rose tree (Hora, 1981) and was described for the first time by Carl Linnaeus in 1753. *Rhododendron* is the only genus of the family Ericaceae that reaches the size of a tree and includes more than 850 species in the world (Mabberely, 2008). Revision of the genus at world level was carried out by Cullen (1980) and Chamberlain (1982). They are mostly distributed at higher elevations in the Sino-Himalayan region. In India, the Rhododendron species are mostly confined to the Himalayan region, particularly in Eastern Himalaya.

J. D. Hooker was the pioneer worker who visited Sikkim Himalaya between 1848-1850 unfold the rhododendron world of this area. In the mean time, he collected and described 34 new species and detail account of 43 species including varieties from the Indian region in his monograph entitled 'Rhododendron of Sikkim Himalaya' (Hooker 1849). It was followed by Clarke (1882) and added some more species in his publication on the 'Indian Rhododendron'.Since then many species have been described and recorded from north-east India by various workers (Calder *et al.*, 1926; Razi 1959; Nayar and Ramamurthy 1973; Nayar and Karthikeyan 1981; Pradhan (1985, 1986); Naithani (1990); Ghosh and Samaddar (1989); Pradhan and Lachunga (1990). In the recent past extensive work on Rhododendron on various aspects was carried out by various workers [Hariharan & Rangaswani (1966), Skidel (1980), Kataki (1983), Naithani & Bahadur (1983), Sastry & Hajra (1983), Mao *et al.* (2001); Maitii & Chauhan (2001), Das & Chauhan (2002), Chauhan *et al.* (2003), Singh (2009), Singh & Chauhan (2000), Sastry and Hajra (2010); Pradhan (2010); Sekar & Srivastava (2010) Das & Singh (2011), Purohit (2014) and Purohit *et al.* (2014), Purohit (2015)] and dealt in their publications.

Distribution: Rhododendrons cover a vast section of south-eastern Asia between the north-western Himalaya through Nepal, Sikkim, eastern Tibet, Bhutan, Arunachal Pradesh, upper Burma and western and central China. More than 90% of the world's natural population of Rhododendrons is from this region. On record, 98% of the Indian species are found in the Himalayan region, of which 72% in Sikkim. In the light of these facts, Sikkim may be considered as the most appropriate location for conservation and propagation of Rhododendrons in India (Singh *et al.*, 2003; Tiwari & Chauhan 2006; Singh 2009).

Study Area

Sikkim (27^0 4' 46" to 28^0 7' 48" N and 88^0 58' to 88^0 5' 25" E) in north-eastern India has an area of 7096 sq. km and its altitude ranges from 100 - 8598 msl. Sikkim is the 22nd state of India, which came into existence on 16th May, 1975 located in the Eastern Himalayas, extending approximately 115 Kms from north to south and 65 Kms from east to west, surrounded by vast stretches of Tibetan Plateau in the North, Chumbi Valley of Tibet and the kingdom of Bhutan in the east, Darjeeling district of West Bengal in the south and the kingdom in Nepal in the west. The state have four district namely, East, West, North and South. The area represented several ecological zones, viz. subtropical, temperate, subalpine and alpine. Altitudinal variations and different ecological zones promote a rich diversity and variations in rhododendron species. This region is listed among the world's ten most critical centers for biodiversity and endemism with 4500 spp. of flowering plants, 36 spp. of rhododendrons, 450 spp. of trees, 430 spp. of orchids, 175 spp. of wild edible plants (Pradhan & Lachungpa, 1990; Singh & Chauhan, 1997; Rai & Rai, 1994). The area experiences a heavy rainfall due to its proximity with the Bay of Bengal. Pre-monsoon rain occurs in April-May and Monsoon operates normally from the month of May and continues up to early October. Average annual rainfall varies from 2000 - 4000 mm at valley to mountain ridges. The humidity remains very high during the rainy season (85-97%).

***Rhododendron* Species Diversity:** The *Rhododendron* habits of the terrestrial species are found to be of dwarf tussocks, small shrub or robust bushes which sometimes form impenetrable thickets at places. A few number of species are epiphytic and mostly are large shrubs or medium height trees attaining 3-15m heights. *Rhododendrons* form dominating species all along the cool temperate, subalpine and alpine zones in the Sikkim Himalayas. For many subalpine species, snow cover is prerequisite for survival during the harsh environmental conditions that prevail during alpine winters (Korner and Larcher, 1988). It is a keystone element and if disturbed can degrade habitats and threaten associated biodiversity. Sekar & Srivastava (2010) reported 87 species, 12 subspecies and 8 varieties of Rhododendrons from Indian Himalayan region, amongst them Arunachal Pradesh is more diverse with having 86% (75 species) of Rhododendrons and only 6 species and one subspecies are reported from Western Himalaya. According to Mao (2010), 120 taxa (73 species, 22 sub-species, 25 varieties) and 3 natural hybrids have been recorded from India, out of which 117 (98% excluding 3 taxa *Rhododendron arboreum* subsp. *nilagiricum* in south India and *R. colletianum* and *R. anthopogon* subsp.

hypenanthum from western Himalaya) are distributed in north-east India. Mao *et al.* (2001) reported that 12 species, 2 subspecies and 5 varieties are endemic to India.

After scrutiny of literature and herbarium it has found most of the Rhododendron species are reported from different reserves and also shows dense population in those areas and a very few population in outskirt areas. After declaration of Biosphere reserves and Sanctuaries by Government of India in Sikkim Himalayas, it plays a very vital role for protection and conservation of important medicinal and threatened plants specially Rhododendrons (Table 1).

Table 1. Protected areas of the Sikkim Himalaya region where rhododendrons are commonly found (Singh *et al.*, 2003; Tiwari, 2004)

Sl.No.	Place/Name	Area (km^2)	Established	Scope
1	Kanchendzonga Biosphere Reserve	2619.92	1977	It is highest altitude National Park in the country (1829 - 8550m amsl), with considerable importance of the flora including rhododendrons, fauna, ecological, geomorphologic locations and wildlife potentiality in the area.
2	Barsey Rhododendron Sanctuary	104.00	1998	It spans over the sharp Singalila Range. The climate of the area favours upon abundance of the Rhododendrons mainly Trees & Shrubs species.
3	Fambanglho Wildlife Sanctuary	51.76	1984	This sanctuary is situated 1280 - 2652m amsl and repository of Himalayan flora fauna including some species of Rhododendrons.
4	Shingba Rhododendron Sanctuary	43.00	1984	It is located in North Sikkim in the Lachung valley and known for its alpine meadow and hot spring. Some rare species like *R. niveum* (state tree of Sikkim) *R. baileyi, R. ciliatum* are found in this sanctuary along with some other Rhododendrons.
5	Maenam Wildlife Sanctuary	35.34	1987	It is located in South Sikkim and is exceedingly rich in *R. griffithianum* and *R. dalhousiae* with some other species.
6	Kyongnosla Alpine Sanctuary	31.00	1977	The sanctuary is located on the way to Nathula and rich in both flora and fauna. Some of the rare and endangered ground orchids and Rhododendrons are reported from here.

Table 2. Indian institutes involved in research and development work on Rhododendron (Tiwari, 2004).

SN	Institution	Location	Activities
1	Botanical Survey of India, Sikkim Himalayan Regional Centre	Gangtok, Sikkim	Survey, collection, characterization, identification, documentation, herbarium preparation and conservation of threatened plants
2	Regional Centre of Institute of Bioresources and Sustainable Development (RCIBSD)	Tadong, Gangtok, Sikkim	Database preparation and documentation
3	Forest Department, Govt. of Sikkim	Gangtok, Sikkim	Rhododendrons conservation through protected area network program
4	Sikkim Rhododendron Society	Gangtok, Sikkim	Rhododendrons status studies
5	World Wildlife Fund for Nature, India-Sikkim Unit	Gangtok, Sikkim	Baseline assessment
6	World Wildlife Fund for Nature, India -Arunachal Unit	Tawang, Arunachal Pradesh	Restoration of Rhododendron community forests in partnership with the locals
7	Ashoka Trust for Research in Ecology and the Environment	Bagdogra, Siliguri, West Bengal	Create awareness among local people and visitors of the problems of nature degradation
8	Tropical Botanical Garden and Research Institute	Palode, Thiruvananthapuram	Genetic variation studies among the population of *R. nilgiricum*
9	High Altitude Plant Physiology Research Centre	H.N.B. Garhwal University, Srinagar, Uttranchal	Studies on phenology, seed germination and polysaccharides estimation, etc.
10	Nainital University	Nainital, Uttranchal	Effect of fertilizers on flower colour and growth in some species
11	Indian Agricultural Research Institute	New Delhi	Genetic diversity studies
12	G.B. Pant Institute of Himalayan Environment and Development (Sikkim Unit)	Pangthang, Gangtok, Sikkim	Status, *ex-situ* and *in-situ* conservation efforts

Importance of *Rhododendron*: At present, nearly 50% of the total species of Rhododendrons are under cultivation worldwide and about 5000-6000 hybrids of Rhododendrons have already been developed (Sally & Greer, 1986). Floriculture of Rhododendrons can play a vital role to enhance the economic growth due to its high cost hybrids. They are mostly grown in the gardens, parks and other important places as avenue plant for their showy and attractive flowers. However, in order to promote cultivation, publicity and conservation of Rhododendrons, many countries have their own societies or organizations, such as Royal Horticultural Society, London; American Rhododendron Society; Rhododendron Society of Canada; Australian

Rhododendron Society; Rhododendron Species Foundations; International Rhododendron Union; Rhododendron, Camellia and Magnolia Group etc.

Ethnic group of Sikkim Himalayan region (*Lepchas, Bhutias, Nepalese* and *Limbus*) are taken in use of Rhododendrons in various ways. Details of economic importance of *Rhododendron* spp in Sikkim Himalaya are given below with their local name, taxonomic characters, phenology, habitat range and specimen examined.

1. ***Rhododendron anthopogon*** D. Don; (Local name: Sunpati, Dhupi, Dhupi Gurans);

   Taxonomic characters: Small erect aromatic, much branched shrub up to 15-60 cm high; Flowers 5-10 in dense subcapitate racemes; Corolla salver-shaped, 5-lobed, white, pink or yellow, glabrous outside, densely white pilose in tube.

   Flowering and Fruiting: April to June.

   Uses: Leaves of this plant are mixed with Juniper to provide incense that is widely used in Buddhist & Tibetans monasteries.

   Habitat range: It grows in open rocky situations and found at an elevation of 2800-5500m. In the Sikkim region, it is found at North district: Lachung, Yomesamdong, Lachen, Zema, Lohnak valley, Thila, Zakophyak, Katao, Yumthang, Tembwa chu, Thangu, Gochung, Momesamdong; East district: Kupup, Thegu, Manju lake, Changu, Gnathang, Tamse, Maryulake, Lungthung; West district: Dzongri, Thangsing, Ghomney, Gomathang, Bakhim, Tsoka, Kosturi;

   Specimen examined: Sikkim, North district, Shingba Rhododendron Sanctuary, $27^0 46^! 16.23$N; $88^0 45^! 55.50$E, 02.05.2014, CS Purohit 38193.

2. ***Rhododendron arboreum*** Sm. subsp. ***arboreum*** (Local name: Laligurans, Gurass, Gurans).

   Taxonomic characters: Small bushy shrubs or tree; Leaves closely matted white or silvery tomentose beneath; Raceme compact, 7-12 cm, bears 7-12 flowers, flowers red; Corolla tubular-campanulate, red colour, dark red dots beneath, dark spots on base of inner side of petals.

Uses: According to common belief, a sip of the juice of the Laligurans flowers dissolves fish bones stuck in the throat. The close, hard-grained wood of this species is used for making "Khukri" handles and pack-saddles. During spring the flowers are harvested and fermented into local wines called GURASE by Sherpas. This wine is believed to be an antidote for altitude sickness. Wood makes good charcoal but cutting and burning trees of this species should be banned. The young leaves are said to be poisonous as well as medicinal and applied on the forehead to alleviate headaches (Watt, 1892). The young leaves are used in headache (Watt, 1892). Local women adorn their hair with the flowers and affectionately call *Lali-Gurans*. The flowers are considered sacred and offered in temples and monasteries.

Poultice of flowers is used in high fever: It is considered as a best remedy for nose bleeding in hills; also used with cow's ghee for blood dysentery. In Homeopathic Materia Medica, the tincture of dried leaves has been used in gout and rheumatism (Skidel, 1980). Ayurvedic preparation "Asoka Aristha," containing *R. arboreum* possesses oxytocic, estrogenic, and prostaglandin synthetase-inhibiting activity (Midlekoop & Labadie, 1983). The fresh and dried corolla is given when fish bones get struck in the gullet (Pradhan & Lachungpa, 1990). Flowers & leaves are fitted in long ropes made of *Saccharum munja* grass and tied around the houses including temples as decorations (Chauhan, 1999). The flowers of *R. arboreum* are used in squash, jams, jellies and local brew. It is a very common and pleasant drink, drunk once daily as refreshing appetizer and also to prevent high altitude sickness. Fresh petals are used to prepare chutney known as barah ki chutney. The juice of the leaves is spread over cots and beds to get rid of bed lice. Wood is used to make charcoal & fuel. The grained wood is used for making 'khukri' handles, packsaddles, gift-boxes, gunstocks and posts (Paul *et al.*, 2005). The dried flowers are used in diarrhoea and blood dysentery (Laloo *et al.*, 2006). Flowers have also been reported for anti-inflammatory and cholinergic activity (Verma *et al.*, 2010).

Habitat range: The plant is a common sight at lower alpine region. In the Sikkim region, it is found at North district: Yumthang, Lachung, Kabi Sacreb Grooves, Thapale, Bey, Chungthang, Bichhu; East district: Pangolakha, Kupup, Rachela,

Karponang, Changu, Lungthang, Rongli, Gnathang; West district: Varsey, Hilley, Kalijhar, Varsey Rhododendron Sanctuary, Geyzing, Pelling, Bakhim, Tsoka; South district: Tendong Reserve forest, Rabong Reserve forest.

Specimen examined: Sikkim, North district, Shingba Rhododendron Sanctuary, $27^{0}43^{'}46.31$N; $88^{0}43^{'}7.10$E, 25.04.2014, CS Purohit 38096.

3. ***Rhododendron arboreum*** Sm. subsp. ***cinnamomeum*** (G. Don) Tagg.

Taxonomic characters: It is similar to subsp. *arboreum* but differs in- Leaves brown-tomentose beneath and rounded or subcordate at the base; occurs above 2700m; flowers variable, rose to white.

Uses: Fresh leaves in combination with Thuja/Juniper/pine leaves are burnt for making smoke that is believed to be sacred and help in purifying surrounding air. Flowers are sweet in taste and are eaten. Wood from large trees is used for making 'kukri' handles, boxes, spoons etc., and as fuel.

Habitat range: The plant is common sight at lower alpine region. In the Sikkim region, it is found at North district: Shingba Rhododendron Sanctuary.

Specimen examined: Sikkim, North Sikkim, Shingba Rhododendron Sanctuary, N $27^{0}45^{'}43.85$; E $88^{0}41^{'}1.47$, 26.04.2014, CS Purohit 38123.

4. ***Rhododendron barbatum*** Wall. ex G. Don

Taxonomic characters: Small tree; young shoot long stiff bristly, covered by bud scales; Bark smooth, purplish red, peeling; Flowers deep red.

Uses: Main stem and branches are used in fuel wood. Its leaves are used as fish poison. Leaves and flowers contain a toxic substance, andromedotoxin. Honey from the flowers is toxic as it contains andromedotoxin. Flowers yield an essential oil (CSIR, 1986).

Habitat range: In the Sikkim region, it is found at West district: Bakhim, Tsoka, Dzongri; South district: Barneli, Maenam Wild Life Sanctuary.

Specimen examined: Sikkim, North district, Yakla, 14 Oct. 1869, CB Clarke 9817.

5. ***Rhododendron cinnabarinum*** Hook.f.; (Local name: Sanu Chimal, Sano Chimal);

   Taxonomic characters: Bushy shrubs; Racemes terminal, compact, 2-6 flowered; Petal- reddish orange, fleshy, tubular-campanulate or campanulate, 3-4cm x 4-5cm; style- green with pinkish tine, hairy on base.

   Uses: The leaves and pollen are poisonous to grazing animals. Corolla is eaten by the local children as it is sour-sweet in taste. Sain (1974) reported that the corolla is used for making jams by the head Lamas and Tibetan aristocrats. In Lachen and Lachung, the local people also fry the corolla to a tasty delicacy. When leaves and wood are employed as fuel, the smoke causes inflammation of the eyes and face. Plant contains andromedotoxin, a toxic compound. Flowers used form preparing a jam, but honey produced from them is poisonous (CSIR, 1986).

   Habitat range: In the Sikkim region, it is found at North district: Shingba Rhododendron Sanctuary, Yumthang, Samthang, Tallam; East district: Kyangnosla Alpine Sanctuary, Karponang, Changu, Kupup; West district: Bakhim, Dzongri, Kalijhar, Varsey, Gurasey dora, Phokte danra, Chewabhanjyang, Tsoka.

   Specimen examined: Sikkim, North district, Shingba Rhododendron Sanctuary, $27^0 45' 43.85$N; $88^0 41' 1.47$E, 26.04.2014, CS Purohit 38121.

6. ***Rhododendron campanulatum*** D. Don; (Local name: Nilo Chimal).

   Taxonomic characters: Large shrubs up to 1.5-4m or small tree to 6m; Racemes 8-15 flowered; Corolla open-campanulate, 2.5-3.5cm, 5-lobed, white or pink, with reddish spots and large blotch at base within.

   Uses: Watt (1892) reports the leaves are exported to the plains, where they are ground up with tobacco and used as snuff, which is said to be useful in cold and hermicrania. They are also said by Baden Powell to be used in chronic rheumatism, syphilis and sciatica. However, leaves are poisonous to livestock. Flowers yield an essential oil. The dried twigs and wood are use in Nepal as medicine in phthisis and chronic fever. The wood makes good fuel, but the smoke is acrid and causes irritation.

Habitat range: In the Sikkim region, it is found at North district: Hot spring water, Yumthang, Jakthang, Thila, Tholung, Kishong; East district: Tamse, Kupup, Mamaichu, Gnathang, Changu, Bitang chu, Karponang; West district: Dzongri, Thangshing, Prekchha river, Ghomney, Tsoka.

Specimen examined: Sikkim, North district- Shingba Rhododendron Sanctuary, 27^{0}49'10.10N; 88^{0}43'11.99E, 01.05.2014, CS Purohit 38186.

7. ***Rhododendron campylocarpum*** Hook.f.; (Local Name: Chimal)

Taxonomic characters: Aromatic shrub of 2-4m or small tree to 6m; Leaves ovate-elliptic, 4-9 x 2.3-5cm, whitish and glabrous beneath; Racemes 5-13 flowered; pedicels Corolla campanulate, 3-4.5cm, 5-lobed, pale-yellow, sometimes with a red basal blotch;

Uses: Used as fuel and widely used for hybridization.

Habitat range: In the Sikkim region, North district: Shingba Rhododendron Sanctuary, Yumthang, Lachung; East Sikkim: Kupup, Gnathang; West district: Phadung, Tsoka, Dzongri, Ghomney, Kasturi, Arralungshok.

Specimen examined: Sikkim, North district, Shingba Rhododendron Sanctuary, N 27^{0}43'36.6; E 88^{0}45'08.5, 29.04.2014, CS Purohit 38157.

8. ***Rhododendron dalhousiae*** Hook.f.; **subsp.** ***tashii*** Pradhan & Lachungpa; (Local Name: Chimal, Pahelochimal)

Taxonomic characters: Shrub up to 1-4m; Racemes 2-3 flowered; Calyx deeply divided into oblong obtuse lobes; Corolla 8-11cm, white or creamy, with or without a red line from base to apex of each lobe.

Uses: It is also called Lahare Chimal is dreaded by the locals as the leaves cause fatal poisoning in cattle. It could have medicinal properties.

Habitat range: In the Sikkim region, North district: Shingba Rhododendron Sanctuary, Lachung, Bitchu, Toona, Swana valley, Lachen, Chungthang, Ramthang, Tingda, Yumthang; East district: Kabi, Phademchen, Famgonglho Wild Life Sanctuary, Rakdong, Pangthang, Rachela, Tashi, Premlakha; West district: Okhrey, Hilley, Ngom, Gorkhey, Deonigala; South district: Ralong Reserve forest, Rabongla, Tendong Reserve forest, Maenam Wild Life Sanctuary, Damthang, Dhatray Reserve forest.

Specimen examined: Sikkim, North district, Shingba Rhododendron Sanctuary, 23-05-2011, Sachin Punekar 35909 (Det.- CS Purohit).

9. ***Rhododendron edgeworthii*** Hook.f.;

Taxonomic characters: Epiphytic, pendulous shrubs, branches straggling; Leaves densely ferruginous tomentose beneath, the hairs completely obscuring the scales; inflorescence 2-4 flowered, white to flushed white with a pale yellow blotch.

Uses: The aromatic smell of mature flower is used for immediate relief in nose block due to cold. Leaves after distillation are used to cure skin diseases.

Habitat range: Sikkim-Yuksum, Bakhim.

Specimen examined: Sikkim: West district, above, Yaksum, on way to Bakhim, 17.05.2002, D. Bhattacharyya 32205.

10. ***Rhododendron falconeri*** Hook.f.; (Local name: Kurling, Korlinga)

Taxonomic characters: Large shrub or tree, 5-15m; bark smooth, branchlets thinly brown tomentose; Racemes dense, subglobose, 12-16cm diameter, 15-20-flowered; Corolla fleshy, obliquely campanulate, 4-5cm, 8-10 lobed, white, creamy or yellow with purple blotch at base, rarely tinged pinkish.

Uses: The rough leaves are used for packaging apples by the people of north Sikkim. It is poisonous to the fish. Leaves are employed as platters and for lining baskets. Wood used for making cups, spoons, ladles and yak-saddles.

Habitat range: In the Sikkim region, North district: Shingba Rhododendron Sanctuary, Lachen; West district: Phakote, Varsey Rhododendron Sanctuary, Hilley, Chewabhanjyang, Phedi; South district: Barneli, Maenam Wildlife Sanctuary.

Specimen examined: Sikkim, North district, Shingba Rhododendron Sanctuary, 18-05-2011, Sachin Punekar 35641 (Det.- CS Purohit).

11. ***Rhododendron fulgens*** Hook.f.

Taxonomic characters: Small tree up to 5m tall; stem: peeling bark with bright, glabrous young shoot; Racemes dense, 10-15 flowers, deep blood-red; pedicels 5-10mm, glabrous; Petal-5, fused, tubular-companulate, 3-4mm x 4-6mm, 5-lobed, lobes rounded, scarlet or blood-red.

Uses: The dense tomentum on the underside of the leaves is scraped and used as wick for lighting fires and as fuel by the inhabitant of north Sikkim.

Habitat range: In the Sikkim region, North district: Shingba Rhododendron Sanctuary.

Specimen examined: Sikkim, North district, Shingba Rhododendron Sanctuary, N $27^{0}43'54.42$; E $88^{0}45'14.64$, 27.04.2014, CS Purohit 38130.

12. ***Rhododendron griffithianum*** Wight; (local Name: Gothale Chimal, Seto Chimal).

Taxonomic characters: Shrubs or tree, 2-10m; bark grey, papery; shoots glabrous; Racemes 3-5 flowered, flowers fragrant; Corolla broadly funnel-shaped, 6-8cm long, 5-lobed, pure white or tinged pinkish.

Uses: Widely used for hybridization.

Habitat range: In the Sikkim region, North district: Lachung, Tholung; West district: Hilley, Ribdi, Varsey, Chewabhanjyang, Okhrey, Kheriperi, Chyjo lake, Bakhim, Tsoka.

Specimen examined: Sikkim, North district, Lachung towards Gompa, 30.04.1955, RS Rao 256, Acc. No. (BSHC)- 14581.

13. ***Rhododendron grande*** Wight.

Taxonomic characters: Tall tree; Bark dark and hard, branches many; petioles very thick, margin recurved; Flowers in head; Flower terminal white; Calyx very small; Corolla large.

Uses: Leaves are used for packing ghee, churpi and butter for transportation. Wood is used as fuel.

Habitat range: In the Sikkim region, North district: Kabi Sacred Grooves, Darmchen, Thaple, Tholung; East district: Lungthung, Rongli, Gnathang; West district: Tsoka, Yoksum, Bakhim; South district: Damthang, Tendong Reserve forest.

Specimen examined: Sikkim, East district, Lungthung, Rongli_Gnathang road, 17.03.1986, DCS Raju & S Singh 5240.

14. ***Rhododendron hodgsonii*** Hook.f.

Taxonomic characters: A small basally branched spreading tree, 3-7 m tall; Bark smooth, pinkish-brown, peeling off in broad membranous flakes; Inflorescence compact rounded leaf, 10-15cm, 10-20 flowers; Petals- 3-4cm, fleshy, tubular-campanulate, 7-8 lobed, pinkish-red or purple.

Uses: The *Bhutias, Lepchas* and *Nepali* use the hard wood to make cup, spoon and ladder and also *khukri* handles. The hard, close-grained and smooth wood makes excellent walking sticks. The thick leaves with glossy surface are used for packing apples and for carrying mashed pulp of *Arisaema griffithii* for bread. Yak butter and cheese are packaged in the attractive foliage for serving and transportation.

Habitat range: In the Sikkim region, North district: Shingba Rhododendron, Yumthang, Momesamdong, Lachen, Kishong, Tholung; East district: Kupup, Mamaichu, Karponang, Changu, Jeelapla; West district: Bakhim, Dzongri, Dentam Varsey, Tsoka.

Specimen examined: Sikkim, North district, Shingba Rhododendron Sanctuary, N $27^0 43^' 54.42$; E $88^0 45^' 14.64$, 27.04.2014, CS Purohit 38133.

15. ***Rhododendron lepidotum*** Wallich ex G. Don; (Local name: Sunpati, Balu sunpate, Bhale sunpate).

Taxonomic characters: Aromatic rounded or mat-forming subshrub 15-60cm; Flowers 1-2, terminal; Corolla campanulate, 1-1.5cm, red, pink, purple, white or yellow, often spotted, scaly but not pubescent outside.

Uses: Leaves are stimulant; yield a volatile oil, used in perfumes and incenses.

Habitat range: In the Sikkim region, North district: Thila, Kishong, Lachung, Katao, Thangu, Gochung, Yumthang, Lachen, Yakchey, Zema, Momesamdong, Chungthang, Muguthang, Logbridge, Jakthang; East district: Suguchu, Kupup, Changu, Lukrip, Kyangnosla Alpine Sanctuary, Zuluk; West district: Dzongri, Bakhim, Kalijhar, Varsey, Tsoka, Bikh Bari, Jorepokhari, Yambung, Ghomney.

Specimen examined: Sikkim, North district, Shingba Rhododendron Sanctuary, N 27^{0}47'24.4; E 88^{0}42'29.3, 14.09.2014, CS Purohit 38283.

16. ***Rhododendron lepidotum*** Wall. ex G. Don var. ***album*** Davidian.

    Taxonomic characters: It is similar to var. *lepidotum* but differs in having white colour flowers with dots on petals.

    Uses: Leaves stimulant, yield a volatile oil used in perfumes and incenses.

    Habitat range: In the Sikkim region, North district: Shingba Rhododendron Sanctuary.

    Specimen examined: Sikkim, North district, Shingba Rhododendron Sanctuary, N 27^{0}43'49.01; E 88^{0}44'52.34, 16.09.2014, CS Purohit 38334.

17. ***Rhododendron nivale*** Hook.f.

    Taxonomic characters: Shrub, much branched, compact, prostrate, cushion like, aromatic plant. Leaves brownish green; Raceme 1-2 flowered; Flower pink to deep purple.

    Uses: The tiny leaves of this species emit fragrance that is comparable to *Eau de Cologne* as mentioned by Sir Joseph Hooker.

    Habitat range: Sikkim, Dzongri,.

    Specimen examined: Sikkim, West district, above Dzongri, 4800m, 18.05.2002, D Bhattacharyya 32235.

18. ***Rhododendron niveum*** Hook.f. (Local name: Hiun-pate Gurans)

    Taxonomic characters: Shrub 1-2m; young shoot puberulous to tomentose; Raceme compact, 10-20 flowers, up to 6 cm long, purple color; Petal- 5, tubular-campanulate, 3-5cm x 7-10cm, mauve or purple colour.

    Uses: Used as fuel.

    Habitat range: In the Sikkim, North district: Shingba Rhododendron Sanctuary, Lachung, Yumthang; East district: Kyangnosla Alpine Sanctuary.

    Specimen examined: Sikkim, North district, Shingba Rhododendron Sanctuary, N 27^{0}43'49.01; E 88^{0}44'52.34, 24.04.2014, CS Purohit 38082.

19. ***Rhododendron setosum*** D. Don (local Name: Tsallu-Gurans in Nepali)

 Taxonomic characters: An aromatic shrubs, 20-50 cm; shoots densely scaly and spreading-bristly; Racemes dense, 2-4 flowered; Calyx deeply 5-lobed, sometimes ciliate; Corolla funnel-shaped, 13-15mm, pale to deep purple, deeply 5-lobed, glabrous outside.

 Uses: The leaves emit a strong heady aroma that causes painful headaches at high altitudes. The leaves yield aromatic oils which can be used in perfumery and cosmetics.

 Habitat range: In the Sikkim, North district: Shingba Rhododendron Sanctuary, Yumthang, Momesamdong, Namthang, Thangu, Gochung, Muguthang, Shivmandir, Katao, Lohnak valley; East district: Kupup, Fambanglho Wildlife Sanctuary; West district: Jorephokhari, Dzongri, Tsoka, Thangshing.

 Specimen examined: Sikkim, North district, Shingba Rhododendron Sanctuary, N $27^0 43' 36.6$; E $88^0 45' 08.5$, 19.07.2013, CS Purohit 37323.

20. ***Rhododendron thomsonii*** Hook.f. subsp. ***thomsonii*** (Bat.) H. Hara.

 Taxonomic characters: Shrub up to 3m tall; stem: Raceme with fewer, 3-8 flowers, deep blood-red; pedicels longer, 1.2-2cm; Calyx larger, red, cup-shaped, with short, shallowly lobed up to 6-15mm; Petal-5, fused, tubular-companulate, 3.5-5mm x 4-6mm, 5-lobed, lobes rounded, deep crimson.

 Uses: Boiled extract of the plant is used as natural insecticide in the Lachen and Lachung villages of north-east Sikkim. Branches and leaves contain poisonous compound, acetylandromedal which probably imparts insecticidal property to plant extract.

 Habitat range: In the Sikkim region, North district: Shingba Rhododendron Sanctuary, Jakthang, Phongi; East district: Lungthung, Rongli, Gnathang, Kupup, Bitonchu lake; West district: Phethang, Ghomney; South district: Tendong Reserve forest.

 Specimen examined: Sikkim, North district, Shingba Rhododendron Sanctuary, N $27^0 48' 55.30$; E $88^0 43' 47.89$, 30.04.2014, CS Purohit 38172.

Conclusion

Rhododendrons act as keystone species in the high altitudinal of the Eastern Himalayan region. The subalpine to alpine transition zone that includes the timberline is the most fragile ecosystem in the Sikkim Himalaya. The Himalayan ecosystem has been greatly affected due to various threats posed by nature as well as by human interferences. Due to the various anthropogenic pressures, 43 species are becoming endangered, rare and threatened (Sastry and Hajra, 1983). The major threats to Rhododendrons are deforestation and unsustainable extraction for firewood. While this is of concern to conservationists, there is requirement of study of its population in natural habitat and counteract the trend by encouraging cultivation at higher altitudes.

Rhododendrons species of Sikkim Himalayas attracts botanists and horticulturist, urging them to unleash new potential as ornamental, avenue plant and dimensions in medical treatment. The detailed survey of literature revealed that most of the species of genus *Rhododendron* have importance as medicine and in commercial uses. The plants exhibited anti-inflammatory, hepato-protective, anti-diarrhoeal, anti-diabetic, anti-oxidant properties due to presence of flavonoids, saponins, tannins and other phyto-chemicals. Because of their over exploitation for fuel wood, incense, medicinal and other economic uses, their population continues to decline in natural habitat. There is requirement of study of *Rhododendron* population in natural habitat before planning any strategy for conservation and further use. To counteract the trend of harvesting the plant form natural environment and for their sustainability it is very important to develop agro-techniques for mass multiplication, propagation and cultivation. This will create source of livelihood for the local people with simultaneously maintain the plant in natural environments. This will prevent their depletion form wild and also be very productive for conservation in the long run.

References

Calder C., Narayanaswami and Ramaswami M.S. 1926. List of species and genera of India phanerogams not included in Sir, J.D. Hooker's Flora of British India. *Record of Botanical Survey of India* 11(1): 1-157.

Chamberlain D.F. 1982. A revision of *Rhododendron* I. Subgenus *Rhododendron*, Sections *Rhododendron* and *Pogonanthum*. *Notes Royal Botanic Garden Edinburgh* 39(2): 209-486.

Chauhan A.S., Pradhan N. and Maity D. 2003. Plant Diversity of the Kanchendzonga Biosphre Reserve in Sikkim Himalaya, *Hymalayan Biosphere Reserve* Vol. 5(1&2).

Chauhan N.S. 1999. *Medicinal and aromatic plants of Himachal Pradesh.* Indus Publishing Company, New Delhi 353.

Clarke, Harold J. 1982. Getting started with Rhododendrons and Azaleas. Portland, Oregon: The American Horticultural Society.

CSIR 1986. The useful plants of India, Publications & Information Directorate, CSIR, New Delhi.

Cullen J 1980. A revision of *Rhododendron* II. Subgenus *Hymenanthes. Notes Royal Botanic Garden Edinburgh* 39(1): 1-207.

Dash S.S. and Chauhan A.S. 2002. Kabi Sacred Grove in Sikkim, Relevance to Conservation, Perspectives of Biodiversity (Eds. Das AP): 713- 322.

Dash S.S. and Singh P. 2011. Trees of Sikkim (in Biodiversity of Sikkim Exploring and Conserving a Global Hot Spots Eds.: ML Arrawatia and Sandeep Tambe), Govt. Sikkim Publication. 89-124.

Ghosh R.B. and Samaddar U.P. 1989. The rhododendrons of the north-east India. *Journal of Economic and Taxonomic Botany* 13(1): 205-220.

Hariharan and Rangaswami S. 1966. Chemical investigation of the bark of *Rhododendron arboreum* Sm. *Current Science* 35: 390-391.

Hooker J.D. 1849. Notes, chiefly botanical, made during an excursion from Darjeeling to Tonglu. *Journal Asiatic Society Bengal* 18: 419-446.

Hora B. 1981, *The Oxford Encyclopedia of Trees of the World*, Oxford University Press, Conscent Books, New York, p. 288.

Kataki S.K. 1983. Some rare plants in Khasi and Jaitia Hills of Meghalaya. In: *An Assessment of Threatened Plants of India* (S.K. Jain and R.R. Rao, eds), pp. 146-150. Botanical Survey of India, Kolkata, India.

Korner C. and Larcher W. 1988. Plant life in cold climates. *In* Plants and Temperature. Eds. S. Long and F.I. Woodward. Company of Biologists Ltd., Cambridge, pp. 25-57.

Laloo R.C., Kharlukhi L., Jeeva S. and Mishra B.P. 2006. Status of medicinal plants in the disturbed and the undisturbed sacred forests of Meghalaya, northeast India: population structure and regeneration efficacy of some important species. *Current Science* 90(2): 225-232.

Mabberley D.J. 2008. Mabberley's Plant-Book. A Portable Dictionary of Plants, Their Classifications and Uses," 2nd Edition, Cambridge University Press, Cambridge.

Maitii D. and Chauhan A.S. 2001. Assessment of plant Diversity of Kangchendzonga Biosphere Reserve in Sikkim. In (Eds.: Sharma JK, Easa PS, Mohahnan C, Sasidharan N, and Rai RK) Biosphere Reserves in India and their Management, Proceeding of the Review meeting Biosphere and their Management 8-11 Sept. 2000 Peechi Kerela.

Mao A.A. 2010. The genus Rhododendron in north-east India. Journal of *Plant Science,* 7: 26–34.

Mao A.A., Singh K.P. and Hajra P.K. 2001. Rhododendron. In: *Floristic Diversity and Conservation Strategies in India,* (Eds.: Singh KP and Singh NP), 5: 2167-2202. Botanical Survey of India, Kolkata, India.

Midlekoop T.B. and Labadie R.P. 1983. Evaluation of 'Asoka Aristha' an indigenous medicine in Sri Lanka. *J Ethnopharmacol.* 8: 13-20.

Naithani H.B. 1990. *Flowering Plants of India, Nepal and Bhutan* (not recorded in Sir J.D. Hooker's Flora of British India). Surya Publications, Dehra Dun, India.

Naithani H.B. and Bahadur K.N. 1983, New and rare taxa of rhododendron from Arunachal Pradesh. In *An Assessment of Threatened Plants of India* (eds Jain, S. K. and Rao, R. R.), BSI, Kolkata, pp. 116–126.

Nayar M.P. and Karthikeyan S. 1981. Fourth list of species of genera of India Phanerogams not included in JD Hooker's Flora of British India. *Record of Botanical Survey of India* 21(2): 131-152.

Nayar M.P. and Ramamurthy K. 1973. Third list of species of genera of India Phanerogams not included in JD Hooker's Flora of British India. *Record of Botanical Survey of India* 15: 204-234.

Paul A., Khan M.L., Arunachalam A. and Arunachalam K. 2005. Biodiversity and conservation of Rhododendrons in Arunachal Pradesh in the Indo-Burma biodiversity hotspot. *Current Science* 89(4): 623-634.

Pradhan K.C. 2010. *The Rhododendrons of Sikkim.* Sikkim Adventure, Sikkim, Nepal.

Pradhan U.C. 1985. A preliminary enumeration of rhododendrons of the Indian Region. Part 1. *Himalayan Plant Journal* 3(8): 123.

Pradhan U.C. 1986. A preliminary enumeration of rhododendrons of the Indian Region. Part 2. *Himalayan Plant Journal* 4(11-12): 73-76.

Pradhan U.C. and Lachungpa S.T. 1990. *Sikkim-Himalayan Rhododendrons.* Primulaceae Books, Kalimpong, West Bengal.

Purohit C.S. 2014. *Rhododendron arboreum* Sm. An economically important tree of Sikkim, Popular Kheti 2(3): 193-198.

Purohit C.S., Agrawala D.K. and Das K. 2014. Ethno-Botanical Plants of Sikkim and their Status- A Review (Eds: Sanjeev Kumar), Deep Publication, 256-277.

Purohit C.S. 2015. Flora of Shingba Rhododendron Sanctuary with GIS mapping of EET species. Project report submitted to BSI headquarter, Kolkata.

Rai T. and Rai L. 1994. *Tress of the Sikkim Himalaya.* Indus Publishing Company, New Delhi p. 94.

Razi B.A. 1959. A second list of species and genera of Indian Phanerogams not included in J.D. Hooker's Flora of British India. *Record of Botanical Survey of India* 18: 1-56.

Sain, M. 1974. *Rhododendrons of Darjeeling and Sikkim Himalaya.* Seattle.

Sastry A.R.K. and Hajra P.K. 1983. Rare and endemic species of rhododendron in India a preliminary study. In: *An Assessment of Threatened Plants of India* (S.K. Jain and R.R. Rao, eds), pp 222-231. Botanical Survey of India, Kolkata, India.

Sastry A.R.K. and Hajra P.K. 2010. *Rhododendrons in India: Floral and Foliar Splendour of the Himalayan Flower.* BSI Publications, Hyderabad, India.

Sekar K. and Srivastava S. 2010. Rhododendrons in Indian Himalayan Region: Diversity and Conservation, *American Journal of Plant Sciences,* 1(2): 131-137.

Salley H.E. and Greer H.E. 1986. Rhododendron hybrids: A guide to their origins. Timber Press, Portland, Ore.

Singh K.K. 2009. Notes on the Sikkim Himalayan rhododendrons: a taxa of great conservation importance. *Turkish Journal of Botany* 33(4): 305-310.

Singh K.K., Kumar S., Rai L.K. and Krishna A.P. 2003. Rhododendron Conservation in Sikkim Himalaya. *Current Science,* 85(5): 602-606.

Singh P. and Chauhan A.S. 2000. Sikkim, Floristic diversity and conservation strategies in India BSI, Howrah, 1419-1450.

Singh P. and Chauhan A.S. 1997. Plant Diversity in Sikkim Himalaya. in: Hajra PK and V Mudgal (ed.), Plant diversity Hotspots of India, An Overview. pp. 137-158. Botanical Survey of India, Kolkata.

Skidel 1980. *Rhododendron.* Text Book of Materia Medica. Sree Bharati Press, Calcutta, p. 540.

Tiwari O.N. and Chauhan U.K. 2006. Rhododendron conservation in Sikkim Himalaya. *Current Science* 90(4): 532-541.

Verma N., Singh A.P., Amresh G., Sahu P.K. and Rao C.V. 2010. Anti-inflammatory and nociceptive activity of Rhododendron arboreum. J. Pharm Res. 3: 1376-1380.

Watt G. 1892. *A Dictionary of the economic products of India.* Supt. of Govt. Prtg, Harvard University, 492-495.

❑❑❑

Chapter 19

Ethno-Conservation of A Sacred Grove Through Tribal Deities

Ram Kumar Bhakat

Department of Botany and Forestry, Vidyasagar University, Midnapore-721102, WB, E-mail : rkbhakat@rediffmail.com/bhakatramkumar@gmail.com

ABSTRACT

Sacred groves are forest patches preserved on religious grounds. This paper highlights the conservation of a sacred grove dedicated to some tribal deities of *Santals* and *Lodhas* of West Midnapore district. The study records 30 species of plants protected in the sacred grove.

Keywords: Sacred grove, tribals and conservation.

Introduction

India is a land of diverse natural resources. It is also a country with the strongest traditions of nature conservation anywhere in the world. It is true that India has suffered an almost unabated devastation of its natural biological heritage, and much of what remains has been preserved through the ages because of a wealth of conservation-oriented cultural and religious traditions. One such significant tradition of nature conservation is that of dedicating patches of forests to some deity by the tribal people.

These pockets of undisturbed natural forests preserved on religious grounds by the local tribes are known as *sacred groves*. These groves represent near-virgin vegetarian preserved in *in situ* form without any outside interference and are indicative of what forest wealth the country once harboured. All forms of life in such a grove are under the protection of the reigning deity of that grove, and the removal of even dead wood is taboo. This phenomenon of grove and forest preservation is quite distinct from the protection of individual and isolated plants and trees like *Tulsi* (*Ocimum sanctum*), *Aswatha* (*Ficus religiosa*), *Bel* (*Aegle marmelos*), *Khejri* (*Prosopis cineraria*) etc. (Gadgil and Vartak, 1975; Bhakat, 2003, 2009, 2011; Bhakat and Sen, 2015).

With this background, this paper is an attempt to study the various ecological and socio-cultural perspectives of Bagdubi sacred grove jointly maintained by *Lodhas* and *Santals* of West Midnapore district in West Bengal. The present study area is selected in view of the fact that these tribals have existing conservation ethos with totemic plants, groves and forests (Hembram, 1988; Bhowmick, 1994; Deb and Malhotra, 1997; Pandit and Bhakat, 2007; Kisku *et al.*, 2009)

The Sacred grove

The Bagdubi sacred grove, named after it's nearby village of the same name, is located 5-km away from Midnapore town along the east-west running Midnapore-Dherua road under Kotwali police station. The grove is spread over a small piece of 0.6 hectare land and represents a 50-year old near-climax forest patch consisting mainly of few old deciduous and evergreen trees (Fig. 1). It stands as an isolated forest island amidst the otherwise paddy field, and is owned by the nearby village communities. *Baram*, *Rankini* and *Shitala* are the three presiding folk deities of the grove. They are anionic, represented by stones smeared with vermillion lying under tall trees. The deities are often presented votive offereing of burnt clay idols of horses and elephants. They demand animal sacrifices (goat, hen and pig) and are supposed to be very ferocious in nature. During the annual *paus sankranti* (a ritual celebrated on the last day of Bengali month *Paus*

Fig. 1. The Bagdubi sacred grove.

(middle of January), local people mainly *Lodhas* and *Santals* gather inside the grove and worship the deities. Additionally, *Sitala* is also worshipped during *chaitra sankranti* (middle of April) annually. Moreover, the deities are also propitiated on every Tuesday and Saturday or any day deemed auspicious by the local people. It is believed that the feminity of the deities is indicative of the cults' origin in the hunting-gathering stage when the society which awed by the miracle of birth was not yet settled in permanent dwelling (Kosambi, 1962). This can also be supported by the fact that the place of worship (tree grove) is away from the village. Since the grove is abode of deities, people do not disturb the area (neither destroy any plant nor remove any plant litter), thus strictly adhering to the taboos and ethics.

Methodology

In the course of investigation for one year, the entire sacred grove in point and the adjoining villages (Bagdubi, Kankabati and Lodhasai) were surveyed during different seasons. Identification of known plants was done on the "spot identification" method. For unknown plants, herbarium sheets of collected specimens were prepared following the standard technique and later identified in consultation with relevant literature. Social and economic values of plants were studied through PRA (Participatory Rural Appraisal) method during which local villagers were interviewed and cross-interviewed. The results thus generated were cross-checked with the relevant available literature (Dhiman, 2003; Pakrashi and Mukhopadhyay, 2004; Paria, 2005).

Plant conservation

Owing to continued protection offered on socio-religious grounds, the Bagdubi sacred grove provides optimum conditions congenial for plant growth and survival. The present study reveals that the grove supports thirty (30) species of plants covering herbs, shrubs, trees and climbers. And among these, 27 species have medicinal value, 5 species are timber yielding, 4 have sacred value and 1 species bears edible fruit (Table 1). Since the sacred grove is located in the tribal zone of West Bengal, a good number of tribals including *Lodhas* and *Santals* depend on the surrounding forest-based plants for home remedies. Out of the 27 medicinal special species of the grove, most of them are also ethnomedicinally significant. The local people normally do not harvest economic plants, but do so only after social sanction from the stakeholders of the grove and that too in a restricted way. Due to high-level of protection, the grove acts as a sanctuary and nursery for a good number of regionally vanishing medicinal plants species. This kind of traditional conservation of disappearing taxa of forest-based medicinal plants in sacred groves of Midnapore district is earlied reported by Bhakat and Pandit (2008), Bhakat and Sen (2008), and Bhakat *et al.* (2010).

Threats

Sacred groves are under various threats everywhere. The Bagdubi sacred grove is no exception. Though it is fairly well-protected by the local villagers, the grove faces mild disturbances from grazing animals and exotic weed invasion. Moreover, erosion of people's traditional ethics along with lack of awareness towards plants and environment is a recent threat to the grove.

Conclusion

The present sacred grove, although small in size, in essence represents the concept of the traditional Indian way of *in-situ* conservation of plants and forests. It acts as an indicator of rich vegetation that had existed in the past, but has now been replaced by crop fields and plantations. The sacred grove also provides social space wherein tribals renew their cultural identities and find community solidarity. Therefore, there is an urgent need not only to protect the grove, but also to study similar other sacred groves which play important roles as socio-cultural institutions and storehouses of germplasms.

Table 1. Plants growing in the Bagdubi sacred grove
(H: Herb, S: Shrub, T: Tree, C: Climber)

S.No.	*Species*	*Habit*	*Local Name*	*Importance*
1	*Abrus precatorius*	C	Kunch	Medicinal
2	*Abutilon indicum*	H	Patari	Medicinal
3	*Achyranthes aspera*	H	Apang	Medicinal
4	*Aegle marmelos*	T	Bel	Medicinal Edible (Fruit), Sacred
5	*Alangium salvifolum*	T	Ankar	Medicinal, Timber
6	*Albizzia lebbeck*	T	Siris	Timber
7	*Andrographis paniculata*	H	Kalmegh	Medicinal

8	*Asparagus recemosus*	C	Satamuli	Medicinal
9	*Azadirachta indica*	T	Neem	Medicinal, Timber
10	*Boerhaavia diffusa*	H	Punarnava	Medicinal
11	*Centella asiatica*	H	Thankuni	Medicinal
12	*Crataeva nurvala*	T	Barun	Sacred
13	*Curculigo orchioides*	H	Talmuli	Medicinal
14	*Cyanodon dactylon*	H	Durba	Medicinal, Sacred
15	*Eclipta prostate*	H	Keshut	Medicinal
16	*Euphorbia hirta*	H	"	Medicinal
17	*Evolvulus alsinoides*	H	"	Medicinal
18	*Gymnema sylvestre*	C	Gurmar	Medicinal
19	*Hemidesmus indicus*	C	Anantamul	Medicinal
20	*Holarrhena antidysenterica*	T	Kurchi	Medicinal
21	*Justicia adhatoda*	S	Basak	Medicinal
22	*Phyllanthus amarus*	H	Bhuiamla	Medicinal
23	*Rauvolfia tetraphylla*	H	Barachan-drika	Medicinal
24	*Shorea robusta*	T	Sal	Timber, Sacred
25	*Solanum nigrum*	H	Kakmachhi	Medicinal
26	*Streblus asper*	T	Sheora	Medicinal
27	*Strychnos nux-vomica*	T	Kuchila	Medicinal
28	*Terminalia arjuna*	T	Arjun	Medicinal, Timber
29	*Tridax procumbens*	H	Tridakha	Medicinal
30	*Vitex negundo*	S	Nishinda	Medicinal

References

Bhakat, R.K., 2003. Socio-religious and ecological perspective of a sacred grove from Midnapore District, *Science and Culture* 69: 371-374.

Bhakat, R.K., 2009. Chilkigarh Kanak Durga Sacred Grove, West Bengal, *Current Science* 96: 185.

Bhakat, R.K., 2011. Tribal culture and plant conservation – case study of village sacred grove. In: P. K. Bhowmik, S. Manna and S., Ghatak (eds.), *Tribal life* Institute of Social Research and Applied Anthropology, Bidisha, West Bengal, pp. 56-53.

Bhakat, R.K. and P.K. Pandit, 2008. Conservation of Vanishing Ethnomedicinal Plants of a tribal Area of Midnapore District through a Sacred Grove, In: A. K. Haler, P. K. Singh and P. C. Coomar (eds.),

Tribal Ethnomedicine: Problem and Prospect, Institute of Social Research and Applied Anthropology, Bidisha, West Bengal, pp. 123-128.

Bhakat, R.K. and U.K. Sen, 2008. Ethnomedicinal Plant Conservation through a Sacred Grove, *Tribes and Tribals* (Spl. Vol.) 2: 55-58.

Bhakat, R.K., U. K. Sen, 2015. Plant resources of Swarga Bauri - a sacred grove in West Midnapore district, *W.B. Journal of Bot. Soc. Bengal*, 69: 71-74.

Bhakat, R.K., S., Koley, N.S., Kundu and U.K. Sen, 2010. Culture and Conservation: Lessons from a Sacred Grove, In: A. K. Danda and I. Talwar (eds.), *Environmental Issues and Other Essays*, Indian National Confederation and Academy of Anthropologists, Jhargram and Indira Gandhi National Centre for the Arts, New Delhi, pp. 58-63.

Bhowmick, P.K., 1994. *The Lodhas of West Bengal*, Institute of Social Research and Applied Anthropology, Kolkata.

Deb, D. and K.C. Malhotra, 1997. Interface between biodiversity and tribal cultural heritage: an exploratory study, *Journal of Human Ecology*, 8: 157-163.

Dhiman, A.K., 2003. *Sacred Plants and their Medicinal uses*, Daya Publishing House, Delhi.

Gadgil, M. and V.D. Vartak, 1975. Sacred groves of India: a plea for continued conservation, *Journal of Bombay Natural History Society*, 72: 314-321.

Hembram, P.C. 1988. *Sari-Sarna (Santal Religion)*, Mittal Publications, Delhi.

Kisku, A.K., R.K., Bhakat and A., Guha, 2009. Culture and Conservation: Case Study of a Sacred Grove, *Science and Culture*, 75: 213-217.

Kosambi, D.D., 1962. *Myth and Reality*, Popular Book Depot, Bombay.

Malhotra, K.C.Y., Gokhale & S., Chatterjee, 2007. *Sacred Grove in India*, IGRMS, Bhopal and Aryan Books International, New Delhi.

Pakrashi, S.C. and S., Mukhopadhyay, 2004. *Medicinal and Aromatic plants of Red lateritic regions of West Bengal (Bakara, Medinipur and Purulia)*, W.B. Academy of Science and Technology, Kolkata.

Pandit, P.K. and R.K., Bhakat, 2007. Conservation of Biodiversity and Ethnic culture through sacred groves in Midnapore district, West Bengal, India, *Indian Forester*, 133: 323-344.

Paria, N. 2005. *Medicinal Plant Resources of South West Bengal*, Directorate of Forests, Govt. of West Bengal, Kolkata.

❑❑❑

Chapter 20

An Efficient Micropropagation Protocol for *Melissa Officinalis* and Assessment of Ursolic Acid in Regenerated Plants Using HPLC Analysis

M. Anuradha[1,2], S. Balasubramanya[1], S.R. Doreswamy[1] and M. Kumara Swamy[1,2*]

[1] RISHI Foundation, #234, 10th C main, 1st Block, Jayanagar, Bangalore- 560011

[2] Padmashree Institute of Management and Sciences, Kengeri, Bangalore -560060, India. * Corresponding author E-mail : swamy.bio@gmail.com

---

ABSTRACT

An efficient protocol for rapid multiplication of *Melissa officinalis* has been described in the present investigation. Axillary and terminal buds were inoculated on to MS medium supplemented with different concentrations and combinations of 6-benzylaminopurine(BA) and kinetin(KN). The best morphogenic response was observed on MS media supplemented with0.5mg/lBA and 0.5mg/lKN, which exhibited highest regeneration frequency(88%), maximum number of shoots /explants (60.1±1.40) and shoot length (3.3±1.21) within 4 weeks.Fortification with 1.0mg/lGA3 enhanced shoot elongation by 2.3 foldin 89% of shoot cultures within 2 weeks, whereas shoot elongation was completely absent in medium devoid of GA3.High percent frequency of rooting (92%) was achieved within 15 days of shoot implantation on ½ strength MS basal media fortified with 100 mg/l and 200mg/l activated charcoal. Rooted plantlets were successfully acclimatized with 95% survival rate. Cost of the media was reduced to 50% by using tap water, table sugar and exvitro rooting. Further, high performance liquid chromatography (HPLC) was carried out to further confirm the existence of qualitative and quantitative differences in the major secondary metabolite (ursolic acid) between the mother plant and in vitro propagated plants. The present results evidently showed comparable chemical profiles. Thus the present protocol can be used for clonal mass propagation of true-to-type elite *M. officinalis* plants.

Keywords: *Melissa officinalis*, ursolicacid, invitro propagation, HPLC, lemon balm.

Introduction

Melissa officinalis (Lemonbalm) belonging to the family Lamiaceae is an herbaceous, perennial plant, a native of the northern Mediterranean region has been used in a variety of practical applications in medical science.The aerial part of plant comprises 0.05 to 0.15% of volatile oil (that contains citronellal, citral, geraniol, linalool), polyphenols, tannins (3 to 6%),mucilages (12%), bitter substances (Tavares et al., 1996). Essential oil of lemon balm is used as an anti-tumeral agent (Turhan 2006). *M. officinalis* is used in folk medicine for nervous complaints, lower abdominal disorders and more recently, for treating Herpes Simplex lesions (Mazzanti *et al.*, 2008; Schnitzler *et al.*, 2008).

The leaves of *M. officinalis* are often used in herbal teas. *M. officinalis* contains some phenolic and flavonoid compounds such as rosmarinic acid (Herodez *et al.*, 2003). Rosmarinic acid and ursolic acid are the major constituents in the plant extracts of lemon balm. These constituents exhibit various pharmacological activities including prevention of oxidation of low density lipoprotein, inhibition of murine cell proliferative activity and of cyclooxygenase, and anti-allergic action. Rosmarinic acid found to have antibacterial, antiviral and antioxidative properties (Szabo *et al.*, 1999; Chen *et al.*, 1999; Hras *et al.*, 2000). Its activity especially against rheumatic and inflammatory conditions makes it a sought-after substance for use in phytotherapy (Pabsch *et al.*, 1991). Rosmarinic acid and the triterpenoids oleanolic acid and ursolic acid, inhibit gamma-aminobutyric acid transaminase (GABA-T) activity and also increase GABA levels in the brain (Awad *et al.*, 2007; Awad *et al.*, 2009).

M. officinalis is declared as herb of the year during 2007 by International Herb Association Horticulture Committee (www.iherb.org). This wonder herb is currently imported by Indian companies from Medittarian region. Owing to its commercial value, this plant is cultivated in high altitudes and in selected geographical locations in India. Flowering and seed set is not witnessed when grown under south Indian conditions. Hence, growers need to depend on rooting the vegetative cuttings. However, vegetative propagation through cuttings is hampered by variation of plants chemical constituents among population and also limited availability of cuttings from elite selected mother plants (Kumara swamy *et al.*, 2010; Swamy *et al.*, 2010a). Erratic rooting pattern, season dependency in establishment of plants adds to the problem. High humidity, shade and low temperature are required for rooting and development of planting stocks and can be considered as potent constraints for large scale cultivation in India. A successful in vitro regeneration protocol should be able to produce uniform plantlets which are genetically and biochemically uniform (Mohanty *et al.*, 2014; Kaushik *et al.*, 2015). Hence in the present study cost effective protocol was optimized to mass propagate true to type under in vitro conditions from selected high yielding lines of *M. officinalis*.

Though several authors have envisaged feasibility of mass propagation of Melissa through tissue culture (Gbolade and Lockwood 1992; Binder and Abou 2000; Meftahizade *et al.*, 2010; Tantos *et al.*, 1999; Tavares *et al.*, 1996), there are limited efforts to study direct organogenesis, which supports cultivation by providing true to type plants in large numbers. Hence, the present study was undertaken to establish a rapid and reliable cost effective regeneration protocol by direct organogenesis using nodal explants of *M. officinalis* under Indian climatic conditions for the first time as per the best of the published prior art. Also, the clonally propagated plants were analyzed by using HPLC analysis for confirming true to type nature of the plants.

Materials and Methods

Plant material

Nodal segments were procured from elite mother plants (Fig. 1) maintained at Rishi Herbal Garden, Bangalore, India.

Fig. 1. (A) Mother plants of *Melissa officinalis*. (B) Shoots formed from nodal segments. (C) Callus induced from nodal segments. (D) Shoot elongation.

In-vitro shoot multiplication

Leaves and petioles were excised and discarded. The explants were cut into 1-2 cm length and washed 3-4 times in tap water and treated with liquid detergent for 15 min followed by thorough washing under running tap water. These were then surface sterilized with 0.5% $HgCl_2$ for 10 min. Explants were rinsed five times with sterile distilled water to remove traces of $HgCl_2$ completely. Under aseptic conditions, explants were inoculated on to MS (Murashige andSkoog 1962) medium, containing 2% (w/v) sucrose, supplemented with different concentrations and combinations of BA (0.25, 0.5, 1.0 and 2.0 mg/L) and KN (0.25, 0.5 and 1.0, 2.0 mg/L) for shoot proliferation and multiplication. The pH of the medium was adjusted to 5.8 prior to the addition of 0.8% (W/V) agar and autoclaved at 121°C, 15 lb pressure for 15 min. All the cultures were incubated at 25 ± 2°C under a 16 h light and 8 h dark regimes with a light intensity of 3000 lux provided by cool-white fluorescent tubes. In vitro derived shoots were excised after 30 days and sub cultured on to fresh medium with the same concentrations of growth regulators unless otherwise mentioned. Control cultures were maintained by implanting explants on to MS basal medium. Data with respect to percent shoot multiplication, number of shoots/explant, and shoot length per culture was recorded after 30 days of subculture. Each treatment, including the control had a total of 20 explants. Cultures were incubated under the same growth conditions as stated above.

To derive cost effective multiplication MS media is prepared by modifying its components. The conventional carbon source i.e., sucrose was replaced by table sugar and distilled water was replaced by autoclaved tap water. The prevailing costs of carbon sources (sucrose and market sugar), water, activated charcoal and other components of medium (MSsalts, vitamins, growth regulators and ascorbic acid) at the time of conducting the experiment were taken to calculate cost of each medium. The cost was expressed in Indian rupees.

Shoot elongation

The explants with shoot clusters, produced after 4 weeks of culture on MS medium containing 0.5 mg/L BA and 1.0 mg/L KN, were transferred to MS medium supplemented with varied concentrations of gibberellic acid (GA3) (0.25, 0.5, 1.0, 1.5, and 2.0 mg/L) for 3 weeks to allow elongation of shoots. The mother explants were repeatedly sub cultured on to shoot multiplication medium after each harvest of the elongated shoots.

In-vitro Rooting

After multiplication, regenerated shoots were separated and transferred to rooting medium. The regenerated shoots (5-6cm) bearing at least 3-4 internodes were excised and cultured on to freshly prepared rooting half strength or full strength MS medium supplemented with different concentrations of activated charcoal (100 and 200 mg/L), IAA and NAA (0.5 and 1.0 mg/L). After 4 weeks of culture, the frequency of root formation, number of roots produced per cultured shoot and length of the root were recorded.

Acclimatization

Plantlets with well-formed roots were removed from culture medium, washed gently from running tap water and transferred to net pots containing sterile soil, sand, and manure (1:1:1).These plantlets were covered with a polythene cover ensuring high humidity (80%) and watered for every three to four days with quarter strength MS salt solution without sucrose and vitamins. After twenty days' polythene cover was removed and maintained in net pots, and subsequently transferred to field.

HPLC profiling and ursolic acid content

Mother plant material and tissue culture grown plant (whole plant) material of M. officinalis were collected, washed and dried at 55^0C in an air drier. Dried biomass was defatted with petroleum ether and filtered. The residual mass was then extracted repeatedly with methanol for five times and evaporated to dryness in a rotoflash evaporator and methanolic extract was dissolved in water. Aqueous methanolic extract thus obtained was then repeatedly extracted with chloroform for 4-5 times. The combined chloroform extract was evaporated to dryness using rotoflash evaporator. Quantitative HPLC was done out on water associate using Novapak C18 column (150nm) and KH_2PO_4: CH_3CN: MeOH (7: 2: 1) solvent system. The flow rate was adjusted to 1.2mL/min and the detector was set at 254 nm. Ursolic acid was quantified by standard calibration curves of authentic samples.

Results and Discussion

Shoot multiplication

The effect of different cytokinin types and concentrations on morphogenetic response of nodal explants after 30days of culture was explored and summarized in table 1. Direct organogenesis was exhibited by nodal explants when cultured on to MS medium supplemented with varied concentrations of BA and KN (0.25, 0.5 and 1.0 and 2.0 mg/L) separately or in combinations. Shoot regeneration frequency ranging from 18 to 88 % was obtained in all the treatments. All cytokinin concentrations generally increased shoot production when compared to the control. MS medium supplemented with 0.5 mg/L BA was best suitable for bud break (Fig. 1B) and resulted maximum number of shoots/explant (47.6±1.53), higher shoot length (2.5±1.42cm) and 65% of regeneration frequency. Among different concentrations of cytokinins tested, explants cultured on BA supplemented media exhibited superior response when compared to explants cultured on KN fortified media. Similarly, Sato *et al.* (2005) and Meftahizade *et al.* (2010) have reported that BA as the most effective cytokinin for induction of shoots in *M. officinalis*. The stimulating effect of BA on multiple shoot formation has been reported earlier in *Artemisia absinthium* (Zia *et al.*, 2007), *Mentha viridis* (Raja and Arockiasamy 2008) and *Pogostemon cablin* (Swamy *et al.*, 2010b). The dosage of cytokinin in the culture medium is known to be critical for shoot organogenesis (Sun *et al.*, 2009).

The result is contradictory to the findings of Tavares et al. (1996), Meftahizade *et al.* (2010) and Sato *et al.* (2005), who reported the use of higher concentration of BA (1-3 mg/L) for better multiplication rate. However, in our study, the increased concentrations of cytokinins beyond 0.5 mg/L, exhibited decreased number of shoot buds coupled with callus proliferation.This may be due to faster cell division leading to profuse callus proliferation and resulting in the hindrance of morphogenesis. Similar findings were reported by Chaudhari *et al.*, (2004) in *Tylophora indica*, Ahmad et al. (2008) in Vitex negundo, Nikam *et al.* (2009) in *Momordica cymbalaria* and Paul *et al.* (2010) in *Pogostemon cablin*. Though KN is less effective on multiple shoot regeneration, it played a role in increasing the length and strength of shoots (Table 1). On KN supplemented media, the regenerated shoots produced roots as well. Though KN was inferior for shoot proliferation in M. officinalis, efficacy of KN either alone or in combination with BA /auxins in direct shoot organogenesis has been demonstrated in

Asparagus maritimus (Stagner *et al.,* 2002), *Bixa orellana* (De Pairva *et al.,* 2003). At higher concentrations of cytokinins (above 0.5 mgL), nodal segments resulted in callus formation (Fig. 1C).

Table 1. Effect of different concentrations and combinations of BA and KN on regeneration frequency (%), number of shoots/explant, and shoot length (cm) from nodal segments of *Melissa officinalis* cultured on MS medium after 30 days.

Cytokinins (mg/l)		Regeneration frequency (%)	Number of shoots/explants ±SD[a]	Shootlength (cm) ± SD[a]	Callus formation
BA	KN				
0	0	18	08.1±0.29	1.2±2.10	–
0.25	0	45	25.3±1.22	1.9±1.01	–
0.5	0	65	47.6±1.53	2.5±1.42	–
1.0	0	53	45.0±2.05	2.2±1.13	+
2.0	0	28	34.1±0.86	0.8±1.42	+
0	0.25	64	22.5±1.56	2.2±1.88	–
0	0.5	69	43.6.±1.28	2.6±1.52	–
0	1.0	54	44.8±0.98	2.9±0.18	–
0	2.0	38	30.5±2.11	2.5±1.79	+
0.25	0.25	73	46.5±1.35	2.6±0.95	–
0.25	0.5	75	51.7±0.97	2.9±1.03	–
0.25	1.0	73	48.1±0.89	3.0±2.05	–
0.25	2.0	34	31.2±1.21	2.0±1.29	+
0.5	0.25	61	53.7±0.48	2.2±1.91	–
0.5	**0.5**	**88**	**60.1±1.40**	**3.3±1.21**	**–**
0.5	**1.0**	**83**	**56.8±0.75**	**2.7±1.10**	**–**
0.5	2.0	38	56.6±1.55	2.6±0.89	+
1.0	0.25	56	46.1±0.73	2.4±1.99	–
1.0	0.5	59	44.8±0.92	1.8±1.37	+
1.0	1.0	57	40.0±0.78	1.6±1.41	+
1.0	2.0	45	34.4±2.10	1.1±1.82	+
2.0	0.5	33	26.6±1.33	0.6±1.21	+
2.0	0.5	35	24.1±1.28	0.5±1.91	+
2.0	1.0	30	25.0±1.32	0.5±2.33	+
2.0	2.0	26	11.3±2.10	0.4±2.10	+
F-value			*	*	
CDat 5% level			6.32	2.05	

* Significant at 5% level.+: Callus induction, –: No callus

[a] Data indicate mean ± standard deviation. Ten replicates were used per treatments and experiment was repeated trice.

The effect of BA in combination with KN at different concentrations was assessed. The combination treatment increased the frequency of multiple shoot formation, average number of shoots per explant and shoot length (Table 1). The best morphogenic response was observed on media supplemented with 0.5 mg/l BA and 0.5mg/L KN, which exhibited highest regeneration frequency (88%). The maximum number of shoots/explant (60.1±1.40) and shoot length (2.8±1.15) was also evidenced on the same treatment. The regenerated shoots developed fragile roots following further culture. Enhancement in the induction of shoots by BA and KN in combination may be due to the synergy of cytokinins as reported in *Rollinia mucosa, Solanum surrattense* and *Pogostemon cablin* (Figueiredo, 2001; Pawar, 2002; Kumara swamy *et al.*, 2010; Swamy *et al.*, 2009; Swamy *et al.*, 2010a). Also Meftahizade *et al.* (2010) and Sato *et al.* (2005) accomplished better shoot proliferation by using combinations of cytokinins. Similarly, the present study clearly indicates that a combination of BA and KN is a better choice for *M.officinalis* as it significantly demonstrated the best morphogenetic response.

Shoot elongation

In the present study, MS medium supplemented with BA alone or in combination with KN induced both multiple shoot bud induction and proliferation, the regenerated shoots obtained failed to elongate on the same. This may be because of regeneration of shoot buds in clumps. BA has often been reported to stimulate shoot proliferation while, inhibiting shoot elongation (Figueiredo, 2001; Purkayastha *et al.*, 2008). The shoot elongation could not be accelerated even after transferring the cultures to hormone-free MS medium. Hence, it was necessary to develop a suitable media for proliferation and elongation of shoot clumps. GA3 at 0.25–2.0 mg/L was used in efforts to stimulate shoot elongation. Inclusion of 1.0 mg/l GA3 enhanced the shoot elongation by 2.3 fold in 89% of shoot cluster cultures within 2 weeks (Fig 1D), whereas the shoot elongation was not significant in cultures regenerated on medium devoid of GA3 (Table 2). Use of 1.0 mg/L GA3 produced mean shoot length of 5.6±0.85. However, stimulatory effect was less significant with increased concentration of GA3. Harsha *et al.* (2012) reported in *Rimina humilis* that the addition of BAP and GA to the nutritive medium favored multiple shooting and inhibited root formation of the new shoots that originate in nodes and shoot tips. Similarly, the promotive effect of GA3 on elongation of stunted shoots has been reported in many plant species (Purkayastha *et al.*, 2008; Sugla *et al.*, 2007 and Paul *et al.*, 2010; Kaushik *et al.*, 2015).

Table 2. Effect of gibberellic acid (GA_3) on elongation of shoots from nodal explants of *Melissa officinalis* on MS medium supplemented with 1mg/IBA and1.5mg/l KN after 3 weeks of culture.

GA3 (mg/L)	Shoot elongation response (%)	Meanshoot length (cm) ± SD[a]	Fold increase in shoot length ± SD[a]
0	13	**3.2±0.88**	0
0.25	54	4.1±1.09	0.8±1.38
0.5	66	5.2±0.70	1.9±1.42
1.0	89	5.6±0.85	2.3±0.79
1.5	79	5.4±1.47	2.1±1.23
2.0	67	4.7±1.27	1.4±2.21
F-value		*	*
CDat 5% level		0.94	3.15

* Significant at 5% level.

[a] Data indicate mean ± standard deviation.Ten replicates were used per treatments and experiment was repeated thrice.

In-vitro rooting of plantlets and acclimatization

The effect of various factors such as strength of media, activated charcoal, auxin type and concentrationss on rhizogenesis of *M. officinalis* were evaluated (Table 3).

Table 3. Effect of various concentrations of IAA, NAA and Activated charcoal on rooting of proliferated shoots of *M. officinalis* cultured on MS medium after 30 days.

Medium (strength) + Auxin (mg/l)	Root induction (%)	Number of roots/shoot (cm) ± SD [a]	Root length (cm) ± SD [a]
MS(½)	85	06.41±1.02	15.12±1.94
MS	78	08.82±1.21	3.93±1.42
MS(½) + IAA(0.5)	76	12.14±1.10	5.27±2.07
MS(½) + IAA(1.0)	72	14.36±0.51	5.42±1.03
MS(½) +NAA(0.5)	66	14.20±0.09	5.57±1.32
MS(½) +NAA(1.0)	58	15.72±1.02	5.84±0.22
MS (½) + Activatedcharcoal (100)	92	17.25±1.14	6.31±1.92
MS (½) + Activatedcharcoal (200)	92	17.25±1.14	6.10±1.53
MS+IAA (0.5)	62	11.96±0.73	5.23±1.21
MS+IAA (1.0)	65	12.83±1.30	5.53±1.31
MS+NAA (0.5)	69	15.12±1.30	4.83±0.19
MS+NAA (1.0)	51	15.82±1.82	5.32±2.21
MS + Activatedcharcoal (100)	89	15.32±1.51	5.85±1.31
MS + Activatedcharcoal (200)	90	15.12±1.94	5.63±0.89
F-value		*	*
CDat 5% level		3.01	6.48

* Significant at 5% level.

[a] Data indicate mean ± standard deviation.

Ten replicates were used per treatments and experiment was repeated thrice.

Among the treatments tried, ½ strength MS medium was enough to get better rooting. Use of ½ strength MS medium for rooting has been reported by in *O. prostrata* (Beegum *et al.*, 2007). However, high percent frequency of rooting (92 %) was achieved within 15 days of shoot implantation on ½ strength MS media fortified with 100 mg/l activated charcoal. Mean number of roots/shoot (16.21±0.44) and root length (6.31±1.92cm) was found to be superior on the same treatment. Among the different auxins tested, NAA was found most effective in inducing roots. The use of ½ MS medium containing 1.0 mg/l NAA produced 15.82±1.82 roots with 5.32±2.21cm length (Table 3). Activated charcoal is an anti-oxidant and known to induce rhizogenesis in *Maerua oblongifolia* (Rathore and Shekhawath, 2011) and Bauhinia cheilantha (Gutiérrez *et al.*, 2011). The encouraging effects of activated charcoal may be endorsed due to creation of darker environment and adsorption of undesirable/inhibitory substances (Thomas 2008). The result obtained by using half strength MS medium and activated charcoal is superior to the results obtained by using auxins. Though auxins (IAA and NAA) were shown to induce rooting with varying degrees, facilitated callus formation in a small number of shoots. This suggests that although the addition of auxins is beneficial for rooting, their use is not essential in *M. officinalis*. Plantlets with four to six fully expanded leaves and well-developed roots were successfully acclimatized in the greenhouse, in pots containing soil, sand and manure (1:1:1) within 2 weeks and eventually established in a nursery with a survival rate of 95%. The established plants were apparently uniform and did not show any morphological variation. About 84% of in vitro derived plants were directly acclimatized suggesting that the formation of in vitro roots prior to acclimatization is not needed and this can reduce time and cost.

HPLC analysis of ursolic acid content

The genetic control of biochemical character is also very important to guarantee stability in the drug yield, for the commercial production of elite M. officinalis plant by tissue culture. So evaluation of the compound, ursolic acid yielding potential of *in vitro* generated plants and mother plant was assessed through HPLC analysis. Our results showed homogeneity in the chemical profile and, ursolic acid content of the mother plant and the randomly selected micropropagated clones of *M. officinalis* (Fig. 2). Similarly, a matching profile of representative chromatograms of in vitro generated plants, market sample and field-grown plants of *Bacopa monnieri* was observed by Binita *et al.* (2005) and Muthiah *et al.* (2013). Likewise, Mohanty *et al.* (2014) reported the similarity found in micropropagated and naturally grown plants of *Leptadenia reticulata*.In an another study, Kaushik *et al.* (2015) reported similarity in the camptothecin content of micropropagated plants of *Ophiorrhiza mungos* compared to mother plants. The level of ursolic acid content in the mature leaves of tissue culture raised clones of *M. officinalis* was comparable in all the clones (0.14%± 0.15%, n = 24) and to that of the mother plant (0.12% ± 0.09 %, n = 5).

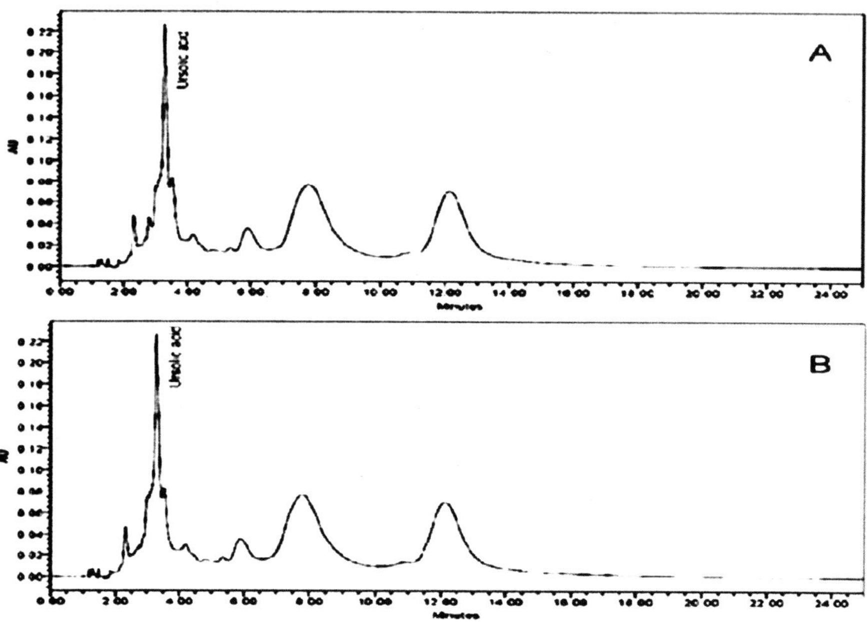

Fig. 2. HPLC analysis of Ursolic acid present in mother plant (A) and a randomly selected micropropagated plants of *Melissa officinalis* (B).

Cost effective medium for regeneration

The cost of the in vitro propagation protocol depends on the composition of the culture media used for shoot proliferation and rooting. Sucrose, agar and distilled water add significantly to the media cost. Many laboratories and researchers have reported the use of table sugar in plant propagation medium (Swamy *et al.*, 2010b; Sudipta *et al.*, 2013). In the present study, for the first time table sugar and tap water was used instead of sucrose and distilled water to propagated *M. officinalis*. The results showed no significant differences in the overall morphogenetic response of the explants (Table 4).

Table 4. Plantlet regeneration in *Melissa officinalis* on various media tested for cost calculation.

Media	No. of shoot per explant (±SE)	Mean Shoot length (±SE)	Medium cost/ liter (Rs)
MS+0.5 BA+0.5Kn+ sucrose+ distilled water	58.2±0.23	3.7±0.65	54.15
MS+ 0.5 BA+ 0.5 Kn + sucrose+tap water	54.5±0.19	3.3±0.58	43.43
MS+ 0.5 BA+ 0.5 Kn + table sugar+ distilled water	56.3±0.26	2.9±0.34	40.21
MS+ 0.5 BA+ 0.5 Kn + table sugar+ tap water	55.63±0.33	3.0±0.52	34.19

Computed F value 124.3. $P<0.05$

Each mean is based on three replicates, each of which consist of 20 individual culture tubes

Data are recorded after 4 weeks of culture initiation.

Similarly, Zapata (2001) has successfully reduced the cost of banana tissue culture by 90% by replacing the tissue culture grade sucrose with a commercial sugar. According to Prakash (1993) distilled water produced through electrical distillation is expensive. In some cases, alternative water sources can be used to lower the cost of the medium. If tap water is free from heavy metals and contaminants, it can be substituted for distilled water. The cost on media ingredients was reduced by using household sugar instead of laboratory grade sucrose and tap water instead of double distilled water (Raghu *et al.*, 2007). The use of tap water has been used for in vitro propagation of banana (Ganapathi *et al.*, 1995) and ginger, *Zingiber officinale* (Sharma and Singh 1995), *Leptadenia reticulate* (Sudipta *et al.*, 2011). Hence, the present study offers a low cost medium for rapid multiplication of *M. officinalis* and holds a great promise in the future for commercial multiplication of many other economically important plants.

Conclusion

In conclusion, a simple, efficient and high fidelity protocol for mass propagation of *M. officinalis* has been established under Indian conditions. In this study, the true to the type nature of the in vitro raised clones was confirmed using DNA-based markers and chemical profiling. No variability was detected among the tissue culture-raised plantlets. Hence this protocol can be successfully employed for the commercial multiplication of M. officinalis without much risk of genetic instability.

Acknowledgements

The authors gratefully acknowledge the financial support received from Department of Science and Technology, (DST) towards this project. The authors also acknowledge the management of Padmashree Institute of Management and Sciences, Bangalore, India, for providing the necessary facilities to carry out this project.

References

Ahmad N., Wali S.A., Anis M. 2008. In vitroproduction of true-to-type plants of *Vitex negundo*L. from nodal explants. *JHort Sci Biotech* 83(3): 313-317.

Awad R., Levac D., Cybulska P., Merali Z., Trudeau V.L., Arnason J.T. 2007. Effects of traditionally used anxiolytic botanicals on enzymes of the gamma-aminobutyric acid (GABA)system. *Can J Physiol Pharmacol* 85: 933-942.

Awad R., Muhammad A., Durst T., Trudeau V.L., Arnason J.T. 2009. Bioassay-guided fractionation of lemon balm (*Melissa officinalis* L.) using an in vitro measure of GABA transaminase activity. *Phytother Res* 23: 1075-1081.

Beegum A.S., Martin K.P., Zhang C.L., Nishitha I.K., Ligimol, Slater A., Madhusoodanan P.V. 2007. Organogenesis from leaf and internode explants of *Ophiorrhiza prostrata*, an anticancer drug (camptothecin) producing plant. *Electronic J Biotechnol* 10(1): 114-123.

Binder G., Abou M.A.A. 2000. Regeneration of plants and production of volatiles from callus cultures of *Melissa officinalis* L. *J Appl Bot* 74(1-2): 26-31.

Binita B.C., Ashok M.D., Yogesh T.J. 2005. *Bacopa monnieri* (L.) Pennell: A Rapid, Efficient and Cost Effective Micropropagation. *Plant Tissue Cult & Biotech* 15(2): 167-175.

Chaudhari K.N., Ghosh S., Jha S. 2004. The root: A potential source of competent cells for high frequency regeneration in *Tylophora indica*. *Plant Cell Rep* 22: 731-740.

Chen H., Chen F., Zhang Y.L., Song J.Y. 1999. Production of lithospermic acid B and rosmarinic acid in hairy root cultures of *Salvia miltiorrhiza*. *J Ind Microbiol Biotechnol* 22: 133-138.

De Paiva Neto, Vespasiano Borges, Da Mota Tiago Ribeiro, Otoni Wagner Campos 2003. Direct organogenesis from hypocotyl-derived explants of annatto (*Bixa orellana*). *Plant Cell Tissue Organ Cult* 75(2): 159–167.

Figueiredo S.F.L. 2001. Micropropagation of *Rollinia mucosa* (Jacq.) Baill. *In vitro Cell Dev Biol Plant* 37: 471–475.

Ganapathi T.R., Mohan J.S.S., Suprasanna P., Bapat V.A., Rao P.S. 1995. A low-cost strategy for in vitro propagation of banana. *Curr Sci* 68: 646-665.

Gbolade A.A., Lockwood G. 1992. Metabolic studies of volatile constituents in tissue culture of *Melissa officinalis*. *Plant Physiol* 140: 28-32.

Gutiérrez I.E.M., Nepomuceno C.F., Ledo C.A.S., Santana J.R.F. 2011. Micropropagation and acclimatization of *Bauhinia cheilantha* (an important medicinal plant). *Afr J Biotechnol* 10(8): 1353–1358.

Harsha P.S.C.S., Khan M.I., Giridhar P., Ravishankar G.A. 2012. In vitro propagation of *Rivina humilis* L. by through proliferation of axillary shootsand shoot tips of mature plants. *Indian J Biotechnol* 11: 481-485.

Herodez S.P., Hadolin M., Kerget M., Knez E. 2003. Solventextractionstudy of antioxidants from Balm (*Melissa officinalis* L.) leaves. *Food Chem* 80(2): 275-282.

HrašA.R., Hadolin M., Knez Z., Bauman D. 2000. Comparison of antioxidative and synergisticeffectsofrosemaryextractwith-tocopherol, ascorbylpalmitate and citricacidinsunfloweroil. *Food Chem* 71: 229-233.

Kaushik P.S., Swamy M.K., Balasubramanya S., Anuradha M. 2015. Rapid plant regeneration, analysis of genetic fidelity and camptothecin content of micropropagated plants of *Ophiorrhiza mungos* Linn.- a potent anticancer plant. *J Crop Sci Biotechnol* 18(1): 1-8.

Kumaraswamy M., Anuradha M. 2010. Micropropagation of *Pogostemon cablin* Benth. through direct regeneration for production of true to type plants. *Plant Tissue Cult Biotechnol* 20(1): 81-89.

Mazzanti G., Battinelli L., Pompeo C., Serrilli A.M., Rossi R., Sauzullo I., Mengoni F. & Vullo V. 2008. In hibitoryactivity of *Melissaofficinalis* L. extracton *Herpessimplex*virus type replication. *Nat Prod Res* 22(16): 1433-40.

Meftahizade H., Lotfi M., Moradkhani H. 2010. Optimization of Micropropagation and establishment of cellsuspensi on culture in *Melissa officinalis* L.*AfrJ Biotechnol* 9(28): 4314-4321.

Meftahizade H., Moradkhani H., Naseri B. Lofti M., Naseri A. 2010. Improved in vitro culture and micropropagation of different *Melissa officinalis* L. genotypes. *J Med Plant Res* 4(3): 240-246.

Mohanty S.K., Mallappa K.S., Godavarthi A., Subbanarasiman B., Maniyam A. 2014. Evaluation of antioxidant, in vitro cytotoxicity of micropropagated and naturally grown plants of *Leptadenia reticulata* (Retz.) Wight & Arn.-an endangered medicinal plant. *Asian Pac J Trop Med* 7: S267-S271.

Murashige T., Skoog F. 1962. A revised medium for rapid growth and bioassay with tobacco tissue cultures. *Physiol Plant* 15, 473-497.

Nikam T.D., Ghane S.G., Nehul J.N., Barmukh R.B. 2009. Induction of morphogenic callus and multiple shoot regeneration in *Momordica cymbalaria* Fenzl. *Indian J of Biotechnol* 8: 442–447.

Pabsch K., Petersen M., Rao N.N., Alfermann A.W., Wandrey C. 1991. Chemo-enzymaticsynthesis of rosmarinic acid. *Recl Trav Chim Pays-Bas* 110: 199-205.

Pawar P.K. 2002. A technique for rapid propagation of *Solanum surrattense* Burm. F. *Indian J Biotechnol* 1: 201–204.

Prakash S. 1993. Production of ginger and turmeric through tissue culture methods and investigations into making tissue culture propagation less expensive. Ph.D. Thesis. Bangalore Univ. Bangalore.

Purkayastha J., Sugla T., Paul A., Solleti S., Sahoo L. 2008. Rapid in vitro multiplication and plant regeneration from nodal explants of *Andrographis paniculata*: a valuable medicinal plant. *In Vitro Cell Dev Biol Plant* 44: 442–447.

Raghu A.V., Martin G., Priya V., Geetha S.P., Balachandran I. 2007. Low Cost Alternatives for the Micropropagation of *Centella asiatica*. *J Plant Sci* 2: 592-599.

Raja H.D., Arockiasamy D.I. 2008. In vitro Propagation of *Mentha viridis* L. from Nodal andShoot tip Explants. *Plant Tissue Cult & Biotech* 18(1): 1-6.

Rathore M.S., Shekhawat N.S. 2011. Micropropagation of *Maerua oblongifolia*: A rare ornamental from semi arid regions of Rajasthan, India. *J Dev Biol Tissue Eng* 3(8): 92–98.

Sato A., Dasilva S., Celso L.S.L., Rosane A.S.S.G., Maria A.E. 2005. Essential Oil Composition of *Melissa officinalis* L. in vitro Produced under the Influence of Growth Regulators. *J Braz Chem* 16: 1387-1390.

Schnitzler P., Schuhmacher A., Astani A., Reichling J. 2008. *Melissa officinalis* oil affectsinfectivity of enveloped herpesviruses. *Phytomedicine* 15(9): 734-740.

Sharma T.R., Singh B.M. 1995. Simple and cost-effective medium for propagation of ginger (*Zingiber officinale*). *Indian J. Agricul. Sciences* 65: 506-508.

Stajner Nataša, Bohanec Borut and Jakše Marijana 2002 In vitro propagation of *Asparagusmaritimus* – A rare Mediterranean salt-resistant species. *Plant Cell Tissue and Organ Cult* 70(3): 269–274.

Sudipta K.M., Kumara Swamy M., Balasubramanya S., Anuradha M. 2011. Cost effective approach for in vitro propagation of (*Leptadenia reticulata* Wight & Arn.) – a threatened plant of medicinal importance. *Journal of Phytology* 3: 72–79.

Sudipta K.M., Swamy M.K., Anuradha M. 2013. Influence of various carbon sources and organic additives on in vitrogrowth and morphogenesis of *Leptadenia reticulate* (Wight & Arn), a valuable medicinal plant of India. *Int J Pharm Sci Rev Res* 21(2): 174-179.

Sugla, T., Purkayastha J., Singh S.K., Solleti S.K., Sahoo L. 2007. Micropropagation of *Pongamiapinnata* through enhanced axillary branching. *In Vitro Cell Dev Biol* 43: 409–414.

Sun Y., ZhaoY., Wang X., Qiao G., Chen G., Yang Y., Zhou J., Jin L., Zhuo R. 2009. Adventitiousbud regeneration from leaf explants of *Platanus occidentalis* L. and genetic stability assessment. *Acta Physiol Plant* 31: 33–41.

Sun YL., Hong S.K. 2010. Effects of plant growth regulators and L-glutamic acid on shootorganogenesis in the halophyte *Leymus chinensis* (Trin.). *Organ Cult Plant Cell Tiss* 100: 317–328.

Swamy M.K., Balasubramanya S., Anuradha M. 2009. Germplasm conservation of patchouli (*Pogostemon cablin* Benth.). *Int J Biodvers Conserv*1(8): 224-230.

Swamy M.K., Balasubramanya S., Anuradha M. 2010a. In vitro multiplication of patchouli through direct organogenesis. *Afr J Biotechnol* 9(14): 2069-2075.

Swamy M.K., Sudipta K.M., Anuradha M. 2010b. Effect of different carbon sources on *in vitro* morphogenetic response of patchouli (*Pogostemon cablin* Benth.). *J Phytol* 2(8): 11-17.

Szabo E., Thelen A., Petersen M. 1999. Fungal elicitor preparations and methyl Jasmonateenhance rosmarinic acid accumulation in suspension cultures of Coleus blumei. *Plant Cell Rep* 18: 485-489.

Tantos A., Meszaros A. 1999. The effect of triacontanol on micropropagation of balm, *Melissaofficinalis* L. *Plant Cell Rep* 19(1): 88-91.

Tavares A.C., pimento M.C., Goncalves M.T. 1996. Micropropagation of *Melissa officinalis* L. through proliferation of axillary shoots. *Plant cell Rep* 15: 441-444.

Thomas T.D. 2008. The role of activated charcoal in plant tissue culture. *Biotechnol Adv* 26: 618–631.

Turhan M. 2004. Hand book of herbal plants, *Melissa officinalis*, 3: 184-245.

Zapata A. 2001. Cost reduction in tissue culture of banana. (Special leaflet), Int. Atom EnergyLabs. Agric. AndBiotech. Lab. Austria.

Zia M., Riaz-ur-Rehman, Chaudhary M.F. 2007. Hormonal regulation for callogenesis and organogenensis of *Artimissia absinthium* L. *Afe J Biotech* 6(16): 1874-1878.

❑❑❑

Chapter 21

Bombax ceiba L. (Semal)- A Multidimensional Folk Medicinal Tree

Vartika Jain*[1]

Department of Botany, Govt. Meera Girl's College, Udaipur-313001, Rajasthan, India, E-mail: vartikajain31@gmail.com

ABSTRACT

Bombax ceiba L. (*Semal* tree) possesses immense significance in lives of indigenous communities all over the country. The plant plays an important role in their many rituals, customs and socio-cultural events and almost every part of the tree is being utilized to treat various animal and human ailments. Many of these folk medicinal claims have also been scientifically validated in various experimental studies in animal and human beings and researches on its phytochemical profiling are being carried out world over. Besides this, its wood, silk-cotton and seed-oil are commercially very valuable. The present paper is an attempt to discuss briefly about multidimensional characters and properties of this medicinal tree in view of history as well as recent researches.

Keywords: Ethnomedicine, Phytochemistry, Silk-Cotton tree, Match-stick.

Introduction

India has a large population of indigenous communities and majority of them are still dwelling in forests which provide almost everything what they need to live sprightly. *Semal*; scientifically known as *Bombax ceiba* L. (Malvaceae) is one such important tree of tribal communities all over the country. It is known by different names such as *Shalmali, Semal, Simal, Simbal, Shimul,* Silk Cotton Tree, Indian Red Kapok Tree etc. in various languages. It is distributed in temperate and tropical Asia, America, Africa and Australia. In India, it is widely distributed throughout the country and the Andamans (Chadha, 1972; Jain, 2012a).

Semal is a large, deciduous tree, armed with hard, conical spines possessing beautiful red flowers and pentafoliate leaves. It reaches a height up to 40 meters with its characteristic horizontally spreading branches. Due to its large size and beautiful scarlet red flowers, it is sometimes called as 'King of the Forest'. It flowers during January-March and fruiting starts in April providing silk-cotton and black color seeds in the month of May (Fig. 1-3). It remains leafless from November-April and new leaves starts appearing in May. It requires relatively less water to survive and therefore, can be planted at most of the places very easily (Jain & Verma, 2012).

Semal tree has left its impact on almost every sphere of life for example, social customs, folk songs, folk tales, materials, medicine, environment balance and spiritual advancement of human beings (Jain, 2012 b,c,d). It provides food, fuel, fibre and fooder and therefore, comes in the category of Multipurpose Tree Species (MPT'S) and thus an integral part of social and agro forestry programs in India. Following few paragraphs describe its popularity, variety of uses and importance in almost every field of human life.

Fig. 1. *Semal* tree in full bloom

Fig. 2. A Flower

Fig. 3. Silk-cotton and seeds

Spiritual Importance

Semal tree is revered by many indigeonous communities across the country and also possesses historical importance. It has been mentioned in epics and ancient scriptures such as *Rigveda, Mahabharta, Guru Granth Sahib* and in many '*Puranas*'. The great epic '*Mahabharata*' mentions that this tree is sacred because Brahma, the creator, after creating the world had taken rest under this tree. It has been given the name of 'Dev-Vriksha' (God Tree) and also known as 'Nakshtra Vriksha' (Constellation Tree) of people born in 18th constellation- 'Jyestha'. It is believed to attract positive microvita and therefore, it was one of the five sacred trees of 'Panchwati'- a place used to do psycho-spiritual meditation in ancient times. Jain scriptures describes that many Jain Tiirthankars have attained salvation under this tree (Jain & Borthakur, 1980; Sane & Ghate, 2006; Jain, 2012 b,c).

Medicinal Importance

Semal tree is an important part of Indian (Ayurveda), Chinese and Taiwanese traditional systems of medicine. Every part of the tree possesses medicinal potential and indigenous communities use it for treatment of various human and animal ailments. The plant is very popular for its use in seminal weakness in man and genital infections and disorders in females (Gupta *et al*., 2004). Many of the folk claims have also been scientifically validated in various parts of world (Jain & Verma, 2014). Medicinal potential of each part as described in indigenous medicine along with results of various scientific investigations has been briefly discussed below:

Root

Tap root of Semal is known as '*Semar-kanda*'/ '*Semal-musli*'. It is rich in calcium, protein and phenolic content and also used as an edible. It is used for treatment of diabetes, heart disease, leucorrhoea, impotence, debility, diarrhea and dysentery in folk medicine (Gupta *et al.*, 2004). Recently, it has been shown to decrease blood sugar in type 2 diabetics and improve fibrinolysis and antioxidant status along with reduction in lipid contents in patients of ischemic heart disease (Verma *et al.*, 2006; Verma *et al.*, 2008; Jain *et al.*, 2012). It has also shown to possess anti-hepatitis B virus, antibacterial, antioxidant, anti-inflammatory, androgenic and anabolic activities in various scientific studies (Jain & Verma, 2012; Wang *et al.*, 2013).

Stem & Stem-bark

Powder of its thorns is mixed with milk and used to remove acne of face. Bark is used for treatment of gastro-intestinal diseases, boils, kidney stone, headache, heart disease, snake bite etc. in tribal medicine (Jain, 1991). Recent researches worldwide have shown its hypoglycemic, hypotensive, hepato-protective, antimicrobial, anti-angiogenic, anti-ulcerogenic, antisecretory, cytoprotective and anti-inflammatory potential (Saleem *et al.*, 2003; You *et al.*, 2003; Jain & Verma, 2012; Hussain *et al.*, 2015).

Leaf

Its protein rich leaves are used for treatment of diabetes, leucorrhoea, arthritis, anemia, body pain and dysentery (Jain, 1991). Scientific investigations have revealed that leaves possess antipyretic, analgesic, antioxidant, anthelmintic, larvicidal, hypoglycemic, hypotensive, and heapto-protective activities (Saleem *et al.*, 1999; Dar *et al.*, 2005; Hossain *et al.*, 2011).

Flower

Young buds are cooked to prepare vegetable. Beautiful red colored flowers are found to be useful in anemia, bleeding piles, leucorrhoea, gonorrhoea, menorrhagia, colitis, skin disease, internal bleeding, ulcers and boils (Jain, 1991). Different extracts of flowers have also been shown to demonstrate hypoglycemic, hypotensive, hepato-protective, cardio-protective, antioxidant, antimicrobial, analgesic, antiviral, anti-proliferative and anti-inflammatory properties in scientific researches

(Vieira *et al.*, 2009; Ravi *et al.*, 2010; Said *et al.*, 2011; Jain & Verma, 2012; Tundis *et al.*, 2014; Zhang *et al.*, 2015).

Fruit

Young fruits are utilized for the treatment of urinary and kidney stones, gonorrhoea, leucorrhoea, menorrhagia, uterus prolapse and snake bite (Jain, 1991). Fruits have shown to possess maximum hypotensive potential among all other parts of *Semal* tree (Saleem *et al.*, 1999). Recent scientific studies have validated its ability to cure urolithiasis in animals (Gadge & Jalalpure, 2012).

Gum

Gum exuding from its stem is known as 'Mochrasa'. Medicinally it is found to be useful against asthma, bleeding piles, debility, diarrhea and dysentery (Gupta *et al.*, 2004). Scientific research has shown that gum possesses antioxidant, analgesic and anti-inflammatory potential (Jagtap *et al.*, 2011; Jain & Verma, 2012).

Phytochemistry

It is an important medicinal plant widely popular in folk medicine and other traditional systems of medicine and therefore, has been a focus of scientific research all over the world. Recent research investigations have determined its various health beneficial properties in experimental and clinical studies and many bioactive compounds such as bombamalones A-D, bombamaloside, lacinilene, bombasinol A, lupeol, b-sitosterol, 5,6-dihydroxymatairesinol, pinoresinol, matairesinol, bombaxquinone, shamimin, mangiferin, taraxeryl acetate, squalene, taraxerone, b-sitosterol palmitate, taraxerol, 4-methyl stigmast-7-en-3-ol, 1H-indole-3-carboxylic acid, 6-O-palmitoylsitosteryl-D-glucoside, 12beta-hydroxyl-pregnane-4, 16-diene-3, 20-dione, loliolide, and 5-(hydroxymethyl) furfural, phenolic compounds, 2-*O*-(3,4-dihydroxybenzoyl)-2,4,6-trihydroxyphenylaceticacid-4-*O*-â-D-glucopyranoside, 4-*epi*-bombalin, *N*-[(2*E*)-3-(4-hydroxyphenyl)-1-oxo-2-propen-1-yl]-L-tyrosine ethyl ester, sesquiterpene lactones and napthoquinones have been isolated from various parts of the plant (Rastogi & Mehrotra, 1995, Zhang *et al.*, 2007, Wu *et al.*, 2008, Jain & Verma, 2012, Wang *et al.*, 2013; Wang *et al.*, 2014; Zhang *et al.*, 2015).

Commercial Importance

Semal tree possesses immense commercial potential and has been included in the list of World's economic plants for its valuable wood and fibre (Wiersema & Leon, 1999). Timber of the tree is an important source for match-stick production besides for planking ceilings, making canoes, catamarans, shingles, toys, pencils, pen-holders, veneers, scabbards, coffins, brush-handles, picture frames, wooden sculptures, ladles, as cushions for mine-props, inside partition of opium-chests, paper pulp and artifact production (Chadha,1972; Pandey, 2005).

Kapok (Silk-cotton), the vermin-proof fiber obtained from its fruits, is used to fill life-belts, life-saving appliances, sleeping bags, mattresses, cushions, pillows, upholstery, wadded cloth quilts etc. Floss can be dry sterilized at 110°C and therefore, used for padded surgical dressings. It is also used as an insulating material for refrigerators, soundproof covers and walls. Edible oil from seeds is also obtained which is used for soap making and as an illuminant (Chadha, 1972).

Socio-cultural Importance

This tree has been part of many customs, rituals and socio-cultural life of various tribal communities in India. In Rajasthan, tribal children play with its numerous stamens for fun. Due to its thorny stem, broilers are kept in a basket which is tied high on the tree and thus broilers get protection from cats and other pets. Tribal communities of Rajasthan don't use its wood for fuel as they believe that it will bring bad luck to them. Garasia tribe sing a song 'Hemlo ropalo re…' and urges to plant this tree in surroundings and take care of this tree as their family. Tribal communities of Deogudis, Chhattisgarh believe that God 'Baskoddo' and Goddess 'Kali Mata' reside on *B. ceiba* tree. It is also believed that it is home of 'Yakshis'; the female tree spirits which are worshipped by women for gift of children. Even many riddles are based on this tree in Rarh Bengal area. It has also been part of marriage or child naming ceremonies and many auspicious occasions of tribal communities in India. Interestingly, many of these beliefs are directly and/or indirectly helping in conservation of this tree (Joshi, 1995; Gupta, 1995; Nonhare *et al.*, 2002; Jain, 2012b).

Besides, all other customs and rituals, one devastating tradition of burning this tree in thousands of quantity in Holi festival as a symbol of mythological character 'Prahlad' has led to serious attention in Udaipur district, Rajasthan (Jain & Verma, 2009). Looking to this,

Society for Microvita Research and Integrated Medicine (SMRIM), Udaipur has initiated 'Semal Conservation Mission' in 2007 which has really helped in conservation of this important medicinal tree in the area. Society has developed saplings of *Semal* and planted them in and around Udaipur district. Not only this, Society has also suggested a novel concept of burning an iron-pole instead of Semal tree in Holika-dahan and implementing it since 2011 in Udaipur, Rajasthan. Both are the historical steps taken for the first time in Udaipur which has got full support from intellectuals and print media (Jain & Verma, 2012).

Conclusion

Semal tree is a common species of tropical dry deciduous forests of India. Being large in size with beautiful blooming as well as as a provider of food, fodder and shelter to many birds and animal species, it can be used to add beauty to cities where it can support simultaneous survival of many animal species dependent on it. It is also popular for its medicinal properties and many new biological properties and bioactive compounds are being discovered world over; however, a thorough pharmological exploration is required to utilize its complete medicinal potential.

Looking to few traditional survival dangers, mass scale *in situ* and *ex situ* conservation strategies should also be initiated and implemented in time, so that future generation can also utilize beneficial properties of this multi-purpose tree. It is cultivated mostly for commercial purpose as one of the best Match-wood resources besides for medicine. In view of, its promising multi-potential, ecological benefits and easy cultivation requirements, it must be promoted as revenue generating plant species in Indian forests. In those regions, where it is present in abundance but not utilized, its economic potential must be highlighted among the dwelling population so that it can become a beneficial source of livelihood. However, techniques of sustainable harvesting must always be incorporated in order to achieve long term benefits and conservation of biodiversity.

References

Chadha Y.R. 1972. The Wealth of India, Raw material. Vol. IX. Publications and Information Directorate, New Delhi.

Dar A., Faizi S., Naqvi S., Roome T., Zikr-ur-Rehman S., Ali M., Firdous S., Moin S.T. 2005. Analgesic and antioxidant activity of mangiferin and its derivatives: the structure activity relationship. *Biol Pharm Bull.* 28(4): 596-600.

Gadge N.B. and Jalalpure S.S. 2012. Curative treatment with extracts of *Bombax ceiba* fruit reduces risk of calcium oxalate urolithiasis in rats. *Pharm Biol.* 50(3): 310-317.

Gupta A.K., Sharma M., Tandon N. 2004. Reviews on Indian Medicinal Plants, Vol IV, pp. 319-331, Indian Council of Medical Research, New Delhi.

Hossain E., Rawani A., Chandra G., Mandal S.C., Gupta J.K. 2011. Larvicidal activity of *Dregea volubilis* and *Bombax malabaricum* leaf extracts against the filarial vector *Culex quinquefasciatus. Asian Pac J Trop Med.* 4(6): 436-441.

Hussain L., Akash M.S., Naseem S., Rehman K., Ahmed K.Z. 2015. Anti-ulcerogenic effects of *Salmalia malabarica* in gastric ulceration-Pilot Study. *Adv Clin Exp Med.* 24(4): 595-605.

Jagtap A.G., Niphadkar P.V., Phadke A.S. 2011. Protective effect of aqueous extract of *Bombax malabaricum* DC on experimental models of inflammatory bowel disease in rats and mice. *Indian J Exp Biol.* 49(5): 343-351.

Jain S.K. 1991. Dictionary of Indian Folk Medicine and Ethnobotany. Deep Publications, New Delhi.

Jain S.K. and Borthakur S.K. 1980. Ethnobotany of Mikirs of India. *Econ Bot.* 34: 264-272.

Jain V., Verma S.K., Katewa S.S. 2012. Effect of *Bombax ceiba* root on some cardiovascular parameters in patients with ischemic heart disease. *Asian J Biol Sci.* 5: 351-357.

Jain V., Verma, S.K., Katewa S.S. 2009. Myths, traditions and fate of multipurpose *Bombax ceiba* - An appraisal. *Indian J Trad Knowledge.* 8(4): 638-644.

Jain V. 2012a. On the occurrence and conservation status of an ethnobotanically important plant *Bombax ceiba* Linn. in India. Project Report, submitted to S.K. Jain Institute of Ethnobiology, Jiwaji University, Gwalior (MP).

Jain V. 2012b. Bahudeshiya aushadhiya vriksha- Semal (In Hindi). Vigyan Pragati. 61 (10): 45-47.

Jain V. 2012c. *Bombax ceiba* (Semal) in Indian folklore and traditions. *Ethnobot.* 24 (1&2): 79-82.

Jain V. 2012d. Dantkathaao mei Shalmali (Semal) (In Hindi). Bull Microvita *Med Res Integr Med.* 4 (2&3): 10-11.

Jain V and Verma S.K. 2012. Pharmacology of *Bombax ceiba* Linn. Springer, Heidelberg.

Joshi P. 1995. Ethnobotany of Primitive tribes in Rajasthan. Printwell, Jaipur, India.

Nonhare B.P., Sahu T.R., Sahu P. 2002. Deogudis: unrecognized centres for biodiversity conservation in Bastar (Chhattisgarh). *Ethnobot.* 14(1&2): 93-99.

Pandey B.P. 2005. Economic Botany. S. Chand and Company limited, New Delhi.

Rastogi R.P. and Mehrotra B.N. 1995. Compendium of Indian Medicinal Plants. Vol 4. PID, New Delhi.

Ravi V., Patel S.S., Verma N.K., Dutta D., Saleem T.S.M. 2010. Hepatoprotective Activity of *Bombax ceiba* Linn against Isoniazid and Rifampicin-induced toxicity in experimental rats. *Int J Appl Res Nat Prod.* 3(3): 19-26.

Said A., Aboutable E.A., Nofal S.M., Tokuda H., Raslan M. 2011. Phytoconstituents and bioactivity evaluation of *Bombax ceiba* L. flowers. *J Trad Med.* 28: 55-62.

Saleem R., Ahmad M., Hussain S.A., Qazi A.M., Ahmad S.I., Qazi M.H., Ali M., Faizi S., Akhtar S., Hussain S.N. 1999. Hypotensive, hypoglycaemic and toxicological studies on the Flavonol C-glucoside Shamimin from *Bombax ceiba. Planta Med.* 65: 331-334.

Saleem R., Ahmad S.I., Ahmad M., Faizi Z., Rehman S., Ali M., Faizi S. 2003. Hypotensive activity and toxicology of constituents from *Bombax ceiba* stem bark. *Biol Pharm Bull.* 26(1): 41-46.

Sane H. and Ghate V. 2006. Sacred conservation practices at species level through tree worship. *Ethnobotany.* 18: 46-52.

Tundis R., Rashed K., Said A., Menichini F., Loizzo M.R. 2014. *In vitro* cancer cell growth inhibition and antioxidant activity of *Bombax ceiba* (Bombacaceae) flower extracts. *Nat Prod Commun.* 9(5): 691-694.

Verma S.K., Jain V., Katewa S.S. 2006. Fibrinolysis enhancement by *Bombax ceiba*- a new property of an old plant. *South Asian J Preventive Cardiology.* 10(4): 212-219.

Verma S.K., Jain V., Katewa S.S. 2008. Potential antihyperglycemic activity of *Bombax ceiba* in type 2 diabetes. *Int J Pharmacol Biol Sci.* 2(1): 79-86.

Vieira T.O., Said A., Aboutabl E., Azzam M., Creczynski-Pasa T.B. 2009. Antioxidant activity of methanolic extract of *Bombax ceiba*. *Redox Rep.* 14(1): 41–46.

Wang G.K., Lin B.B., Qin M.J. 2014. Study on chemical constituents from leaf of *Bombax ceiba* (II). *Zhong Yao Cai.* 37(2): 240-242.

Wang G.K., Lin B.B., Rao R., Zhu K., Qin X.Y., Xie G.Y., Qin M.J. 2013. A new lignan with anti-HBV activity from the roots of *Bombax ceiba*. *Nat Prod Res.* 27(15): 1348-52.

Wiersema J.H. and Leon B. 1999. World Economic Plants. p. 79. CRC Press, USA.

Wu J., Zhang X.H., Zhang S.W., Xuan L.J. 2008. Three novel compounds from the flowers of *Bombax malabaricum*. *Helvetica Chimica Acta.* 91(1): 136-143.

You Y.J., Nam N.H., Kim H.M., Bae K.H., Ahn B.Z. 2003. Antiangiogenic activity of lupeol from *Bombax ceiba*. *Phytother Res.* 17(4): 341-344.

Zhang X., Zhu H., Zhang S., Yu Q., Xuan L. 2007. Sesquiterpenoids from *Bombax malabaricum*. *J Nat Prod.* 70: 1526-1528.

Zhang Y.B., Wu P., Zhang X.L., Xia C., Li G.Q., Ye W.C., Wang G.C., Li Y.L. 2015. Phenolic compounds from the flowers of *Bombax malabaricum* and their antioxidant and antiviral activities. *Molecules.* 20(11): 19947-57.

❑❑❑

Chapter 22

Utilization of Plant Biodiversity for Treatment and Cure of Diseases of Human Beings in Thalamalai hills, Namakkal District, Tamil Nadu, India

Vikas Kumar*

Ph.D. Scholar, Department of Silviculture and Agroforestry, College of Forestry, Vellanikkara, Kerala Agricultural University, Thrissur- Kerala-680656, Corresponding author E-mail: vkskumar49@gmai.com

ABSTRACT

Demand for medicinal plant is increasing today in both developing and developed countries due to growing recognition of natural products, being non-toxic, having no side-effects, easily available at affordable prices. An ethnomedicinal survey was conducted for gathering information to treat and cure diseases of human beings in surrounding area of Thalamalai hills, Namakkal district, Tamil Nadu during April 2014 to March 2016. The information was collected through personal interviews with traditional healers and village elders. The investigation revealed that 120 plant species belong to 100 genera and 56 families out of which trees (29 spp.), shrubs (16 spp.), climbers (21 spp.) and herbs (54 spp.).

Keywords: Ethnomedicinal, Plant Biodiversity, Thalamalai hills, Namakkal, Tamil Nadu, treatment and cure of disease.

Introduction

Since the beginning of civilization, human being has been using plants for different aspects of life. Communities engaging with the forest are manifestation of trust on the forests as a source of life. Utilization of forest plant itself has been applied in filling out and meets the needs of the people living in particular to complementary and accessories. By maximizing its potency, various pattern of utilization and other forms are very likely to be generated (Plowden *et al.*, 2003). Indian sub-continent has a very rich diversity of plant species and having wide range of ecosystems. There are about 17,000 species of higher plants, of which approximately 8,000 species are considered medicinal and used by village communities, particularly tribal communities, or in traditional medicinal systems, such as the Ayurveda (Pei, 2001; Abu-Rabia, 2005; Kumar *et al.*, 2014). The conservation of environment, natural resources and biological diversity has been deeply rooted in the Indian tradition and culture (Kumar, 2015). The term Ethnobotany was coined by John Harshburger to delimit a specific field of botany and described plant used (Harshburger, 1896). Particularly traditional knowledge has been described as a cumulative body of knowledge, practice and belief, evolving through adaptive processes and handed over through generations by cultural transmission (Berkes *et al.*, 2003). The wide spread use of Ethnomedicinal plants could be attributed to cultural acceptability, economic affordability and efficacy against certain type of diseases as compared to modern medicines. Thus, different local communities in countries across the world have indigenous experience in various medicinal plants where they use their perceptions and experiences to categorize plants and plant parts to be used when dealing with different ailments (Omoruyi *et al.*, 2012). Demand for medicinal plant is increasing in both developing and developed countries due to growing recognition of natural products, being non-toxic, having no side-effects, easily available at affordable prices (Kumar and Mano, 2015).

Recently considerable attention has been paid to utilize eco-friendly and bio-friendly plant based product for the preservation and cure of different human diseases. Medicinal and aromatic plants (MAPs) are looked upon not only as a source of affordable health care products but also as a source of income. It is documented that 80% of the world's population have faith in traditional medicine, particularly plant drug for their primary healthcare (Dubey, 2004; Ganeshan, 2004). Many plants have become endangered or vulnerable or threatened as 90% collection of MAPs is from wild source and 70% collection involved destructive harvesting (GOI, 2000). With dwindling supplies from natural resources and increasing global demand, expanding the cultivation of MAPs appears to be an important strategy (Rao *et al.*, 2004). Documentation of the indigenous knowledge through ethno botanical studies is important for the conservation and utilization of biological resources (Macia *et al.,* 2005; Muthu *et al.,* 2006; Kumar, 2015). Therefore, determining the local names and indigenous uses of plants have significant potential societal benefits (Kargioglu *et al.,* 2008; Jayakumar, 2013; Kumar, 2015). The objective of this study was to assess the diversity of ethnomedicinal plant species used in Thalamalai hills area in Namakkal district of Tamil Nadu and to document the ethnomedicinal practices for treatment and cure of diseases of human beings.

Materials and Methods

Study Area

Thalamalai hills are situated in Southern Eastern Ghats comes under Rasipuram Taluk, Namakkal district. Thalamalai is at the elevation of 3,937 ft mountain in the Eastern Ghats of South India. It lies between 11^0 14′46" - 12^0 53′30" North latitude and between 77^0 32′52" - 78 ^{0}53′05" East longitude and it has an elevation of 881 meters above sea level (Fig. 1). Thalamalai is in an area with a humid subtropical climate, only Hindu Malayali tribes residing in this area. Namakkal District comes under the North Western Agro climatic zone of Tamil Nadu. It is situated in the dividing portion of two watersheds between Cauvery and the Vellar System with the Taluks of Attur, Rasipuram and Namakkal on the East and Salem, Omalur and Mettur on the West. The area is receives maximum rainfall from north east monsoon with amount of 600-700 mm per annum with 35-50% humidity. It enjoys maximum temperature of 25-32°C and minimum of 15-20°C. The study is carried out during May 2014 to Feb 2015 in the weekends.

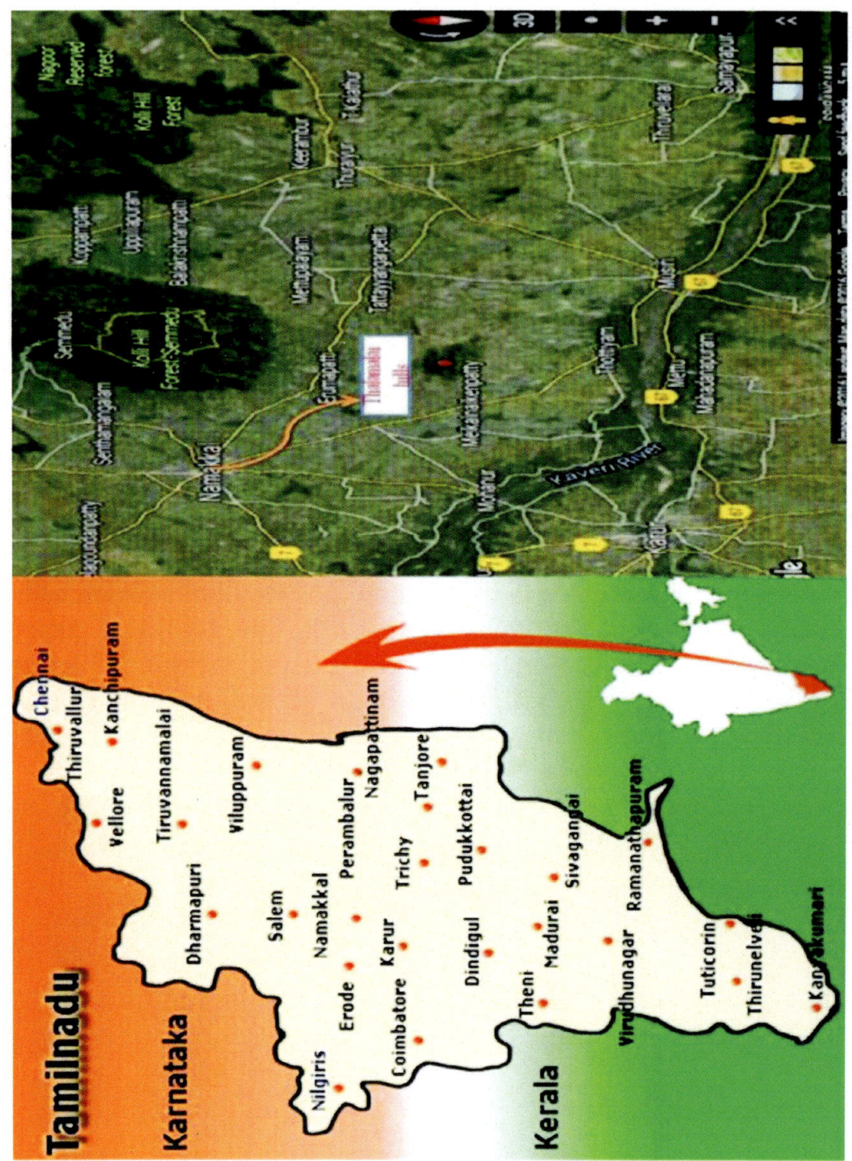

Fig. 1. Location map of study area of Thalamalai hills, Namakkal District, Tamil Nadu, India

Methods

Several field trips were carried out in Thalamalai hills from April 2014 to March 2016, covering different seasons, in order to know the phenology of the plants. Extensive field survey was made in Thalamalai hills and villages in Namakkal district. Standard method was followed from record to collection of plant materials, drying, mounting, preparation and preservation of plant specimens (Jain, 1964). Voucher specimens of medicinal plants in triplicate were collected, prepared and identiûed. Plants with their correct nomenclature were arranged alphabetically by family name, vernacular name and ethno medicinal uses. Identiûcation and nomenclature of the listed plants were based on the Flora of Presidency of Madras (Gamble, 1935), Flora of Tamil Nadu Carnatic (Matthew, 1983) and available monographs, taxonomic revisions (Henry *et al.*, 1989) and by using field keys. The data were collected through repeated field visits and the careful interaction with the village peoples and by participating rural appraisal. The specimens were deposited in the Department Silviculture and Agroforestry, College of Forestry, Vellanikkara, Kerala for future reference. The data's were obtained from the informed constants of interviewed individuals. Interview of minimum 5 and maximum of 10 traditional healers, and village elders who have been using the medicinal plant for curing the various health problems were conducted. The collected data were confirmed and compiled by repeated visits and general talk with the patients.

Results and Discussion

The present study revealed that diversity of ethnomedicinal plants in Thalamalai hills, Tamil Nadu has 120 plant species for treating and healing the diseases in their traditional way. Data obtained from the survey is compiled in Table 1. All plant species are arranged in alphabetical order. Among these 120 plant species, there were trees (29 spp.), shrubs (16 spp.), climbers (21 spp.) and herbs (54 spp.) (Table 1). The 120 ethnomedicinally valuable plant species belonged to 100 genera and 56 families and included. For each species scientific name with local name, family and method of drug preparation with dosage are provided. The most dominating families were Euphorbiaceae and Fabaceae with 7 species each followed by Convolvulaceae (06), Acanthaceae, Cucurbitaceae, Mimosaceae, Solanaceae with 5 species each (Fig. 2). Twenty nine species (24%)

are trees, 16 species (13%) are shrubs, 54 species (45%) are herbs and 21 species (18%) are climbers (Fig. 3). Different plant parts were used for the treatment of ethnomedicinal perspectives in Thalamalai hills, Namakkal district, Tamil Nadu. Leaves were the most used (41.79%) followed by roots (18.65%), fruits (11.94%), bark (6.71%), flowers, seeds, stem and whole plant (3.73%), latex (2.98%), Rhizomes and tuber (1.49%) in decreasing order (Fig. 4).

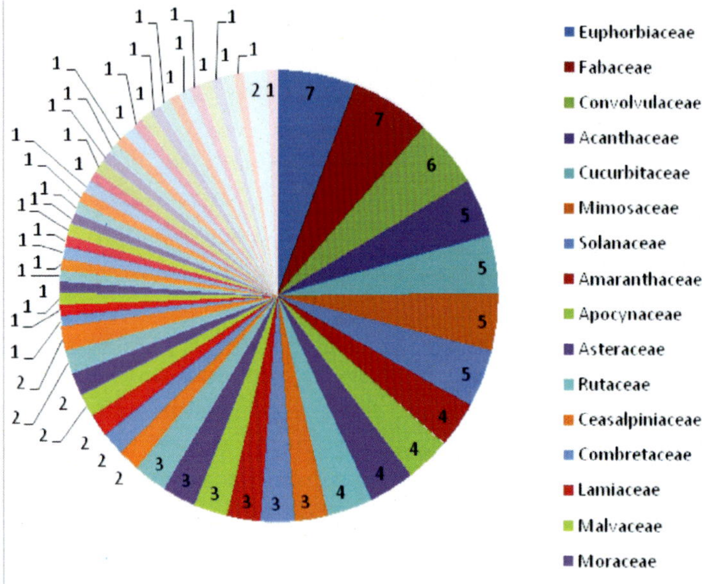

Fig. 2. Family basis ethnomedicinal plants categories in Thalamalai hills, Namakkal district, Tamil Nadu, India

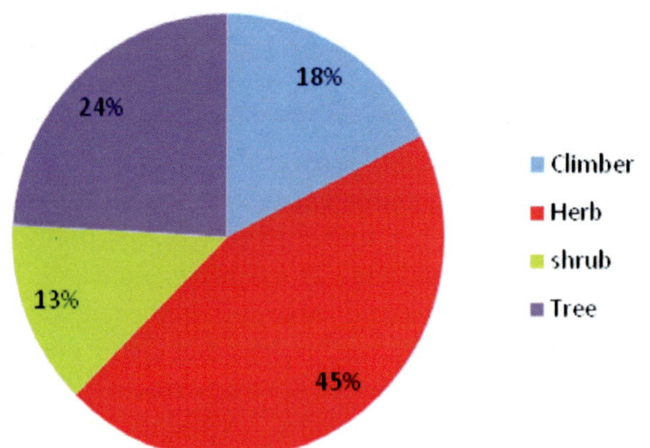

Fig. 3. Habit of ethnomedicinal plants in Thalamalai hills, Namakkal district, Tamil Nadu, India

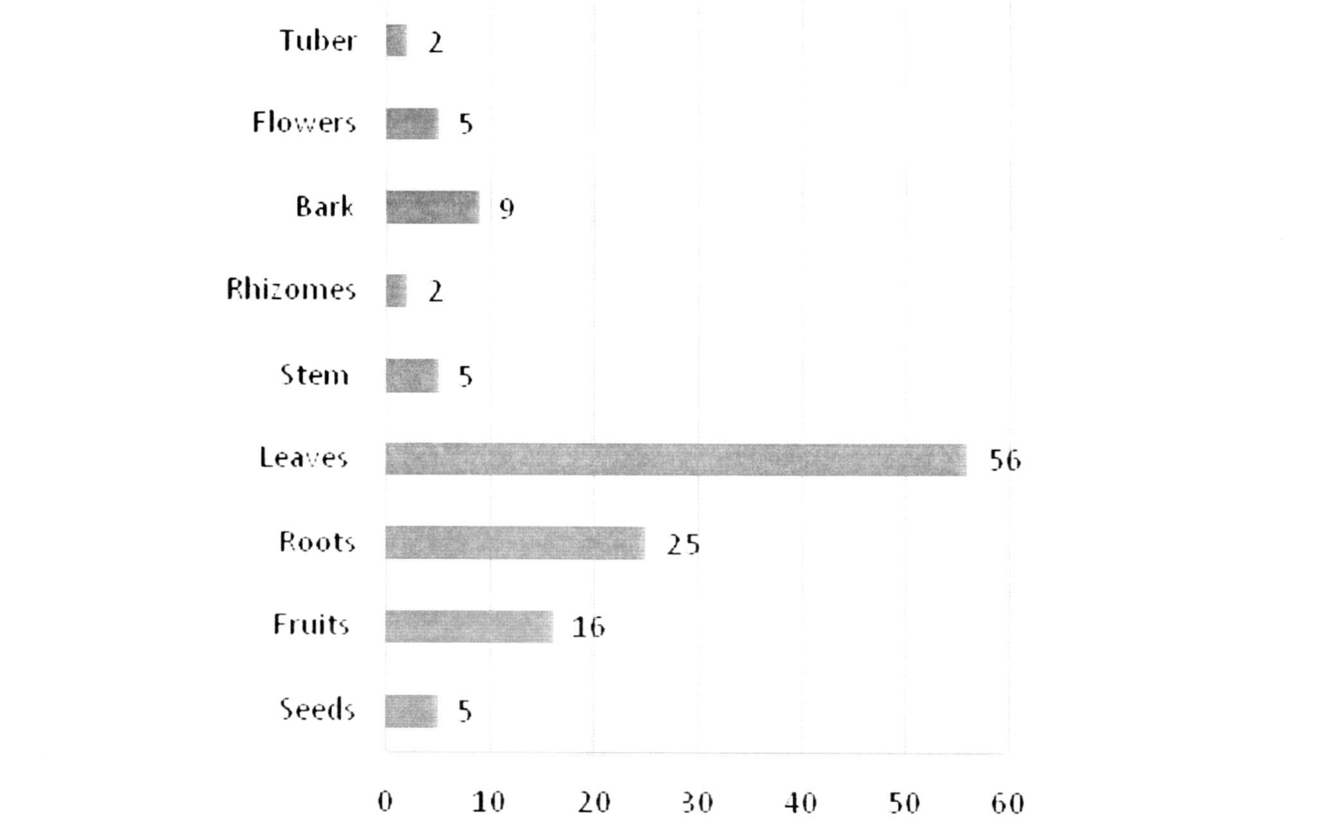

Fig. 4. Plants parts used in different ethnomedicinal perspectives in Thalamalai hills, Namakkal district, Tamil Nadu, India

Table 1. Important ethnomedicinal uses of angiosperms of Thalamalai hills, Namakkal district, Tamil Nadu

Sl. No.	Scientific name	Local name	Family	Uses
1.	*Abrus precatorius* L.	Kundumani	Fabaceae	Dried seeds decoction was taken orally to induce abortion and also to cure in tuberculosis, painful swellings.
2.	*Abutilon indicum* (L.) Sweet.	Thuththi	Malvaceae	Crushed leaves were applied on skin diseases like eczema.
3.	*Acacia chundra* (Roxb. ex Roth) Willd.	Karungali	Mimosaceae	Bark juice were taken twice in day for three days orally to cure stomach ache.
4.	*Acacia concinna* (Willd.) DC	Shikakai	Fabaceae	Fruit paste/powder applied to head 30 minutes before bath to promote hair growth.
5.	*Acacia leucophloea* (Roxb.) Wild.	Velvalan	Mimosaceae	Crushed bark were applied Skin disease like eczema.
6.	*Acacia nilotica* (L.) Willd.	Karuvelam	Mimosaceae	Young branches were used as toothbrush daily will give whitish teeth. Leaves paste consumed to cure stomach disorder twice a day for three days.
7.	*Acalypha indica* L.	Kuppaimeni	Euphorbiaceae	Leaves paste were applied on throat once a day for two days to cure the severe cough.
8.	*Achyranthes aspera* L. var. aspera.	Nayuruvi	Amaranthaceae	Leaves paste were applied for wounds and dog bite. Roots of a plant boiled with glass of milk, reduced to half and taken orally by mentally retarded people in a single dose.
9.	*Acorus calamus* L.	Vasambu	Acoraceae	Rhizomes paste is sedative, analgesic in nature used to cure the skin disease, indigestion, blood pressure and respiration.
10.	*Adhatoda vasica* Nees.	Adathoda	Acanthaceae	Leaves were grind with *Zingiber officinale* and *Piper betle* L. for fever twice in a day.
11.	*Aegle marmelos* (L.) Corr.	Vilvam	Rutaceae	Leaves paste with curd are used in dysentery. Bark juice is taken thrice a day for one week in intermitted fever. Fruit extract will reduce the diabetics.
12.	*Aerva lanata* (L.) Juss. Ex Schult.	Cerupulapoo	Amaranthaceae	Decoction of the root taken orally in empty stomach once a day for a month to cure diabetics.
13.	*Ailanthus excelsa* Roxb.	Peeyamaram.	Simaroubaceae	Leaves decoction with *Aristolochia indica* and goat milk Scorpion sting and snake bite.
14.	*Alangium salvifolium* L. f.	Alangimaram	Alangiaceae	Leaf paste/ powder applied to head before 30 minutes to get good hair growth. Bark decoction were used for fever and jaundice.
15.	*Albizia amara* (Roxb.)B. Boivin.	Usilmaram	Fabaceae	Leaves paste/powder were applied to head for good hair growth.
16.	*Aloe vera* (L.) Burm. Fil.	Chottukathalai	Liliaceae	Leaves juices were applied externally swelling due to itching and burns.
17.	*Alternanthera sessilis* (L.) R. Br. ex. Dc.	Ponnakanni	Amaranthaceae	Cooked leaves taken regularly for a month to cure night blindness.

18.	*Amaranthus tristis* Roxb.	Thandangeerai	Amaranthaceae	Decoction of the leaves was taken orally in thrice in day to reduce pain during mensuration.
19.	*Andrographis paniculata* Nees.	Nilavembu	Acanthaceae	Leaves paste along with *Andrographis alata* leaves paste for Scorpion sting and snake bite. Whole plant decoction was used for fever.
20.	*Argemone mexicana* L.	Nari mirati	Papavaraceae	Yellow latex use to cure the wounds, pimples and ulcer in the lips.
21.	*Aristolochia indica* L.	Eswaramooligai	Aristolochiaceae	Fresh juice of the leaves is a popular antidote to snake poison. The leaves and bark are used in bowel complaints of children, diarrhoea and in intermittent fever.
22.	*Asparagus racemosus willd.*	Thanneervitankizhngu	Asparagaceae	Roots of plant are considered to be effective as antisplasmodic, aphrodisiac, antidiarrheal, antidysentiric, blood purifier, and also in night blindness, kidney and urinary troubles.
23.	*Azadirachta indica* A. Juss.	Vembu.	Meliaceae	Leaves paste were mixed with *Ricinus communis* oil applied for the Small pox for 7 days. Fresh juice of leaves with salt is remove intestinal worms and applied externally skin disease.
24.	*Azima tetracantha* Lam.	Mulluchangu	Salvadoraceae	Leaves juice were applied with *Piper longum* to cure the Cold and cough. Root paste was applied to wound.
25.	*Boerhavia diffusa* L.	Padarmookirattai	Nyctaginaceae	Leaves decoction used to cure fever, urinary troubles and asthma.
26.	*Boerhavia erecta* L.	Mookirattai	Nyctaginaceae	Plant powder was smoked to cure the asthma.
27.	*Cardiospermum halicacabum* L.	Mudakkathaan	Sapindaceae	The cooked leaves were eaten for week to cure rheumatism.
28.	*Carissa carandas* L.	Kalakkai	Apocynaceae	Latex used to cure the mouth cancer. Fruit were eaten raw to reduce the diabetics.
29.	*Cassia auriculata* L.	Avaaram	Ceasalpiniaceae	Leaves paste were applied on the head to cool the body and good hair growth.
30.	*Cassia obtusifolia* L.	Nilavagai	Ceasalpiniaceae	Leaves paste with *Zingiber officinale* powder to cure anti tumour and throat pain.
31.	*Cassine glauca* (Rottb.) Kuntze	Karuvaali	Celastraceae	Paste of leaves externally applied to cure and to give relief in stomach pains.
32.	*Catharanthusroseus* (L.) G. Don	Sudukattumalli	Apocynaceae	Decoction of white flowers and leaves taken for loose motion.
33.	*Cissus quadrangularis* L.	Pirandai	Vitaceae	Stem were cooked as chatini and eaten as a food to strong bone and indigestion.
34.	*Citrullus colocynthis* (L.) Schrad.	Kumattikkaai	Cucurbitaceae	Fruits were pungent, purgative antipyretic, anthelmintic cure asthma, tumors, ulcer and jaundice.
35.	*Cleome gynandra* L.	Nallavelai	Capparidaceae	Leaves paste were applied to cure wounds daily.
36.	*Cleome viscosa* L.	Naaikkadukhu	Capparidaceae	Juice of leaves is used in curing ear pain.
37.	*Clitoria ternatea* L.	Sankupu	Fabaceae	Root paste is diuretic, anthelmintic, depurative and Aphrodisiac in nature.

38.	*Coccinia indica* (L.) J. Voigt.	Kovai.	Cucurbitaceae	Decoction of the leaves was mixed with *Acalyphaindica* and *Piper nigrum* to cure the jaundice andulcer. Leaves juice applied externally to cure itching.
39.	*Cocculus hirsutus* Diels.	Kattukkodai.	Menispermaceae	Decoction of leaves used externally in sunstroke. Paste of root is externally used in rheumatism.
40.	*Corallocarpus epigaeus* Rottl. Ex. Wild.	Kolankovakilang hu	Cucurbitaceae	Tuber paste applied to snake bite to remove the poison. Tuber powder with hot milk will reduce the blood sugar.
41.	*Croton sparsiflorus* Morong.	Rail poondu	Euphorbiaceae	Leaves juice were applied to the wounds to stop bleeding.
42.	*Curcuma longa* L.	Manjal	Zingiberaceae	Rhizome powder is mixed with *Azadirachta indica leaf* powder applied daily to heal wound.
43.	*Cuscuta chinensis* Lam.	Manjapullaruvi	Convolvulaceae	Stem paste applied to fractured bone in order to promote the joining.
44.	*Cynodon dactylon* (L.) Pers. var. dactylon.	Arugambullu	Poaceae	Leaves juice are made used as drink to cool the body heat.
45.	*Cyperus rotundus* L.	Koraipul	Cyperaceae	Tuber paste taken for the stomach ache.
46.	*Datura metel* L.	Oomaththai	Solanaceae	Leaves paste were applied to cure swelling asthma.
47.	*Delonix elata* (L.) Gamble.	Vadanaraayam	Ceasalpiniaceae	Leaves juice was mix Juice mix with *Piper longum* to cure cough and cold.
48.	*Desmodium gangeticum* (L.) DC.	Pullati	Fabaceae	Leaves decoction used for Diarrhoea and dysentery. Whole plant used as antipyretic.
49.	*Dioscorea bulbifera L.*	Kaivallikizhangu	Dioscoreaceae	Paste of the root is used in piles and syphilis.
50.	*Dodonaea viscosa* Jacq.	Virali	Sapindaceae	Powder of the flower was used to cure the toothache, sore throats, haematoma and wounds.
51.	*Eclipta alba* (L.) Hassk.	Karisalanganni	Asteraceae	Root paste used for spleen, problems bronchitis, asthma and leucoderma.
52.	*Eclipta prostrata* (L.) L.	Karisalankakki	Asteraceae	Leaves paste are applied with *Ricinus communis* oil in head for hair growth and applied on the wounds between toes.
53.	*Euphorbia hirta* L.	Ammanapachari si	Euphorbiaceae	Decoction of whole plant to cure the diarrhea, ulcers, enteritis and asthma.
54.	*Evolvulus alsinoides* L.	Vishnukarandhi	Convolvulaceae	Leaves paste applied regularly to promote hair growth.
55.	*Feronia elephantum* Correa.	Vila	Rutaceae	Fruit juice was drunk orally to cure hiccough, sore throat and diseases of the gums.
56.	*Ficus benghalensis* L.	Aalamaram	Moraceae	Milky juice was tapped in the early morning applied over swelling area, and ulcers. Aerial roots were used as toothbrush.
57.	*Ficus religiosa* L.	Arasamaram	Moraceae	Dry fruits were eaten as raw for diabetes. Tender leaves paste applied externally to cure itching.

58.	*Ficus retusa* L.	Athi	Moraceae	Latex was tapped from the tree applied in bone fracture area daily for months. Fruits were able to reduce blood pressure and purifier.
59.	*Fimbristylis cymosa* R.Br.	Kothuppullu	Cyperaceae	Root extract with hot milk in empty stomach to cure dysentery.
60.	*Gloriosa suberba* L.	Kanvilikizhangu	Liliaceae	The paste of seeds and roots is applied on skin diseases and arthritis. Rhizome paste applied to the wounds.
61.	*Gymnema sylvestre* (Retz.) Schult.	Sirukurinjan	Asclepiadaceae	Leaves decoction taken orally twice a day to cure the fever. Root decoction will reduce diabetics.
62.	*Hemidesmus indicus* (L.) R.Br.	Nannari	Apocynaceae	Syrup made with an infusion of the roots is used to make a Sherbet. The decoction of root is used to cure the fever, skin diseases and blood purifier.
63.	*Hybanthus enneaspermus* L.	Oridhazhthamar ai	Violaceae	Leaf extract with cow milk in the early morning to increases sexual vigour.
64.	*Indoneesiella echioides* (L.) Sreemadh.	Koburandhaangi	Acanthaceae	Leaves paste were applied to externally to cure the Itches and skin disease.
65.	*Ipomoea aquatica* Forsskal	Veelaikkeerai	Convolvulaceae	Flower juice applied around the eye to Cure black ring.
66.	*Ipomoea obscura* (L.) Ker Gawler.	Siruthaalikkodi	Convolvulaceae	Leaves juice were applied to Snake bite.
67.	*Jatropha gossypifolia* L.	Kattamanakku	Euphorbiaceae	Leaves paste with coconut oil applied externally to cure the itching and skin diseases.
68.	*Justicia adhatoda* L.	Adathodai	Acanthaceae	Leaves were boiled with pepper to cure cough and fever.
69.	*Leucas aspera* Spreng.	Thumbai.	Lamiaceae	Leaves decoction with hot milk fever and headache. Whole plant is used as an insect repellent.
70.	*Madhuca longifolia* (Koen) Macler.	Iluppai	Sapotaceae	Flowers and stem were crushed and applied to control the haemorrhage. Leaves paste were applied externally to secreting milk in feeding mothers.
71.	*Marsilea minuta* L.	Arakeerai	Marsileaceae	Leaves powder were given to reduce diabetics daily
72.	*Melochia corchorifolia L.*	Yennaichedi	Sterculiaceae	Leaves decoction taken orally in empty stomach to cure the dysentery.
73.	*Merremia emarginata* Burm.	Elikaadhukeerai	Convolvulaceae	Leaves juice with honey for 3 days to cure cold and cough.
74.	*Merremia tridentata* (L.) Hallier.	Mudhiyaarkoond hal	Convolvulaceae	Root decoction taken orally for the 40 - 60 days to cure the diabetics.
75.	*Mimosa pudica* L.	Thottaalsunugki	Mimosaceae	Decoction of whole plant taken orally to reduce the excess menstrual bleeding, swelling and rheumatism.
76.	*Momordica charantia* L.	Pagal	Cucurbitaceae	Fruit juice or matured fruit were fried in groundnut oil taken daily to kill the Intestinal worms.
77.	*Mukia maderaspatana* (L.) M. Roemer.	Musumusukkai	Cucurbitaceae	Root paste applied twice a day to cure the tooth-ache.
78.	*Murraya koenigii* (L.) Spreng.	Kariveppilai	Rutaceae	Fried leaves were made as chatini taken in food for good hair growth.

79.	*Musa paradisiaca* L.	Vazhai.	Musaceae	Stem juice is extracted in the early morning taken orally to cure nephrolith.
80.	*Ocimum basilicum* L.	Thiruneetrupatch ilai	Lamiaceae	10-15 matured leaves taken raw to cure cold and cough. Root extract taken orally ton cure malarial fever in once day for a week.
81.	*Ocimum tenuifiorum* L.	Thullasi.	Lamiaceae	Leaves were boiled with pepper and taken daily to cure cough cold and fever.
82.	*Oldenlandia umbellata* L.	Muthkkaasu	Rubiaceae	Leaves decoction orally taken daily to cure asthma and bronchitis.
83.	*Passiflora foetida* L.	Siruppunakkai	Passifloraceae	Fruit powder with *Piper longum* and milk to cure the asthma.
84.	*Pedalium murex* L.	Aanainerungil	Pedaliaceae	Fruit decoction with hot water for 2-3 days to cure gonorrhea, diarrhea and dysentery.
85.	*Pergularia daemia* (Forsk). Choir.	Veliparuthi.	Asclepiadaceae	Leaves paste was applied externally to cure the insect bites, Snake bite.
86.	*Phoenix pusilla* Roxb.	Icham	Palmaceae	Fruit were eaten raw to cure the diabetics.
87.	*Phyllanthus amarus* L.	Keelanelli	Euphorbiaceae	Root paste with hot milk for the 10 days to cure the jaundice.
88.	*Phyllanthus emblica L.*	Malainelli	Phyllanthaceae	Whole fruits with sugar syrup fruits are used in preparation of hair oil, which has hair darkening properties. Dry fruits used in bronchitis. dysentery, asthma.
89.	*Plumbago zeylanica* L.	Chitramoolam	Plumbaginaceae	Cup of fresh cow milk is stirred with a piece of the root and taken internally to stop pregnancy.
90.	*Polygonum plebeium* R.Br.	Kanganichedi	Polygonaceae	Root paste applied externally twice a day to cure inflammations
91.	*Pongamia pinnata* L.	Pungam	Fabaceae	Seed oil used in rheumatism. Stem bark with *Piper longum* to cure stomach ulcer.
92.	*Portulaca oleracea* L.	Paruppukkeerai	Portulacaceae	Leaves of the plant were cooked and eaten to cure the mouth wounds.
93.	*Rhinacanthus nasutus* (L.) Kurz.	Nagamalli.	Acanthaceae	Leaves juice with cow's milk to cure the fever and cough. Root juice was applied to cure skin disease and ring worm.
94.	*Ricinus communis* L.	Amanakku.	Euphorbiaceae	Seed oil were taken orally for remove the stomach worms and applied to head for hair growth and also body pain.
95.	*Santalum album* L.	Chandhanam	Santalaceae	Wood paste applied in the forehead in religious customs. Also cure pimples in the face.
96.	*Sapindus emarginatus* Vahl	Poochamaram	Sapindaceae	Matured fruit pulp was used as detergent soap for washing cloths and disinfects the home.
97.	*Sida cordifolia* L.	Nilathuthi	Malvaceae	Leaf extract were taken twice a day for the 2 -3 days to cure the itching and headache.
98.	*Sida rhombifolia* L.	Pazhambaasi	Malvaceae	Decoction of the leaves with hot water to cure the Stomachache and fever.
99.	*Solanum nigrum* L.	Manathakkaali	Solanaceae	Leaf juice was consumed orally to cure the stomach ulcer and high fever. Leaves were cooked and eaten to get good vision.

100.	*Solanum surattense* Burm.f.	Kandankatthari	Solanaceae	Fruit powder/paste was used to cure cough asthma and tooth problems.
101.	*Solanum torvum* Sw.	Sundakkai	Solanaceae	Fruits extraction was used in Asthma, tuberculosis,
102.	*Stachytarpheta indica* (L.) Vahl.	Seemainaayuruv i	Verbenaceae	Root extract were with hot water to cure diarrhoea dysentery and intestinal worms.
103.	*Strychnos nux-vomica* L.	Etti	Loganiaceae	Fruit paste is applied externally to cure the wounds.
104.	*Syzygium cumini* L.	Naval	Myrtaceae	Fruit raw /with salt to cure diabetics. Stem bark decoction with hot water to cure indigestion and stomach ulcer.
105.	*Tephrosia purpurea* (L.) Pers.	Kozhinji	Fabaceae	Root decoction with pepper to cure jaundice, diarrhea, rheumatism, asthma and urinary disorder.
106.	*Terminalia bellarica* Roxb.	Thaanikai	Combretaceae	Fruit powder leaves were applied to cure skin disease, small pox and tooth ache. Fruit extracts were applied to cure the dysentery.
107.	*Terminalia chebula* Retz.	Kadukkai	Combretaceae	Fruit powder with *Phyllanthus emblica* L and *Terminalia bellarica* Roxb. Digestive disorder.
108.	*Terminalia tomentosa* (Roxb. Ex DC.) Wight & Arn.	Karimaruthu	Combretaceae	A decoction of the bark is taken internally for intestinal problems.
109.	*Tinospora cordifolia* (willd.) Miers.	Seendhil	Menispermaceae	Leaf extract with honey ton cure the jaundice.
110.	*Toddalia asiatica* Lam.	Kindumullu	Rutaceae	Fruits along with *Solanum torvum* and goat`s milk taken internally to strengthen the body for 10-20 days.
111.	*Tragia involucrata* L.	Sendhatti	Euphorbiaceae	Root decoction with *Piper longum* L. seed paste orally to cure the bronchitis and asthma.
112.	*Trianthema decandra* L.	Sirusaaranai	Aizoaceae	Root extract with honey to cure the asthma.
113.	*Trianthema portulacastrum* L.	Vellaichaaranai	Aizoaceae	Leaf decoction with hot water to cure Rheumatism.
114.	*Tribulus terrestris* L.	Nerunchi.	Zygophyllaceae	Whole plant juice with hot water for thrice a day for 2 -3 days to cure the cough and cold.
115.	*Trichodesma indicum* (L.) R.Br.	Kavizhthumbai	Boraginaceae	Root paste twice a day to cure swelling in the joints.
116.	*Tridax procumbens* L.	Thaathapoothala i	Astraceae	Juice and paste of leaves were used to cure the haemorrhage and antitumour.
117.	*Vernonia cinerea* L.	Sirudhevisengel uneer	Astraceae	Juice of roots is used to dissolve kidney stone and diarrhoea.
118.	*Withania somnifera* Dun.	Amukramkizhan gu	Solanaceae	Root powder with *Piper longum* seed paste drunken orally to cure the fever and stomachache. Root was aphrodisiacs in nature.
119.	*Wrightia tinctoria* (Roxb.)R. Br.	Paalai	Apocynaceae	Bark along with jeera and garlic used against jaundice.
120.	*Zingiber officinale* Roscoe	Inji	Zingiberaceae	Rhizome used as such and also added in different food preparation and headache, stomachache and cure the severe cough.

During the last few decades there has been an increasing interest in the study of medicinal plants and their traditional use in different parts of India and there are many reports on the use of plants in traditional healing by either tribal people or indigenous communities of Namakkal district (Subramaniyan *et al.*, 2003; Udayan *et al.*, 2005; Santhya *et al.*, 2006; Sankaranarayanan *et al.*, 2010; Kumari Subitha *et al.*, 2011; Kishor Kumar and Satheesh Kumar, 2011; Vethanarayanan *et al.*, 2011; Francisca and Rajendran, 2012; Prabu and Kumuthakalavalli, 2012; Sindhu *et al.*, 2012; Dhayapriya and Senthil, 2014; Ramanathan *et al.*, 2014; Kumar and Mano, 2015). All ethnomedicinal plants documented in the present study have continuously been used and the results also revealed that some of them are less known and some of them supplements to the available earlier data.

Conclusion

For development and domestication of new drugs study of Traditional medicines are very important. The unique traditional system of healthcare progressed from generation to generation within the society is still prevalent within the remote rural areas of the country. Important medicinal plants need immediate conservation in order to avoid degradation and the deforestation in this area. The cultivation and establishment should be encouraged to prevent the extinction of potentially valuable species.

References

Abu-Rabia, A. 2005. Urinary diseases and Ethnobotany among pastoral nomads in the middle East. *J. Ethnobiol. and Ethnomedicine,* 1: 4-18.

Berkes, F., Colding, J., Folke, C. 2003. Navigating Social-Ecological Systems: Building Resilience for Complexity and Change. Cambridge University Press, Cambridge, UK.

Dhayapriya, R.G. and Senthil K.S. 2014. Studies on Ethnomedicinal Plants of Malayali Tribes in Bodamalai Hills of Southern Eastern Ghats, Tamil Nadu, India. *Inter. J. Pharma. Res. & Devel.,* 6(3): 115-118.

Dubey, N.K. 2004. Global promotion of herbal medicine: India opportunity. *Curr. Sci.,* 86(1): 37-41.

Francisca G.B. and Rajendran, A. 2012. Ethnobotany of Irulartribes in Redhills, Tamil Nadu, India. *Asian Pacific J. Trop. Dis.,* 874-877.

Gamble, J.S. 1935. The Flora of the Presidentcy of Madras, Bishen singh Mahendra Pal Singh, Dehradun, Vol. 1-III, 1986, 2017p.

Ganesan, S. 2004. Traditional oral care medicinal plants survey of Tamil Nadu. *Nat. Prod. Radi.*, 7(2): 166-172.

GOI, 2000. Report of the Taskforce on Medicinal Plants in India. Planning Commission, Government of India, Yojna Bhawan, New Delhi, India.

Harshburger, J.W. 1896. Purpose of Ethnobotany, *Botanical Gazette*, 21: 146-154.

Henry, A.N., Chitra, V. and Balakrishnan, N.P. 1989. Flora of Tamil Nadu, India (Series I. Vol. 3) Bot. Surv. India, Southern Circle, Coimbatore.

Jain, S.K. 1891-1936. Dictionary of Iaolar folk medicine and Ethanobiology. Deep Publishers, New Delhi, India.

Jain, S.K., 1964. The role of Botanist in folklore research. *Folklore,* 5: 145-150.

Jayakumar, K. 2013. Traditional Medicinal Values of *Adhatoda vasica. International Journal of Traditional and Natural Medicines,* 3(1): 26-31.

Kargioglu, M., Cenkci, S., Serteser, A., Evliyaoglu, N., Konuk, M., Kok, M.S. and Bagci, Y. 2008. An ethnobotanical survey of inner-West Anatolia, Turkey. *Hum. Ecol.,* 36: 763-777.

Kishor Kumar, V. and Satheesh Kumar, P. 2011. Ethnomedicinal Plants from Vattamalai Hills of Namakkal District, Tamil Nadu, India. Inter. *J. Phar. Sci. Revi. Res.,* 11(2): 100-106.

Kumar, V. 2015a. Ethno-medicinal plants in five forest ranges in Dang district, south Gujarat, India. *Indian Journal of Tropical Biodiversity,* 23(2): 148-156.

Kumar, V. 2015b. Ethnomedicines of Malayan tribes of southern region of Kerala, India. In book: Recent advances in Ethnobotany, Sanjeev Kumar, Deep Publication, New Delhi, pp: 173-185.

Kumar, V. and Mano, G. 2015. Ethno-Medico-Biology of Cuddalore District, Tamil Nadu, India. In Sanjeev Kumar (Eds), Recent advances in Ethnobotany. Deep Publication, New Delhi, pp. 209-214.

Kumar, V., Babu, S., Revale, A.K., Meena, R.K., Ranjan, M.K. and Desai, B.S. 2014. Cultivation of medicinal plants in natural ecosystem in Gujarat (India): constraints and conservation need. *Journal of Plant Development Sciences*, 6(3): 425-435.

Kumari Subitha, T., Ayyanar, M., Udayakumar, M. and Sekar, T. 2011. Ethnomedicinal plants used by Kani tribals in Pechiparai Forests of Southern Western Ghats, Tamil Nadu, India. *Inter. Res. J. Plan. Sci.* 2(12): 349-354.

Macia, M.J., Garcia, E. and Vidaurre, P.J. 2005. An ethnobotanical survey of medicinal plants commercialized in the markets of La Paz and El Alto, Bolivia. *J. Ethnopharmacol.,* 97: 337-350.

Matthew, K.M., 1983. The Flora of the Tamil Nadu Carnatic. The Rapinat Herbarium, vol. 3. St. Josephs College, Tiruchirapalli, India, IXXXIV (2154).

Muthu, C., Ayyanar, M., Raja, N. and Ignacimuthu, S. 2006. Medicinal plants used by traditional healers in Kancheepuram district of Tamil Nadu, India. *J. Ethnobiol. Ethnomed.,* 2: 43-48.

Omoruyi, B.E., Bradley, G. and Afolayan, A.J. 2012. Ethnomedicinal survey of medicinal plants used for the management of HIV/AIDS infection among local communities of Nkonkobe Municipality, Eastern Cape, South Africa. *J. Med. Plants Res.* 6: 3603-3608.

Pei, S.J., 2001. Ethnobotanical approaches of traditional medicine studies some experiences form Asia. *Pharma Bio.,* 39: 74-79.

Plowden, C., Uhi, C. And Oliveira, F.A. 2003. The ecology and harvest potential of titica vine roots (*Heteropsis flexuosa*: Araceae) in the eastern Brazilian Amazon. *For. Ecol. Manage. J.,* 182(1,3): 59-73.

Prabu, M. and Kumuthakalavalli 2012. Folk remedies of Medicinal plants for snake bites, scorpion stings and dog bites in Eastern Ghats of Kolli hills, Tamil Nadu, India. *IJRAP,* 3(5): 696-700.

Ramanathan, R., Bhuvaneswari, R., Indhu, M., Subramanian, G. and Dhandapani, R., 2014. Survey of ethnobotanical observation on wild tuberous medicinal plants of Kolli hills, Namakkal district, Tamil Nadu. *J. Med. Plan. Stud.,* 2(4): 50-58.

Rao, M.R., Palada, M.C. and Becker, B.N. 2004. Medicinal and aromatic plants in agroforestry systems. *Agroforest. Syst.,* 61: 107-122.

Sankaranarayanan, S., Bama, P. and Ramachandran, J. 2010. Ethnobotanical study of medicinal plants used by traditional users in Villupuram district of Tamil Nadu, India. J. *Med. Pla. Res.,* 4(12): 1089-1101.

Santhya, B., Thomas, S., Isabel, W. and Senbagarathi, R. 2006. Ethnomedicinal Plants Used by the Valaiyan community of Piranmalai Hills (Reserved Forest), Tamil Nadu, India. A Pilot Study. *Afr. J. Tradit. Complement. Altern. Med.,* 3(1): 101-114.

Sindhu, S., Uma, G. and Kumudha, P. 2012. Survey of medicinal plants in Chennimallai Hills, Erode Districts, Tamilnadu. *Asian J. Pla. Sci. and Res.*, 2(6): 712-717.

Subramaniyan, A., Mohan, V.R., Kumaresan, S. and Chelladurai, V. 2003. Medicinal plants used by the Valaiyan of Madurai district, Tamil Nadu. *J. Econ. Taxon. Bot.*, 27(4): 785-787.

Udayan, P.S., George, S., Thushar, K.V. and Balachandran, I. 2005. Ethnomedicine of the chellipale community of Namakkal district, Tamil Nadu. Ind. *J. Tra. Knowl.*, 4 (4): 437-442.

Vethanarayanan, P., Unnikannan, P., Baskaran, L. and Sundaramoorthy, P. 2011. Survey on traditional medicinal plants used by the village peoples of Cuddalore district, Tamil Nadu, India. *Asian J. Bioche. and Pharma. Res.*, 3(1): 351-36.

❑❑❑

Chapter 23

Conservation and Utilization of High Attitude Valuable Medicinal Plant

Dhiman Mukherjee*

Directorate of Research, Bidhan Chandra Krishi Viswavidyalaya, Kalayani, Nadia, W.B.-741235, India, E.mail: dhiman_mukherjee@yahoo.co.in.

* Ex-officer Incharge of AICRP on Medicinal and Aromatic Plant, Kalimpong, Dist: Darjeeling, West Bengal.

ABSTRACT

Sustaining the natural resource viz. valuable medicinal plant base is a must, but is a great challenge for India, especially in the highland areas where alternate opportunities of livelihood and income generation for the communities is less and agriculture production situation is poor. Planning and budgeting research and extension promotion strategies and activities for economically important medicinal plants is a priority issue for developing suitable cultivation technologies aimed at domestication and commercial cultivation. The available technologies need to be updated, tested, verified and documented. Likewise, sustainable harvesting techniques should be identified and adopted by grass-root users for in situ conservation of high altitude especially endangered medicinal herb. Special attention need to be given on *Aconitum ferox, Acorus calamus, Dioscorea composite, Dichroa febrifuga, Drymaria diandra, Digitalis purpurea, Gentiana kurro, Heracleum wallichii, Litsaea cubeba, Nardostachys grandiflora, Oroxylum indicum, Panax pseudo-ginseng, Picrorhiza kurrooa, Podophylum hexandrum, Przewalskia tangutica, Pteris biaurita, Rheum modi, Swertia chirata* and *Valeriana jatamansi.* Research and management are necessary for processing and improving quality of products and increasing organic production. Medicinal plant should be included in the income generation activities of the community forestry to enhance rural income supporting livelihoods and reduce poverty. Capacity of the medicinal plant collectors and cultivators needs to build up in areas such as cultivation, sustainable harvesting, post-harvest/primary processing through training, visits, interactions etc.

Keywords: Conservation, Endangered, Medicinal plant, Utilization.

Introduction

The use of valuable plants in curing various human ailments is known since time immemorial. In recent past, despite considerable progress made in synthetic drugs, plant constituents are still considered as major sources of valuable medicines, as no substitute for chirayita, kuth etc. It is estimated that close to 15% of the 70,000 known plant species have medicinal properties. As per different report about 70-80% of people worldwide rely chiefly on traditional, largely herbal medicine to meet their primary healthcare needs. India has tradition and long history on use of valuable medicinal plants as medicine, health hygiene and food supplements in improving the quality of life (Bhusan *et al.*, 2005). India has written records of the use of plants for curing human and animal diseases which can be traced back to the earliest scriptures of Hindus (4500-1600 BC). India has a rich heritage of traditional system of medicine that is mostly based on formulations derived from herbs, shrubs, trees, vines and some animal products. The country has 45,000 different plant species and 15000 medicinal plants that include 7000 plants used in Ayurveda, 700 in Unani, 600 in Siddha, 450 in Homeopathy and 30 in modern medicines. (Ravishankar and Shukla, 2007). The Indian system of medicines (ISM) comprises of Ayurveda, Siddha and Unani which have their long roots in the society. Ayurveda is about 5,000 years old and predominantly uses medicinal and aromatic plants for their preparations and formulations. Modern pharmacopeia also listed about 25 per cent of drugs derived from plants and vast majority of synthetic analogues built on prototype compounds isolated from plants. The Himalayas are exceptionally rich in medicinal plants. In India, the rich plant diversity of the Himalaya – over 8000 angiosperms, 44 gymnosperms, 600 pteridophytes, 1737 bryophytes, 1159 lichens, etc. – has been a source of medicine for millions of people in the country and elsewhere in the world (Dhar *et al.*, 2002, Mukherjee, 2008). Plant based herbal medicines particularly high altitude (> 1800 m asl) plant have been used for thousands of years in many parts of the world. The therapeutic use of herbal medicines is gaining considerable momentum in the world during the last decade. Much of the tropical plants like *Adhatoda zeylanica, Centella asiatica, Emblica officinalis, Piper mullesua, Terminalia bellirica, T. chebula. Piper longum* and many others are found in lower and mid hill regions play crucial role in medicinal plant industry. The sub-tropical species such as *Acorus calamus, Terminalia chebula, Sapindus pinnatus, Melia azadiracta, Solanum khasianum,*

Aspagarus racemosus, Rauvolfia serpentina, Dioscorea bulbifera, Aristolochia indica, etc. temperate species such as *Swertia chirayata, Taxus baccata, Valerina jatamansi, Viola odorata, Sapindus pinnatus, Rubia cordifolia* and various species of *Artemesia, etc.* and alpine species like *Aconitum ferox, Podophyllum hexandrum, Picrorhiza kurrooa, Bergenia purpurascens,* etc. use as human and animal cure of various disease. Out of these, namely *Glycyrrhiza glabra, Swertia chirayita, Commiphora mukul, Plantago ovata, Aloe barbadensis* and *Azadirachta indica* are even used in modern medicine (Mukherjee, 2009). The plant *Glycyrrhiza glabra, Piper longum, Adhatoda vasica, Withania somnifera, Tinospora codifolia, Berberis aristata,* and *Boerhavia diffusa* have been used in more than 200 herbal formulation and triphala (*Terminalia chebula, Terminalia belerica* and *Emblica officinalis*) alone have been used in 253 formulation (Mukherjee *et al.*, 2015).

Eastern and western Himalaya of India is known for its rich biological and cultural diversity and the unique river system. Based on its physiography and biological composition, the region can broadly be differentiated into the high to low altitude regions. Its unique situation at the confluence of the Indo-Malayan, Indo-Chinese and Indian bio-geographical regions coupled with its diverse physiography has generated a profusion of habitats, which harbours diverse biota with high-level endemism. Particularly in Darjeeling - Sikkim Himalaya, a wide range of physiography, eco-climatic conditions, varied vegetation and forest types adequately expressed themselves in giving rise to rich gene pool both of wild, endangered and cultivated plant species. The region exhibits innumerable varieties and kinds of medicinal plant. The region is home of over one hundred tribal communities with distinct socio-cultural settings and a large percentage of the population is dependent on traditional natural resource-based livelihood sustenance. Its strategic location at the confluence of South, Southeast and East Asia made it an important gateway between peoples of the region as well as isolates it from the rests part of the country. A survey conducted during 2011-14, by the All India Coordinated Research Project on Medicinal and Aromatic Plants, Kalimpong centre (Darjeeling) more than 2561 of wild plants used by the tribals and other traditional communities in India for treating various health problems. Few of these medicines have high demand in International Market.

Threatened medicinal plant from medium to high altitude range: *Acacia catechu* (T), *Aconitum balangrense* (EN), *Aconitum bisma*

(DD), *Aconitum ferox* (DD), *Aconitum gammiei* (T), *Aconitum heterophyllum* (V,T), *Aconitum laciniatum* (T), *Aconitum spicatum* (V,T), *Allium hypsistum* (V), *Alstonia neriifolia* (EN,R), *Alstonia scholaris* (V,R), *Arisaema costatum* (LC), *Arnebia benthamii* (V), *Bergenia ciliate* (T), *Butea monosposperma* (V,EN), *Corydalis megacalyx* (EN), *Curculigo orchioides* (V), *Dactylorhiza hatagirea* (EN), *Dalbergia latifolia* (V), *Delphinium himalayai* (V), *Dioscorea deltoidea* (EN,T), *Elaeocarpus sphaericus* (V), *Ephedra intermedia* (EN), *Ephemerantha macraei* (V), *Fritillaria cirrhosa* (V), *Gloriosa superb* (EN), *Heracleum lallii* (EN), *Jurinea dolomiaea* (NT), *Lilium nepalense* (DD), *Maharanga bicolor* (DD), *Maharanga emodi* (DD,K), *Meconopsis dhwojii* (NT), *Michelia champaca* (CR,EN), *Nardostachys grandiflora* (V,V), *Neopicrorhiza scrophulariiflora* (V), *Operculina turpethum* (EN), *Oroxylum indicum* (EN), *Otochilus porrectus* (EN), *Paeonia emodi* (V), *Panax pseudo-ginseng* (V), *Paris polyphylla* (V,V), *Piper longum* (V), *Pistacia chinensis* (R), *Podophyllum hexandrum* (V), *Pongamia pinnata* (DD,K), *Pterocarpus marsupium* (CR), *Rauvolfia serpentina* (CR,EN), *Rheum austral* (V), *Rheum moorcroftianum* (NT), *Rheum nobile* (V,R), *Rubia manjith* (V), *Swertia angustifolia* (EN), *Swertia chirayita* (V), *Swertia multicaulis* (DD), *Taxus wallichiana* (EN), *Tinospora sinensis* (V), *Valiriana jatamansi* (V).

(Legend: CR: Critically endangered; DD: Data deficient; EN: Endangered; K: Insufficiently known; LC: Little cared; NT: Nearly threatened; R: Rare; T: Threatened; V: Vulnerable)

High altitude: *Taxus baccata, Saussurea lappa* (kuth), *Picrorrihiza scrofularifolia* (kutki), *Swertia chirayita* (chiraito), *Rheum australe* (padamchal), *Valeriana wallichii* (sugandhawal) *and Zanthoxylum armatum* (timur).

Mid altitude: *Asparagus racemosus* (satawari/kurilo), *Valeriana wallichii* (sugandhawal) and *Anacyclus pyrethrum* (akarkara).

The World Health Organization (WHO) estimates that herbal medicine is still the mainstay for primary healthcare of about 80 per cent of world population, mainly in the developing countries (Chapman and Chomchalow, 2004). The use of botanical raw material is in many cases much cheaper than to use chemical alternate substances. As a consequence, there is an enormous demand for botanicals in domestic use and commercial trade in the international market (Mukherjee, 2008). The chemical ingredients that make a plant valuable for medicinal use are: i) alkaloids (additives, pain killing effects); ii) glycosides (heart stimulants, purgative, better sexual health);

iii) tannins (used for gastro-intestinal problems, like diarrhea, dysentery, ulcer, skin diseases); iv) essential oils (facilitates digestion, insect repellents); gum resins and mucilage (analgesic, suppress inflammation, mild purgatives); and v) vitamins and minerals (fruits, vegetable sources). With respect to therapeutic category, the uses of botanical medicines are mainly to treat: i) cardiovascular disorders, ii) respiratory disorders, iii) GUT disorders, iv) CNS disorders, and v) skin disorders (Chakraborty *et al.*, 2015). Also, botanical products are being used as dietary supplements. Modern pharmacopoeia contains about 25 per cent plant drugs. China has made very good progress in utilizing herbal wealth by promoting its use in the developed world. In China, the annual herbal production is worth US$ 48 billion, with export of about US$ 3.6 billion (Kumari *et al.*, 2011). Global market for all categories of herbal products estimated at US$ 62 billion. European Union (EU) countries, USA, Japan, Hongkong, Korea, and Singapore are the major export markets (Source: International Trade Centre, Market News Service, http://www.intracen.org/trade support/medicinal plants). Few of the medicinal plant have high export value and quite observed in Indian Himalaya (Table 1).

Fact – We Need to Understand

Frankly, not all that much has changed when it comes to a deeper understanding of trade in high altitude medicinal plants. This sector has not been overhauled in a manner that we have seen in food related agro-produce being formatted and the clarity we see in 'user' and 'grower' profiles of countries. Technology and science have been employed to maximize productivity, food security being a national priority attention is explainable. Food sector and agro-produce for the world's largest business segment - Agriculture. A similar concerted and strong effort is needed in the high altitude medicinal plant sector to ensure sustainable supply for generations to come. It would be simple to format and place this database, as 'agriculture' related data collection dragnet is already in place. It is a matter of priority and attention denied so far particularly with context of high altitude medicinal plant. An ideal medicinal plant can be described as "Sustainably sourced from organically certified forests and farms, documentarily supporting traceability and legal procurement, in full compliance of a fair trade guideline and to complete satisfaction of desired quality monographs". It may be pertinent to note here that, with effect from July 2013, EU regulations require good agricultural practice, good laboratory practice and good manufacturing practice certification for all botanicals imported by member countries intended for 'medicinal use' (EU Guidelines, 2008). This indicated that all aspects of medicinal plants have to be addressed to achieve the object of sustainable utilization of quality medicinal plants. For traditional trade channel, the critical and neglected factors that need to be looked into are: i) produce sourced either from forests primarily through tribal communities and or from farms through farmers subsequently moving to weekly markets; ii) produce from both collected by collectors at district level; iii) produce transported to urban market place via traditional/rural/urban transport system; and iv) eventually the produce is transported to pharma companies, perfumeries, herbal extract industries, etc. for value addition purpose (http://en.wikipedia.org/wiki 2011-12). Basically, this is a great challenge to maintain the balance between three factors, viz., 'Forest - Farm - Pharma' so that sustainable supply and utilization of quality medicinal plants is assured.

Table 1. Medicinal plant being exported from India

S.No.	Botanical name	Part of the plant
1.	*Aconitum sp.*	Root
2.	*Acorus calamus*	Rhizome
3.	*Adhatoda vasia*	Whole plant
4.	*Berberis aristata*	Root
5.	*Colchicum luteum*	Rhizome and seed
6.	*Hedychium spicatum*	Rhizome
7.	*Juniperus communis*	Fruit
8.	*Picrorhiza kurrooa*	Root
9.	*Podophyllum hexandrum*	Rhizome
10.	*Punica granatum*	Flower, root and bark
11.	*Swertia chirayita*	Seed, Root and bark
12.	*Valerianal jatamansi*	Rhizome, root

India is setting on a gold mine of well-recorded and traditionally well-practised knowledge of herbal medicine. This country is perhaps the largest producer of medicinal herb and is rightly called the botanical garden of the world. India has 16 Agro climatic zones, 45000 different plant species out of which 20, 000 are medicinal plants (Mukherjee, 2008 a). The Indian Systems of Medicine have identified 3000 plants for their medicinal values, of which 500 species are mostly used in the preparation of drugs. It is generally estimated that over 8000 plants in India are in use in traditional, folk and herbal medicine, representing about 75% of the medicinal needs of the "Third World" countries (Mukherjee, 2013). Presently India is being imported many of plant products for their biotech and pharma company (Table 2). Few of the medicinal plant have been used by local tribe, vaidaya for their day to day disease curing (Table 3).

Table 2. Medicinal plant being imported in India.

S.No.	Botanical name	Native name
1.	*Cuscuta epithymum*	Aftimum vilaiyti
2.	*Glycyrrhiza glabra*	Mullathi
3.	*Lavendula stoecheas*	Ustukhudus
4.	*Operculina turpethum*	Turbud
5.	*Pimpinella anisum*	Anise frui
6.	*Smilax china*	Chobchini
7.	*Smilax ornate*	Ushba
8.	*Thymus vulgaris*	Hasha

Table 3. Some important valuable wild plant used by local vaidays.

Botanical name	Vernacular name	Part used/mode of application	Uses
Aconitum atrox	Mitha bish	Paste of rhizome fried in ghee (clarified butter) is applied externally	Rheumatism, neuralgia, paralysis, rheumatic and puerperal fever.
A. heterop-hyllum	Atibish	Root powder mixed with honey and galls of *Pistacia ichinjuk* (kakarsingi) is applied externally.	Bronchial inflammation
Actaea acuminata	Mamira	Decoction of roots	Flatulence, colic
Asparagus filichinus	Jhirni	Powder of dried tuberous roots	Sexual debility and urinogenital disorder

Bergenia straqcheyi	Shilpari	Root decoction and juice of leaves	Kidney stone, sores, swellings and jaundice
Dactylorrhiza hatagirea	Salampanja	Powder of roots	Cuts and wounds
Dioscorea bulbifera	Genthi	Tuber	Bronchial cough
Diocorea deltoids	Tairu	Rhizomes	Spermetonorrhoea
Euphorbia hirta	Dudhibari	Entire plant with curd	Pile
Hedychium accuminatum	Kapoor-kachri	Purified root powder	Dyspepsia and piles
Megacarpea polyandra	Barmoola	Roots	Fever, stomach disorder
Picrorrhiza kurroo	Kutki	Root powder	Severe coughing, fever and stomach disorder.
Swertia chirayita	Chirota	Whole plant	Fever, malaria remedies
Valerianal jatamansi	Sameva	Roots, rhizome	Epilepsy, hysteria
Zanthozylum acanthopodium	Timru	Seed powder and stem bark	Toothache, tooth decay.

Approaches for meeting emerging challenges

1. Constrains for proper conservation

Conservation and cultivation of medicinal plant is a comparatively new occupation for farmers and the risk of failure is particularly high. In addition to the risk of crop failure, the farmers face serious market-related risks and difficulties in getting right price. Industry has the upper hand in deciding the price. Unlike major crops, limited research and development (R&D) work has been done on cultivation and improvement of medicinal plants in the country. Lack of proper cultivation methods for the local species contributes towards improper cultivation and harvesting, and hence, is the major cause of stagnation in the development of medicinal plants and herbs in India and its adjoining country. Poor crop husbandry practices including poor harvesting and post-harvest practices pose serious problems. Medicinal plants require specific soil and climate for expression of their best quality. However, there is no policy to regulate the cultivation of right species in right location which is standing in the way of assuring quality of raw drug even it is cultivated. Long gestation period and high risk are serious constraints. Many medicinal plants can be harvested only after three years or more (ex. *Swertia chirayita*). This is particularly true of the plants grown in high-altitude areas. As most farmers are poor, have small landholdings and lack credit, they cannot wait so long for returns. Understandably, they are reluctant to convert a significant part of their land to production of medicinal plants. Candidate species for cultivation are so large, and therefore, there is a clear need to develop technologies related to cultivation, harvesting, storage, transportation and quality control for many more species that are not available. At present, there are no proper management plans in place for the collections from forest or reserve biosphere zone. It is a common property resource and people tend to overharvest using unsustainable practices. This could endanger some of the rare species such as *Swertia chirayata, Taxus baccata, Valerina jatamansi, Viola odorata, Sapindus pinnatus, Rubia cordifolia, Aconitum ferox, Podophyllum hexandrum, Picrorhiza kurroo* and *Bergenia purpurascens*. Non-availability of quality seeds of high yielding varieties with desired quality is a big constraint. There is lack of awareness among the farmers regarding commercial cultivation of high altitude plants. Poor post-harvest technologies and inefficient processing techniques lead to low yields and poor quality products. Non-availability of sufficient information on physiochemical parameters of cultivated medicinal plants would lead to its extinction from earth. Lack of education and training of collectors, growers and traditional medicinal practitioners create problem. The untrained collectors harvest natural resources mindlessly to subsidize their meager income. A large quantity (15%) of the material collected is wasted. Lack of coordination and linkages among various (such as) local communities, Government organizations, research institutes, private sector, NGOs and international organizations. There is lack of real statistics and database concerning the area and production of valuable plant species. There is uncertainty and discriminating prices to the growers and collectors by market forces.

2. Opportunities

Indian Himalaya zone is gifted with rich flora of medicinal plants due to its varied ecological conditions. These plant have immense pharmaceutical potential which need to be exploited. Traditional Unani medicine system heavily relies on use of different plant parts in prevention and cure of diseases. More than 80 per cent population of the hill people resorts to such crude drugs of vegetable origin. The hot spots of medicinal plants are spread over the regions from Darjeeling-Sikkim Himalaya to whole of north eastern state of India. There is good scope to exploit these hotspots appropriately for collection and utilization of medicinal plant species. Implementation of R&D programs involving Govt. agencies, institutions and private sector is likely to pay good dividends to this sector. Public–private collaboration can play an important role in removing many of the bottlenecks described above. Successful model of tripartite partnerships among the farmers' group of Gheshe village of Uttarkhand, HAPPRC and Dhawan International, a Delhi-based firm for cultivation of *Picrorhiza kurrooa* (kutki) and *Saussurea lappa* (kuth) could be replicated. In Nepal, Dabour industry use *Swertia chirayita* plant under PPP model, this help to improves livelihood of local people and industry as well. The public–private collaboration supporting the cultivation of medicinal plant is in its infancy. What can be done to promote large number of collaborations will depend upon a number of conditions that have to be met before the private sector get attracted to join program to support the cultivation of medicinal plants.

Conservation of genetic resources

Plant genetic resources have made substantial contributions to the domestication, utilization and improvement of medicinal plants. Collection, characterization and their efficient utilization are keys to efficient management of any kind of genetic resource including those of medicinal plants. Modern techniques offer the opportunity for collecting, rapid propagation, medium and long-term storage and distribution of germplasm. Complementary strategies are significant for conservation, particularly of medicinal and aromatic plants as we come across a wide spectrum of species with orthodox or recalcitrant or intermediate seed storage behaviour or exclusively vegetatively propagated plants. Collections from different and widely placed areas will greatly enhance the existing collections in genebanks by providing back-ups in case of losses through diseases, insects and environmental stresses and weather changes. The major objectives of conservation programmes are to provide safety against loss of genetic resources and to make these resources available for crop improvement at present and in the future. Each strategy for conservation has to offer relatively greater safety and cost effectiveness. Any useful plant can be considered for conservation but medicinal plants with known biological activities and chemical constituents responsible for such activities if influenced by agro-ecological situations needs to be conserved in ideal situations to avoid loss of essential compounds responsible for biological actions. However, prioritization of species is essential to make full use of any particular strategy with justification. Modification of the environment and particularly associations combined with any treatment exercised for survival or increasing productivity needs to be approached cautiously. According to the estimates of Medicinal Plants Specialist Group of the IUCN Species Survival Commission, the number of medicinal plants which are threatened world wide is at least 10,000 species (Leaman, 1998). Conservation approach should meet both i) future supply, and ii) genetic resource conservation. According to IUCN, WHO and WWF (1993) the cultivation of medicinal and aromatic plants is the best effective way to satisfy the market demand. Good agricultural and collection practices (GACP) for medicinal plants have accordingly been developed through WHO and other initiatives. Medicinal plant species are of high priority for conservation action, as wild collections are likely to play a significant role in the future trade. In more recent times, the conservation strategies have been drawn in many countries, including India. The basic principles include: The protection of species (in the wild) through restrictions in collection for trade. In case of critically endangered species, a complete ban for trade and export is must. Development of adequate genera and species specific management programs, including *in-situ* and *ex-situ* conservation strategy and guidelines for sustainable collections, may be supported by an effective certification system. The ex situ conservation program could be extended to herbal gardens, botanic gardens and arboreta, medicinal plant gardens, nurseries and gene banks. The ex situ conservation of threatened and endangered species of medicinal plants could be taken in priority. Tissue culture protocols for micro-propagation and in vitro long-term storage should also be attempted to RET species of medicinal plant. Increasing number of medicinal plant species are threatened or endangered due to unsustainable collection methods in the wild. In world, China and India has rich genetic resources and a great variety of medicinal plants. It is reported that there are 15,361 kinds of medicines, out of which about 9,325 (64%) are of plant origin. In general, 320 medicinal plants are the most commonly used. China's annual herbal drug production is reported to be around US $ 48 billion with export of about US $ 3.6 billion (Kumari *et al.*, 2011). Japan, Hong Kong, Korea, and Singapore are the major importers of herbal drugs making 66 per cent share of China's botanical drug exports. The ethnic groups in China, namely, Han, Zang, Meng, Wei and Dai have traditionally utilized medicines. According to statistics, 3,781 ethnic medicines are in use among the 25 minority nationalities in Yunnan Province alone (http://english.biodiv.govt.or/imagesbiodiv/resources/medicineal-en.htm). Great diversity of cultivated species and strains exists and a good number of varieties of medicinal plants have been developed locally. For example, Panax ginseng has many local cultivars such as, Dmaya, Ermaya Changbo, Yuanbang and Yuanly. There are abundant wild relatives of cultivated plants. For example, the cultivated Panax ginseng has many wild species, e.g. Panax stipuleanthus, Pijaponica var. major, *P. zingibarensis, P. natoginseng, P. pseudoginseng* and others. Among the 32 most prioritized medicinal plants having commercial / industrial potential and identified by the Government of India for developmental work, 21 species are known to occur in the Darjeeling Himalaya to North Eastern Himalayan region of India. Valuable plants identified for integrated development, are being investigated for their medicinal value (both traditional as well as western medicines) and functional foods (for heart health, gut health, bone health and immune functions). Out of 8,000 plant species, more than 1,000 are reported to enjoy commercial trade potential as botanical raw drugs. The 2004 IUCN Red List (endangered threatened species) included 297 plant species (angiosperms and gymnosperms) in India, of which 16 are medicinal plants. Some of the threatened species are *Aconitum ferox, Picrorhiza kurroa, Nardostachys grandiflora, Panax pseudo-ginseng, Coptis teeta, Dioscorea deltoidea,* and *Costus* sp (Mukherjee *et al.*, 2015). Loss of these genetic resources is happening through excessive picking and digging of wild medicinal plants and many valuable species are confronted with imminent danger. Deforestation for cultivation and industrialization, large scale felling of forest trees resulted in wrecking of valuable medicinal plant resources. Out of the 5 Panax species, China cultivates 4 species (*Panax ginseng, P. natoginseng, P. pseudoginseng, P. quirquefallia*) on commercial scale and thus dominates the international market. However this plant, faces tremendous pressure due to its high demand in medicinal plant industry resulting in reduced population. In Nepal The top 5 species traded were *Nardostachys grandiflora, Swertia chirayita, Neopicrorhiza scrophulariflora, Zanthoxylum armatum* and *Sapindus mukorossi*, together they made up more than 50 per cent of the total value (Olsen, 2005). Conservation of these plants is becoming very crucial because of high demand in international market. Rare and high priced medicinal herbs need commercial cultivation by following good agricultural practices. Elite clones need to be identified, multiplied, and maintained in field gene banks. It is crucial to identify respective roles and responsibilities of various stake holders ranging from collectors to end users, local traders to exporters, traditional healers to profession practices, small formulators to industrial manufacturers and the government agencies. Two of the most important aspect of genetic resource conservation include are germplasm collection and characterization.

1. Germplasm collection

Plant genetic resources (PGR) are the basic raw material for genetic improvement and varietal development. In medicinal plants wild species play an important role. About 90-95 per cent species are collected from forests or its wild habitat. Utilization of national heritage of medicinal plant resources should, therefore, be planned in such way that it must conserve the valuable herbs as well as it should bring out the maximum potential of resources to ensure a healthy and prosperous future generation. India will play a key role at the international level also, if medicinal plant genetic resources are to be properly managed. Collection of vast diversity of germplasm is the key to the success of development of superior variety of medicinal plants. Chomchalow (1980) reported the importance of genetic resources in the improvement of medicinal plants. The potential of the germplasm has not yet been fully utilized like any other crop, cultivation of medicinal plants started with the direct introduction of the species from the wild habitat to agriculture.

2. Germplasm characterization

The second and foremost step after collection is characterization of the germplasm. Value of any germplasm is known once it is characterized. Characterization and evaluation of the valuable high altitude germplasm have been done in limited areas. Thorough survey/ expedition of forests and rangelands may identify additional medicinal plant resources. However, there is need to verify and characterize employing different methods including molecular techniques (Maity and Geetha, 2013). Characterization is needed for better identification and confirmation of the species. Conservation of valuable medicinal plant in situ has become a challenge for high altitude hill people particularly in Eastern himalaya. The word conservation has been popular in media and published materials from time to time. But, Himalayan medicinal plant diversity is waiting for effective conservation plan/ program in the days to come. There is positive impact of community forestry user groups in conserving medicinal plant in most areas. However, the most demanded and high value plant in the high hills such as Chiretta, Gentian, Spikenard, Valerian are not conserved but threatened. Their population is declining *in situ* due to overharvesting, theft and other reasons. There are 60 high value species already in various degree of threat as categorized by IUCN and CAMP. Severe genetic erosion and vulnerability of the priority species has occurred. Advocacy advertisements and publications only could not mitigate the situation. Despite education on conservation and awareness on sustainable harvesting, the trend was found to be continued. Role of plant breeding to select best plants within the variable population as a potential cultivar to increase yield many fold. The potential productivity of cereal crops has traditionally been increased by modifying its morphological characters such as number of branches, number of kernels per year, test weight of seeds, etc. but in medicinal plant improvement, it is entirely different from other agricultural crops, because quality in terms of alkaloid, steroid or essential oil is not apparently dependant on morphological characters. All the breeding objectives and selection criteria should, therefore, be directed to enhance the total secondary metabolite content and biomass production in a particular crop species. Simple introduction of a variety for successful cultivation in a new environmental condition may sometime be futile due to chemical changes and infra-specific chemical modifications, which may take place due to, altered ecological and geographical conditions (Tetenyi, 1992).

Conservation, utilization and value addition

Medicinal plants are large in number under Indian subcontinent and hence 5-10 species need to be prioritized based on the medicinal value and market demand for efficient R&D management. The RET species should be given top priority for conservation efforts (Mukherjee *et al.*, 2009). At least one field genebank needs to be established in each climatic zone in each country for conservation and utilization of valuable medicinal plant species. In situ conservation is the cost-effective method for conservation. About 112 species in Southern India, 74 species in Northern and Central India and 42 species in high altitude of Himalayas are threatened in the wild (*Source:* BSI data (http://164.100.52.111/search1species.asp?cc=2)). Conservation and sustainable utilization of medicinal plants is a major concern in the present situation. The best way to conserve the medicinal plants as in the case of any other taxa is in situ conservation and in some rare cases if the habitat is threatened, selected species can be saved through ex situ conservation. Ex situ conservation of medicinal plants has been initiated at national and state levels. National Bureau of Plant Genetic Resources (NBPGR) acts as nodal agency for the collection and maintenance of medicinal plants. Under All India Coordinated Research Project on Medicinal and Aromatic Plants and Betelvene (AICRP-MAP), since its inception, constant efforts are being made for collection and evaluation of germplasm in selected medicinal plant. In India, no separate policy or regulation exists for conservation of medicinal plants. Their conservation is mainly covered under the India Forest Act (1927) and Wild Life Protection Act (1972), which are enforced by the State Forest Departments and Indian Government's Directorate of Wildlife Preservation.

The following Acts influence the medicinal plant sector in India:

- The Indian Forest Act, 1927 and its subsequent amendments, 1930
- The Drugs and Cosmetics Act, 1940 and its amendments time to time
- The Seed Act, 1966
- The Wild Life Protection Act (1972)
- The Foreign Trade Development and Regulation Act of 1992
- The Protection of Plant Varieties and Farmers' Rights Act, 2001
- The Biological Diversity Act, 2002

The first *in-situ* conservation project for the conservation of medicinal plants in India was initiated in the states of Karnataka, Kerala and Tamil Nadu involving the State Forest Departments, leading NGOs and Research Institutes in 1993 in collaboration with Foundation for Revitalization of Local Health Traditions (FRLHT) - a Bangalore based NGO group. This project, sponsored by the Govt. of India, developed a network of medicinal plant conservation sites (MPCS), medicinal plant conservation parks (MPCP) and medicinal plant development sites (MPDS). Apart from, there are national parks, biosphere reserves and world heritage sites. Indigenous communities develop their own ways for the conservation of natural resources. One of such conservation practices is the 'sacred groves'. Development of varieties with improved quality should be given greater thrust for enhancing their conservation. Exchange of improved varieties of medicinal plant species between the countries on bilateral basis needs to be encouraged. To ensure the sustainable harvesting of species from forest and systematic cultivation, each country should develop the GACP as per the guidelines of WHO and implement faithfully at all steps from collection to production in order to produce quality raw material. The species collected most from the forests may also be brought under cultivation and this needs to be organized following the cluster approach to solve the problems of smallholder and small volumes. The tree medicinal plant species are more vulnerable to extinction as their bark and root are used. Therefore, a sustainable harvesting method by staggered harvesting and replanting in accordance with the harvesting cycle need to be adopted. Tree selection from the available diversity and their vegetative propagation method should be standardized for tree and shrub species improvement.

Guidelines on good agricultural and collection practices (GACP) need to be developed and implemented at all stages from collection to cultivation in each country for production of quality raw materials. Encourage involvement of Self-Help Groups/local entrepreneurs/ community groups for value addition (cleaning, grading, packaging, labelling etc.) at farm level to enhance the income of farmers and collectors. Value addition means, anything that is done to raise the value of the product in the market. The value added practices are going to be the key for the future of sustainable farming, since it enables the growers to advance economically without targeting the unsustainable increase in productivity from the land. Value added medicinal and aromatic plant products may range from simple processes such as drying, grading, grinding, and cleaning to very sophisticated processes such as spray drying, vacuum drying, fractionation of molecules, etc (Maity and Geetha, 2013). Possible primary processing and value addition activities for selected individual crops are as follows:

1. *Psyllium* (*Plantago ovata*): Usually the seeds are sold as such by the farmers. The usual value addition practice for export is as follows: (i) cleaning of seeds from clay and other foreign material, and (ii) separation of seed coat and grading

2. Aswagandha (*Withania somnifera*): Root is sold in the market. Processing and value addition include: (i) making into pieces, (ii) cleaning and grading based on diameter, and powder making.

3. Aloe (*Aloe barbadensis*): Gel present in the leaf is of major demand in the market. Extraction of the gel, stability maintenance of the gel and powder making are important steps.

4. Senna (*Senna angustifolia*): Dry leaves are sold in the market. Processes and value addition include: (i) mechanical drying through hot draft of air or solar drying, and (ii) grading of leaf ased on size and colour.

5. *Safed musli* (*Chlorophytum borivilanum*): Dried root is sold in the market. The steps for processing and value addition include: (i) peeling of the fresh root, and (ii) drying.

6. Satavary (*Asparagus racemosus*): Dried root is sold in the market. The steps for processing and value addition include: (i) cleaning, and (ii) drying and chopping.

7. Kalmegh (*Andrographis paniculata*): Dried herbage is sold in the market. The steps for processing and value addition include: (i) cleaning, (ii) drying and chopping, and (iii) making powder/ herbal concentrate.

8. Glory lily (*Gloriosa superba*): Dried seeds and roots are sold in the market. The steps for processing and value addition include: (i) cleaning, (ii) drying, and (iii) colchicine extraction and concentration.

This kind of value addition is very much urgent for high altitude medicinal plant so this will help to improve rural livelihood and farmer take part in its cultivation in large scale. This help to reduce pressure on plant population of *Sausaria lappa, Podophylum hexandrium, Valeriana jatamansi* and *Swertia chirayita* etc. The cluster-based approach needs to be followed to facilitate higher investment for betterment of community based value chain programs. Efforts are needed to develop the database on production, processing, trade of raw material and the finished products of medicinal plant species in each country to facilitate the development of regional database. At present, the researcher-farmer-industry linkage does not exist in most of the countries. This type of linkage needs to be developed with involvement of SHGs and contract farming groups. Mechanism for buy-back arrangement should be developed between the producer and user industry. Globally acceptable certification (organic produce with defined quality) needs to be linked with the local certifying agency through credibility building for end user along with good labeling code with complete disclosure of details to facilitate transparent export and trade. This help to maintain value addition trust to the consumers. Knowledge exchange related to trade along with the other regulatory procedures needs to be promoted to facilitate the effective and fair trade of high altitude medicinal plant in Indian and other part of world. Leading research efforts on medicinal plants could unveil the curative measures of dreadful diseases that can be therapeutically significant in drug discovery. There is an urgent need to have strong linkage between research institutions and pharmaceutical industry which needs to be established through appropriate linkage model. The public-private partnership (PPP) model needs to be adopted for implementing R&D programs for overall conservation and development of medicinal plant species. Regulatory measures for conservation, cultivation and marketing as well as quality standards of commercial species need to be developed by all herb growing countries in the region.

Patent issue of high altitude medicinal plant related to India

A number of medicinal plants and their uses have been patented by foreign countries. There has been criticism by the people and in the press on this growing trend of patenting of our medicinal plants and their uses. Some of the well-known plants *Picrorhiza kurroa, Nardostachys jatamansi, Swertia chirata, Valeriana jatamansi, Aconitum heterophyllum* and *Sassurea lappa* etc. are well use by various industry (Chakraborty *et al.*, 2015b). Some of the patents have been successfully contested by India at various forum regarding patent and other issue.

Various product developed from these valuable medicinal plant also claimed by other nation. Patents have been granted because the knowledge about the uses of these plants is not available in the format and manner which the patent examiners can have easy access. Therefore, it was considered necessary to bring the knowledge contained in ancient texts and in public domain in patent compatible format to prevent patenting by others. Now it is important we should conserve our rich heritage of herbal source side by side we should document in proper way. This will help in future to show our dominance in herbal sector with various high altitude endangered medicinal plant of Indian subcontinent , and this will also help to enhance country income, which ultimately shares large proportion in country GDP. There is thus an enormous scope for India to emerge as a major player in the global valuable medicinal plant product based medicines. But unfortunately various lacunae pertaining to quality of herbal drugs do exit which are the major hindrance to come up to the expected level of trade of these medicines both within and outside the country.

The action plan for development of medicinal plant with reference to high altitude

1. Encouragement for cultivation of selected medicinal plants backed by buyback arrangements (*Podphyllum hexandrium, Valeriana jatamansi* etc.).
2. Simplification of Transit permit/legal procurement certificate for transportation of raw drugs (*Swertia chirayita* etc.).
3. Few selected priority medicinal plants, like Atis, Chirayita, *Podophyllum, Plantago, Taxus* sp etc., which are in great demand both in domestic and international market to be brought into cultivation status for the overall development of the medicinal plants sector.
4. General and specialized surveys of the international market for medicinal plants and products to be undertaken for identifying niche areas.
5. Registration of farmers/cultivators and traders of medicinal plants to be entrusted to the respective State Medicinal Plants Board.

6. Research and development studies in the areas of post harvest management shelf life, storage and simple agro techniques to be taken up through CSIR, NBRI, CIMAP, ICFRE, RRLs, DBT, Horticulture and Forest Department.
7. Constitution of State Medicinal Plants Board in every State/UT of the country for overall development of medicinal plants sector.
8. Efforts to create mass awareness about the importance of medicinal plants among the people and publish distribution material for the purpose.
9. Extension activities: These extension activities should be done either through KVKs or by University level with involvement of Private Sector Company.

- Training/seminar/workshop
- Visit of growers to demonstration spots and research institutes.
- Extension material on agro-techniques.

The following research topics in phytomedicine research suggested by Handa (2013) need immediate consideration

- The high-tech methods for chemical analysis of plant extracts and standardization
- The search for bioactive botanical constituents and their use as templates to develop new drug for diseases which at present cannot be effectively treated
- The integration of new molecular biology methods into screening of plant extracts and their constituents
- The good clinical practice (GCP) confirming studies on the efficacy proof and bioavailability of standardized plant extracts

Cultivation Practices

The growing requirements for medicinal plants are understandably more stringent than the other agricultural crops. The medicinal plants must be cultivated at least 1 km away from major roads and farms where chemical pesticides are sprayed. The soil should be loose and free from toxic metals. A clean water source should be available. There is ample scope of medicinal cultivation in the cropping and arable lands and in community forests (Mukherjee, 2016). Research and development works in propagation, cultivation are limited. But, there is increased awareness among farmers, cooperatives, government agencies and NGOs in cultivating medicinal plant and increase production and income. Emphasis is given to commercial cultivation for additional income and employment generation of subsistence farmers, in situ conservation of the threatened plant, and utilization of the unproductive land, etc. Despite the advocacy for cultivation and conservation, cultivation of indigenous/endemic plant has been taking place only few year back. Cultivation and farming of medicinal plants require better understanding of the agro-climatic and edaphic condition, propagation and other management techniques. The indigenous communities have a tradition of practicing mixed farming systems that includes herbal species. Few of the high value medicinal observed in Darjeeling himalaya which immense need for improved agrotechnique because of high demand in export market (Table 4).

Table 4. List of medicinal plants having high export market demand

S.No.	Medicinal Plants
1.	*Picrorhiza kurroa*
2.	*Nardostachys jatamansi*
3.	*Swertia chirata*
4.	*Valeriana jatamansi*
5.	*Aconitum heterophyllum*
6.	*Sassurea lappa*
7.	*Asparagus racemosus*
8.	*Rauvolfia serpentina*
9.	*Dactylorhiza hatagarea*
10.	*Withania somnifers*
11.	*Costus specious*
13.	*Bacopa monnieri*
14.	*Taxus wallichiana*
15.	*Digitallis purpurea*

Cultivation of medicinal plants in a limited scale and in limited species started in the beginning of the 19th century but large scale cultivation with large number of species is a recent phenomenon. Cultivation of medicinal species gives scope to improve the quality of the drugs. Merits of commercial cultivation of medicinal plant is the outcome of implementation of number of critical factors like good genetically stable planting materials; good agro-technological practices; nutrient input; harvesting management and implementation of suitable post-harvesting techniques to preserve the end product till smart and effective marketing arrangements are made. At present, National Medicinal Plants Board (NMPB) is promoting 32 species for large scale cultivation which include various threatened plant also. List of medium to high altitude medicinal plants identified by National Medicinal Plants Board for development and cultivation on priority are as follows:

- Kuth (*Saussurea costus* C. B. Clarke.)
- Atis (*Aconitum heterophyllum* Wall.)
- Kutki (*Picrorhiza scrophulariflora* Pennell. /P. Kurrooa)
- Psyllium (Plantago ovata Forsk.)
- Saffron (*Crocus sativus* Linn.)
- Chiraita (*Swertia chirata* Buch-Ham.)
- Sarpgandha (*Rauvolfia serpentina* Benth. Ex. Kurz.)
- Glory lily (*Gloriosa superba* Linn.)
- Gymnema (*Gymnema sylvestre* R. Br.)
- Indian aconite (*Aconitum ferox* Wall)
- Tinospora (*Tinospora cordifolia* Miers.)
- Indian bellidium (*Commiphora wightii* (Arn.) Bhant.)
- Jatamansi (*Nardostachys jatamansi* DC.)

Direction and planting of medicinal plant play crucial role in proper development of plant. A study on the population structure of *Persicaria amplexicaule*, *Valeriana jatamansi* and *Viola serpens* was conducted during 2002 – 2004 in different ecologically important sites of Malam Jabba, Swat, Pakistan. The results revealed that overall population density of each plant showed a significant decrease after the collection period. However, all the investigated parameters generally showed two to four-fold increase in protected sites as compared to the unprotected areas. The growth performance of the tested species increased with rise in elevation. The abundance/ distribution and overall population of each targeted plant was generally high in North facing slopes as compared to South facing features. The air and soil temperatures were slightly higher in open areas as compared to the protected sites and showed a decline with increasing elevation. Both air and soil temperatures were relatively higher in Southern slopes as compared to North-eastern slopes in Malam Jabba. By using cluster analysis, dynamic, normal and regressive populations were distinguished in the population of selected plants. The three population types differed with respect to their population size and total plant density. The differences were maximum in dynamic, intermediate in normal, and lowest in regressive populations. The study concluded that the conservation of the remaining populations of targeted plants would best be achieved by proper time of sustainable harvesting (Alyemeni and Sher, 2010). The medicinal plants have to be grown without chemical fertilizers and use of pesticides. Organic manures like, farm yard manure (FYM), vermi-compost, green manure etc. (Table 5) may be used as per requirement of the species (Mukherjee, 2014).

Table 5. Organic inputs recommendation on observation basis at field level.

Input	Quantity required(t/ ha)	Remarks
Neem/ Pongamia cake/Green leaf manure	1.5 0	Protection against termite and other soil borne pest and pathogen.
FYM	10 .00	There is possibility of weed infestation and poor establishment of the crop.
Compost	10.00	Possibility of weed and pathogen infestation and poor establishment of the crop.
Vermicompost	0.50	This source is totally weed free and help in better performance of the crop.

Development of varieties

The variety development work has practically not been undertaken for even the most valuable medicinal plant species contrary to agro-crops. Local landraces are being grown, promoted and expanded in a few areas only known as herbal farms. Improved varieties are the products of plant breeding efforts. However, it did not find acceptance in totality for improvement of medicinal plant. The obvious reason for low acceptance is the lack of understanding on mode of action of majority of medicinal plants that are used in Ayurvedic preparations. In most of the cases, there is no single bio-active molecule but a battery of various groups of chemicals such as flavoinoids, alkaloids, glycosides, terpinodes, phenols, etc. which synergistically produce the curative result (Maity and Geetha, 2013). Their proportion is also important. Plant breeding for quality improvement in terms of increasing the percentage of these chemicals are possible only when the medicinal plants are used for extraction of specific chemicals such as sennoside from senna; morphine from opium poppy; quinine from Cinchona; artemisin from Artemisia. But, when breeding of medicinal plants used in Ayurveda, Siddha, and Unani medicines is concerned, it becomes difficult to set the breeding objectives. Problem of setting the breeding objectives in terms of increasing the quality of a particular constituent will not be accepted because it will upset the dosages prescribed and also efficacy will be altered. Therefore, every new cultivar of particular species has to be supported by further standardization of either dosage or quantity of the species in a codified Ayurvedic drug which will jeopardise the entire Ayurvedic, Siddha and Unani medicines preparation. Therefore, at present the scope of plant breeding efforts in these groups of plants are limited to disease and pest resistance and yield improvement without compromising the initial quality in terms of chemical constituents. This is also a grey area to the researchers since there is no base value in terms of bio-active molecules available in various species of plants that have been considered as quality material in Ayurvedic preparations. So far, Ayurvedic industry demands only right species with clean material as quality standard. Therefore, there is a need for concerted efforts to identify species-wise quality parameters before taking up any serious breeding programme are used for Ayurvedic medicines. At present, research organizations are identifying the elite cultivars and releasing either at their institute level or through Central Varietal Release Committee. Recently Protection of Plant Varieties and Farmers' Rights Authority (PPVFRA) has started developing Distinctiveness, Uniformity and Stability (DUS) descriptors and is ready for registration of few of the endangered and valuable medicinal plant of India. Efforts are continuing to include more and more species under PPVFRA registration domain.

Production of quality seed and planting material

Quality seed and planting material are mostly based on natural resource. Good quality of seed and planting can be produced through good agricultural practices (Mukhrjee, 2013). As most of the requirements are met from the wild collection, not much effort has been made to upscale the production of seeds and planting materials of medicinal plant. The production of seeds and planting materials are currently limited to the few pocket. Primary processing, value addition and product development is one of the important factor for high altitude temperate medicinal production. Involvement of self-help groups/local entrepreneurs/community groups for value addition at farm level to enhance the income of farmers and collectors and help to preserve endangered medicinal herb of high altitude himalaya. Basic processing and value addition of medicinal plant raw or crude products are the areas on which the work has been initiated and will continue to expand upon in the areas like bio-extraction. Primary processing is done locally in the field and at collectors' home. The process undergoes a number of operations after harvesting: washing/cleaning, drying, sorting/grading, bundle making and storing of all plant products. Almost all medicinal plants are sold and exported in this form. Cleaning is done for the rhizome/roots of spikenard, valerian, atis, gentian, etc. whereas drying, curing, are done to chiretta (Mukherjee, 2008) and other plants such as *Valeriana jatamansi* (Mukherjee and Chkraborty, 2014). The essential oil producing species are allowed for 2-3 days for drying/weathering in the field. Spikenard, valerian and other essential oil species are further processed in a boiler plant by the distillation of rhizomes/roots. The farmers are not well trained to perform all practices safely maintaining quality. Care should be taken while harvesting, e.g. uprooting chiretta plants (Mukherjee, 2009). The farmers who are engaged in processing, collect their raw materials from the forest with the help of state and community forests personals. The processed medicinal plant product is sold directly by the farmers through various marketing channels. Product diversification and value addition has been one of the key focus areas of the future high altitude medicinal plant production and conservation. Farmers have been trained and technologies have been passed on to the farmer groups for making different herbal products such as liver tonic (*Swertia chirayita, Aconitum heterophyllum* etc use), pain killer (*Saussurea costus, Picrorhiza scrophulariflora*). The role of the various government and state added institute has been seen as very pivotal in developing different herbal and natural products. The role of farmers, scientist and various research and development wings, in product development and value addition will be very critical in the overall development of the medicinal plant sector particularly high altitude ones, in India. Market exploration and intelligence were found to be the crucial factors in the continuation and the long-term sustainability of the program activities. This very aspect has also been underscored as a major activity amongst others in the program's of coming five year plan. R&D efforts need to be reoriented so as to keep this sector free from the ill effects of chemical fertilizers. The efforts are required to generate eco-friendly agro-technology to support commercial cultivation under organic environment with buy-back arrangements with user industry. An effective mechanism needs to be in place for certification of organically produced medicinal plant species in individual country with its recognition for international trade. In order to meet the growing demand of quality raw material for industry, the organized cultivation provide an answer and, therefore, the production and processing technologies must be worked out and effect of climate change on the quality of these resources may also be studied by the various scientist through R& D wings.

Quality standards

No system of medicine can achieve a high degree of credibility and mass acceptance unless some degree of quality assurance is maintained. In many countries, herbal products are marketed through health stores as dietary supplements. For quality assurance, a minimum documentation of the product needs to be done indicating: i) that the products are manufactured by following good manufacturing practices (GMP), ii) certificate of analysis of the batch detailing the tests performed, limits, actual value recorded, and iii) minimum labeling of contents. The drugs are generally required to be labeled to comply with the label provision of the concerned country into which imported. In Europe, there were different pharmacopoeias to define medicinal plants/extracts. European pharmacopoeia is a legal binding in EEC countries. Specific tests on quality and regulatory requirements in respect of packing, labeling, testing, fumigation and storage environment do exist for export/import of herbal medicine products. The main difficulty in prescribing the standards for herbal medicines is due to the fact that most of the products use whole plant or extracts of parts of plants and in some cases a mixture of number of plants (polyherbal drugs). It is challenging to develop suitable standards because the preparation of the drugs based on medicinal plants is regarded as one active entity in its entirety. For herbal medicines, the qualitative and quantitative variations in the contents of bioactive phytochemicals should be considered in fixing standards and specifications. Standardization, optimization and control of growing conditions should guarantee quality controlled production of many plant derived compounds. Good agricultural practices need to be prescribed for enabling production of quality products free from pesticide residues, heavy metals and microbial contaminations. The quality control must start with the raw materials and the next stage is the quality control during processing and manufacturing. In India, concerned nodal department have already notified elaborate good manufacturing practices. The guidelines of good agricultural and collection practices for medicinal plants documented in a publication of ICAR, IFAD and FAO, (2010) should be adhered in order to ensure quality control of raw material derived from cultivated medicinal plant species. Drug testing laboratories engaged exclusively, for traditional herbal medicines may also play important role in ensuring quality of herbal medicines. Enforcing uniformity in processing may also reduce the variation in the quality of same product of different manufacturers.

Also, there is a need to strengthen safety evaluation centers as some high altitude plant drugs are toxic if not properly procured and used judiciously. Safety evaluation should be mandatory before marketing for human consumption. Clinical validation for confirming efficacy of the drug by following standard protocols prescribed by WHO will help in building confidence in the minds of consumers about quality standards of plant based medicines. In recent times, there had been some published reports of consumers suffering adverse health effects caused by poor quality medicines, from valuable plant species of high range of himalaya. Quality of raw material used is often indicated as the cause of the problems. As a result, herbal medicine industry has been under pressure for quality assurance and some countries have developed/created new laws/stricter regulations both for manufacturing the medicine as well as sourcing the raw materials. The guidelines for good agricultural and collection practices (GACP) for medicinal plants developed in 2003 by the WHO aimed at improving the quality of medicinal plant material being used in the herbal medicines in the market.

Screening medicinal plants for desirable biological activities and identification of compound

Screening medicinal plants for desirable biological activities and identification of compound with selective and specific biological effects for target diseases. For isolation and structure elucidation of active constituent(s), a system of prioritization and de-replication process need to be developed. Development of reliable methods of de-replication may reduce the laborious, time consuming and expensive steps in natural product drug discovery system. The goal of de-replication is to select only the extracts that are likely to yield novel chemotypes. Structural characterization of isolated active compounds is accomplished mostly using spectroscopic and chromatographic techniques such as LC-MS, LC-MS-MS. Therapeutically potential new compounds (based on biological activity) considered as promising pharmaceutical candidates are to be selected for pre-clinical studies and clinical trials (Mukherjee, 2008). Synthesis and isolation from plant material have been the most consistent in providing large quantities of selected plant derived compounds. The botanical sources identified to provide the following classes of NCEs for drug discovery process included: i) herbal extracts as botanical drug e.g. green tea extract; ii) bioactive compounds for direct use e.g. dioxin, and iii) lead compounds (e.g. paclitaxol from *Taxus* sp.) as potent drug.

Anti-malarial drug: Discovery of Artemisin from *Artemisia annua* as anti-malarial compound, especially for cerebral malaria. One third of world population is exposed to the risk of malaria.

Anti-bird flu respite: The avian-influenza resulting in bird flu caused by HSNI virus can be managed through Tamilflu drug Oseltamivir produced through a 12 steps synthesis, starting from a plant product, the shikimic acid, obtained from the fruits of Illicium verum (Chinese star anise) and Illicium griffithi (Indian star anise). At present, this is the only drug available for bird flu (Handa, 2013).

Anti inflammatory property: Turmeric (*Curcuma longa*) rhizome is being increasingly used for remedy for a number of inflammatory conditions. The pigment curcumin in turmeric rhizome has multiple uses. Non-steroidal, non-aspirin like anti-inflammatory drug is reported from gum resin of *Boswellia serrata.*

Other leads: Oleogum resin from *Commiphora wightii* (gugul) for use as anti-hyperlipidemic drug (standardized on the basis of bioactive gugulosterone E and Z); some liver protecting drugs form *Picrorhiza kurooa, Andrographis paniculata, Schizandra chinensis* are showing good promise. Compounds such as 'reserpine' from Rauvolfia serpentina and ' paclitaxol' from Himalaya yew (*Taxus wallichiana*) have important pharmaceutical uses in Europe and USA.

Case study of Darjeeling Himalaya

The variation in elevation with diverse climatic conditions has favored the occurrence of variety of plant species and making Darjeeling Himalaya ecologically one of the most diverse in the world. Major diversity rich areas include mid hill, to high hill and mountains. Darjeeling district is a region characterized by diverse physiography ranging from plains, plateaus to hills and valleys of various dimension. The area is rich in flora and fauna and it lies in one of the most biodiversity rich regions of the world. The forest is of sub-tropical to temperate type which harbors different varieties of endangered plants. Most of the tribal villagers greatly depend on forest based products for their livelihood. Wild edible plants are a necessary supplement in their daily diet. The tribal villagers also have great faith in their traditional system of medicine and prefer them to allopathic medicine. From the enumeration of the plant species collected it can be inferred that many threatened plants are being routinely used by the tribal villagers of the area for the treatment of ailments and also for other purposes. A perusal of the available literature has revealed that at least 19 species of plants have been reported elsewhere and also mentioned in the Red Data Book of Indian Plants to be rare, endangered or vulnerable Example include *Gloriosa superba* is extinct in the wild in Darjeeling Himalaya, or endangered (EN) in Africa and other parts of Asia [32] but is least concerned (LC) in the study area. On the contrary, species like *Swertia chirata* while vulnerable (VU) in Darjeeling Himalaya[] and endangered (EN) in Himachal Pradesh [35] are now extinct in the wild. Several critically endangered (CR) species like *Renanthera inschootiana, Taxus baccata* needs immediate conservation in order to prevent extinction in the district. Many near threatened (NT) species like *Terminalia chebula, Ipomoea coccinea, Pinus roxburghii* and vulnerable species like *Ipomoea quamoclit, Mesua ferrea* needs conservation (Mukherjee, 2009 a). The soil in these hilly zones is mostly categorized as brown forest soil due to their characteristic reddish brown or brownish colour (Table 5). Accumulation of clay particles increased with the decrease in elevation in hilly areas, whereas, a reverse trend of results was obtained for sand particles. Both bulk and particle densities decreased with decreased in elevation. Decrease in pH was recorded with decrease in elevation. Further soil sample study revealed that organic carbon and available N increased with decrease in elevation. Increase in organic matter and nitrogen in lower hill because of organic matter present in leaf litter and decomposed materials were carried out by seasonal springs at lowest elevation. The amount of both available P and K were also found to decrease with the decrease in elevations.

Table 5. Physico-chemical status of soil sample under different altitude.

Parameters		**Elevation from Mean sea level**			
		2500 m	**2000 m**	**1500m**	**1000m**
Mechanical separates	Sand	78.37	52.66	39.34	27.81
	Silt	17.31	31.36	27.19	14.94
	Clay	13.91	23.14	33.26	61.31
	BD (g cc^{-1})	1.07	1.06	0.98	0.87
	PD (g cc^{-1})	2.10	1.94	1.87	1.79
pH	5.1	4.9	4.6	4.3	
Organic C (g 100 g^{-1})	0.95	1.26	1.34	1.41	
Available N (mg Kg^{-1})	289.91	306.99	391.32	489.61	
Available P (mg Kg^{-1})	2.61	2.49	2.29	1.92	
Available K (mg Kg^{-1})	362.34	230.1	169.78	129.34	

The study has shown that continuous exploitation of several individual plant species from the wild, legally or illegally and substantial loss of their habitats has resulted in the population decline of many high value medicinal plant species in the district. As for example, it was learnt from the Takda and Sukhiapokhri adjoining village that in the early 1980s different establishments paid money to the local people to collect indiscriminately the raw biomass of *Taxus baccata* which was abundant at that time. Presently the natural population of *T. baccata* is critically endangered and has almost been obliterated due to that practice (Mukherjee, 2009a). Most of the villagers in the district are illiterate and they have no idea or consequences or loss if a species become extinct. The forest department and a few NGOs have been working to preserve the rich flora and fauna with little success. The present study has also revealed some information on a few conservation strategies that can be applied to give a more or at least some effective result. Another thing that's comes to highlight to safe guard the forest is by rehabilitation of the biodiversity of the area. Those species on which the tribals mostly depend such as, wild edibles and timber trees, firewood species etc. can be planted in large scale or cultivated. This would rejuvenate the eco-system with maximum indigenous floral and faunal elements. This will also bring in several associated species which will automatically sustain the traditional lifestyle of the tribals as well as protect the forest. There is also several forests which are related to myth and some of them are sacred to the tribals and these forests are preserved by them. Scientific and modern methods are not known to them and very often difficult to explain to them. Therefore, the message of conservation would be more meaningful to them through myth, faith and traditions rather through scientific approach. Distribution and diversity of a species varies from location to location. Few areas have been roughly estimated through resource mapping for particular species. Due attention is not paid to estimate area of the medicinal plants. Detailed survey was conducted during 2011 to 2013 to study about the valuable medicinal herbs and its conservation aspect in Darjeeling himalaya, under AICRP on Medicinal and Aromatic plants, under the aegis of Uttar Banga Krishi Viswavidayalaya. After a preliminary survey of the study area, 25 representative sites were selected on the basis of physiognomic contrast, altitude, habitat, species composition and stages of degradation (deforestation, grazing, over harvesting and erosion). The general vegetation data in these two aspects was collected at five altitudinal zone viz: 1000 m (foot hill), 1800-2000 m (mid hills) and 2200-2500 m (top hill). The data of each targeted species was collected by using 10 x 10 m quadrate in each site of the study. In each site ten square, each of 10 x 10 m were fixed with 4 wooden pegs at 4 corners following the methods described by Shreshta *et al.* (1998) and marked for later re-identification. The quadrates were laid randomly in a nested manner. Day to day survey of all kinds of medicinal plants was made in different land situations with the help of local people, vaidya and scientist. Thirty five medicinal plants were identified. The species of medicinal plants found in the hilly tracts of Kalimpong, Darjeeling and Sikkim are as follows : *Aconitum ferox, Acorus calamus, Artemisia vulgaris, Astilbe rivularis, Bergenia ciliate, Cephaelis ipecacuanha, Ceritella asiatica Clematis buchanaria Dioscorea composite, Dichroa febrifuga, Drymaria diandra, Digitalis purpurea, Eupatorium cannabinum, Ficus semicordatus, Fraxinus floribunda, Gentiana kurro, Heracleum wallichii, Litsaea cubeba, Nardostachys grandiflora, Oroxylum indicum, Panax pseudo-ginseng, Paederia foetida, Phytolacca acinosa, Picrorhiza kurrooa, Podophylum hexandrum, Rheum modi, Rhus semialata, Rumex nepalensis, Swertia chirata, Thysamolaena maxima, Urtica dioica, Viscum articulatum, Valeriana officinalis*. This plant categorized into various families as per taxonomy (Table 6). Diversification within various plants particularly in *Swertia chirayita* and *Valeriana jatamansi* observed (Chakraborty *et al.*, 2016).

Table 6. Number of economic medicinal species found in each family.

Botanical family	No. of species present for economic use.
Anacardiaceae	1
Apiaceae	2
Araceae	1
Araliaceae	1
Asteraceae	2
Bignoniaceae	1
Caryophyllaceae	1
Dioscoreaceae	1
Gentianaceae	2
Hydrangeaaceae	1
Lauraceae	1
Loranthaceae	1
Moraceae	1
Oleaceae	1
Petridiaceae	1
Phytolaccaceae	1
Poaceae	1
Podophyllaceae	1
Polygonaceae	2
Ranunculaceae	2
Rubiaceae	2
Saxifragaceae	1
Scrophulariaceae	2
Solanaceae	1
Valerianaceae	2

Table 7. List of endangered and vulnerable medicinal plant of Darjeeling Himalaya.

S.No.	Species	Ecological regions
Endangered medicinal plants of high altitude		
1.	*Aconitum heterophyllum*	Temperate Himalayas
2.	*Atropa acuminata*	Temperate Himalayas
3.	*Acorus calamus*	Temperate Himalayas
4.	*Bergenia ciliata*	Temperate Himalayas
5.	*Commophora wightii*	Deserts
6.	*Dioscoria deltoidea*	Temperate Himalayas
7.	*Picrorrhiza kurroo*	Alpine Himalayas
8.	*Paeonia emodi*	Temperate Himalayas
9.	*Podophyllum hexandrum*	Temperate Himalayas
10.	*Rheum emodi*	Temperate Himalayas
11.	*Saussurea costus*	Alpine Himalayas
12	*Valeriana wallichii*	Temperate Himalayas
Most vulnerable medicinal plant of high altitude zone		
1.	*Artemisia spp.*	Tropical to subtropical
2.	*Cochicum luteum*	Sub-tropical Himalayas
3.	*Pistacia integerrima*	Subtropical Himalayas
4.	*Plantago ovata*	Cold arid hills
5.	*Glycyrrhiza glabra*	Hindukush, Karakorum

Disease and Pest Management

This area is still in infancy. Diseases and insect pest surveys and surveillances have not been conducted even for the commercially cultivated species. However, little information is available on this aspect. Some diseases and insect pests of particular species are known but their management practices have not been developed. Very few studies had been carried out but were not conclusive. The cultivated medicinal plant such as *Swertia chirayita, Valeriana jatamansi* are seriously affected by fungal disease. Roy *et al.* (2008) reported infection of root not nemoted (*Melodogyna* sp) in *Swertia chirayita.* The cultivation of these species is seriously threatened in most of the production sites. As these crops are cultivated by the farmers on limited scale Directorate of Medicinal Plant, Gujrat, Anand, India has initiated number of programs through various AICRPs to address the issues of pests and diseases. Few endangered plant that are collected from the wild, not much work and initiatives have been taken up on the aspects of pests and disease management. Lack of technical capacity and limited human resource are the key limitations to start activities under this program. Use of synthetic chemicals to control pests and diseases and chemical fertilizers to increase productivity of medicinal plants is discouraged by few states (Sikkim, Kerala, West Bengal) under Indian subcontinent. This is because to avoid adverse residual effect and toxicity to human health. Farmers are encouraged to use biopesticides and biofertilizers to produce quality raw material as demanded by pharmaceutical industry. To avoid pesticide residual hazards in the first place of occurrence at field level, pharmaceuticals are encouraging contract farming and necessary field visits to control the use of unapproved chemicals. For example, pyrethrin- a natural insecticide derived from the Pyrethrum plant is used to control pest like cut-worms of *Artemisia annua*, a plant known to cure malaria. Considering the little information available on organic cultivation practice, biological control of pests with sustainable cultivation techniques, emerge as new arena for hilly area. In order to address the pest and disease management, the cultivation sites were attacked by pests and pathogens in different stages and seasons. Specimens of attacking pests and pathogens were collected from the site for rearing and identify the nature and extent of damage. After rearing, it was identified that most of the affected parts of the medicinal plants were attacked by mole cricket, psyllid and red mite which are responsible for the damage at seedlings and growing stages in the field. Rate of infestation and symptoms were recorded. As a preventive measure, neem (*Azadirachta indica*) oil was applied. The use of weed killers and pesticides may result in a risk to the environment, to the growers and the medicinal and aromatic plant itself. Therefore the use of chemicals (if use at all) should be reduced to an absolute minimum, and where chemicals are used, necessary regulations should be in place and rigidly followed.

The pest (insect pests and diseases) management in medicinal plant is aimed to minimize the pressure of pesticide residues in the herbal products. The integrated pest management (IPM) strategies are the best choice for and include cultural practices (*eg*: change of sowing date, crop rotation to break pest life cycle, soil pulverisation), mechanical control measures (*e.g*: collection and destruction of adults and young ones, use of different types of traps for monitoring and management of insect pests), cultivation of pest resistant varieties under field condition, use of botanicals (eg: neem based pesticides), use of biocontrol agents (e.g. Trichogramma sp. as egg parasitoid of lepidopteran pests, antagonistic microorganisms against plant pathogens, *Trichoderma* spp.), use of microbial insecticides (e.g. use of *Bacillus thuriengenesis*, nuclear polyhedrosis virus against various caterpillar pests and use of entomopathogenic fungi like *Beauveria bassiana, Verticillium lecanii* against insect pests) and other biorational approaches like use of insect growth regulators (IGR), pheromone traps, etc. But under GAP cultivation, pesticides are allowed in extreme cases to save the crop from total failure. However, care should be taken to harvest the crop or sending the produce to the market only after the prescribed safe waiting period is over (Maity and Geetha, 2013). To prevent diseases, bio-pesticides could be prepared (either single or mixture) from neem (kernel, seeds & leaves), chitrakmool, dhatura, cow's urine etc. Observation revealed that under field condition this plant is mostly attacked by ant. To overcome this problem applies phorate as per need and necessity.

Marketing strategy of high altitude medicinal plant

Proper and genuine market is one of the important aspects for medicinal plant growers. If market availability is poor then definitely growth and production will be directly hamper. This will leads to endangered many medicinal plant which are found to be frequent in present situation. However, we should think in following aspect :

- Marketing information service on medicinal plants for domestic as well as global market.
- Ensure supply of quality planting material in bulk to the farmers by way of appropriate technology viz. vegetative propagation, tissue culture etc.
- Production of medicinal plants in bulk as per demand and supply position of most preferred species.
- Area expansion for selected species in the specific agro-climatic zones.
- Develop proper harvesting techniques.
- Semi-processing of produces viz. collection, grading, drying, packing etc.
- Develop innovative marketing mechanism.

Conclusion

High altitude medicinal plant is one of the most neglected part of research. This sector needs proper conservation and utilization as per need by the medicinal plant sector. Our future strategies would be focused on collecting as many plants as possible, which have medicinal values in the present context. The alkaloids, metabolites and other complex chemical substances that are responsible for curing diseases and have aesthetic importance need to be identified for the use of pharmaceutical industry and for domestic purpose also.

References

Alyemeni, M.N. and Sher, H. 2010. Impact of human pressure on the population structure of Persicaria amplexicaule, *Valeriana jatamansi* and *Viola serpens* the naturally growing medicinal plants in Malam Jaba, Swat, Pakistan. *Journal of Medicinal Plants Research,* 4(20): 2080-2091.

Bhushan, P., Warude, D., Pushpangadan, P. and Bhalt, N. 2005. Ayurveda and traditional chinese medicine: A comparative Overview. CAM, 2005, 2(4): 465-473.

Chakraborty, S., Mukherjee, D. and Baskey, S. 2016. Morphological diversity and nomenclature of *Swertia chirayita* (Gentianaceae)-Recovery of endangered medicinal plant population in north eastern Himalaya. *American Journal of Plant Sciences,* 7: 741-755.

Chakraborty, S., Mukherjee, D. and Baskey, S. 2015. Selection of lines of Valeriana jatamansi Jones, a high value medicinal plant in North Eastern Himalayan region. *Indian Journal of Genetics and Plant Breeding,* 75(3): 1-4.

Chakraborty, S., Mukherjee, D. and Baskey, S. 2015a. Indian Valerian, a Highly Endangered Medicinal Plant in North Eastern Himalayan Region. *Advances in Plants and Agriculture Research,* 2(4): 1-7 (DOI: 10.15406/apar.2015.02.00058).

Chakraborty, S., Mukherjee, D. and Baskey, S. 2015b. Paradigm of demographic stochasticity way to extinction of *Valeriana jatamansi* Jones, a valuable medicinal plant in North Eastern Himalaya region. *Ecology, Environment and Conservation,* 21(1): 521-528.

Chapman, K.R. and Chomchalow, N. 2004. Production of Medicinal Plants in Asia In: Medicinal Plant Research in Asia. Vol.I: The Framework and Project Work Plans. P. Batugal, J.Kanniah, Lec SY and J.T. Oliver

(Eds). IPGRI-Regional office for the Asia, Pacific and Occania, Serdong, Selangor DE, Malaysia.

Chomchalow, N. 1980. Medicinal plants and spices in Asia. *Plant Genetic Resources Newsletter*, 44: 2-11.

Dhar, U., Manjkhola, S., Joshi, M.,Bhatt, A., Bisht, A.K. and Joshi, M. 2002. Current status and future strategy for the development of medicinal plant sector in Uttaranchal, India. *Current science*, 83(8): 64-69.

Handa, S.S. 2013. Research trends in drug development: Exploration of Medicinal & Aromatic Plants. Guest Lecturer in National Conference at Shoolini University, Bajhol, Solan, H.P., India (20 June, 2013) p. 65.

ICAR I.F.A.D. and F.A.O. 2010. Good agricultural and Collection Practices for medicinal plants. FAO of UN, New Delhi, India.

IUCN W.H.O. and W.W.F. 1993. Guidelines on conservation of medicinal plants. IUCN, WHO and WWF, Gland, Switzerland.

Kumari, S., Shukla, G. and Rao, A.S. 2011. The present status of Medicinal Plants-Aspects and Prospects. *International Journal of Research in Pharmaceutical and Biomedical Sciences*, 2(1): 64-68.

Leaman, D. 1998. How many medicinal plants are threatened with extinction? *Plants Talk* 14: 4.

Maity, S. and Geetha, K.A. 2013. Country status report on medicinal and aromatic plant in India. Paper presented in expert consultation on promotion of medicinal and aromatic plant in the Asia pacific region, Bangkok, Thailand (2-3 December, 2013), pp. 101-123.

Mukherjee, D. 2016. Studies on ecology and outcome of various seed and organic input treatments on yield attribute and economic yield of *Valeriana jatamansi*. *The Himalayan Review*, 1(2): 193-200.

Mukherjee, D., Chakraborty, S. and Baskey, S. 2015. Threatened medicinal plants biodiversity of Eastern Himalaya and its conservation. *Journal of Agriculture and Technology*, 2(1): 101-107.

Mukherjee, D. and Chkraborty, S. 2014. Studies on ecology, habitats diversification and seed germination behavior of *Valeriana jatamansi* Jones : A critical endangered plant. *International Journal of Agricultural Sciences*, 4(5): 203-209.

Mukherjee, D. 2013. Improved agricultural practices for high value medicinal plants under mid to high altitude situation. *Journal of Crop and Weed*, 9(1): 201-06.

Mukherjee, D. 2014. Nutrient and its management: Prospect and challenges under the changing environment scenario. In Advances in Plant Physiology, (eds. Hemantaranjan, A.). Scientific Publishers, Jodhpur, India, 15: 413-442.

Mukherjee, D., Chakraborty, S., Roy, A and Mokthan, M.W. 2009. Differential approach of germplasm conservation of high value of medicinal plants in north eastern himalayan region. *International Journal of Agriculture Environment & Biotechnology,* 2(4): 332-340.

Mukherjee, D. 2009. *Swertia chirayita* an endangered herbs in Darjeeling hill. In : Krishi Sandesh (eds. Rai, S., Mokthan, M.W. and Ali, S). Miazik International Volunteer Center, Japan. pp 122-132.

Mukherjee, D. 2009a. Current status, distribution and ethno-medicinal values of medicinal plant in hilly region of Darjeeling district of West Bengal. *Journal of Crop and Weed,* 5(1): 316-320.

Mukherjee, D. 2008. Association of medicinal plants with important tree species in hills of Darjeeling. *Environment and Ecology,* 26(4A): 1697-99.

Mukherjee, D. 2008a. Cultivation of *Swertia chirayita* : a high value medicinal herb in high altitude. *Journal of Medicinal & Aromatic Plant Sciences,* 30(4): 350-355.

Olsen, C.S. 2005. Quantification of the trade in medicinal and aromatic plants in and from Nepal. *Acta Horticulture,* 678: ISHS, 2005.

Ravishankar, B. and Shukla, V.J. 2007. Indian system of medicine: A brief profile. *African Journal of Traditional, Complementary and Alternative Medicines,* 4(3): 319–337.

Roy, K., Jash, S., Mukherjee, D., Pramanik, A. and Mukhopadhyay, A.K. 2008. A note on the incidence of Melodogyna javanica in *Swertia chirayita* (Family : Gentianaceae) from West Bengal. *Indian Journal of Nematology,* 38(2): 258.

Shrestha, K.K., Ghimire, S.K., Gurung, T.N., Lama, Y.C. 1998. Conservation of plant resources, Community Development and training in applied ethnobotany in Nepal. Final Report Prepared by WWF-Nepal Programme and people and plants initiative.

Tetanyi, P. 1992. Plant breeding and variety control of medicinal plants. In cultivation and processing of Medicinal plants (ed.) Hornok L. John Wiley and Sons. NY. Pp. 10-18.

❑❑❑

Chapter 24

Sacred Grove for *In-situ* Conservation of Ethnomedicinal Plants

R.K. Bhakat* and U.K. Sen

Department of Botany & Forestry, Vidyasagar University, Midnapore-721 102, West Bengal, India, E. mail: rkbhakat@rediffmail.com/bhakatramkumar@gmail.com

ABSTRACT

Sacred groves are forest patches protected on socio-religious grounds. *It* acts as an ideal center for biodiversity *conservation*. Several *plants* that are threatened in the forest areas are still well conserved in some of the *sacred groves*. It has been observed that several *ethnomedicinal plants* that are vanishing in nearby forests are abundant in the *sacred groves*. This paper attempts to highlight the role of a sacred grove in *in-situ* conservation of 163 species of ethnomedicinal plants in West Midnapore district of West Bengal. It also discusses some prevailing threats to the grove and recommends certain remedial measures.

Keywords: Conservation, Ethnomedicinal plant, Sacred grove.

Introduction

The practice of nature conservation through worship is a very ancient tradition in India. One such important practice of environmental conservation is that of dedicating patches of forest to some local deities and or spirits. These forest pockets, commonly known as sacred groves, are protected from all human interferences on grounds of socio-religious beliefs and taboos associated with them. Owing to social protection, these groves support a rich collection of wild ethnomedicinal plants, including some rare and endemic taxa (Gadgil and Vartak, 1981; Khan *et al.*, 2008). In India, sacred groves are found mainly in tribal dominated areas and are known by different names in ethnic terms (Bhakat, 1990). About 4,215 sacred groves covering an area of 39,063 hectares are estimated to be distributed in India (Malhotra, 1998). The sacred groves are considered to be a rich source of ethnomedicinal, rare and endemic plants as refugia for relic flora of a region and as centre of seed dispersal (Jeeva *et al.*, 2006). Ethnomedicinal plants constitute the base of the health care system in many societies. Globally, about 85% of the traditional medicines used in primary health care derived from plants (Farnsworth, 1988). Today, according to the World Health Organization (WHO), as many as 80% of the world's people depend on traditional medicine and in India, 65% of the population in the rural areas use Ayurvedic and ethnomedicinal plants to help meet their primary health care needs (WHO, 2002). In India, more than 43% of the total flowering plants are reported to be of medicinal importance (Pushpangadan, 1995).

But due to easy availability and much dependence of the rural folk on the surrounding ethnomedicinal taxa, people often harvest them mercilessly from the countryside. As a result, population of these taxa, are gradually shrinking in the wild. It is in this context, the nearby sacred groves act a sanctuary to shelter them because the groves are themselves protected. Under this background, the paper attempts to discuss the *in-situ* way of ethnomedicinal plant conservation through a sacred grove situated in a tribal area of West Midnapore district in West Bengal.

Methodology

The sacred grove

The sacred grove under study, popularly known as *Swarga Baurir* than (named after the presiding deity *Swarga Bauri* of the grove) is situated in the Jamboni block (latitude 22° 26′00.09″- 22° 26′ 01.48″N and longitude 86° 50′00.90″- 86° 50′01.56″E, average altitude 86.70 masl.) of Jhargram sub-division in West Midnapore district of West Bengal. The grove houses a brick-made small temple, and is spread over a 3.5 acre public land at the outskirts of villages *Chhotopindara, Dochakhuria* and *Ranijhor* along the south-western bank of a perennial rivulet *Palpala.* It represents a 400-500 year old isolated forest patch consisting of evergreen, deciduous and semi-deciduous plants. The vegetation of the grove is multi-tiered and multi-specific in structure (Fig. 1).

Fig. 1. *Swarga Bauri* temple inside the sacred grove.

Most of the adjoining village inhabitants belong to tribal communities like Sabar and Santal. Out of these tribes, Sabar stand high in the majority. Traditionally, the tribal communities in West Midnapore district have been preserved small patches of virgin forest since time immemorial due to their religious belief. Sabar and Santal tribes live close to the sacred forest and are largely dependent on the wild biological resources for their livelihood. They consume leaves, flower, corm, fruits, rhizome and bulbs in a restricted manner, which are considered by the tribesmen as highly nutritious with high medicinal value. Since, the grove is an abode of deity, the entire area along with its plants and other creatures is considered as sacred. Owing to this socio-religious tag on the grove, the local villagers do not cut or disturb the flora thus, strictly adhering to the local taboos and sentiments.

Survey and inventorisation of ethnomedicinal plants

During the course of investigation for a period of five years (2008-2012), the sacred grove was surveyed in different seasons for the estimation of floral wealth and its role in conservation. Through questionnaire survey the information was collected in the name of the sacred grove, its locality, size and area of the grove, deity worshipped, history or folklores and gender issues associated and occurrence of ethnomedicinal plants at the sacred grove site. Besides, the local peoples were encouraged to give their views and perceptions on the sacred grove with respect to the ethnomedicinal plant conservation perspective. During the field work, the sacred grove site was also visited by the local knowledgeable people for preparing the list of species and associated knowledge with such sacred grove. Participant observations were also employed and information was collected through PRA (Participatory Rural Appraisal) method and through local literature. A brief ethnomedicinal plant survey has been carried out through "spot identification" basis. For unknown plants, samples of plants with flowers or fruits were collected. After collection, the specimens were processed, preserved and mounted on herbarium sheets following the standard and modern herbarium techniques (Jain and Rao, 1977). Photographs were taken for some of the more common, very rare, endemic and ethnomedicinally valuable plant species in the sacred grove. Abbreviations of author's names of plant species strictly follow Brummitt and Powell (1992). The herbarium sheets have been identified by matching with correctly annotated materials available at the Vidyasagar University Herbarium. For identification purpose, different relevant floras, monographs, revision works and other literature were consulted (Hooker, 1872-1897; Prain, 1903a and b; Haines, 1921–1925; Sanyal, 1994).

In the systematic enumeration of the taxa (herbs, shrubs, trees and climbers), scientific name, family, vernacular name, angiosperm type, habit and medicinal uses/ general properties are arranged alphabetically (Table 1). Information about ethnomedicinal plants was collected through interviewing and cross-interviewing the local people, traditional healers and by literature (Kirtikar and Basu, 1935; Cunningham, 2001; Pakrashi and Mukhopadhyay, 2004; Paria, 2005).

Results and Discussion

In the present ethnobotanical study, a total of 163 species (dicots 147 and monocots 16) belonging to 147 genera distributed in 56 families were recorded from the sacred grove. Among these, 68 (41.72%), 44 (26.99%), 22 (13.5%), 29(17.79%) are herbs, shrubs, trees and climbers including lianas respectively of the 56 families, dicots are represented by 43 (76.79%) families; monocots consist of 13 (23.21%) families. Amongst the total dicots 147 (90.18%) and monocots

16 (9.82%); herbs, shrubs, trees and climbers represent 59, 42, 22, 24 and 9, 2, 0, 5 species respectively, representing 36.2%, 25.76%, 13.5%, 14.72% and 5.52%, 1.23%, 0%, 3.07% of the total species (Table 1).

The ten well represented families, according to descending species number (≥4) are Fabaceae 18 (11.04%), Apocynaceae 17 (10.42%), Asteraceae 15 (9.2%), Malvaceae 11 (6.74%), Acanthaceae 10 (6.13%), Lamiaceae 9 (5.52%), Euphorbiaceae 7 (4.29%), Solanaceae 5 (3.06%), Amaranthaceae 4 (2.45%), and Rubiaceae 4 (2.45%); which represent 100 (61.3%) of the total flora (Table 1).

The twelve well represented genera are *Sida, Solanum* and *Terminalia* (3 species each); *Andrographis, Barleria, Bauhinia, Jatropha, Justicia, Mimosa, Ocimum, Rauvolfia* and *Senna* (2 species each) respectively. The present study revealed that the local people living nearby the sacred grove were using all the representative species of ethnomedicinal plants to cure various ailments (Table 1). This kind of traditional conservation of ethnomedicinal plants in sacred groves of West Midnapore district is also earlier reported by Bhakat (2014), Bhakat and Pandit (2003, 2007, 2008) and Bhakat and Sen (2008a and b, 2013).

The frequent use of herbaceous species among the tribal communities could be a result of their relative abundance as compared to trees and shrubs as also witnessed by investigators of this study. The study area experiences wet tropical monsoon with moderate humidity for most months of the year, creating favorable conditions for the growth of herbs. The common use of herbs as sources of medicine was also indicated by studies conducted elsewhere in the world (Muthu *et al.*, 2006; Yineger *et al.*, 2007). In the tropical regions, there is a tendency for fewer families of flora to dominate the pharmacopoeia; Fabaceae is such family. In the study of Upadhyay *et al.* (2010), Fabaceae is the most used family in the ethnomedicine system of Rajasthan. Musa *et al.* (2011) reported Fabaceae is the most represented ethnomedicinal family in the savannah of Sudan. The plant of Fabaceae is generally herbaceous in habit and found in abundance in the nutrient rich soils of sacred groves.

Table 1. Ethnomedicinal plants in *Swarga-bauri* sacred grove.

Sl. No.	Scientific Name	Family	Vernacular Name	Angiosperm Type	Habit	Medicinal Use (s)/General properties
1.	*Abelmoschus moschatus* Medik.	Malvaceae	*Kasturi*	D	S	Aphrodisiac, constipating, demulcent, stomachic and tonic
2.	*Abroma augusta* (L.) L.f.	Malvaceae	*Olatkambal*	D	S	Antispasmodic and anti-inflammatory
3.	*Abrus precatorius* L.	Fabaceae	*Lal Kunch*	D	C	Substitute of liquorices; cures body pain and skin diseases
4.	*Abutilon indicum* (L.) Sweet	Malvaceae	*Patari*	D	S	Cures dysentery, hepatitis and diabetes
5.	*Acalypha indica* L.	Euphorbiaceae	*Muktojhuri*	D	H	Laxative, cures asthma and bronchitis
6.	*Achyranthes aspera* L.	Amaranthaceae	*Apang*	D	H	Diuretic, purgative and insect bites
7.	*Aegle marmelos* (L.) Corrêa	Rutaceae	*Bel*	D	T	Diarrhea and constipation.
8.	*Aerva lanata* (L.) Juss.	Amaranthaceae	*Chaldhowa*	D	H	Anthelmintic, diuretic and cure headache
9.	*Ageratum conyzoides* (L.) L.	Asteraceae	*Dochunti*	D	H	Diarrhea, dysentery and leukemia
10.	*Alangium salviifolium* (L.f.) Wangerin	Cornaceae	*Ankar*	D	T	Laxative, astringent, pungent and purgative
11.	*Alocasia macrorrhizos* (L.) G.Don	Araceae	*Mankachu*	M	H	Diuretic and laxative
12.	*Aloe vera* (L.) Burm.f.	Xanthorrhoeaceae	*Ghritakumari*	M	H	Astringent, cooling, vermifuge and diuretic
13.	*Alstonia scholaris* (L.) R. Br.	Apocynaceae	*Chhatim*	D	T	Cures leprosy, skin diseases, malaria, diarrhea and asthma
14.	*Alternanthera sessilis* (L.) R.Br. ex DC.	Amaranthaceae	*Sanchi*	D	H	Used as febrifuge and cures night blindness
15.	*Amaranthus spinosus* L.	Amaranthaceae	*Kantanotey*	D	H	Used as febrifuge, cures sore and eczema

16.	*Andrographis echioides* (L.) Nees	Acanthaceae	*Kalmegh*	D	H	Antibiotic, stomachic and anti-typhoid
17.	*Andrographis paniculata* (Burm.f.) Nees	Acanthaceae	*Kalmegh*	D	H	Cures dysentery, worm infection and used as liver tonic
18.	*Anisochilus carnosus* (L.f.) Wall.	Lamiaceae	*Panjiri-Ka-Pat*	D	S	Anti-inflammation in liver
19.	*Anisomeles indica* (L.) Kuntze	Lamiaceae	*Gopali*	D	S	Astringent, carminative and uterine affections
20.	*Aristolochia indica* L.	Aristolochiaceae	*Ishermul*	D	C	Antidiabetic, blood pressure and leucoderma
21.	*Artemisia vulgaris* L.	Asteraceae	*Nagdamani*	D	H	Emmenagogue, anthelmintic, antiseptic and stomachic
22.	*Asparagus racemosus* Willd.	Asparagaceae	*Satamuli*	M	C	Hyperacidity, health tonic and uterine tonic
23.	*Ayapana triplinervis* (Vahl) R.M.King & H.Rob.	Asteraceae	*Ayapan*	D	H	Astringent, acrid, stimulant, diaphoretic and cardio tonic
24.	*Azadirachta indica* A.Juss.	Meliaceae	*Neem*	D	T	Antimicrobial, cooling, appetizer, laxative, analgesic, epilepsy and hypertensive
25.	*Bacopa monnieri* (L.) Wettst.	Plantaginaceae	*Bramhi*	D	H	Anti-inflammatory, digestive, diuretic; cures asthma and skin diseases
26.	*Barleria cristata* L.	Acanthaceae	*Swet jhanti*	D	S	Stomachache, tonic and febrifuge
27.	*Barleria lupulina* Lindl.	Acanthaceae	*Kanta bishalyakarani*	D	S	Decoction used to treat dropsy and glandular swelling
28.	*Bauhinia acuminata* L.	Fabaceae	*Swet Kanchan*	D	S	Biliousness, leprosy and stone in bladder
29.	*Bauhinia vahlii* Wight & Arn.	Fabaceae	*Chihurlata*	D	C	Aphrodisiac, tonic and cures dysentery
30.	*Bixa orellana* L.	Bixaceae	*Sindure*	D	T	Antimicrobial, antitumor and anticancer
31.	*Blumea lacera* (Burm.f.) DC.	Asteraceae	*Kukurshoka*	D	H	Antipyretic, anthelmintic, febrifuge, stimulant and diuretic
32.	*Boerhavia diffusa* L.	Nyctaginaceae	*Punarnova*	D	H	Jaundice, child birth and liver complaints
33.	*Breynia vitis-idaea* (Burm.f.) C.E.C.Fisch.	Phyllanthaceae	*Kali Sitki*	D	S	Headache, leaves are used to relieve skin inflammation and meningitis.

34.	*Butea superba* Roxb.	Fabaceae	*Latapalash*	D	C	Bark/gum cures piles, anthelmintic and sedative
35.	*Caesalpinia bonduc* (L.) Roxb.	Fabaceae	*Natakaranj*	D	C	Fever, skin diseases and worm infestation
36.	*Calotropis gigantea* (L.) Dryand.	Apocynaceae	*Dhusar Akanda*	D	S	Diaphoretic, expectorant, purgative and indigestion
37.	*Capparis zeylanica* L.	Capparaceae	*Rohini*	D	C	Bark sedative, stomachic and cholagogue
38.	*Cardiospermum halicacabum* L.	Sapindaceae	*Shibjhul*	D	C	Diuretic, laxative and nervous disease
39.	*Carissa spinarum* L.	Apocynaceae	*Ban Karamcha*	D	S	Rheumatism and purgative
40.	*Cascabela thevetia* (L.) Lippold	Apocynaceae	*Kalke*	D	T	Antihaemohagic and antipyretic
41.	*Catharanthus roseus* (L.) G.Don	Apocynaceae	*Nayantara*	D	S	Leukemia, breast cancer, sedative and stomachache
42.	*Cayaponia laciniosa* (L.) C.Jeffrey	Cucurbitaceae	*Mala*	D	C	Aphrodisiac, increase masculinity and enhance youthfulness
43.	*Cayratia trifolia* (L.) Domin	Vitaceae	*Amallata*	D	C	Astringent and skin diseases
44.	*Celastrus paniculatus* Willd.	Celastraceae	*Kujri*	D	C	Abortifacient and depurative
45.	*Centella asiatica* (L.) Urb.	Apiaceae	*Thankuni*	D	H	Used in jaundice and dysentery
46.	*Centratherum anthelminticum* (L.) Kuntze	Asteraceae	*Somraj*	D	H	Acrid, anthelmintic, astringent, anti-inflammatory, depurative and diuretic
47.	*Cheilocostus speciosus* (J.Koenig) C.D.Specht	Costaceae	*Kew*	M	H	Anthelmintic, depurative and snake bite
48.	*Cinnamomum camphora* (L.) J.Presl	Lauraceae	*Karpur*	D	T	Stimulant, carminative, rheumatism and sexual diseases
49.	*Cissus quadrangularis* L.	Vitaceae	*Harjora*	D	C	Anti-osteoporotic, anti-diabetic and anti-metabolic syndrome
50.	*Cleome gynandra* L.	Cleomaceae	*Brahma Surbbachala*	D	H	Earache, eye troubles and skin diseases
51.	*Clerodendrum infortunatum* L.	Lamiaceae	*Ghetu*	D	S	Antiseptic, antipyretic and anti-inflammatory
52.	*Codariocalyx motorius* (Houtt.) H.Ohashi	Fabaceae	*Ban Chandal*	D	H	Antidote, anti-inflammatory and cardiac tonic

53.	*Colocasia esculenta* (L.) Schott	Araceae	*Altikachu*	M	H	Astringent and hemorrhages
54.	*Commelina benghalensis* L.	Commelinaceae	*Kanchira*	M	H	Diuretic and febrifuge
55.	*Cotula anthemoides* L.	Asteraceae	*Babun*	D	H	Rheumatic inflammation
56.	*Crateva nurvala* Buch.-Ham.	Capparaceae	*Barun*	D	T	Antipyretic, diuretic, stomachic and laxative
57.	*Crotalaria pallida* Aiton	Fabaceae	*Jhunjhunia*	D	S	Astringent and expectorant
58.	*Croton bonplandianus* Baill.	Euphorbiaceae	*Churchuri*	D	H	Diarrhea, irruption and laxative
59.	*Cryptolepis dubia* (Burm.f.) M.R.Almeida	Apocynaceae	*Karilata*	D	C	Anorexia, bowel and skin diseases
60.	*Curculigo orchioides* Gaertn.	Hypoxidaceae	*Talmuli*	M	H	Filarial, venereal disease and anticancer
61.	*Curcuma aromatica* Salisb.	Zingiberaceae	*Banhaldi*	M	H	Anti-inflammatory, anthelmintic, laxative and diuretic
62.	*Cynodon dactylon* (L.) Pers.	Poaceae	*Durbaghas*	M	H	Used for treating leprosy, fever, dysentery, vomiting and skin diseases
63.	*Cyperus rotundus* L.	Cyperaceae	*Mutha*	M	H	Stomach disorders, heal wounds, indigestion and bronchitis
64.	*Datura stramonium* L.	Solanaceae	*Sadadhutra*	D	S	Narcotic, anodyne and antispasmodic
65.	*Dioscorea alata* L.	Dioscoreaceae	*Khamalu*	M	C	Tonic and used in dispensing swellings
66.	*Ecbolium viride* (Forssk.) Alston	Acanthaceae	*Neel kantha*	D	H	Jaundice, menorrhea and rheumatism
67.	*Eclipta prostrata* (L.) L.	Asteraceae	*Keshut*	D	H	Hair growth and the black color of hair
68.	*Elephantopus scaber* L.	Asteraceae	*Samdulum*	D	H	Antipyretic, astringent and diuretic
69.	*Enydra fluctuans* DC.	Asteraceae	*Hinche*	D	H	Laxative, antibilious and demulcent
70.	*Euphorbia antiquorum* L.	Euphorbiaceae	*Bajbaran*	D	T	Anodyne, digestive, purgative, stomachic and useful in wounds and ulcer
71.	*Euphorbia hirta* L.	Euphorbiaceae	*Baradudhe*	D	H	Used in dysentery, cough, asthma and worm infections
72.	*Evolvulus alsinoides* (L.) L.	Convolvulaceae	*Lataghas*	D	H	Useful in bronchitis, epilepsy, falling and graying of hair

73.	*Ficus racemosa* L.	Moraceae	*Jaggya Dumur*	D	T	Anti-diabetes, astringent, aphrodisiac and purgative
74.	*Flacourtia indica* (Burm.f.) Merr.	Salicaceae	*Benchi*	D	S	Diuretic, skin diseases and poisonous bites
75.	*Flemingia strobilifera* (L.) W.T.Aiton	Fabaceae	*Ghora Chabuk*	D	S	Vermifuge and tonic
76.	*Gardenia resinifera* Roth	Rubiaceae	*Hing*	D	S	Astringent and anti-worms
77.	*Gloriosa superba* L.	Colchicaceae	*Ulat-chandal*	M	C	Abortifacient, anthelmintic, stomachic, purgative, antipyretic and gastrointestinal disorders
78.	*Glycosmis pentaphylla* (Retz.) DC.	Rutaceae	*Ban Jamir*	D	T	Astringent, vermifuge, febrifuge, expectorant and rheumatism
79.	*Grangea maderaspatana* (L.) Poir.	Asteraceae	*Namuti*	D	H	Useful in headache and earache
80.	*Grewia helicterifolia* Wall. ex G.Don	Malvaceae	*Pohalsa*	D	T	Astringent, antipruritic, aphrodisiac and expectorant
81.	*Gymnema sylvestre* (Retz.) R.Br. ex Sm.	Apocynaceae	*Gurmar*	D	C	Astringent, anti-inflammatory, stomachic and anthelmintic
82.	*Helicteres isora* L.	Malvaceae	*Atmora*	D	S	Used in diarrhoea, dyspepsia and gastrointestinal disorders
83.	*Heliotropium indicum* L.	Boraginaceae	*Hatisur*	D	H	Cures impotency, fever and wounds
84.	*Hemidesmus indicus* (L.) R. Br. ex Schult.	Apocynaceae	*Anantamul*	D	C	Used as tonic and cures poisonous bites
85.	*Hemigraphis hirta* (Vahl) T.Anderson	Acanthaceae		D	H	Used in dysentery
86.	*Hibiscus vitifolius* L.	Malvaceae	*Bala*	D	S	Jaundice, fractured bones and sprained muscles
87.	*Holarrhena pubescens* Wall. ex G.Don	Apocynaceae	*Kurchi*	D	T	Astringent and antidysenteric
88.	*Hybanthus enneaspermus* (L.) F.Muell.	Violaceae	*Nunbora*	D	H	Aphrodisiac and sexual disorders
89.	*Hygrophila auriculata* (Schumach.) Heine	Acanthaceae	*Kulekhara*	D	H	Anti-inflammatory, aphrodisiac and diuretic
90.	*Ichnocarpus frutescens* (L.) W.T.Aiton	Apocynaceae	*Shyam*	D	C	Used in leprosy and skin diseases

91.	*Indigofera tinctoria* L.	Fabaceae	*Neel*	D	H	Diuretic and wound healing
92.	*Ipomoea aquatica* Forssk.	Convolvulaceae	*Kalmishak*	D	H	Used as purgative; cures nervous and general debility
93.	*Jatropha curcas* L.	Euphorbiaceae	*Sada Bherenda*	D	S	Galactogogue, ulcer and tumor
94.	*Jatropha gossypiifolia* L.	Euphorbiaceae	*Lal Bherenda*	D	S	Laxative and appetite
95.	*Justicia adhatoda* L.	Acanthaceae	*Basak*	D	S	Used in cold and cough
96.	*Justicia gendarussa* Burm.f.	Acanthaceae	*Bishalyakarani*	D	S	Coughs, colds, asthma, skin infections and inflammations
97.	*Leonotis nepetifolia* (L.) R.Br.	Lamiaceae	*Bhutbhairab*	D	S	Ringworm and skin diseases
98.	*Leucas cephalotes* (Roth) Spreng.	Lamiaceae	*Ghalghasa*	D	H	Diaphoretic, laxative and anthelmintic
99.	*Limnophila indica* (L.) Druce	Plantaginaceae	*Karpur*	D	H	Antiseptic and carminative
100.	*Lippia javanica* (Burm.f.) Spreng.	Verbenaceae	*Ban Nebu*	D	S	Expectorant and skin diseases
101.	*Litsea glutinosa* (Lour.) C.B.Rob.	Lauraceae	*Leda*	D	T	Bark paste as emollient to treat body pain
102.	*Ludwigia octovalvis* (Jacq.) P.H.Raven	Onagraceae	*Banlavanga*	D	H	Anti-dysentery and anthelmintic
103.	*Martynia annua* L.	Martyniaceae	*Baghnakh*	D	H	Antiseptic, epilepsy and inflammation
104.	*Meyna spinosa* Roxb. ex Link	Rubiaceae	*Maina Kanta*	D	S	Diphtheria and dysentery
105.	*Mimosa pudica* L.	Fabaceae	*Lajjabati*	D	S	Used in iron deficiency
106.	*Mimosa rubicaulis* Lam.	Fabaceae	*Kunchikata*	D	S	Piles and burns
107.	*Morinda citrifolia* L.	Rubiaceae	*Ach*	D	T	Dysentery, dyspepsia and fever
108.	*Mucuna pruriens* (L.) DC.	Fabaceae	*Alkushi*	D	C	Diabetes and high blood pressure
109.	*Nerium oleander* L.	Apocynaceae	*Karabi*	D	S	Leprosy and insect or snake bite
110.	*Ocimum americanum* L.	Lamiaceae	*Krishnatulsi*	D	S	Antimicrobial, antispasmodic and expectorant
111.	*Ocimum basilicum* L.	Lamiaceae	*Mistitulsi*	D	H	Anti-inflammatory, alexipharmic, antispasmodic, anthelmintic and expectorant

112.	*Oldenlandia corymbosa* L.	Rubiaceae	*Papartak*	D	H	Used in jaundice and liver diseases
113.	*Oxalis corniculata* L.	Oxalidaceae	*Aamrul*	D	H	antiseptic and stomachic
114.	*Pandanus odorifer* (Forssk.) Kuntze	Pandanaceae	*Keya*	M	S	Emollient, depurative, antiseptic, stomachic. astringent and tonic
115.	*Parkinsonia aculeata* L.	Fabaceae	*Gandal*	D	T	Used to treat cuts and wounds
116.	*Pergularia daemia* (Forssk.) Chiov.	Apocynaceae	*Chagalbati*	D	C	Used in rheumatism and skin diseases
117.	*Phyllanthus fraternus* G.L.Webster	Phyllanthaceae	*Bhuiamla*	D	H	Diabetes, jaundice and gonorrhea
118.	*Physalis minima* L.	Solanaceae	*Bantepari*	D	H	Appetizing, diuretic. expectorant and laxative
119.	*Plectranthus amboinicus* (Lour.)	Lamiaceae	*Patharchur*	D	H	Febrifuge and sexual diseases
120.	*Plumeria rubra* L.	Apocynaceae	*Garurchampa*	D	T	Astringent and laxative
121.	*Polygala arvensis* Willd.	Polygalaceae	*Nilkantha*	D	H	Asthma, fevers and catarrhal affections
122.	*Portulaca oleracea* L.	Portulacaceae	*Baralonia*	D	H	Antibacterial, diuretic and antidysenteric
123.	*Premna mollissima* Roth	Lamiaceae	*Brihat Agnimantha*	D	T	Diuretic, anti dropsy, astringent, stomachic and cardiotonic
124.	*Rauvolfia serpentina* (L.) Benth. ex Kurz	Apocynaceae	*Sarpagandha*	D	S	Used to treat hypertension, insomnia, nervous disorders
125.	*Rauvolfia tetraphylla* L.	Apocynaceae	*Gandhagokul*	D	S	High blood pressure, insanity and insomnia
126.	*Ricinus communis* L.	Euphorbiaceae	*Rerhi*	D	S	Anthelmintic, diuretic and skin diseases
127.	*Rungia pectinata* (L.) Nees	Acanthaceae	*Pindi*	D	H	Febrifuge, refrigerant and aperients
128.	*Scindapsus officinalis* (Roxb.) Schott	Araceae	*Gajpipul*	M	C	Aphrodisiac, diaphoretic and anthelmintic
129.	*Scoparia dulcis* L.	Plantaginaceae	*Bandhane*	D	H	Fevers, cough, bronchitis and toothache
130.	*Senna occidentalis* (L.) Link	Fabaceae	*Kalkasunda*	D	S	Anti-inflammatory, depurative and whooping cough
131.	*Senna tora* (L.) Roxb.	Fabaceae	*Chakunda*	D	H	Anthelmintic, expectorant and snake-bite
132.	*Sesbania sesban* (L.) Merr.	Fabaceae	*Jayanti*	D	S	Anthelmintic, astringent and skin eruption

133.	*Sida acuta* Burm.f.	Malvaceae	*Kureta*	D	S	Disorders, headache, leucorrhoea, tuberculosis, diabetes, fever and uterine disorders
134.	*Sida cordata* (Burm.f.) Borss.Waalk.	Malvaceae	*Latjoka*	D	H	Astringent, diuretic and tonic
135.	*Sida cordifolia* L.	Malvaceae	*Berela*	D	S	Astringent, diuretic and tonic
136.	*Smilax ovalifolia* Roxb. ex D.Don	Smilacaceae	*Kumarica*	M	C	Dysentery, rheumatism and urinary complaints
137.	*Solanum americanum* Mill.	Solanaceae	*Kakmachi*	D	H	Antiseptic, anti-inflammatory, anodyne, expectorant and laxative
138.	*Solanum surattense* Burm. f.	Solanaceae	*Kantikari*	D	H	Anthelmintic, anti-inflammatory, carminative, depurative, febrifuge and diuretic
139.	*Solanum torvum* Sw.	Solanaceae	*Sada Kantikari*	D	H	Diuretic, digestive and sedative
140.	*Sphaeranthus senegalensis* DC.	Asteraceae	*Sarbani*	D	H	Useful in epilepsy, hemicranias, hepatopathy and gastropathy
141.	*Sphagneticola calendulacea* (L.) Pruski	Asteraceae	*Bhringaraj*	D	H	Emollient, febrifuge and purgative
142.	*Stephania japonica* (Thunb.) Miers	Menispermaceae	*Rajpatha*	D	C	Cures fever and urinary diseases
143.	*Strychnos nux-vomica* L.	Loganiaceae	*Kunchila*	D	T	Appetizer, digestive, dyspepsia and nervous disorders
144.	*Synedrella nodiflora* (L.) Gaertn.	Asteraceae	*Synedrela*	D	H	Antiseptic and anti-inflammatory
145.	*Tephrosia purpurea* (L.) Pers.	Fabaceae	*Ban Neel*	D	S	Elephantiasis, dyspepsia, anemia and skin diseases
146.	*Terminalia arjuna* (Roxb. ex DC.) Wight & Arn.	Combretaceae	*Arjun*	D	T	Astringent, aphrodisiac, cardiotonic, antidysenteric febrifuge and anticancerous
147.	*Terminalia bellirica* (Gaertn.) Roxb.	Combretaceae	*Bahera*	D	T	Astringent, digestive and insomnia
148.	*Terminalia chebula* Retz.	Combretaceae	*Haritaki*	D	T	Astringent, cardio tonic, digestive and insomnia
149.	*Tinospora sinensis* (Lour.) Merr.	Menispermaceae	*Padma Gulancha*	D	C	Emetic, liver disorders and rheumatoid arthritis
150.	*Trianthema portulacastrum* L.	Aizoaceae	*Swet Punarnova*	D	H	Asthma, edema and dropsy

151.	*Tribulus terrestris* L.	Zygophyllaceae	*Kantagokhur*	D	H	Diuretic, anthelmintic, anti-inflammatory, cardiotonic and skin diseases
152.	*Trichosanthes tricuspidata* Lour.	Cucurbitaceae	*Makal*	D	C	Liver disorder and skin infection
153.	*Tridax procumbens* (L.) L.	Asteraceae	*Targanda*	D	H	Cataract, dysentery and diarrhea
154.	*Triumfetta rhomboidea* Jacq.	Malvaceae	*Banokra*	D	S	Antidiabetic and sexual diseases
155.	*Tylophora indica* (Burm. f.) Merr.	Apocynaceae	*Antamul*	D	C	Acrid, cathartic, purgative, expectorant and stomachic
156.	*Urena sinuata* L.	Malvaceae	*Kunjia*	D	S	Diuretic and externally applied to treat rheumatism
157.	*Vallaris solanacea* (Roth) Kuntze	Apocynaceae	*Haparmali*	D	C	Diarrhea and rheumatic pain
158.	*Ventilago denticulata* Willd.	Rhamnaceae	*Raktapita*	D	C	Dyspepsia, general debility and skin diseases
159.	*Vitex negundo* L.	Verbenaceae	*Nishinda*	D	S	Antiseptic, tonic, anthelmintic and tranquillizer
160.	*Xanthium strumarium* L.	Asteraceae	*Okra*	D	H	Diuretic, leucorrhoea and hyperglycemia
161.	*Yucca gloriosa* L.	Asparagaceae	*Betal*	M	S	Purgative, rheumatism and bronchitis
162.	*Ziziphus oenopolia* (L.) Mill.	Rhamnaceae	*Shia Kul*	D	C	Antiseptic, stomachic and antiworm
163.	*Zornia gibbosa* Span.	Fabaceae	-	D	H	Antioxidant and anti-inflammatory

Abbreviation:

***In Angiosperm Type*: D-**Dicot, **M-**Monocot

***In Habit*: T-**Tree, **S-**Shrub, **H-**Herb, **C-**Climber.

Conclusion

The *Swarga-Bauri* sacred grove though fairly well protected by the villagers is facing mild threats by grazing animals, invasion of exotic weeds and erosion of traditional values towards plants among the younger generations. Therefore, the groves need care and attention. Steps should be taken to promote awareness among the villagers. Presently, there is no provision by the Government of West Bengal to protect sacred groves. Thus, a sacred grove conservation programme may be initiated taking the concerned scientists, local people and local administrative bodies into confidence.

References

Bhakat, R.K. 1990. Tribal ethics of forest conservation. *Yojana* (14-31 March): 24-27.

Bhakat, R.K. 2014. Ethnobotanical value of a sacred grove. *In*: Kumar, S. (Ed.). *Ethnobotanical studies in India*. Deep Publications, New Delhi. pp. 118-126.

Bhakat, R.K. and Pandit, P. K. 2003. Role of a sacred grove in conservation of medicinal plants. *Indian Forester*. 129: 224-232.

Bhakat, R.K. and Pandit, P.K. 2007. Role of a sacred grove in conservation of medicinal plants. *In*: Negi, S. S., Srivastava, R. K. and Bisht, N. S. (Eds.). *Medicinal and aromatic plants. Indian Forester*, Dehra Dun, India. pp. 332-339.

Bhakat, R.K. and Pandit, P.K. 2008. Conservation of vanishing ethnomedicinal plants of a tribal area of Midnapore district through sacred grove. *In*: Halder, A. K., Singh, P. K. and Coomar, P. C. (Eds.). *Tribal ethnomedicine: Problem and prospects*. Institute of Social Research and Applied Anthropology, Bidisha, Paschim Medinipur. pp. 123-128.

Bhakat, R.K. and Sen, U.K. 2008a. Ethnomedicinal plant conservation through sacred groves. *Tribes and Tribals* (Spl. Vol.). 2: 55-58.

Bhakat, R.K. and Sen, U.K. 2008b. *In–situ* conservation of medicinal plants through a sacred grove. *In*: *Proceedings of Present Status of Scientific and Technological Interventions on the Development of Medicinal and Aromatic Plants in India*. Kalyani, West Bengal. pp. 171-175.

Bhakat, R.K. and Sen, U.K. 2013. Conservation value of a sacred grove. *Indian Journal of Forestry*. 36(2): 217-226.

Brummit, R.K. and Powell, C. E. 1992. *Authors of plant names*. Royal Botanic Gardens, Kew. pp. 1-732.

Cunningham, A.B. 2001. *Applied ethnobotany: People, wild plant use and conservation*. Earthscan Publications Ltd., London.

Farnsworth, N.R. 1988. Screening plants for new medicines. *Biodiversity*. 3: 81-99.

Gadgil, M. and Vartak, V.D. 1981. Sacred groves of Maharashtra: An inventory. *In*: Jain, S. K. (Ed.). *Glimpses of ethnobotany*. Oxford University Press, Bombay. pp. 279-294.

Haines, H.H. 1921–1925. *The Botany of Bihar and Orissa*. Vol 1-6 (Reprint 1961). Botanical Survey of India, Calcutta.

Hooker, J.D. 1872-1897. *The Flora of British India*. Vols. 1-7 (Repr. 1973). Bishen Singh Mahendra Pal Singh, Dehra Dun.

Jain, S.K. and Rao, R.R. 1977. *A Handbook of field and herbarium methods*. Today and Tomorrow's Printers and Publishers, New Delhi.

Jeeva, S., Mishra, B.P., Venugopal, N., Kharlukhi, L. and Laloo, R.C. 2006. Traditional knowledge and biodiversity conservation in the sacred groves of Meghalaya. *Indian Journal of Traditional Knowledge*. 5(4): 563-568.

Khan, M.L., Khumbongmayum, A.D. and Tripathi, R.S. 2008. The sacred groves and their significance in conserving biodiversity: An overview. *International Journal of Ecology and Environmental Sciences*. 34(3): 277-291.

Kirtikar, K.R. and Basu, B.D. 1935. *Indian medicinal plants* (Vols. 1-4). Bishen Singh Mahendra Pal Singh, Dehra Dun.

Malhotra, K.C. 1998. Anthropological dimensions of sacred groves in India: An overview. *In*: Ramakrishnan, P.S., Saxena, K.G. and Chandrashekara, U.M. (Eds.). *Conserving the sacred for biodiversity management*. Oxford and IBH Publ. Co. Pvt. Ltd., New Delhi. pp. 423-438.

Musa, M.S., Abdelrasool, F.E., Elsheikh, E.A., Ahmed, L.A., Mahmoud, A.L.E. and Yagi, S.M. 2011. Ethnobotanical study of medicinal plants in the Blue Nile State, south-eastern Sudan. *Journal of Medicinal Plants Research*. 5(17): 4287-4297.

Muthu, C., Ayyanar, M., Raja, N. and Ignacimuthu, S. 2006. Medicinal plants used by traditional healers in Kancheepuram District of Tamil Nadu, India. *Journal of Ethnobiology and Ethnomedicine*. 2(1): 43.

Pakrashi, S.C. and Mukhopadhya, S. (Eds.). 2004. *Medicinal and aromatic plants of red laterite region of West Bengal (Bankura, Medinipore and Purulia)*. West Bengal Academy of Science and Technology, Kolkata. pp. 1-508.

Paria, N. (Ed.) 2005. *Medicinal plant resources of south West Bengal*. Vol.-1. Research Wing, Directorate of Forests in collaboration with Department of Environment, Government of West Bengal, Kolkata. pp. 1-198.

Prain, D. 1903a. *Bengal plants* Vol. 1 (Ind. Repr. 1963). Botanical Survey of India, Calcutta. pp. 1-487.

Prain, D. 1903b. *Bengal plants* Vol. 2 (Ind. Repr. 1963). Botanical Survey of India, Calcutta. pp. 488-1013.

Pushpangadan, P. 1995. *Ethnobiology in India: A status report*. Government of India. New Delhi.

Sanyal, M.N. 1994. *Flora of Bankura district*. Bishen Singh Mahendra Pal Singh, Dehra Dun, India.

Upadhyay, B., Dhaker, A.K. and Kumar, A. 2010. Ethnomedicinal and ethnopharmaco-statistical studies of eastern Rajasthan, India. *Journal of Ethnopharmacology*. 129(1): 64-86.

WHO. 2002. *WHO traditional medicine strategy 2002-2005*. Document WHO/EDM/TRM/2002.1

Yineger, H., Kelbessa, E., Bekele, T. and Lulekal, E. 2007. Ethnoveterinary medicinal plants at bale mountains national park, Ethiopia. *Journal of Ethnopharmacology*. 112(1): 55-70.

❑❑❑

Chapter 25

Utilization and Conservation of Ethnomedicinal Plants

Dawn C.P. Ambrose*

*Central Institute of Agricultural Engineering – Regional Centre, Coimbatore-7

ABSTRACT

Herbals are considered as green medicines which are safe to consume. These firms depend on the forests for their raw material which has led to the extinction of some of the herbal species. To prevent their extinction, it is necessary to ensure their survival by germplasm protection, multiplication and conservation. Domestication of ethnomedicinal plants is of utmost importance. The paper deals with conservation the biodiversity of ethnomedicinal plants and their utilization.

Keywords: Laxatives, Alkaloids, Traditional Medicinal.

Introduction

Medicinal plants have been used since time immemorial for the treatment of various ailments. Ethno medicinal plants are being used as folklore medicines by the ethnic group of a particular location. It belongs to the group of alternative medicine which uses natural or plant-based fixes. The application of plants for medicinal purposes is primitive and existing before human history. Plants are used in medicines for the physical, mental and spiritual sustenance of mankind. Even the animals used to graze on these medicinal plants to self treat their ailments. Mankind used many herbs and spices having special compounds to ward off the microbes contaminating the food. However, such plants were also found to prevent illness. Even the common weeds that are found in human habitats are found to possess medicinal properties. Many of the plants are used in fork lore medicines and also some have been significantly used in the preparation of modern drugs. The ancient people had acquired knowledge on these medicinal plants. At times the plants were used as medicine based on superstition, since they believed that illness were caused due to evil forces. It was thought that the evil spirits could be driven out from the body using toxic plants which would render an undesirable habitat to the spirits. Hence the person of a tribe was of the knowledge of such plants often guided by previous observations and beliefs.

History of Medicinal Plants

Medicinal plants draws special interest and have been used in the Chinese medicines in the early days in 5000 B.C. It has been mentioned in Sanskrit writings on the preparation of drugs from these medicinal plants. Even the ancient Babylonians, Hebrew and the Assyrians used plant herbs as medicines. The present medicines from plants were well known by the Greeks as quoted in the works of Hippocrates, Aristotle and Pythagoras. The work of Dioscorides is still valued in Turkey and North Africa on the medicinal properties of plants present those time. There existed a period of herbalists after the Dark Ages. Plants were believed to cure sickness based on their leaf shapes as a superstitious doctrine (Doctrine of Signatures). The early botanists practised medicine due to their interest in medicinal plants.

Ethnomedicinal plants as a source of drugs

Pharmacognosy is a branch of medicine dealing with the medicinal drugs from plants or natural sources. Medicinal plant comprises of primary and secondary metabolites responsible for their medicinal effect. There are several thousands of plants that are being used for the preparation of drugs across the globe. Many such plants are restricted in use by the native people residing in that particular region for a long time. The medicinal plants for drugs are obtained from the wild especially in tropical areas and very few are cultivated. The drug plants are collected and processed in a crude way before shipment and eventually processed in the drug trade centers. Countries play a monopoly of particular drug. The presence of chemical substance in the tissue of a plant producing a physiological action on the body contributes to the medicinal value of that particular plant. Alkaloids, glucosides, essential oils, tannins and gums are those phytochemicals which attribute to the healing properties of the plants. Some of these compounds are poisonous and hence the administration of such drugs needs to be under the supervision of the medical practitioner.

Importance of ethno medicinal plants

Ethnomedicinal plants play a major role in the lives of the poorer people, being not only cheaper but also as the only medicine available in the remote areas. In the recent years, there is a growing awareness on healthy lifestyle among individuals due to the alarming rise of diseases world around. People are more reclined to the natural system of medicine, wherein the traditional medicine has gained importance. The term medicinal plants include a variety of herbals having curative properties. Due to their curative nature, they are used as a raw material or extraction of their active ingredient in the development of herbal drugs, viz., laxatives, blood thinners, antibiotics and antimalarial medications, contain ingredients from plants. Traditional medicine is of primitive age practised by the sages. People living in the remote regions of the world have a unique knowledge of the natural resources they depend upon. They are the keepers of the cumulative knowledge for generations. The plants used by them are the store houses of potential medicines. However, the knowledge on such traditional plants has started to dwindle with the passage of time due to insufficiency of records/documents for reference.

Forests are a large source of ethno botanical plants. Any plant is useful because it provides oxygen to breathe. Man on his quest for food has identified the edible and non edible plants. Similarly he has found the medicines suitable for illness during his search for the same. The health of rural and tribal people living in the remote places is dependent on ethno medicinal plants.

The ethno botanical uses of the medicinal plants are of great importance in the drug industry. The following table shows the categories of herbal drugs and their usage

Table 1. Medicinal term and use of herbal drugs

S.No.	Medicinal term	Use
1.	Abortifacient	Inducing abortion
2.	Anthelmintic	Expelling or destroying parasitic worms, especially of the intestines
3.	Antipsoric	Relieving itches
4.	Aperient	Gently moving the bowels
5.	Astringent	Drawing together soft organic tissues
6.	Cardiotonic	Tending to increase the tonus of heart muscle

7.	Diaphoretic	Increasing perspiration
8.	Dysenteric	Dispelling dysentery or severe diarrhoea
9.	Emetic	Inducing vomiting
10.	Expectorant	Promoting discharge of mucus from the respiratory tract
11.	Febrifugic	Mitigating or removing fever
12.	Hypotensive	Causing low blood pressure or a lowering of blood pressure
13.	Purgative	Purging or tending to purge
14.	Rubefacient	Inducing redness of the skin by external application
15.	Vulnerary	Speeding the healing of wounds

Source: Robin Levingston and Rogelio Zamora, 1983

Uses of Ethnomedicinal Plants

Ethno medicinal plants have been used as herbal medicine, documented in the history of all civilisations. With modernisation, research in medicine concluded that the active principles present in plants are responsible for curative action against diseases. The scientists were even able to identify the toxic substances present in plants. The used part may be the seeds, berries, leaves, barks, roots, fruits, or other parts of a plant. The vital information on the utilisation of medicinal plants by the local people is collected through ethnobotanical data. This data is important in natural resource management. The following table presents the plant based remedies with botanical name of species followed by local name, parts used and ethno medical uses.

Table 2. Uses of ethnomedicinal plants as traditional medicine

S No.	Name of the plant	Common name	Family name	Part used	Use	Source
1.	*Abutilom indicum*	Khangi	Malvaceae	Roots, seed and leaves	Gonnorhea, Asthma & Constipation	2
2.	*Acacia catechu (L.f.)*	Khair	Fabaceae	Bark	eczema	9
3.	*Acacia nilotica Linn.*	Babool	Fabaceae	Stem, Bark	Tooth Problem, Skin Diseases	9
4.	*Acalypa indica*	Kuppi	Euphorbiaceae	Whole plant	Brain weakness and asthma	2
5.	*Adathoda zylamica*		Acanthaceae	Seeds and whole plants	Diarrohea, cough	2
6.	*Aegle marmelos Linn.*	Bilpatra	Rutaceae	Roots, Leaves and Fruit	Digestive problem	9
7.	*Ageratum conyzoides L.*	Lensa	Asteraceae	Leaves	Nose bleeding	10
8.	*Alangium salvifolium*	Ankol	Alangaceae	roots	Piles and seasonal fever	2
9.	*Aloe barbadensis Mill.*	Aloe vera	Liliaceae	Leaves	Jaundice and liver disorders	10
10.	*Alstonia scholaris R. Br.*	Thum riat	Apocynaceae	latex	Wounds, boils, ear ache	12
11.	*Amebia absinthium L.*	Tethwan	Boraginaceae	Leaves	Stomach pain, worm infections	1
12.	*Annona squamosa* l.	Ata	Annonaceae	Bark	diabetes	7
13.	*Aralia montana Blume*	Kotabell	Aralliaceae	Leaves	Malarial fever	9
14.	*Argemone mexicana L.*	Satyanashi	Papaveraceae	Root, latex	Gout, Dysentery, Liquid film in the eye	8
15.	*Asparagus racemosus*	Satawari	Lilliaceae	Pods and roots	Gout and rheumatism	2
16.	*Azadirachta indica A. Juss.*	neem	Meliaceae.	Whole plant	Insecticidal, liver tonic and urinary astringent, leprosy, skin diseases, leucoderma, dyspepsia, ulcers, tuberculosis, eczema, malarial fever	6

17.	*Baccaurea ramiflora lour.*	Khusmai	Euphorbioceae	Fruit	jaundice	7
18.	*Bauhenia variegata L.*	Vaufavang	Caesalpiniaceae	Bud	Piles and dysentry	9
19.	*Berberis lyceum Royle*	Kawdach	Beriberidaceae	Root	Indigestion, constipation	1
20.	*Bombax ceiba L.*	Semal	Bombacaceae	Root	Wound healing	8
21.	*Buchanania lanzan Spreng.*	Chironji	(Fabaceae	Bark and Seeds	Wounds, skin diseases, snake bite and rheumatism	8
22.	*Cojanus cajan (L)*	Orol	Papillionaceae	Leaf	jaundice	7
23.	*Calicarpa arborea Roxb.*	Hnah kiah	Verbanaceae	Bark	wound	9
24.	*Callicarpa arborea Roxb.*	Priyangu Hnahkiah	Verbenaceae	Leaves	Gastric ulcer	12
25.	*Carissa spinarum L*	Karaunda	Apocynaceae	Fruits and Roots	Rheumatic pain, fever and wound healing	8
26.	*Cassia angustifolia Vahl.*	Senna	Caesalpiniaceae	leaves	gout and rheumatism, effective purgative	5
27.	*Centella asiatica*	Brahmi	Apeaceae	Leaves	Blood pressure & urinary diseases	2
28.	*Clerodendram glandulosum*	Misimao	Verbenaceae	Tender leaves	High blood pressure	7
29.	*Clerodendrum phlomidis Linn.*	Arni	Verbenaceae	Leaf juice	to cure syphilis	5
30.	*Crotolaria juncea L.*	Sana	Fabaceae	Flower	Worm repellent	
31.	*Croton tiglium*	Jamalgota	Euphorbiaceae	Seed	asthma	11
32.	*Cuscuta reflexa*	Amarbel	Convolvulaceae	Leaves	leucoderma	11
33.	*Cynodon govianiana Wall*	Sangi-barb	Fumariaceae	Aerial part	Whooping cough, respiratory disorder	1
34.	*Dalbergia sissoo Roxb.*	Sheesham	Fabaceae	Leaves, Bark and Roots	Eye diseases and gonorrhea, scabies, leprosy, diarrhea and dysentery.	8

35.	*Datura metal*	Datura	Solanaceae	Seeds and leaves	Head ache, scorpion stings	12
36.	*Dillenia indica L.*	Kawsthinda ng	Dilleniaceae	Leaves, bark and fruit	dysentry	2
37.	*Dioscorea deltroiden Wall*	Kraeth	Dioscoreaceae	Leaves	Eye problem	1
38.	*Echinops echinatus Roxb*	Oont-kanti	Asteraceae	roots	quick delivery in women, to cure cough and cold	5
39.	*Emblica officinales Gaertn.*	Sinhlu	Euphorbiaceae	Bark	dysentry	2
40.	*Euphorbia hirta L.*	Dudh Ghas	Euphorbiaceae	Leaves	arthritis	8
41.	*Ficus religiosa Linn.*	Pipal	Moraceae	Whole Plant	Gonorrhoea, scabies and snake bite.	8
42.	*Gelsemium elegans Benth*	Hnam tur	Logianaceae	Leaves, root	Wound, stomach ulcer	12
43.	*Helicteres isora*	Marorphali	Sterculialaceae	Fruits and seeds	Ameobiosis	11
44.	*Hemidesmus indicus*	Annantmul	Asclepiadeceae	Roots	arthritis	
45.	*Houlfuynia cardata Thunb.*	Mosokmao	Sauraraceae	leaf	Heart problem	7
46.	*Hyoscyamus niger L.*	Bazarbang	Solanaceae	Seed	Tooth ache	1
47.	*Hyptis suaveolens (l.) Poit.*	Tukhma	lamiaceae	Seed	Urinary track infection	7
48.	*Indigofera prostrata Willd.*	Sekhupthur	Fabaceae	Seed	piles	2
49.	*Leucas aspera* (Willd.) Spreng	Goma	Lamiaceae	Leaves	Skin diseases, snake bites	5
50.	*Malva sylvestris L.*	Sotsal	Malvaceae	Seeds	Cough & fever	1
51.	*Mentha arvensis L.*	Pudna	Lamiaceae	Aerial part	Cough, sore throat, indigestion and constipation	1
52.	*Mjcheliachampoca 1.*	champa	Magnoliaceoe	Seed	jaundice	7

53.	*Mikania mickrantha*	Japan hlo	Asteraceae	leaves	Wound healing,	12
54.	*Mimosa hamata* Wild.	Jinjani,	Mimosaceae	leaves	diarrhoea and dysentery, bronchitis	5
55.	*Moringa oleifera* Lamk.	Sanjna	Moringaceae	leaves	To heal wounds caused by snake-bite and dog-bite	5
				Roots	Swellings, tumour and rheumatic pain	
56.	*Mucuna pruriens*	Kivach	Fabaceae	Seed	Male sterility	11
57.	*Nasturtium officinate L.*	KulHak	Brassicaceae	Leaf	Stomach ulcer, intestinal disorder	1
58.	*Oroxylum indicum (L.) Benth.*	Archangkon	Bignoniaceae	Leaves	tonsilitis	2
59.	*Pedalium murex Linn*	Bara Gokhru	Pedaliaceae	Sap of fresh leaves	gonorrhoea and dysuria	5
60.	*Peganum harmala Linn.*	Harmal,	Zygophyllaceae	leaves	to cure rheumatism	5
61.	*Pranella vulgaris L.*	Kulwanth	Lamiaceae	Flowering top	Headache, fever and muscular pain	1
62.	*Rheum emodi Wall*	Pam Tsalem	Polygonaceae	Rhizome	Body wounds	1
63.	*Salix wallichiana Andrs.*	Danthiveer	Salicaceae	Leaves	Fever and general body pain	1
64.	*Sambucus wightiana*	Gandula	Sambucaceae	Fruits	Stomach disorder	1
65.	*Sida cordifolia* Linn	Bal	Malvaceae	Flowers and unripe fruits	against painful urination	5
				roots	to cure the effect of sun-stroke, promoting healing of wounds	
66.	*Solanum viarum Dunal*	Khimkatai	*Solanaceae*	Fruit	Heart problem	7
67.	*Stellaria media L.*	Losdhi	Caryophyllaceae	Seed	Skin allergy	1
68.	*Taraxacum offianale Weber*	Haend	Asteraceae	Leaves	Bone fractures	1

69.	*Tectona grandis Linn. f.*	Sagun	Fabaceae	Bark, flowers, seeds and oil	Headache, toothache, inflammation and irritation of skin	8
70.	*Terminalia chebula Retz.*	Harra	Combretaceae	Fruits	Astringent, digestive, laxative, cardio tonic	8
71.	*Thunbergia gramndiflora Roxb.*	Vako	Acanthaceae	Root	Kidney stone and jaundice	2
72.	*Tinospora wrdifo/ia*	Daothulu	Menispermaceae	Shoot	Jaundice & diabetes	7
73.	*Triticum aestivum L.*	Kaenak	Poaceae	Seed	Treatment of worms in children	1
74.	*Urtica dioica L.*	Soi	Urticaceae	Roots	Cysts, joint pain	1
75.	*Viola odorata L.*	Banafsha	Violaceae	Flowers	Cough, fever and sore throat	1
76.	*Vitis vinifera L.*	Daech	Vitaceae	Leaves	Skin rashes	1
77.	*Vitex negundo L.*	Nirgudi	Verbenaceae	Leaves	Rheumatism	
78.	*Wede/ia (hinensis (Osb.) Merr.*	Vringaraj	Asteraceoe	Leaf	High blood pressure	7
79.	*Woodfordia fruticosa (L.) Kurz*	Dhawai	Lytharaceae	Leaves	Arthritis	8
80.	*Zizyphus mauritriana La.*	Brag Kund	Rhamnaceae	Leaves	Skin rashes	1
81.	*Zizyphus mauritiana Lamk.*	Ber, Beri	Rhamnaceae	Fruits	Cold and Cough	
82.	*Withania somnifera*	Asgandh	Solanaceae	Root	Heamotological problems	11

Conservation of Ethnomedicinal Plants

Throughout history, civilizations have moved in around the plants. Seeds of wild plants were stored in clay pots by man in earlier days which resulted in the advent of regular cultivation i.e. agriculture. As the search for species led to colonialism so also the search for medicinal plants led to search the remote.

In earlier times, the medicinal plants were collected from the nearby forests or planted in the home yards based on the traditional knowledge passed on to generations by the elders. But in recent times, deforestation, afforestation as well by monoculture and civilization in cultivation, the use and knowledge on ethnomedicinal plants are found to decline. Apart from these, natural calamities create a threat to these plants.

However in the present days, herbal medicines are gaining momentum due to wide numbers of firms involved in the production of herbal medicines to counteract the side effects of allopathy medicines. These herbals are considered as green medicines which are safe to consume. These firms depend on the forests for their raw material which has led to the extinction of some of the herbal species. To prevent their extinction, it is necessary to ensure their survival by germplasm protection, multiplication and conservation. Domestication of ethnomedicinal plants is of utmost importance. To conserve the biodiversity of ethnomedicinal plants, the following strategies need to be adopted:

- Farmers should be encouraged in the cultivation of medicinal plants at least in their barren and fallow lands. This would help in raising their income and in turn help in the conservation of the species. Also the farmers need to cultivate medicinal plants by integrating into the traditional crops grown which would pave way for income generation throughout the year.
- Creating nurseries for multiplication.
- Establishing seed banks.
- Scientific research on standardising the protocols for use in various diseases and to find suitable plants for curing the emerging diseases.
- Safe guarding the traditional knowledge on ethnomedicinal plants by the traditional healers through intellectual property rights.
- Creation of linkage between the producers and processers on the promotion of medicinal plants cultivation and processing.
- Documenting the traditional knowledge on medicinal plants.
- Herbal practitioners can motivate the farmers on suitable cultivation practises through the knowledge on farming practises.

Hence conservation of medicinal plants in their natural habitat needs to be explored. Efforts are being taken by various researchers across the globe in reviving and conserving the heritage of these traditional medicines, which in turn preserves the cultural heritage of the traditional/tribal people depending on the ethnomedicinal plants for their well being.

Conclusion

The role of ethnomedicinal plants for human well being is infinite. The weaker sections living in remote areas are greatly benefited by the use of such plants. Modern era also demand the natural plant raw material for safer and stress free living. The traditional knowledge needs to be safe guarded and efforts to be made in the conservation of valuable natural resource; ethnomedicinal plants by cultivation and preservation to protect the bio diversity. The age old adage, "is not health dearer than wealth?" is a truth fulfilled by nature.

References

Akhtar H., Malik, Anzar A., Khuroo, Dar, G.H. & Kha, Z.S. 2011, Ethnomedicinal uses of plants in Kashmir, *Indian J. Traditional Knowledge,* 10(2): 362-366.

Bhanwar Singh Hada and Katewa, S.S. 2015. Ethnomedicinal plants used against various diseases in Jhalwar district in Rajasthan, India, *Journal of Global Biosciences,* 4(4): 2077-2086.

Chhetri, D.R., Deewa Basnet, Po Fong Chiu, Sujata Kalikotay, Gagan Chhetri and Sippy Parajuli, 2005. Current status of ethnomedicinal plants in the Darjeeling Himalaya, *Current Science,* 89(2): 264-268.

Jean Giblette, 2006. The Role of Cultivation in Conserving Medicinal Plants, Call, Elizabeth *et al.,* 2006. Mending the Web of Life: Chinese Medicine and Species Conservation. Published by the International Fund for Animal Welfare (IFAW) and the Foundation for Education and Research on Botanicals, American Herbal Products Association (AHPA-ERB).

Kapoor , B.B.S. and Swati Lakhera, 2013. Ethnomedicinal plants of Jodhpur District, Rajasthan used in herbal and folk remedies, *Indian J. Pharm. Biol. Res.* 1(4): 71-75.

Kumkum Agarwal and Ranjana Varma, 2012. Some ethnomedicinal plants of Bhopal district used for treating stone diseases, *Int. J. of Pharm. & Life Sci.* 3(1): 1356-1362.